Integrated Study of Probability and Statistics

Volume I

Integrated Study of Probability and Statistics
Volume I

Edited by **Calanthia Wright**

NY RESEARCH PRESS
New York

Published by NY Research Press,
23 West, 55th Street, Suite 816,
New York, NY 10019, USA
www.nyresearchpress.com

Integrated Study of Probability and Statistics: Volume I
Edited by Calanthia Wright

International Standard Book Number: 978-1-63238-296-2 (Hardback)

Printed in China.

Contents

	Preface	**VII**
Chapter 1	**Bayesian Estimation and Prediction for Flexible Weibull Model under Type-II Censoring Scheme** Sanjay Kumar Singh, Umesh Singh and Vikas Kumar Sharma	**1**
Chapter 2	**Masses of Negative Multinomial Distributions: Application to Polarimetric Image Processing** Philippe Bernardoff, Florent Chatelain and Jean-Yves Tourneret	**17**
Chapter 3	**Note on Qualitative Robustness of Multivariate Sample Mean and Median** Evgueni Gordienko, Andrey Novikov and J. Ruiz de Chávez	**30**
Chapter 4	**Weighted Kappas for 3 × 3 Tables** Matthijs J. Warrens	**38**
Chapter 5	**Bayesian and Non-Bayesian Inference for Survival Data Using Generalised Exponential Distribution** Chris Bambey Guure and Samuel Bosomprah	**47**
Chapter 6	**On the Study of Transience and Recurrence of the Markov Chain Defined by Directed Weighted Circuits Associated with a Random Walk in Fixed Environment** Chrysoula Ganatsiou	**54**
Chapter 7	**On the Generalized Lognormal Distribution** Thomas L. Toulias and Christos P. Kitsos	**59**
Chapter 8	**The Complementary Exponentiated Exponential Geometric Lifetime Distribution** Francisco Louzada, Vitor Marchi and James Carpenter	**74**
Chapter 9	**Using Time Deformation to Filter Nonstationary Time Series with Multiple Time-Frequency Structures** Mengyuan Xu, Wayne A. Woodward and Henry L. Gray	**86**
Chapter 10	**Growth Estimators and Confidence Intervals for the Mean of Negative Binomial Random Variables with Unknown Dispersion** David Shilane and Derek Bean	**101**

Chapter 11 **Exact Group Sequential Methods for Estimating a Binomial Proportion** **110**
Zhengjia Chen and Xinjia Chen

Chapter 12 **On-Line Selection of *c*-Alternating Subsequences from a Random Sample** **134**
Robert W. Chen, Larry A. Shepp and Justín Ju-Chen Yang

Chapter 13 **Testing for Main Random Effects in Two-Way Random and Mixed Effects Models: Modifying the *F* Statistic** **139**
Trent Gaugler and Michael G. Akritas

Chapter 14 **Nonlinear Survival Regression Using Artificial Neural Network** **150**
Akbar Biglarian, Enayatollah Bakhshi, Ahmad Reza Baghestani, Mahmood Reza Gohari, Mehdi Rahgozar and Masoud Karimloo

Chapter 15 **Estimation of Extreme Values by the Average Conditional Exceedance Rate Method** **157**
A. Naess, O. Gaidai and O. Karpa

Chapter 16 **A Survey Design for a Sensitive Binary Variable Correlated with Another Nonsensitive Binary Variable** **172**
Jun-Wu Yu, Yang Lu and Guo-Liang Tian

Chapter 17 **Bayesian Estimation of the Scale Parameter of Inverse Weibull Distribution under the Asymmetric Loss Functions** **183**
Farhad Yahgmaei, Manoochehr Babanezhad and Omid S. Moghadam

Chapter 18 **A Note on Functional Averages over Gaussian Ensembles** **191**
Gabriel H. Tucci and Maria V. Vega

Permissions

List of Contributors

Preface

Probability and Statistics are two related yet different fields of study that are a part of the world of mathematics. The measure of the likelihood of an event occurring is given by probability. We can usually quantify the attitude of the mind towards some proposal of whose veracity we are not certain through the science of probability. Such a proposition is usually in the form of "Will a specific event occur or not". The field of probability has many theories and axioms that are used widely in disciplines such as mathematics, finance, statistics, science and AI as well as even philosophy and gambling. Even the underlying regularities and mechanics of complex systems can be described using probability theory. Probability can be applied to everyday life in fields such as risk assessment and financial markets. Probability is even used on a global scale to assess environmental and economic issues and future trends. Statistics on the other hand is a discipline focused on the study of the collection, organisation, interpretation, analysis and presentation of data. Such aspects of data like planning of data collection for the reasons of designing of surveys and experiments. There are basically two methodologies in statistics and any one can be used to analyse data - descriptive statistics or inferential statistics. It is thus a branch of mathematics that pertains to the collection and analysis of data into a more understandable presentation, though some may consider Statistics to be an entirely different science rather than a branch of mathematics. Such rigorous methods to use probability and statistics have changed society.

This book is an attempt to collate all current available data on the disciplines of probability and statistics and consequently I am thankful to those who put their effort and hard work behind this compilation.

Editor

1

Bayesian Estimation and Prediction for Flexible Weibull Model under Type-II Censoring Scheme

Sanjay Kumar Singh, Umesh Singh, and Vikas Kumar Sharma

Department of Statistics and DST-CIMS, Banaras Hindu University, Varanasi 221005, India

Correspondence should be addressed to Vikas Kumar Sharma; vikasstats@rediffmail.com

Academic Editor: Shein-chung Chow

We have developed the Bayesian estimation procedure for flexible Weibull distribution under Type-II censoring scheme assuming Jeffrey's scale invariant (noninformative) and Gamma (informative) priors for the model parameters. The interval estimation for the model parameters has been performed through normal approximation, bootstrap, and highest posterior density (HPD) procedures. Further, we have also derived the predictive posteriors and the corresponding predictive survival functions for the future observations based on Type-II censored data from the flexible Weibull distribution. Since the predictive posteriors are not in the closed form, we proposed to use the Monte Carlo Markov chain (MCMC) methods to approximate the posteriors of interest. The performance of the Bayes estimators has also been compared with the classical estimators of the model parameters through the Monte Carlo simulation study. A real data set representing the time between failures of secondary reactor pumps has been analysed for illustration purpose.

1. Introduction

In reliability/survival analysis, generally, life test experiments are performed to check the life expectancy of the manufactured product or items/units before products produced in the market. But in practice, the experimenters are not able to observe the failure times of all the units placed on a life test due to time and cost constraints or due to some other uncertain reasons. Data obtained from such experiments are called censored sample. Keeping time and cost constraints in mind, many types of censoring schemes have been discussed in the statistical literature named as Type-I censoring, Type-II censoring and progressive censoring schemes, and so forth. In this paper, Type-II censoring scheme is considered. In Type-II censoring scheme, the life test is terminated as soon as a prespecified number (say, r) of units have failed. Therefore, out of n units put on test, only first r failures will be observed. The data obtained from such a restrained life test will be referred to as a Type-II censored sample.

Prediction of the lifetimes of future items based on censored data is very interesting and valuable topic for researchers, engineers and reliability practitioners. In predictive inference, In predictive Inference, we can infer about the lifetimes of the future items using observed data. The future prediction problem can be classified into two types: (1) one-sample prediction problem. (2) two-sample prediction problem. In one-sample prediction problem, the variable to be predicted comes from the same sequence of variables observed and dependent of the informative sample. In the second type, the variable to be predicted comes from another independent future sample. Reference [1] has developed the Bayesian procedure to the prediction problems of future observations and use the concept of Bayesian predictive posterior distribution. Many authors have focussed on the problem of Bayesian prediction of future observations based on various types of censored data from different lifetime models (see [2–9], and references cited therein).

The flexible Weibull distribution is a new two-parameter generalization of the Weibull model which has been proposed

Table 1: Average estimates and MSEs (in the brackets) of the estimators of α (in the first row of each cell) and β (in the second row of each cell) under different Type-II censoring schemes for fixed values of $\alpha = 2$ and $\beta = 2$.

(n, r)	MLE	Bayes		
		Jeffrey's prior	Gamma-1	Gamma-2
20, 20	2.16201 (0.19478)	2.12381 (0.18568)	2.11852 (0.16520)	2.12323 (0.18297)
	2.15231 (0.25599)	2.11832 (0.24964)	2.11214 (0.21910)	2.11751 (0.24560)
20, 16	2.25425 (0.38927)	2.19812 (0.37041)	2.18709 (0.31676)	2.19695 (0.36310)
	2.20897 (0.35759)	2.16973 (0.34807)	2.15957 (0.30027)	2.16858 (0.34157)
20, 12	2.42308 (0.86955)	2.33505 (0.82782)	2.31067 (0.66996)	2.33284 (0.80561)
	2.29886 (0.55647)	2.25176 (0.54174)	2.23387 (0.45669)	2.24973 (0.52982)
30, 30	2.10084 (0.10210)	2.07541 (0.09830)	2.07318 (0.09054)	2.07517 (0.09720)
	2.09362 (0.13855)	2.07067 (0.13589)	2.0680 (0.123778)	2.07039 (0.13420)
30, 24	2.16669 (0.22014)	2.12924 (0.21169)	2.12430 (0.189528)	2.12873 (0.20866)
	2.13586 (0.20569)	2.10924 (0.20143)	2.10460 (0.18104)	2.10875 (0.19861)
30, 18	2.25330 (0.43040)	2.19459 (0.41323)	2.18513 (0.35242)	2.19370 (0.40483)
	2.17914 (0.27758)	2.14681 (0.27137)	2.13951 (0.23802)	2.14599 (0.26667)
50, 50	2.05641 (0.05175)	2.04115 (0.05045)	2.04046 (0.04793)	2.04111 (0.05013)
	2.05189 (0.07560)	2.03798 (0.07468)	2.03712 (0.07043)	2.03793 (0.07413)
50, 40	2.08371 (0.09921)	2.06117 (0.09674)	2.05986 (0.08995)	2.06105 (0.09583)
	2.06881 (0.10032)	2.05256 (0.09900)	2.05119 (0.09234)	2.05243 (0.09811)
50, 30	2.14568 (0.18827)	2.11044 (0.18191)	2.10701 (0.16367)	2.11011 (0.17946)
	2.09982 (0.13099)	2.08009 (0.12866)	2.07740 (0.11804)	2.07982 (0.12722)

by [10]. The density function of the flexible Weibull distribution is given by

$$f(x) = \left(\alpha + \frac{\beta}{x^2}\right) \exp\left(\alpha x - \frac{\beta}{x}\right) \exp\left(-e^{\alpha x - \beta/x}\right), \quad \alpha, \beta > 0. \tag{1}$$

The corresponding CDF is given by

$$F(x) = 1 - \exp\left(-e^{\alpha x - \beta/x}\right). \tag{2}$$

They have shown that this distribution is able to model various ageing classes of lifetime distributions including IFR, IFRA, and MBT (modified bathtub). They have also checked the goodness-of-fit of this distribution to the failure time between secondary reactor pumps data in the comparison of various extensions of Weibull model and found that this model gives the better fit. Therefore, this distribution can be considered as an alternative lifetime model of the various well-known generalizations of the Weibull distribution. Some statistical properties and the classical estimation procedure for the flexible Weibull distribution have been discussed by [10]. It is to be mentioned here that this distribution has not been considered under Bayesian setup in the earlier literature.

It is well-known that the squared error loss function is the most widely used loss function in Bayesian analysis. This loss is well justified in the classical paradigm on the ground of minimum variance unbiased estimation procedure. In most of the cases, Bayesian estimation procedures has been developed under the same loss function. For Bayesian estimation, we also need to assume a prior distribution for the model parameters involved in the analysis. In this paper, Bayesian analysis have been preformed under the squared error loss function (SELF) assuming both Jeffreys scale invariant and Gamma priors.

A major difficulty to the implementation of Bayesian procedure is that of obtaining the posterior distribution. The process often requires the integration which is very difficult to calculate especially when dealing with complex and high-dimensional models. In such a situation, Monte Carlo Markov chain (MCMC) methods, namely, Gibbs sampling [11] and Metropolis-Hastings (MH) algorithms [12, 13], are very useful to simulate the deviates from the posterior density and produce the good approximate results.

The rest of the paper is organized as fallows. In Section 2, we have discussed the point estimation procedures for the parameters of the considered model under classical set-up. The confidence/bootstrap intervals have been constructed in Section 3. In Section 4, we have developed the Bayesian estimation procedure under the assumption that model parameters have the gamma prior density function. We have also derived the one-sample and two-sample predictive densities and corresponding survival functions of the future observables in Sections 5 and 6, respectively. The predictive bounds of future observations under one-sample and two-sample predictions have also been discussed in respective sections. For comparing the performance of the classical and Bayesian estimation procedures, the Monte Carlo simulation study has been presented in Section 7. To check the applicability of the proposed methodologies, a real data set has been analysed in Section 8. Finally, the conclusions have been given in Section 9.

Table 2: Average estimates and MSEs (in the brackets) of the estimators of α (in the first row of each cell) and β (in the second row of each cell) for complete sample $n = 20$ with varying parameters.

(α, β)	MLE	Bayes		
		Jeffrey's prior	Gamma-1	Gamma-2
1, 1	1.1139 (0.07556)	1.0684 (0.06575)	1.0656 (0.06067)	1.0680 (0.06508)
	1.0900 (0.08537)	1.0493 (0.07775)	1.0462 (0.07101)	1.0490 (0.07688)
1, 2	1.1034 (0.06502)	1.0709 (0.05950)	1.0681 (0.05440)	1.0706 (0.05881)
	2.1769 (0.30904)	2.1195 (0.29285)	2.1093 (0.24159)	2.1183 (0.28551)
1, 3	1.0993 (0.06207)	1.0739 (0.05822)	1.0703 (0.05164)	1.0734 (0.05726)
	3.2631 (0.66390)	3.1976 (0.64043)	3.1758 (0.47965)	3.1949 (0.61568)
2, 1	2.2067 (0.26010)	2.1418 (0.23801)	2.1317 (0.20114)	2.1407 (0.23297)
	1.0884 (0.07726)	1.0597 (0.07321)	1.0561 (0.06690)	1.0594 (0.07235)
2, 2	2.1942 (0.24329)	2.1522 (0.23141)	2.1449 (0.20424)	2.1512 (0.22769)
	2.1746 (0.28703)	2.1396 (0.27953)	2.1318 (0.24468)	2.1387 (0.27465)
2, 3	2.1894 (0.23938)	2.1580 (0.23120)	2.1510 (0.20434)	2.1571 (0.22745)
	3.2608 (0.62525)	3.2237 (0.61496)	3.2093 (0.51498)	3.2219 (0.60058)
3, 1	3.2979 (0.55861)	3.2216 (0.52396)	3.1977 (0.40431)	3.2187 (0.50676)
	1.0877 (0.07376)	1.0659 (0.07116)	1.0610 (0.06349)	1.0653 (0.07006)
3, 2	3.2841 (0.53861)	3.2370 (0.52021)	3.2220 (0.44077)	3.2352 (0.50919)
	2.1739 (0.27789)	2.1491 (0.27331)	2.1411 (0.24128)	2.1481 (0.26884)
3, 3	3.2979 (0.55861)	3.2216 (0.52396)	3.1977 (0.40431)	3.2187 (0.50676)
	1.0877 (0.07376)	1.0659 (0.07116)	1.0610 (0.06349)	1.0653 (0.07006)

Table 3: Average estimates and MSEs (in the brackets) of the estimators of α (in the first row of each cell) and β (in the second row of each cell) for fixed $n = 20$, $r = 16$ with varying parameters.

(α, β)	MLE	Bayes		
		Jeffrey's prior	Gamma-1	Gamma-2
1, 1	1.1601 (0.13900)	1.0873 (0.12006)	1.0832 (0.10606)	1.0870 (0.11811)
	1.1051 (0.09747)	1.0605 (0.08798)	1.0565 (0.07969)	1.0600 (0.08695)
1, 2	1.1393 (0.11004)	1.0928 (0.10056)	1.0886 (0.08998)	1.0924 (0.09915)
	2.2098 (0.36960)	2.1461 (0.34912)	2.1335 (0.28565)	2.1446 (0.34025)
1, 3	1.1317 (0.10126)	1.0968 (0.09494)	1.0916 (0.08277)	1.0962 (0.09321)
	3.3147 (0.81320)	3.2416 (0.78322)	3.2143 (0.58046)	3.2382 (0.75242)
2, 1	2.2786 (0.44015)	2.1857 (0.40223)	2.1691 (0.31750)	2.1840 (0.39033)
	1.1049 (0.09240)	1.0731 (0.08728)	1.0681 (0.07812)	1.0725 (0.08607)
2, 2	2.2551 (0.38836)	2.1990 (0.36937)	2.1879 (0.31593)	2.1978 (0.36213)
	2.2099 (0.35695)	2.1708 (0.34734)	2.1605 (0.29969)	2.1696 (0.34086)
2, 3	2.2463 (0.37240)	2.2055 (0.35965)	2.1952 (0.31102)	2.2044 (0.35296)
	3.3157 (0.79237)	3.2738 (0.77911)	3.2552 (0.64353)	3.2716 (0.75988)
3, 1	3.3950 (0.91132)	3.2905 (0.85450)	3.2509 (0.60166)	3.2864 (0.81711)
	1.1049 (0.09035)	1.0805 (0.08702)	1.0736 (0.07525)	1.0798 (0.08538)
3, 2	3.3695 (0.83789)	3.3083 (0.80919)	3.2850 (0.65786)	3.3059 (0.78822)
	2.2105 (0.35216)	2.1825 (0.34626)	2.1716 (0.29991)	2.1813 (0.33992)
3, 3	3.3598 (0.81499)	3.3160 (0.79548)	3.2972 (0.65221)	3.3139 (0.77860)
	3.3167 (0.78472)	3.2879 (0.77676)	3.2710 (0.66096)	3.2859 (0.76072)

2. Classical Estimation

Let $x_1, x_2, \ldots, x_n$ be the IID random sample from (1) and let $x_{(1)}, x_{(2)}, \ldots, x_{(r)}$, $(r \leq n)$ be the ordered sample obtained under Type-II censoring scheme. Then, the likelihood function for such type of censored sample can be defined as

$$L = \frac{n!}{(n-r)!} \prod_{i=1}^{r} f\left(x_{(i)}\right) \left[1 - F\left(x_{(r)}\right)\right]^{n-r}. \quad (3)$$

TABLE 4: Average estimates and MSEs (in the brackets) of the estimators of α (in the first row of each cell) and β (in the second row of each cell) for fixed $n = 20$, $r = 12$ with varying parameters.

(α, β)	MLE	Bayes		
		Jeffrey's prior	Gamma-1	Gamma-2
1, 1	1.2957 (0.37835)	1.1122 (0.35851)	1.1380 (0.26430)	1.1250 (0.33656)
	1.1456 (0.13641)	1.0903 (0.12315)	1.0856 (0.10866)	1.0902 (0.12098)
1, 2	1.2447 (0.27307)	1.1609 (0.25275)	1.1552 (0.21467)	1.1611 (0.24655)
	2.2777 (0.50948)	2.2017 (0.48238)	2.1838 (0.38908)	2.1996 (0.46883)
1, 3	1.2222 (0.23392)	1.1652 (0.21985)	1.1553 (0.18599)	1.1642 (0.21487)
	3.4460 (1.24114)	3.3585 (1.19561)	3.3153 (0.87861)	3.3528 (1.14642)
2, 1	2.4823 (1.07353)	2.3141 (0.99717)	2.2810 (0.67931)	2.3164 (0.93999)
	1.1468 (0.13669)	1.1086 (0.12925)	1.0996 (0.11056)	1.1078 (0.12651)
2, 2	2.4243 (0.86825)	2.3363 (0.82612)	2.3117 (0.66887)	2.3340 (0.80407)
	2.3000 (0.55574)	2.2530 (0.54086)	2.2349 (0.45605)	2.2509 (0.52898)
2, 3	2.4028 (0.80078)	2.3416 (0.77287)	2.3197 (0.64588)	2.3392 (0.75508)
	3.4559 (1.26492)	3.4054 (1.24372)	3.3731 (1.01013)	3.4014 (1.20982)
3, 1	3.6598 (2.05251)	3.4886 (1.92783)	3.3923 (1.13282)	3.4815 (1.79894)
	1.1472 (0.13534)	1.1180 (0.13032)	1.1048 (0.10602)	1.1165 (0.12669)
3, 2	3.6041 (1.80173)	3.5124 (1.73891)	3.4586 (1.31761)	3.5071 (1.67923)
	2.3039 (0.56221)	2.2702 (0.55276)	2.2498 (0.46441)	2.2679 (0.54033)
3, 3	3.5805 (1.69795)	3.5168 (1.65595)	3.4757 (1.33982)	3.5125 (1.61182)
	3.4617 (1.28014)	3.4267 (1.26723)	3.3959 (1.05514)	3.4230 (1.23701)

Substituting (1) and (2) in (3), we have

$$L = \frac{n!}{(n-r)!}\prod_{i=1}^{r}\left(\alpha + \frac{\beta}{x_{(i)}^2}\right)\exp\left(\sum_{i=1}^{r}\left(\alpha x_{(i)} - \frac{\beta}{x_{(i)}}\right)\right) \times \exp\left(-\sum_{i=1}^{r}\exp\left(\alpha x_{(i)} - \frac{\beta}{x_{(i)}}\right) - (n-r)\exp\left(\alpha x_{(r)} - \frac{\beta}{x_{(r)}}\right)\right). \tag{4}$$

The log-likelihood function is given by

$$\log L = \ln\left(\frac{n!}{(n-r)!}\right) + \sum_{i=1}^{r}\ln\left(\alpha + \frac{\beta}{x_{(i)}^2}\right) + \sum_{i=1}^{r}\left(\alpha x_{(i)} - \frac{\beta}{x_{(i)}}\right) - \sum_{i=1}^{r}\exp\left(\alpha x_{(i)} - \frac{\beta}{x_{(i)}}\right) - (n-r)\exp\left(\alpha x_{(r)} - \frac{\beta}{x_{(r)}}\right). \tag{5}$$

The MLEs $\hat{\alpha}$ and $\hat{\beta}$ of α and β can be obtained as the simultaneous solution of the following two nonlinear equations:

$$\sum_{i=1}^{r}\frac{1}{\left(\alpha + \beta/x_{(i)}^2\right)} + \sum_{i=1}^{r}x_{(i)} - \sum_{i=1}^{r}x_{(i)}e^{(\alpha x_{(i)} - \beta/x_{(i)})} - (n-r)\,x_{(r)}e^{(\alpha x_{(r)} - \beta/x_{(r)})} = 0, \tag{6}$$

$$\sum_{i=1}^{r}\frac{1/x_{(i)}^2}{\left(\alpha + \beta/x_{(i)}^2\right)} - \sum_{i=1}^{r}\frac{1}{x_{(i)}} + \sum_{i=1}^{r}\frac{e^{(\alpha x_{(i)} - \beta/x_{(i)})}}{x_{(i)}} + (n-r)\frac{e^{(\alpha x_{(r)} - \beta/x_{(r)})}}{x_{(r)}} = 0 \tag{7}$$

It can be seen that the above equations cannot be solved explicitly and one needs iterative method to solve them. Here, we proposed the use of the fixed point iteration method, which can be routinely applied as follows:

The equation (6) can be rewrite as

$$\alpha = \phi_1(\alpha, \beta) = \frac{\sum_{i=1}^{r}\left(1/\left(1 + \beta/\alpha x_{(i)}^2\right)\right)}{h_1(\alpha, \beta)}. \tag{8}$$

Similarly, from (7), we have

$$\beta = \phi_2(\alpha, \beta) = \frac{\sum_{i=1}^{r}\left(\left(1/x_{(i)}^2\right)/\left((\alpha/\beta) + \left(1/x_{(i)}^2\right)\right)\right)}{h_2(\alpha, \beta)}, \tag{9}$$

TABLE 5: Average 95% confidence/HPD/bootstrap intervals (in brackets) along with their width and shape (in square brackets) for bootstrap method under different Type-II censoring schemes for fixed $\alpha = 2$ and $\beta = 2$.

(n, r)		Asymptotic	Bootstrap	Bayes		
				Jeffrey	Gamma-1	Gamma-2
20, 20	α	1.42104	1.672185 [2.492175]	0.59304	0.57805	0.59072
		(1.452209, 2.873245)	(1.676250, 3.348435)	(1.82722, 2.42026)	(1.82973, 2.40778)	(1.82818, 2.41890)
	β	1.66567	1.892912 [2.134547]	0.68685	0.66402	0.68298
		(1.320337, 2.986004)	1.548423, 3.441335	(1.77583, 2.46268)	(1.78165, 2.44567)	(1.77731, 2.46029)
20, 16	α	1.96999	2.426357 [2.773682]	0.74758	0.72096	0.74375
		(1.270139, 3.240126)	(1.611282, 4.037639)	(1.82464, 2.57222)	(1.82727, 2.54823)	(1.82559, 2.56934)
	β	1.93131	2.301952 [2.467296]	0.72063	0.69357	0.71606
		(1.244336, 3.175642)	(1.545066, 3.847018)	(1.81036, 2.53099)	(1.81423, 2.5078)	(1.81135, 2.52741)
20, 12	α	2.81147	3.749764 [3.347512]	0.97546	0.92306	0.96713
		(1.018547, 3.830021)	(1.560572, 5.310336)	(1.84771, 2.82317)	(1.85060, 2.77366)	(1.85008, 2.81721)
	β	2.28314	2.936211 [2.990892]	0.77081	0.73686	0.76509
		(1.158481, 3.441619)	(1.563132, 4.499343)	(1.86734, 2.63815)	(1.86716, 2.60402)	(1.86832, 2.63341)
30, 30	α	1.12314	1.247883 [2.056709]	0.4791	0.47062	0.47771
		(1.539796, 2.662935)	(1.692596, 2.940479)	(1.83586, 2.31496)	(1.83784, 2.30846)	(1.83645, 2.31416)
	β	1.32406	1.438015 [1.840961]	0.56098	0.54809	0.55861
		(1.432405, 2.756466)	(1.587448, 3.025463)	(1.79086, 2.35184)	(1.79496, 2.34305)	(1.79186, 2.35047)
30, 24	α	1.54929	1.7729 [2.272754]	0.60329	0.58807	0.60117
		(1.392709, 2.942003)	(1.624975, 3.397875)	(1.82805, 2.43134)	(1.83084, 2.41891)	(1.82825, 2.42942)
	β	1.52657	1.706546 [2.06264]	0.58835	0.57302	0.58567
		(1.373484, 2.900058)	(1.578646, 3.285192)	(1.81574, 2.40409)	(1.81937, 2.39239)	(1.81669, 2.40236)
30, 21	α	2.15385	2.59599 [2.624895]	0.78178	0.75125	0.77734
		(1.176378, 3.330229)	(1.537144, 4.133134)	(1.80412, 2.58590)	(1.81047, 2.56172)	(1.80539, 2.58273)
	β	1.76257	2.071998 [2.388763]	0.62933	0.60969	0.62606
		(1.297852, 3.060424)	(1.567708, 3.639706)	(1.83298, 2.46231)	(1.83617, 2.44586)	(1.83368, 2.45974)
50, 50	α	0.84922	0.902724 [1.740127]	0.36791	0.36381	0.36732
		(1.631801, 2.481019)	(1.726964, 2.629688)	(1.85716, 2.22507)	(1.85882, 2.22263)	(1.85731, 2.22463)
	β	1.00564	1.054286 [1.595255]	0.43445	0.42822	0.43321
		(1.549076, 2.554712)	(1.645654, 2.699940)	(1.82095, 2.25540)	(1.82367, 2.25189)	(1.82162, 2.25483)
50, 40	α	1.1557	1.250972 [1.873862]	0.46222	0.45477	0.46123
		(1.505854, 2.661558)	(1.648413, 2.899385)	(1.83026, 2.29248)	(1.83315, 2.28792)	(1.83062, 2.29185)
	β	1.14531	1.222437 [1.735241]	0.45574	0.44818	0.45429
		(1.496150, 2.641461)	(1.621884, 2.844321)	(1.82516, 2.2809)	(1.82774, 2.27592)	(1.82558, 2.27987)
50, 30	α	1.6002	1.784637 [2.088423]	0.59843	0.58372	0.59636
		(1.345588, 2.945790)	(1.567841, 3.352478)	(1.81112, 2.40955)	(1.81576, 2.39948)	(1.81225, 2.40861)
	β	1.31418	1.442373 [1.937533]	0.48696	0.47735	0.48517
		(1.442738, 2.756921)	(1.608814, 3.051187)	(1.83724, 2.32420)	(1.83956, 2.31691)	(1.83778, 2.32295)

where

$$h_1(\alpha,\beta) = \sum_{i=1}^{r} x_{(i)} e^{(\alpha x_{(i)} - \beta / x_{(i)})} + (n-r)\, x_{(r)} e^{(\alpha x_{(r)} - \beta / x_{(r)})} - \sum_{i=1}^{r} x_{(i)},$$
$$h_2(\alpha,\beta) = \sum_{i=1}^{r} \frac{1}{x_{(i)}} - \sum_{i=1}^{r} \frac{e^{(\alpha x_{(i)} - \beta / x_{(i)})}}{x_{(i)}} - (n-r)\frac{e^{(\alpha x_{(r)} - \beta / x_{(r)})}}{x_{(r)}}. \quad (10)$$

The following steps followed to obtain the solution of the normal equations (6) and (7).

Step 1. Start with initial starting points, say α^0 and β^0.

Step 2. By using α^0 and β^0, obtain $\widehat{\alpha} = \phi_1(\alpha^0, \beta^0)$ from (8).

Step 3. Then, obtain $\widehat{\beta} = \phi_2(\alpha^0, \beta^0)$ from (9).

Step 4. If $|\alpha^0 - \widehat{\alpha}| \le \epsilon$ and $|\beta^0 - \widehat{\beta}| \le \epsilon$, where ϵ is some preassigned tolerance limit, then $(\widehat{\alpha}, \widehat{\beta})$ will be the desired solution of (8) and (9).

TABLE 6: Average 95% confidence/HPD intervals (in brackets) and bootstrap method along with their width and shape (in square brackets) for bootstrap method for complete sample $n = 20$ with varying parameters.

(α,β)		Asymptotic	Bootstrap	Bayes Jeffrey	Bayes Gamma-1	Bayes Gamma-2
1, 1	α	0.84323	1.024972 [2.695397]	0.51464	0.506	0.51346
		(0.6923179, 1.5355495)	(0.836569, 1.861541)	(0.81088, 1.32552)	(0.81256, 1.31856)	(0.81082, 1.32428)
	β	0.94998	1.086861 [2.203492]	0.56244	0.5504	0.56069
		(0.6150548, 1.5650366)	(0.750772, 1.837632)	(0.76999, 1.33243)	(0.77297, 1.32337)	(0.77060, 1.33129)
1, 2	α	0.79894	0.9567911 [2.582233]	0.40899	0.40252	0.40818
		(0.7039037, 1.5028432)	(0.836280, 1.793071)	(0.86641, 1.27540)	(0.86717, 1.26969)	(0.86644, 1.27462)
	β	1.80953	2.068489 [2.192656]	0.90842	0.86205	0.9015
		(1.272122, 3.081648)	(1.528995, 3.597484)	(1.66685, 2.57527)	(1.68082, 2.54287)	(1.66961, 2.57111)
1, 3	α	0.78582	0.9352116 [2.53492]	0.3523	0.3457	0.35143
		(0.7063879, 1.4922066)	(0.834733, 1.769945)	(0.89762, 1.24992)	(0.89777, 1.24347)	(0.89765, 1.24908)
	β	2.65059	3.034498 [2.197503]	1.17065	1.07699	1.15653
		(1.937839, 4.588432)	(2.314114, 5.348612)	(2.61358, 3.78423)	(2.63980, 3.71679)	(2.61865, 3.77518)
2, 1	α	1.59788	1.913583 [2.582236]	0.81799	0.78679	0.81391
		(1.407806, 3.005689)	(1.672561, 3.586144)	(1.73282, 2.55081)	(1.73930, 2.52609)	(1.73389, 2.54780)
	β	0.90477	1.034245 [2.19266]	0.45421	0.44617	0.45288
		(0.6360592, 1.5408260)	(0.764498, 1.798743)	(0.83341, 1.28762)	(0.83447, 1.28064)	(0.83390, 1.28678)
2, 2	α	1.56103	1.851625 [2.507755]	0.63	0.6128	0.6276
		(1.413654, 2.974686)	(1.666304, 3.517929)	(1.8373, 2.4673)	(1.8391, 2.4519)	(1.8379, 2.4655)
	β	1.74157	1.996533 [2.203399]	0.6938	0.6703	0.6904
		(1.303866, 3.045431)	(1.551394, 3.547927)	(1.7936, 2.4874)	(1.7980, 2.4683)	(1.7945, 2.4849)
2, 3	α	1.55385	1.836112 [2.47468]	0.5332	0.52	0.5314
		(1.412461, 2.966312)	(1.660960, 3.497072)	(1.8912, 2.4244)	(1.8915, 2.4115)	(1.8919, 2.4233)
	β	2.56731	2.948256 [2.214166]	0.8724	0.8288	0.866
		(1.977191, 4.544503)	(2.343578, 5.291834)	(2.7883, 3.6607)	(2.7967, 3.6255)	(2.7900, 3.6560)
3, 1	α	2.35746	2.805634 [2.534927]	1.05692	0.99444	1.04846
		(2.119162, 4.476625)	(2.504204, 5.309838)	(2.69286, 3.74978)	(2.70190, 3.69634)	(2.69436, 3.74282)
	β	0.88353	1.0115 [2.197506]	0.39022	0.38322	0.389
		(0.6459449, 1.5294797)	(0.771372, 1.782872)	(0.87118, 1.26140)	(0.87031, 1.25353)	(0.87132, 1.26032)
3, 2	α	2.33077	2.754159 [2.474687]	0.7998	0.7697	0.7957
		(2.118684, 4.449453)	(2.491433, 5.245592)	(2.8368, 3.6366)	(2.8381, 3.6078)	(2.8379, 3.6336)
	β	1.71154	1.965495 [2.214164]	0.5816	0.5654	0.5791
		(1.318121, 3.029657)	(1.562379, 3.527874)	(1.8589, 2.4405)	(1.8589, 2.4243)	(1.8591, 2.4382)
3, 3	α	2.32879	2.743817 [2.44709]	0.6713	0.6509	0.6685
		(2.114429, 4.443215)	(2.482841, 5.226658)	(2.9086, 3.5799)	(2.9071, 3.5580)	(2.9090, 3.5775)
	β	2.53128	2.910998 [2.225123]	0.7242	0.6965	0.7202
		(1.994582, 4.525857)	(2.357619, 5.268617)	(2.8734, 3.5976)	(2.8744, 3.5709)	(2.8739, 3.5941)

Step 5. If $|\alpha^0 - \widehat{\alpha}| > \epsilon$ and $|\beta^0 - \widehat{\beta}| > \epsilon$, then set $\alpha^0 = \widehat{\alpha}$ and $\beta^0 = \widehat{\beta}$ and repeat Steps 2–5, until tolerance limit is achieved.

3. Confidence Intervals

3.1. Asymptotic Confidence Intervals. The exact distribution of MLEs cannot be obtained explicitly. Therefore, the asymptotic properties of MLEs can be used to construct the confidence intervals for theparameters. Under some regularity conditions, the MLEs $(\widehat{\alpha}, \widehat{\beta})$ are approximately bivariate normal with mean $(\widehat{\alpha}, \widehat{\beta})$ and variance matrix $I^{-1}(\widehat{\alpha}, \widehat{\beta})$, where $I(\widehat{\alpha}, \widehat{\beta})$ is the observed Fishers information matrix and is defined as

$$I\left(\widehat{\alpha}, \widehat{\beta}\right) = \begin{bmatrix} -\dfrac{\partial \log L}{\partial \alpha^2} & -\dfrac{\partial \log L}{\partial \beta \partial \alpha} \\ -\dfrac{\partial \log L}{\partial \alpha \partial \beta} & -\dfrac{\partial \log L}{\partial \beta^2} \end{bmatrix}_{(\alpha=\widehat{\alpha}, \beta=\widehat{\beta})}, \tag{11}$$

TABLE 7: Average 95% confidence/HPD intervals (in brackets) and bootstrap method along with their width and shape (in square brackets) for bootstrap method for complete sample $n = 20$, $r = 16$ with varying parameters.

(α,β)		Asymptotic	Bootstrap	Bayes Jeffrey	Bayes Gamma-1	Bayes Gamma-2
1, 1	α	1.13535 (0.5924266, 1.7277759)	1.441475 [3.029834] (0.802400, 2.243875)	0.6736 (0.7505, 1.4241)	0.6537 (0.7568, 1.4105)	0.6704 (0.7517, 1.4221)
	β	1.0079 (0.6011738, 1.6090738)	1.195897 [2.431916] (0.756660, 1.952557)	0.5714 (0.7766, 1.3480)	0.5585 (0.7786, 1.3371)	0.5693 (0.7772, 1.3465)
1, 2	α	1.03659 (0.6210150, 1.6576012)	1.290073 [2.861497] (0.805222, 2.095295)	0.5053 (0.8404, 1.3457)	0.4943 (0.8421, 1.3364)	0.5036 (0.8409, 1.3445)
	β	1.96599 (1.226844, 3.192829)	2.336601 [2.444149] (1.531410, 3.868011)	0.9327 (1.6816, 2.6143)	0.8825 (1.6950, 2.5775)	0.9249 (1.6844, 2.6093)
1, 3	α	1.00217 (0.6305741, 1.632746)	1.238806 [2.801578] (0.805794, 2.044599)	0.4245 (0.8848, 1.3093)	0.4145 (0.8848, 1.2993)	0.423 (0.8849, 1.3079)
	β	2.91598 (1.856743, 4.772727)	3.471131 [2.45426] (2.309851, 5.780982)	1.2101 (2.6382, 3.8483)	1.10717 (2.66343, 3.7706)	1.194 (2.6429, 3.8369)
2, 1	α	2.07318 (1.242028, 3.315206)	2.580142 [2.861492] (1.610445, 4.190587)	1.0107 (1.6808, 2.6915)	0.9546 (1.6934, 2.6480)	1.0023 (1.6835, 2.6858)
	β	0.983 (0.6134203, 1.5964171)	1.168299 [2.444144] (0.765705, 1.934004)	0.4663 (0.8408, 1.3071)	0.4571 (0.8407, 1.2978)	0.4646 (0.8413, 1.3059)
2, 2	α	1.96999 (1.270139, 3.240126)	2.426357 [2.768508] (1.611282, 4.037639)	0.7477 (1.8255, 2.5732)	0.7211 (1.8280, 2.5491)	0.7439 (1.8264, 2.5703)
	β	1.93131 (1.244336, 3.175642)	2.301952 [2.461984] (1.545066, 3.847018)	0.7208 (1.8113, 2.5321)	0.6938 (1.8151, 2.5089)	0.7162 (1.8123, 2.5285)
2, 3	α	1.93619 (1.278243, 3.214431)	2.37571 [2.731208] (1.609624, 3.985334)	0.6222 (1.8946, 2.5168)	0.6938 (1.8942, 2.4972)	0.6192 (1.8948, 2.5140)
	β	2.87555 (1.877916, 4.753462)	3.434175 [2.475603] (2.327609, 5.761784)	0.9124 (2.8180, 3.7304)	0.8621 (2.8259, 3.6880)	0.9042 (2.8200, 3.7242)
3, 1	α	3.00653 (1.891721, 4.898246)	3.716418 [2.801586] (2.417387, 6.133805)	1.2737 (2.6544, 3.9281)	1.172 (2.6668, 3.8388)	1.258 (2.6585, 3.9165)
	β	0.972 (0.6189131, 1.5909124)	1.157043 [2.454261] (0.769952, 1.926994)	0.40339 (0.8794, 1.28279)	0.39493 (0.8769, 1.27183)	0.40167 (0.8794, 1.28107)
3, 2	α	2.90428 (1.917364, 4.821645)	3.56358 [2.73123] (2.414436, 5.978016)	0.9335 (2.8418, 3.7753)	0.8884 (2.8420, 3.7304)	0.9265 (2.8427, 3.7692)
	β	1.91703 (1.251943, 3.168972)	2.289456 [2.475615] (1.551738, 3.841194)	0.6083 (1.8787, 2.4870)	0.5889 (1.8784, 2.4673)	0.6047 (1.8795, 2.4842)
3, 3	α	2.87151 (1.924054, 4.795567)	3.514358 [2.704448] (2.411125, 5.925483)	0.7726 (2.9299, 3.7025)	0.7435 (2.9263, 3.6698)	0.7682 (2.9299, 3.6981)
	β	2.8594 (1.886973, 4.746374)	3.42188 [2.488985] (2.335907, 5.757787)	0.7619 (2.9074, 3.6693)	0.7293 (2.9074, 3.6367)	0.7564 (2.9086, 3.6650)

where

$$\frac{\partial \log L}{\partial \alpha^2} = -\sum_{i=1}^{r} \frac{1}{\left(\alpha + \beta/x_{(i)}^2\right)^2} - \sum_{i=1}^{r} x_{(i)}^2 e^{(\alpha x_{(i)} - \beta/x_{(i)})} - (n-r)\, x_{(r)}^2 e^{(\alpha x_{(r)} - \beta/x_{(r)})},$$

$$\frac{\partial \log L}{\partial \beta^2} = -\sum_{i=1}^{r} \frac{\left(1/x_{(i)}^2\right)^2}{\left(\alpha + \beta/x_{(i)}^2\right)^2} + \sum_{i=1}^{r} \frac{e^{(\alpha x_{(i)} - \beta/x_{(i)})}}{x_{(i)}^2} + (n-r)\frac{e^{(\alpha x_{(r)} - \beta/x_{(r)})}}{x_{(r)}^2},$$

$$\frac{\partial \log L}{\partial \beta \partial \alpha} = \frac{\partial \log L}{\partial \alpha \partial \beta} = -\sum_{i=1}^{r} \frac{\left(1/x_{(i)}^2\right)}{\left(\alpha + \beta/x_{(i)}^2\right)^2} + \sum_{i=1}^{r} e^{(\alpha x_{(i)} - \beta/x_{(i)})} + (n-r)\, e^{(\alpha x_{(r)} - \beta/x_{(r)})}. \tag{12}$$

TABLE 8: Average 95% confidence/HPD intervals (in brackets) and bootstrap method along with their width and shape (in square brackets) for bootstrap method for complete sample $n = 20$, $r = 12$ with varying parameters.

(α,β)		Asymptotic	Bootstrap	Bayes Jeffrey	Bayes Gamma-1	Bayes Gamma-2
1, 1	α	1.80922	2.494413 [3.704015]	1.04609	0.96179	1.03549
		(0.3894311, 2.1986549)	(0.763769, 3.258183)	(0.58521, 1.63130)	(0.65815, 1.61994)	(0.60219, 1.63768)
	β	1.1317	1.440344 [2.899471]	0.58801	0.57383	0.58606
		(0.5780497, 1.7097506)	(0.774531, 2.214875)	(0.79841, 1.38642)	(0.80093, 1.37476)	(0.79897, 1.38503)
1, 2	α	1.5505	2.090383 [3.461255]	0.7101	0.6779	0.7043
		(0.4658775, 2.0163793)	(0.772565, 2.862947)	(0.8058, 1.5159)	(0.8171, 1.4950)	(0.8094, 1.5137)
	β	2.26918	2.903446 [2.942088]	0.974	0.9167	0.9661
		(1.159005, 3.428187)	(1.557071, 4.460517)	(1.7172, 2.6912)	(1.7286, 2.6453)	(1.7188, 2.6849)
1, 3	α	1.45582	1.948467 [3.381644]	0.5674	0.5473	0.5638
		(0.4942760, 1.9500981)	(0.777498, 2.725965)	(0.8813, 1.4487)	(0.8819, 1.4292)	(0.8827, 1.4465)
	β	3.41489	4.381877 [2.966392]	1.2838	1.1625	1.265
		(1.738521, 5.153412)	(2.341215, 6.723092)	(2.7184, 4.0022)	(2.7368, 3.8993)	(2.7219, 3.9869)
2, 1	α	3.10101	4.180747 [3.461238]	1.4143	1.2678	1.3829
		(0.9317525, 4.0327632)	(1.545131, 5.725878)	(1.6060, 3.0203)	(1.6496, 2.9174)	(1.6259, 3.0088)
	β	1.1346	1.451719 [2.942077]	0.4892	0.4771	0.4871
		(0.5795007, 1.7140960)	(0.778536, 2.230255)	(0.8650, 1.3542)	(0.8622, 1.3393)	(0.8651, 1.3522)
2, 2	α	2.81147	3.749764 [3.341453]	0.9754	0.9233	0.9672
		(1.018547, 3.830021)	(1.560572, 5.310336)	(1.8490, 2.8244)	(1.8515, 2.7748)	(1.8513, 2.8185)
	β	2.28314	2.936211 [2.984448]	0.771	0.737	0.7653
		(1.158481, 3.441619)	(1.563132, 4.499343)	(1.8684, 2.6394)	(1.8681, 2.6051)	(1.8694, 2.6347)
2, 3	α	2.7054	3.594662 [3.297717]	0.7911	0.7575	0.786
		(1.050075, 3.755473)	(1.566362, 5.161024)	(1.9465, 2.7376)	(1.9422, 2.6997)	(1.9466, 2.7326)
	β	3.44036	4.436749 [3.006102]	0.9862	0.9225	0.976
		(1.735746, 5.176104)	(2.348427, 6.785176)	(2.9134, 3.8996)	(2.9127, 3.8352)	(2.9144, 3.8904)
3, 1	α	4.36047	5.837303 [3.381616]	1.7022	1.493	1.6623
		(1.479581, 5.840054)	(2.327943, 8.165246)	(2.6367, 4.3389)	(2.6489, 4.1419)	(2.6515, 4.3138)
	β	1.13673	1.458758 [2.966365]	0.4279	0.4164	0.4257
		(0.5787959, 1.7155273)	(0.779418, 2.238177)	(0.9046, 1.3325)	(0.8975, 1.3139)	(0.9042, 1.3299)
3, 2	α	4.05809	5.391884 [3.297632]	1.1866	1.108	1.1739
		(1.575106, 5.633193)	(2.349532, 7.741416)	(2.9198, 4.1064)	(2.9066, 4.0146)	(2.9209, 4.0948)
	β	2.29357	2.957763 [3.006009]	0.6575	0.6326	0.653
		(1.157161, 3.450727)	(1.565612, 4.523375)	(1.9423, 2.5998)	(1.9347, 2.5673)	(1.9421, 2.5951)
3, 3	α	3.9438	5.227407 [3.269421]	1.0633	1.0157	1.0556
		(1.608629, 5.552424)	(2.356143, 7.583550)	(2.9352, 3.9985)	(2.9191, 3.9348)	(2.9352, 3.9908)
	β	3.45715	4.469468 [3.026892]	0.9317	0.889	0.901
		(1.733091, 5.190240)	(2.351761, 6.821229)	(2.9117, 3.8434)	(2.9027, 3.7917)	(2.9352, 3.8362)

The diagonal elements of $I^{-1}(\widehat{\alpha},\widehat{\beta})$ provide the asymptotic variances for the parameters α, β and respectively. A two-sided $100(1-\gamma)\%$ normal approximation confidence interval of α can be obtained as

$$\left\{\widehat{\alpha} \mp Z_{\gamma/2}\sqrt{\operatorname{var}(\widehat{\alpha})}\right\}. \tag{13}$$

Similarly, a two-sided $100(1-\gamma)\%$ normal approximation confidence interval of β can be obtained as

$$\left\{\widehat{\beta} \mp Z_{\gamma/2}\sqrt{\operatorname{var}(\widehat{\beta})}\right\}, \tag{14}$$

where Z is the standard normal variate (SNV).

3.2. Bootstrap Confidence Intervals. In this subsection, we have discussed another method for obtaining the confidence intervals proposed by [14]. They have developed the computer-based technique that can be routinely applied without any heavy theoretical consideration. The bootstrap method is very useful when an assumption regarding the normality is invalid. For bootstrap procedure, the computational algorithm is given as follows:

Step 1. Generate sample $\{x_1, x_2, \ldots, x_n\}$ of size n form (1) by using inversion method. Then estimated distribution function is given by $\widehat{F}(x, \widehat{\Theta})$, where $\Theta = \{\alpha, \beta; \alpha, \beta > 0\}$.

TABLE 9: Estimates and confidence/HPD/bootstrap intervals for real data set.

r		MLE	Bootstrap	Bayes
Complete	α	0.2071	Shape = 3.316171	0.19774
		0.1225917, 0.2916155	(0.1534971, 0.3848719)	0.1384182, 0.2585010
	β	0.25876	Shape = 2.136357	0.24668
		0.1300606, 0.3874579	(0.1746025, 0.4385483)	0.1613039, 0.3347743
18	α	0.27514	Shape = 7.412105	0.22501
		0.07038155, 0.4799067	(0.1595285, 1.1320991)	0.0554001, 0.4020123
	β	0.2557	Shape = 2.094187	0.24559
		0.1253812, 0.3860173	(0.1708061, 0.4334815)	0.1562725, 0.3323715
15	α	0.38256	Shape = 9.825055	0.26299
		−0.01972 = 0, 0.7848314	(0.1684655, 2.4860048)	0.0002199, 0.5562250
	β	0.25644	Shape = 2.330094	0.24741
		0.1237523, 0.3891245	(0.1741460, 0.4481874)	0.1567716, 0.3352633

TABLE 10: The summary of the one sample predictive densities for real data set when $r = 20$.

s	Mode	Mean	SE	95% predictive bounds	
				Lower	Upper
1	4.08201	4.98539	0.63119	4.08213	6.18207
2	5.06696	5.86145	1.15616	4.1365	8.08115
3	6.50403	7.42144	1.78186	4.49902	10.8474

Step 2. Generate a bootstrap sample $\{x_1^*, x_2^*, \ldots, x_n^*\}$ of size n from $\widehat{F}(x, \widehat{\Theta})$. Obtain bootstrap estimates $\widehat{\Theta}^* = \{\widehat{\alpha}^*, \widehat{\beta}^*\}$ of $\Theta = \{\alpha, \beta\}$ using bootstrap sample.

Step 3. Repeat Step 2, B-times. Obtain the bootstrap estimates $\{\widehat{\alpha}_1^*, \widehat{\alpha}_2^*, \ldots, \widehat{\alpha}_B^*\}$ and $\{\widehat{\beta}_1^*, \widehat{\beta}_2^*, \ldots, \widehat{\beta}_B^*\}$.

Step 4. Let $\{\widehat{\theta}_{(1)}^*, \widehat{\theta}_{(2)}^*, \ldots, \widehat{\theta}_{(B)}^*\}$ be the ordered values of a sequence $\{\widehat{\theta}_1^*, \widehat{\theta}_2^*, \ldots, \widehat{\theta}_B^*\}$ of a variable θ. Then, the empirical distribution function (EDF) of $\{\widehat{\theta}_1^*, \widehat{\theta}_2^*, \ldots, \widehat{\theta}_B^*\}$ is given by $\widehat{G}(t) = \{\text{Number of}\,(\widehat{\theta}^* < t)/B\}$. The $100(1-\gamma)\%$ boot-p confidence intervals for θ can be obtained by the following formula: $(\widehat{\theta}_{B\gamma/2}^*, \widehat{\theta}_{B(1-\gamma/2)}^*)$. By using the above definition, the two-sided $100(1-\gamma)\%$ boot-p confidence intervals for $\widehat{\alpha}$ and $\widehat{\beta}$ are given by $(\widehat{\alpha}_{B\gamma/2}^*, \widehat{\alpha}_{B(1-\gamma/2)}^*)$ and $(\widehat{\beta}_{B\gamma/2}^*, \widehat{\beta}_{B(1-\gamma/2)}^*)$, respectively.

Step 5. The bootstrap measure of symmetry can be defined as

$$\text{Shape} = \left[\frac{\widehat{\theta}_{B(1-\gamma/2)}^* - \widehat{\theta}^*}{\widehat{\theta}^* - \widehat{\theta}_{B(\gamma/2)}^*}\right]. \tag{15}$$

Using the above formula, we can also easily obtain the measure of symmetry (Shape) for α and β. For standard normal approximate confidence intervals, the shape is always equals to one.

4. Bayes Estimation

In Bayesian scenario, we need to assume the prior distribution of the unknown model parameters to take into account uncertainty of the parameters. The prior densities for α and β are given as

$$\begin{aligned} g_1(\alpha) &\propto \alpha^{b-1} e^{-\alpha a}, \quad a, b, \alpha > 0, \\ g_2(\beta) &\propto \beta^{d-1} e^{-\beta c}, \quad c, d, \beta > 0. \end{aligned} \tag{16}$$

Further, it is assumed that the parameters α and β are independent. Therefore, the joint prior of α and β is given by

$$g(\alpha, \beta) \propto g_1(\alpha) \times g_2(\beta), \tag{17}$$

where a, b, c, and d are the hyperparameters. Then, the joint posterior PDF of α and β can be readily defined as

$$\pi\left(\alpha, \beta \mid \underset{\sim}{x}\right) = \frac{L\left(\underset{\sim}{x} \mid \alpha, \beta\right) g(\alpha, \beta)}{\int_0^\infty \int_0^\infty L\left(\underset{\sim}{x} \mid \alpha, \beta\right) g(\alpha, \beta)\, d\alpha\, d\beta}. \tag{18}$$

If $\lambda(\alpha, \beta)$ is the function of α and β, then the Bayes estimates of $\lambda(\alpha, \beta)$ are given by

$$\widehat{\lambda}\left(\alpha, \beta \mid \underset{\sim}{x}\right) = \frac{\int_0^\infty \int_0^\infty \lambda(\alpha, \beta) L\left(\underset{\sim}{x} \mid \alpha, \beta\right) g(\alpha, \beta)\, d\alpha\, d\beta}{\int_0^\infty \int_0^\infty L\left(\underset{\sim}{x} \mid \alpha, \beta\right) g(\alpha, \beta)\, d\alpha\, d\beta}. \tag{19}$$

The above expression cannot be obtained in nice closed form. The evaluation of the posterior mean of the parameters will be complicated and it will be the ratio of two intractable integrals. In such situations, Monte Carlo Markov chain (MCMC) method, namely, Gibbs sampling techniques can be

Table 11: The summary of the two sample predictive densities for real data set.

r	k	Mode	Mean	SE	95% predictive bounds	
					Lower	Upper
23	1	0.06074	0.06507	0.01882	0.03105	0.10218
	2	0.0831	0.08968	0.02478	0.04469	0.13839
	3	0.10216	0.11247	0.03223	0.05422	0.17551
	4	0.12085	0.13377	0.03811	0.06713	0.20994
	23	6.19579	6.54602	1.7254	3.26888	9.81353
20	1	0.06155	0.06477	0.01884	0.03084	0.10226
	2	0.08422	0.08932	0.02484	0.04549	0.13953
	3	0.10355	0.11353	0.03237	0.05431	0.17682
	4	0.1225	0.1351	0.0385	0.06572	0.21009
	23	6.68878	7.39385	2.22182	3.2862	11.7307

effectively used. For implementing the Gibbs algorithm, the full conditional posterior densities of α and β are given by

$$\begin{aligned}\pi_1\left(\alpha \mid \beta, \underset{\sim}{x}\right) &= e^{-\alpha(\sum_{i=1}^{r} x_{(i)}+a)}\prod_{i=1}^{r}\left(\alpha+\frac{\beta}{x_{(i)}^2}\right) \\ &\quad\times \exp\left(-\sum_{i=1}^{r}\exp\left(\alpha x_{(i)}-\frac{\beta}{x_{(i)}}\right)\right) \\ &\quad\times \alpha^{b-1}\exp\left(-(n-r)\exp\left(\alpha x_{(r)}-\frac{\beta}{x_{(r)}}\right)\right), \\ \pi_2\left(\beta \mid \alpha, \underset{\sim}{x}\right) &= e^{-\beta(\sum_{i=1}^{r}(1/x_{(i)}+c))}\prod_{i=1}^{r}\left(\alpha+\frac{\beta}{x_{(i)}^2}\right) \\ &\quad\times \exp\left(-\sum_{i=1}^{r}\exp\left(\alpha x_{(i)}-\frac{\beta}{x_{(i)}}\right)\right) \\ &\quad\times \beta^{d-1}\exp\left(-(n-r)\exp\left(\alpha x_{(r)}-\frac{\beta}{x_{(r)}}\right)\right).\end{aligned} \tag{20}$$

The simulation algorithm consists of the following steps.

Step 1. Start with $j = 1$ and the initial values of $\{\alpha^{(0)}, \beta^{(0)}\}$.

Step 2. Using the initial values $\{\alpha^{(0)}, \beta^{(0)}\}$, generate candidate points $\{\alpha_c^{(j)}, \beta_c^{(j)}\}$ from proposal densities $\{q_1(\alpha^{(j)}, \alpha^{(j)}), q_2(\beta^{(j)}, \beta^{(j-1)})\}$, where, $q(\alpha^{(j)}, \alpha^{(j-1)}) = q(\alpha^{(j)} \to \alpha^{(j-1)})$ is the probability of returning a value of $\alpha^{(j)}$ given a previous value of $\alpha^{(j-1)}$.

Step 3. Generate U uniform variate on range 0-1; that is, $u \sim U(0,1)$.

Step 4. Calculate the ratios at the candidate point $\alpha_c^{(j)}$ and previous point $\alpha^{(j)}$

$$R_1 = \left(\frac{\pi_1\left(\alpha_c^{(j)} \mid \beta^{(j-1)}, \underset{\sim}{x}\right) q_1\left(\alpha_c^{(j)}, \alpha^{(j-1)}\right)}{\pi_1\left(\alpha^{(j-1)} \mid \beta^{(j-1)}, \underset{\sim}{x}\right) q_1\left(\alpha^{(j-1)}, \alpha_c^{(j)}\right)}\right). \tag{21}$$

Step 5. If $u \le \min(1, R_1)$, accept the candidate point with probability $\min(1, R_1)$, that is, $\alpha^{(j)} = \alpha_c^{(j)}$. Otherwise set $\alpha^{(j)} = \alpha^{(j-1)}$.

Step 6. Similarly from Step 4, the ratio

$$R_2 = \left(\frac{\pi_2\left(\beta_c^{(j)} \mid \alpha^{(j)}, \underset{\sim}{x}\right) q_2\left(\beta_c^{(j)}, \beta^{(j-1)}\right)}{\pi_2\left(\beta^{(j-1)} \mid \alpha^{(j)}, \underset{\sim}{x}\right) q_2\left(\beta^{(j-1)}, \beta_c^{(j)}\right)}\right). \tag{22}$$

Step 7. If $u \le \min(1, R_2)$, accept the candidate point with probability $\min(1, R_2)$, that is, $\beta^{(j)} = \beta_c^{(j)}$. Otherwise set $\beta^{(j)} = \beta^{(j-1)}$.

Step 8. Repeat Steps 2–7 for all $j = 1, 2, \ldots, M$ and obtain $(\alpha_1, \beta_1), (\alpha_2, \beta_2), \ldots, (\alpha_M, \beta_M)$.

Note that if the candidate point is independent on previous point, that is, if $q(x \to y) = q(x)$, then the M-H algorithm is called independence M-H sampler. The acceptance function becomes

$$\min\left(1, \left(\frac{\pi_1\left(\alpha_c^{(j)} \mid \beta^{(j-1)}, \underset{\sim}{x}\right) q_1\left(\alpha_c^{(j)}\right)}{\pi_1\left(\alpha^{(j-1)} \mid \beta^{(j-1)}, \underset{\sim}{x}\right) q_1\left(\alpha^{(j-1)}\right)}\right)\right). \tag{23}$$

The Bayes estimates under SELF of the parameters can be obtained as the mean of the generated samples from the posterior densities by using the algorithm discussed previously. The formulae are given by

$$\hat{\alpha} = E_\pi\left(\alpha \mid \underset{\sim}{x}\right) \approx \frac{1}{M - M_0}\sum_{k=M_0+1}^{M}\alpha_k, \tag{24}$$

$$\hat{\lambda} = E_\pi\left(\lambda \mid \underset{\sim}{x}\right) \approx \frac{1}{M - M_0}\sum_{k=M_0+1}^{M}\lambda_k, \tag{25}$$

where M_0 is the burn-in-period of Markov Chain. The HPD credible intervals for α and λ can be constructed by using the algorithm given in [15]. Let $\{(\alpha_{(i)}, \beta_{(i)}); i = 1, 2, \ldots, M\}$ be the corresponding ordered MCMC sample of $\{(\alpha_i, \beta_i), i = 1, 2, \ldots, M\}$. Then construct all the $100(1-\psi)\%$ credible intervals of α and λ as

$$\begin{gathered}\left(\alpha_{[1]}, \alpha_{[M(1-\psi)]}\right), \ldots, \left(\alpha_{[M\psi]}, \alpha_{[M]}\right), \\ \left(\beta_{[1]}, \beta_{[M(1-\psi)]}\right), \ldots, \left(\beta_{[M\psi]}, \beta_{[M]}\right).\end{gathered} \tag{26}$$

Here, $[x]$ denotes the largest integer less than or equal to x. Then, the HPD credible interval is that interval which has the shortest length.

5. One-Sample Prediction

In a Type-II censoring scheme, the life test consists only of few observed items (say r) out of all units/items (say n) under study due to time and cost constraints. In practice, the experimenter may be interested to know the life time of the $(n-r)$ removed surviving units on the basis of informative units. In such a situation, one-sample prediction technique may be helpful to give an idea about the expected life of the removed units. Let $x_{(1)}, x_{(2)}, \ldots, x_{(r)}; (r \leq n)$ the observed censored sample and $y_{(1)}, y_{(2)}, \ldots, y_{(n-r)}$ be the unobserved future ordered sample from the same population. Let $Y_{(s)} = X_{(r+s)}$ represent the failure lifetimes of the remaining surviving units. From [16], the conditional PDF of $Y_{(s)}$ given $X = x$ can be obtained as

$$f(y_{(s)} \mid x_{(r)}) = \frac{(n-r)!}{(s-1)!\,(n-r-s)!} \times \frac{[1-F(y_{(s)})]^{n-r-s}[F(y_{(s)})-F(x_{(r)})]^{s-1} f(y_{(s)})}{[1-F(x_{(r)})]^{n-r}}. \quad (27)$$

Putting (1) and (2) in (27), we get

$$f(y_{(s)} \mid x_{(r)}) = \frac{(n-r)!}{(s-1)!\,(n-r-s)!}\left(\alpha + \frac{\beta}{y_{(s)}^2}\right) \times \sum_{j=0}^{s-1}\sum_{i=0}^{\infty}\sum_{l=0}^{i}\frac{-1^{(i+j)}}{i!}\binom{s-1}{j}\binom{i}{l}(n-r-s+j+1)^i \times \exp\left[(i+l-1)\left(\alpha y_{(s)} - \frac{\beta}{y_{(s)}}\right) + l\left(\alpha x_{(r)} - \frac{\beta}{x_{(r)}}\right)\right]. \quad (28)$$

Then, the predictive posterior density of future observables under Type-II censoring scheme is given by

$$f_1\left(y_{(s)} \mid \underset{\sim}{x}\right) = \int_0^\infty \int_0^\infty f\left(y_{(s)} \mid \alpha, \beta, \underset{\sim}{x}\right) \pi\left(\alpha, \beta \mid \underset{\sim}{x}\right) d\alpha\, d\beta. \quad (29)$$

Equation (29) cannot be evaluated analytically. Therefore, to obtain the consistent estimator for $f_1(y_{(s)} \mid \underset{\sim}{x})$, MCMC sample obtained through Gibbs algorithm is used. The consistent estimate of $f_1(y_{(s)} \mid \underset{\sim}{x})$ is obtained as

$$f_1^*\left(y_{(s)} \mid \underset{\sim}{x}\right) = \frac{1}{M - M_0}\sum_{i=1}^{M-M_0} f\left(y_{(s)} \mid \alpha_i, \beta_i, \underset{\sim}{x}\right). \quad (30)$$

To obtain the estimate of future sample, we used the M-H algorithm, to draw the sample from (29). Similarly, form (24), we can estimate the future observations under SELF as the mean of simulated sample drawn from (29). The survival function of future sample can be simply defined as

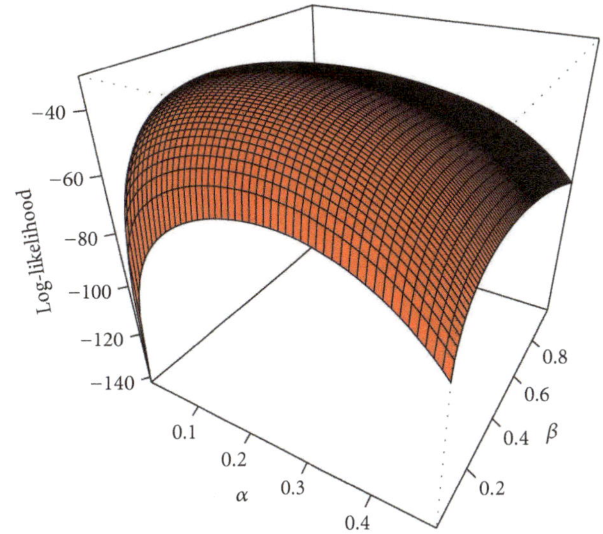

FIGURE 1: Likelihood profile with respect to parameters for real data set.

$$S_{y_{(s)}}(T) = 1 - \int_{y_{(s)}=x_{(r)}}^{T} f_1\left(y_{(s)} \mid \underset{\sim}{x}\right) dy_{(s)} = 1 - \int_{y_{(s)}=x_{(r)}}^{T}\int_0^\infty\int_0^\infty f\left(y_{(s)} \mid \alpha, \beta, \underset{\sim}{x}\right) \times \pi\left(\alpha, \beta \mid \underset{\sim}{x}\right) d\alpha d\beta dy_{(s)}. \quad (31)$$

We can also obtain the two-sided $100(1-\alpha)\%$ prediction intervals (L_s, U_s) for $y_{(s)}$ by solving the following two equations:

$$P\left(y_{(s)} > U_s \mid \underset{\sim}{x}\right) = \frac{\alpha}{2}, \qquad P\left(y_{(s)} > L_s \mid \underset{\sim}{x}\right) = 1 - \frac{\alpha}{2}. \quad (32)$$

Confidence intervals can be obtained by using any suitable iterative procedure as the above equations cannot be solved directly.

6. Two-Sample Prediction

In some situations, only lifetime model is given, and no priori information is available then assumes $p(y_{(k)} \mid \alpha, \beta, \underset{\sim}{x}) = p(y_{(k)} \mid \alpha, \beta)$. This leads to the two-sample prediction problems. That is, the experimenters are interested in the kth failure time in a future sample of size N following the same life

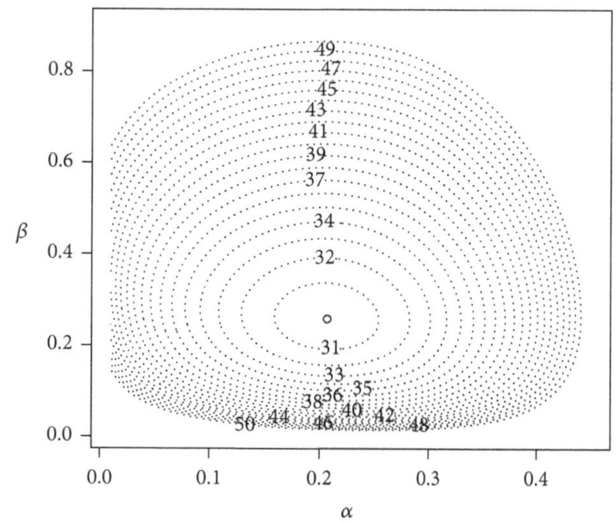

FIGURE 2: Likelihood contour plot with respect to parameters for real data set.

time distribution. From [16], the PDF of kth order statistics is given by

$$p\left(y_{(k)} \mid \alpha, \beta\right) = \frac{N!}{(k-1)!\,(N-k)!} \times \left[F\left(y_{(k)}\right)\right]^{k-1}\left[1 - F\left(y_{(k)}\right)\right]^{N-k} f\left(y_{(k)}\right). \tag{33}$$

Putting (1) and (2) in (29), we get

$$p\left(y_{(k)} \mid \alpha, \beta\right) = \frac{N!}{(k-1)!\,(N-k)!}\left(\alpha + \frac{\beta}{y_{(k)}^2}\right) \times \sum_{j=0}^{k-1}\sum_{i=0}^{\infty}\frac{-1^j}{i!}\binom{s-1}{j}(N-k+j+1)^i \times \exp\left[(i+1)\left(\alpha y_{(k)} - \frac{\beta}{y_{(k)}}\right)\right]. \tag{34}$$

The predictive posterior density of future observables under Type-II censoring scheme is given by

$$p_1\left(y_{(k)} \mid \underset{\sim}{x}\right) = \int_0^{\infty}\int_0^{\infty} p\left(y_{(k)} \mid \alpha, \beta, \underset{\sim}{x}\right)\pi\left(\alpha, \beta \mid \underset{\sim}{x}\right) d\alpha\, d\beta. \tag{35}$$

Equation (35) cannot be evaluated analytically. Therefore, to obtain the consistent estimator for $p_1(y_{(k)} \mid \underset{\sim}{x})$, MCMC sample obtained through Gibbs algorithm is used. The consistent estimate of $p_1(y_{(k)} \mid \underset{\sim}{x})$ is given by

$$p_1^*\left(y_{(k)} \mid \underset{\sim}{x}\right) = \frac{1}{M - M_0}\sum_{i=1}^{M-M_0} p\left(y_{(k)} \mid \alpha_i, \beta_i\right). \tag{36}$$

To obtain the estimate of future sample, again M-H algorithm is used to draw the sample from (35). Similarly, form (24), we can estimate the future observations under SELF as the mean of simulated sample drawn from (35). The survival function of future sample can be simply defined as

$$\begin{aligned} S_{y_{(k)}}(T) &= 1 - \int_{y_{(k)}=0}^{T} p_1\left(y_{(k)} \mid \underset{\sim}{x}\right) dy_{(k)} \\ &= 1 - \int_{y_{(k)}=0}^{T}\int_0^{\infty}\int_0^{\infty} p\left(y_{(k)} \mid \alpha, \beta, \underset{\sim}{x}\right) \\ &\quad \times \pi\left(\alpha, \beta \mid \underset{\sim}{x}\right) d\alpha\, d\beta\, dy_{(k)}. \end{aligned} \tag{37}$$

We can also obtain the two sided $100(1-\alpha)\%$ prediction intervals (L_k, U_k) for $y_{(k)}$ by solving the following two nonlinear equations:

$$\begin{aligned} P\left(y_{(k)} > U_k \mid \underset{\sim}{x}\right) &= \frac{\alpha}{2}, \\ P\left(y_{(k)} > L_k \mid \underset{\sim}{x}\right) &= 1 - \frac{\alpha}{2}. \end{aligned} \tag{38}$$

Confidence intervals can be obtained by using any suitable iterative procedure as the above equations cannot be solved directly.

7. Simulation Study

This section consists of the simulation results to compare the performance of the classical and Bayesian estimation procedures under different Type-II censoring schemes and parameter combinations. The comparison between the MLEs and Bayes estimators of the model parameters made in terms of their mean square errors (MSEs). We have also compared the average lengths of the asymptotic confidence intervals, bootstrap intervals and HPD credible intervals. For this purpose, we generate the sample of sizes $n = 20$ small, 30 medium, and 50 large from (1) for fixed values of $\alpha = 2$ and $\beta = 2$. We have considered the different Type-II censoring schemes for each sample size so that the sample contains 100%, 80%, and 60% of the available information.

The choice of the hyperparameters is the main concerning issue in the Bayesian analysis. Reference [17] argues that when information is not in compact form it is better to perform the Bayesian analysis under the assumption of non-informative prior. If we take $a = b = c = d = 0$, then posterior becomes as obtained under Jeffery's scale invariant prior. For the choice of hyper parameters under the subjectivism, we have taken prior means equals to the true values of the parameters with varying variances. The prior variance indicates the confidence of our prior guess. A large prior variance shows less confidence in prior guess and resulting prior distribution is relatively flat. On other hand, small prior variance indicates greater confidence in prior guess. In this study, we have taken prior variance equals to 1 (small) and 8 (large), and we call these Gamma-1 and Gamma-2 priors, respectively.

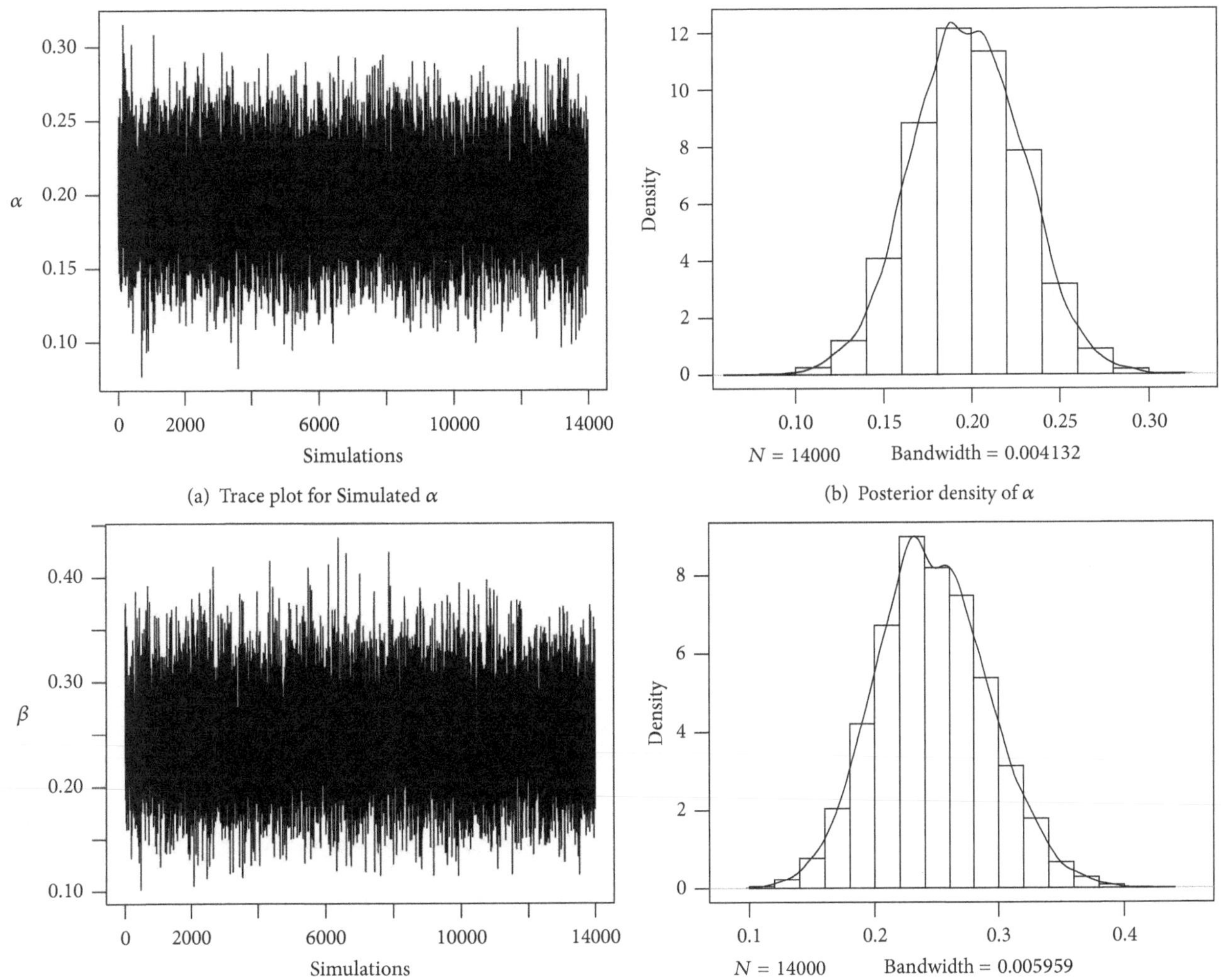

(a) Trace plot for Simulated α

(b) Posterior density of α

(c) Trace plot for Simulated β

(d) Posterior density of β

FIGURE 3: MCMC runs and posterior density plot of α and β for complete real data set.

For obtaining the Bayes estimates, we generate posterior deviates for the parameters α and β using algorithm discussed in Section 4. First thousand MCMC iterations (Burn-in period) have been discarded from the generated sequence. We have also checked the convergence of the sequences of α and β for their stationary distributions through different starting values. It was observed that all the Markov chains reached to the stationary condition very quickly.

For the unknown model parameters, we have computed MLEs and Bayes estimates under informative and non-informative priors along with their asymptotic confidence/bootstrap/HPD intervals. We repeat the process 1000 times, and the average estimates with the corresponding mean square errors (MSEs) of the estimators, and average confidence/bootstrap/HPD intervals are recorded. The simulation results are summarized in Tables 1, 2, 3, 4, 5, 6, 7, and 8. All necessary computational algorithms are coded in R-environment [18] and codes are available with the authors.

On the basis of the results summarised in Tables 1–8, some conclusions can be drawn which are stated as follows:

(i) The MSE of all the estimators decreases as sample increases (i.e., as n and r increases) for fixed values of α and β.

(ii) The MSE of all the estimators increases with increasing the value of the parameters for any fixed values of n and r.

(iii) The MSE of the maximum likelihood and Bayes estimator of α increases with increasing α for given values of β, n and r.

(iv) The MSE of the maximum likelihood and Bayes estimator of β increases with increasing β for given values of α, n and r.

(v) The Bayes estimators have the smaller risks than the classical estimators for estimating the parameters in all the considered cases. Although the Bayes estimates obtained under Gamma-1 prior more efficient than those obtained under Jeffery's and Gamma-2 priors. This indicates that the Bayesian procedure with

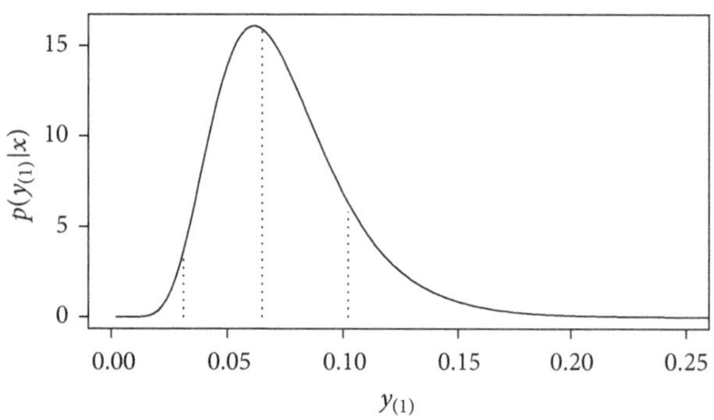

FIGURE 4: Two-sample predictive density function when $k = 1$ for complete real data set.

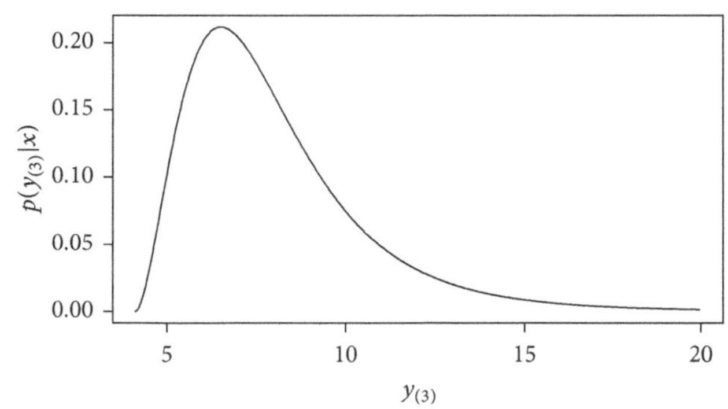

FIGURE 5: One-sample predictive density function when $S = 3$ and $r = 20$ for real data set.

accurate prior information provides more precise estimates.

(vi) The width of the HPD credible intervals is smaller than the width of the asymptotic confidence/bootstrap intervals.

(vii) In all the cases, the bootstrap procedure provides larger width of the confidence intervals for the parameters and the wide range of the confidence intervals helps to cover the asymmetry.

(viii) The shape is greater than one in all the considered cases which indicates that the distribution of the maximum likelihood estimators is positively skewed and becomes more skewed with decreasing sample size n.

8. Real Data Analysis

In this section, we analysed the data set of time between failures of secondary reactor pumps. This data set has been originally discussed in [19]. The chance of the failure of the secondary reactor pump is of the increasing nature in early stage of the experiment and after that it decreases. It has been checked by [10] that flexible Weibull distribution is well fitted model to this data set. The times between failures of 23 secondary reactor pumps are as follows: 2.160, 0.150, 4.082, 0.746, 0.358, 0.199, 0.402, 0.101, 0.605, 0.954, 1.359, 0.273, 0.491, 3.465, 0.070, 6.560, 1.060, 0.062, 4.992, 0.614, 5.320, 0.347, and 1.921.

For analysing this data set under Type-II censoring scheme, we generate two artificial Type-II censored samples from this real data set by considering two different values of $r (= 18, 15)$. In the real applications, we have nothing in our hand other than few observations following any distribution function. Let us sketch the likelihood profile with respect to the parameters. The log-likelihood function is plotted over the whole parameter space in Figure 1. The contour plots of the likelihood function are also plotted in Figure 2. The maximum likelihood estimates, Bayes estimates, and corresponding confidence/bootstrap/HPD intervals are presented in Table 9 for different values of n and r. The simulation runs and the histogram plot of simulated α and β are plotted in Figure 3.

The summary of one-sample predictive densities of future samples is presented in Table 10 for different values of s. Two-sample predictive density functions for different values of r and k are summarised in Table 11. From Tables 10 and 11, it can be observed that the standard error (SE) of future observables increases as the values of s and k increase for one-sample and two-sample prediction, respectively. Two-sample predictive density of the first future ordered sample is plotted in Figure 4. The density of $y_{(3)}$th future sample in case of one-sample prediction is plotted in Figure 5.

The expressions (31) and (37) do not seem to be possible to compute analytically. Therefore, we prosed to use the Monte Carlo technique to solve these two equations. To compute the integral part

$$I_1 = \int_{y=x_{(r)}}^{T} \int_0^{\infty} \int_0^{\infty} f\left(y \mid \alpha, \beta, \underset{\sim}{x}\right) \pi\left(\alpha, \beta \mid \underset{\sim}{x}\right) d\alpha\, d\beta\, dy,$$
$$I_2 = \int_{y=0}^{T} \int_0^{\infty} \int_0^{\infty} p\left(y \mid \alpha, \beta, \underset{\sim}{x}\right) \pi\left(\alpha, \beta \mid \underset{\sim}{x}\right) d\alpha\, d\beta\, dy \tag{39}$$

appearing in the expressions, we follow the following steps.

Step 1. The approximate value of I_1 can be obtained as

$$I_1 = \int_{y=x_{(r)}}^{T} E_{\pi}\left[f\left(y \mid \alpha, \beta, \underset{\sim}{x}\right)\right] dy$$
$$\approx \left(T - x_{(r)}\right) \frac{1}{M} \sum_{i=1}^{M} E_{\pi}\left[f\left(y_i \mid \alpha, \beta, \underset{\sim}{x}\right)\right], \tag{40}$$
$$y \sim U\left(x_{(r)}, T\right).$$

Similarly,

$$I_2 \approx T \frac{1}{M} \sum_{i=1}^{M} E_{\pi}\left[p\left(y_i \mid \alpha, \beta, \underset{\sim}{x}\right)\right], \quad y \sim U(0, T). \tag{41}$$

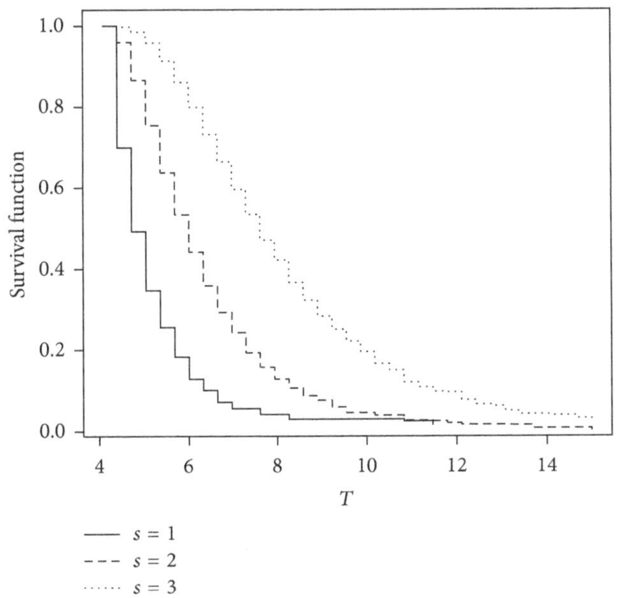

FIGURE 6: One-sample predictive survival function for real data set when $r = 20$.

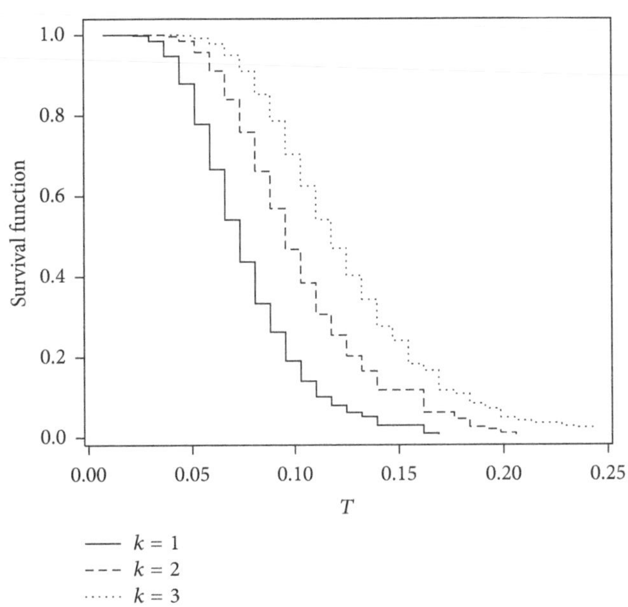

FIGURE 7: Two-sample predictive survival function for real data set.

E_π denotes the expectation of the function with respect to the joint posterior pdf of α and λ.

Step 2. Simulate α and λ from (19) by using the algorithm discussed in Section 2.

Step 3. Generate U_1 from uniform density of size M on range $(x_{(r)}, T)$. That is, $U_1 \sim U(x_{(r)}, T)$.

Step 4. Generate U_2 from uniform density of size M on range $(0, T)$. That is, $U_2 \sim U(0, T)$.

Step 5. By using U_1 and U_2, then calculate the approximate values of integrals as

$$I_1 \approx (T - x_{(r)}) \frac{1}{M} \sum_{i=1}^{M} E_\pi \left[f\left(U_{i:1} \mid \alpha, \beta, \underset{\sim}{x} \right) \right],$$
$$I_2 \approx T \frac{1}{M} \sum_{i=1}^{M} E_\pi \left[p\left(U_{i:2} \mid \alpha, \beta, \underset{\sim}{x} \right) \right]. \qquad (42)$$

By using the above-discussed algorithm, the survival functions (31) and (37) of future samples are plotted in Figures 6 and 7, respectively.

9. Conclusion

In this paper, we have considered the problem of estimation and prediction for flexible Weibull distribution in the presence of Type-II censored sample. We have found that Bayesian procedure provides the precise estimates of the unknown parameters of flexible Weibull model with smaller mean square error. The width of HPD intervals is smaller than that of the asymptotic and the bootstrap confidence intervals. Prediction has been applied in medicine, engineering, business, and other areas, and the Bayesian approach using MCMC methods can be effectively used to solve the prediction problems. The methodology developed in this paper will be very useful to the researchers, engineers, and statisticians where such type of life test is needed and especially where the flexible Weibull distribution is used.

Acknowledgment

The authors would like to thank the editor and revivers for the valuable suggestions to improve this manuscript. The third author (Vikas Kumar Sharma) thanks to the University grant commission (UGC), New Delhi for financial assistance.

References

[1] I. R. Dunsmore, “The Bayesian predictive distribution in life testing models,” *Technometrics*, vol. 16, pp. 455–460, 1974.

[2] E. K. AL-Hussaini, “Predicting observables from a general class of distributions,” *Journal of Statistical Planning and Inference*, vol. 79, no. 1, pp. 79–91, 1999.

[3] E. K. Al-Hussaini, “On Bayes prediction of future median,” *Communications in Statistics*, vol. 30, no. 7, pp. 1395–1410, 2001.

[4] B. Pradhan and D. Kundu, “Bayes estimation and prediction of the two-parameter gamma distribution,” *Journal of Statistical Computation and Simulation*, vol. 81, no. 9, pp. 1187–1198, 2011.

[5] D. Kundu and M. Z. Raqab, “Bayesian inference and prediction of order statistics for a Type-II censored Weibull distribution,” *Journal of Statistical Planning and Inference*, vol. 142, no. 1, pp. 41–47, 2012.

[6] D. Kundu and H. Howlader, “Bayesian inference and prediction of the inverse Weibull distribution for Type-II censored data,” *Computational Statistics & Data Analysis*, vol. 54, no. 6, pp. 1547–1558, 2010.

[7] I. G. Evans and A. M. Nigm, “Bayesian prediction for two-parameter Weibull lifetime models,” *Communications in Statistics*, vol. 9, no. 6, pp. 649–658, 1980.

[8] S. K. Singh, U. Singh, and V. K. Sharma, "Bayesian estimation and prediction for the generalized lindley distribution under assymetric loss function," *Hacettepe Journal of Mathematics and Statistics*. In press.

[9] S. K. Singh, U. Singh, and V. K. Sharma, "Bayesian prediction of future observations from inverse weibull distribution based on type-ii hybrid censored sample," *International Journal of Advanced Statistics and Probability*, vol. 1, pp. 32–43, 2013.

[10] M. Bebbington, C.-D. Lai, and R. Zitikis, "A flexible weibull extension," *Reliability Engineering and System Safety*, vol. 92, no. 6, pp. 719–726, 2007.

[11] A. F. M. Smith and G. O. Roberts, "Bayesian computation via the Gibbs sampler and related Markov chain Monte Carlo methods," *Journal of the Royal Statistical Society B*, vol. 55, no. 1, pp. 3–23, 1993.

[12] W. K. Hastings, "Monte carlo sampling methods using Markov chains and their applications," *Biometrika*, vol. 57, no. 1, pp. 97–109, 1970.

[13] S. P. Brooks, "Markov chain Monte Carlo method and its application," *Journal of the Royal Statistical Society D*, vol. 47, no. 1, pp. 69–100, 1998.

[14] B. Efron and R. Tibshirani, "Bootstrap methods for standard errors, confidence intervals, and other measures of statistical accuracy," *Statistical Science*, vol. 1, no. 1, pp. 54–77, 1986.

[15] M.-H. Chen and Q.-M. Shao, "Monte Carlo estimation of Bayesian credible and HPD intervals," *Journal of Computational and Graphical Statistics*, vol. 8, no. 1, pp. 69–92, 1999.

[16] H. A. David and H. N. Nagaraja, *Order Statistics*, Wiley Series in Probability and Statistics, Wiley, New York, NY, USA, 3rd edition, 2003.

[17] J. O. Berger, *Statistical Decision Theory and Bayesian Analysis*, Springer Series in Statistics, Springer, New York, NY, USA, 2nd edition, 1985.

[18] R Development Core Team, "R: a language and environment for statistical computing," Tech. Rep., R Foundation for Statistical Computing, Vienna, Austria, 2012, http://www.r-project.org/.

[19] M. S. Suprawhardana, Prayoto, and Sangadji, "Total time on test plot analysis for machenical components of the rsg-gas reactor," *Atom Inones*, vol. 25, no. 2, 1999.

2

Masses of Negative Multinomial Distributions: Application to Polarimetric Image Processing

Philippe Bernardoff,[1] Florent Chatelain,[2] and Jean-Yves Tourneret[3]

[1] *Université de Pau et des Pays de l'Adour, Avenue de l'Université, 64012 Pau, France*
[2] *Gipsa-Lab, Département Images et Signal, 961 rue de la Houille Blanche, BP 46, 38402 Saint Martin d'Héres, France*
[3] *Université de Toulouse, IRIT/ENSEEIHT/TESA, 2 rue Charles Camichel, BP 7122, 31071 Toulouse Cedex 7, France*

Correspondence should be addressed to Florent Chatelain; florent.chatelain@gipsa-lab.inpg.fr

Academic Editor: Edwin Ortega

This paper derives new closed-form expressions for the masses of negative multinomial distributions. These masses can be maximized to determine the maximum likelihood estimator of its unknown parameters. An application to polarimetric image processing is investigated. We study the maximum likelihood estimators of the polarization degree of polarimetric images using different combinations of images.

1. Introduction

The univariate negative binomial distribution is uniquely defined in many statistical textbooks. However, extensions defining multivariate negative multinomial distributions (NMDs) are more controversial. Most definitions are based on the probability generating function (PGF) of these distributions. Doss [1] proposed to define the PGF of an NMD as the inverse λth power of a polynomial linear in each of its variables. This definition can also be found in the famous textbook [2, page 93] and the computation of its modes has been investigated in [3]. A more general class of NMDs introduced in [4] was characterized by PGFs of the form $|\mathbf{I} - \mathbf{Q}|^{\lambda}|\mathbf{I} - \mathbf{QZ}|^{-\lambda}$, where $\lambda > 0$, $\mathbf{Q}$ is an $n \times n$ matrix, and $\mathbf{Z} = \operatorname{diag}(z_1, \ldots, z_n)$. In particular, matrices $\mathbf{Q}$ yielding infinitely divisible PGFs were derived. Finally, Bar-Lev et al. [5] introduced NMDs whose PGFs are defined as the inverse λth power of any affine polynomial. Necessary and sufficient conditions on the coefficients of this affine polynomial were derived to obtain the PGF of a multivariate distribution defined on $\mathbb{N}_0^n$ (where $\mathbb{N}_0$ is the set of nonnegative integers) [6]. These very general multivariate NMDs were recently used for image processing applications in [7].

The family of NMDs introduced in [5] can be defined as follows. Let us denote $[n] = \{1, \ldots, n\}$ the set of the n first nonzero integers. We denote $\mathbf{z}^T = \prod_{t \in T} z_t$ as the monomial obtained by multiplying all the entries of the vector $\mathbf{z} = (z_1, \ldots, z_n) \in \mathbb{R}^n$ whose indexes belong to T, where $T \subset [n]$ stands for any subset of the indexes. Let $P_n(\mathbf{z}) = \sum_{T \subset [n], T \neq \emptyset} p_T \mathbf{z}^T$ be an affine polynomial with respect to the n variables $(z_1, \ldots, z_n)$ such that $1 - P_n(\mathbf{1}) \neq 0$. The NMD distribution defined at pair (n, P_n) is represented by its PFG which is given by

$$\mathcal{G}_{\mathcal{NM}(n,P_n)}(\mathbf{z}) = \left(1 - P_n(\mathbf{z})\right)^{-\lambda} \left(1 - P_n(\mathbf{1})\right)^{\lambda}. \tag{1}$$

Such laws are denoted as $\mathcal{NM}(n, P_n)$. However, as explained in [6], all couples (n, P_n) do not provide a valid NMD. More specifically, Bernardoff has derived a finite number of conditions over P_n such that $(1 - P_n(\mathbf{z}))^{-\lambda}(1 - P_n(\mathbf{1}))^{\lambda}$ is the PGF of an NMD for all positive integer n. The corresponding expression of the coefficient of $\mathbf{z}^{\alpha}$ in the Taylor expansion of $(1 - P_n(\mathbf{z}))^{-\lambda}$ is given by the formula

$$c_{\alpha}(\lambda, P_n) = \sum_{k \in K_{\alpha}} (\lambda)_{|k|} \frac{\mathbf{p}^k}{k!}, \tag{2}$$

where $K_\alpha = \{k : \mathscr{P}_n \to \mathbb{N}\}$ and $\mathscr{P}_n$ is the set of all subsets of $[n]$. However, this expression of $c_\alpha(\lambda, P_n)$ does not allow us to explicitly compute the masses of NMDs in the general case.

As a first goal of this paper, we propose a way of computing the masses of multivariate NMDs $\mathscr{N}\mathscr{M}(n, P_n)$ defined above. A specific attention is devoted to bivariate and trivariate cases. In particular, it allows us to retrieve the results of [7] obtained for bivariate NMDs. The second part of the paper is devoted to the application of NMDs to image processing, more specifically to polarimetric image processing [8, 9]. Polarimetric image processing has received a considerable attention in the image processing and optical communities (see for instance [10–12] and references therein). The state of polarization of a polarimetric image is classically characterized by the degree of polarization (DoP) whose estimation is of major importance [13, 14]. The DoP of polarimetric images can be classically estimated by using four images associated to four different polarizations [15]. However, estimating the DoP using less than four images is interesting since it allows one to reduce the acquisition time and the resulting cost of the imaging system. As a consequence, there has been recently an increasing interest in deriving estimators of the DoP based on a reduced number of polarimetric images. Depending on the intensity of the acquired images, polarimetric images are referred to as low flux or high flux images (low flux corresponding to a small intensity and high flux to a larger intensity). DoP estimation based on a single polarimetric image was considered in [16, 17] under high flux and low flux assumptions. DoP estimators derived from two intensity images degraded by fully developed speckle noise were studied in [18, 19]. Finally, imaging systems using three polarimetric images were studied in [20, 21], under high flux and low flux assumptions.

This paper studies the maximum likelihood estimators (MLEs) of the square DoP based on two or three polarimetric images. These estimators are computed by maximizing the masses of bivariate or trivariate NMDs derived in the first part of this work.

The paper is organized as follows. Section 2 recalls important results on NMDs. Section 3 proposes a new way of computing masses of NMDs. A particular attention is devoted to bivariate and trivariate cases. Section 4 addresses the problem of estimating the square DoP of low flux polarimetric images using the maximum likelihood (ML) principle. Different MLEs are constructed depending on the number of available polarimetric images. Simulation results are presented in Section 5.

2. Negative Multinomial Distributions

An n-variate NMD is the distribution of a random vector $\mathbf{N} = (N_1, \ldots, N_n)$ taking its values in $\mathbb{N}_0^n$ whose PGF is

$$G_{\mathbf{N}}(\mathbf{z}) = \mathbb{E}\left(\prod_{k=1}^{n} z_k^{N_k}\right) = [P_n(\mathbf{z})]^{-\lambda}, \tag{3}$$

where $\mathbb{E}$ denotes the mathematical expectation, $\mathbf{z} = (z_1, \ldots, z_n)$, $\lambda > 0$, and $P_n(\mathbf{z})$ is an affine polynomial of order n. (A polynomial $P_n(\mathbf{z})$ with respect to $\mathbf{z} = (z_1, \ldots, z_n)$ is affine if the one variable polynomial $z_j \mapsto P_n(\mathbf{z})$ can be written as $A^{(-j)} z_j + B^{(-j)}$ (for any $j = 1, \ldots, d$), where $A^{(-j)}$ and $B^{(-j)}$ are polynomials with respect to the z_i's with $i \neq j$.) These discrete distributions have received much interest in the literature (see for instance [2] and the references therein). Of course, the affine polynomial P_n has to satisfy appropriate conditions to ensure that $G_{\mathbf{N}}(\mathbf{z})$ is a PGF. These conditions include the trivial equality $P_n(1, \ldots, 1) = 1$. However, determining all pairs (P_n, λ) such that $G_{\mathbf{N}}(\mathbf{z})$ is a PGF is still an open problem (see [6], for discussions related to this problem). As explained in [6], the affine polynomial $P_n(\mathbf{z})$ can be rewritten

$$P_n(\mathbf{z}) = \frac{A_n(a_1 z_1, \ldots, a_n z_n)}{A_n(a_1, \ldots, a_n)}, \tag{4}$$

where $a_1, \ldots, a_n$ are positive numbers and A_n is an affine polynomial such that $A_n(0, \ldots, 0) = 1$. The Taylor expansions of $[A_n(\mathbf{z})]^{-\lambda}$ and $[P_n(\mathbf{z})]^{-\lambda}$ in the neighborhood of $(0, \ldots, 0)$ will be denoted as follows:

$$\begin{aligned} [A_n(\mathbf{z})]^{-\lambda} &= \sum_{\boldsymbol{\alpha} \in \mathbb{N}_0^n} c_{\boldsymbol{\alpha}}(\lambda, A_n) \mathbf{z}^{\boldsymbol{\alpha}}, \\ [P_n(\mathbf{z})]^{-\lambda} &= \sum_{\boldsymbol{\alpha} \in \mathbb{N}_0^n} c_{\boldsymbol{\alpha}}(\lambda, P_n) \mathbf{z}^{\boldsymbol{\alpha}}, \end{aligned} \tag{5}$$

where $\boldsymbol{\alpha} = (\alpha_1, \ldots, \alpha_n)$ and $\mathbf{z}^{\boldsymbol{\alpha}} = \prod_{i=1}^{n} z_i^{\alpha_i}$. Equations (3) and (4) clearly show that the masses of multivariate NMDs denoted as $c_{\boldsymbol{\alpha}}(\lambda, P_n)$ can be expressed as follows:

$$c_{\boldsymbol{\alpha}}(\lambda, P_n) = c_{\boldsymbol{\alpha}}(\lambda, A_n) \frac{\prod_{i=1}^{n} a_i^{\alpha_i}}{A_n(a_1, \ldots, a_n)^{-\lambda}}. \tag{6}$$

3. Masses of Negative Multinomial Distributions

In this section, we derive new expressions for the coefficients $c_{\boldsymbol{\alpha}}(\lambda, A_n)$ that will be used to compute the masses of NMDs. The particular cases of bivariate and trivariate NMDs will play an important role for the estimation of the DoP on polarimetric images. In order to compute the $c_{\boldsymbol{\alpha}}(\lambda, A_n)$, we derive several results summarized in this section whereas all demonstrations are reported in the appendix.

Theorem 1. *Denote $\mathfrak{P}_n^*$ as the set of nonempty subsets of $[n] = \{1, \ldots, n\}$. Any affine polynomial A_n such that $A_n(\mathbf{0}) = 1$ denoted as*

$$A_n(\mathbf{z}) = 1 - \sum_{T \in \mathfrak{P}_n^*} a_T \mathbf{z}^T \tag{7}$$

can be expressed as follows:

$$A_n(\mathbf{z}) = \prod_{i \in [n]} (1 - a_i z_i) - \sum_{T \in \mathfrak{P}_n^*, |T| \geqslant 2} d_T^n \mathbf{z}^T \prod_{i \in [n] \setminus T} (1 - a_i z_i), \tag{8}$$

where $|T|$ is the cardinal of the set T. Moreover

$$A_n(\mathbf{z}) = \left[\prod_{i\in[n]}(1-a_iz_i)\right] \times \left(1 - \sum_{T\in\mathfrak{P}_n^*, |T|\geqslant 2} d_T^n \frac{\mathbf{z}^T}{\prod_{i\in T}(1-a_iz_i)}\right) \quad (9)$$

$$= \left[\prod_{i\in[n]}(1-a_iz_i)\right] \times \left(1 - Q_n\left(\frac{z_1}{1-a_1z_1}, \ldots, \frac{z_n}{1-a_nz_n}\right)\right), \quad (10)$$

where Q_n is the polynomial defined by $Q_n(\mathbf{z}) = \sum_{T\in\mathfrak{P}_n^*, |T|\geqslant 2} d_T^n \mathbf{z}^T$ and d_T^n is related to the $2^{|T|}-1$ variables a_S, $S \in \mathfrak{P}_T^*$ as follows:

$$d_T^n = \sum_{\substack{T\in\mathfrak{P}_n \\ |T|>1}}^{|T|} a_T a^{[n]\setminus T} + (|T|-1)\prod_{i\in T} a_i. \quad (11)$$

Remark 2. In the trivariate case defined by $n = 3$, the polynomial $A_3(\mathbf{z})$ can be expressed as

$$A_3(\mathbf{z}) = (1-a_1z_1)(1-a_2z_2)(1-a_3z_3) \times \left[1 - Q_3\left(\frac{z_1}{1-z_1}, \frac{z_2}{1-z_2}, \frac{z_3}{1-z_3}\right)\right], \quad (12)$$

where the coefficients of the polynomial

$$Q_3(\mathbf{z}) = b_{1,2}z_1z_2 + b_{1,3}z_1z_3 + b_{2,3}z_2z_3 + b_{1,2,3}z_1z_2z_3 \quad (13)$$

can be determined using the relations

$$b_{i,j} = a_{i,j} + a_ia_j, \quad i, j \in \{1,2,3\}, \ i \neq j,$$
$$b_{1,2,3} = a_{1,2,3} + a_1a_{2,3} + a_2a_{1,3} + a_3a_{1,2} + 2a_1a_2a_3. \quad (14)$$

The next theorem provides a relation between the coefficients of the polynomials $A_n(\mathbf{z})$ and $Q_n(\mathbf{z})$ introduced above.

Theorem 3. *Let $A_n(\mathbf{z}) = 1 - \sum_{T\in\mathfrak{P}_n^*} a_T z^T$, $\mathbf{a} = (a_1, \ldots, a_n)$, and Q_n be the affine polynomial defined in (10) and (11). For any α and γ in $\mathbb{N}^n$, denote as $c_\gamma(\lambda, A_n)$ the coefficient of $\mathbf{z}^\gamma$ in the Taylor expansion of $[A_n(\mathbf{z})]^{-\lambda}$ and as $c_\alpha(\lambda, 1-Q_n)$ the coefficient of $\mathbf{z}^\alpha$ in the Taylor expansion $[1-Q_n(\mathbf{z})]^{-\lambda}$. The following relation can be obtained:*

$$c_\gamma(\lambda, A_n) = \sum_{\alpha+\beta=\gamma} c_\alpha(\lambda, 1-Q_n)(\lambda\mathbf{1}+\alpha)_\beta \frac{\mathbf{a}^\beta}{\beta!} \quad (15)$$

$$= \sum_{0\leqslant\beta_i\leqslant\gamma_i,\, i=1,\ldots,n} c_{\gamma-\beta}(\lambda, 1-Q_n) \times \prod_{i=1}^n (\lambda+\gamma_i-\beta_i)_{\beta_i} \frac{a_i^{\beta_i}}{\beta_i!} \quad (16)$$

$$= \sum_{0\leqslant\alpha_i\leqslant\gamma_i,\, i=1,\ldots,n} c_\alpha(\lambda, 1-Q_n) \times \prod_{i=1}^n (\lambda+\alpha_i)_{\beta_i} \frac{a_i^{\gamma_i-\alpha_i}}{(\gamma_i-\alpha_i)!}. \quad (17)$$

The masses of NMDs can be directly obtained from this theorem. The particular cases of bivariate and trivariate NMDs are considered in the following subsections since the corresponding masses will be useful in the application considered in the second part of this paper.

3.1. Bivariate NMDs

Theorem 4. *Consider the affine polynomial of order 2 with variables $\mathbf{z} = (z_1, z_2)$ defined by*

$$A_2(\mathbf{z}) = 1 - \sum_{T\in\mathfrak{P}_2^*} a_T z^T = 1 - a_1z_1 - a_2z_2 - a_{1,2}z_1z_2. \quad (18)$$

The coefficient of $\mathbf{z}^\gamma$ in the Taylor expansion of $[A_2(\mathbf{z})]^{-\lambda}$, can be computed as follows

$$c_\gamma(\lambda, P_2) = (\lambda)_{\max(\gamma_1,\gamma_2)} \sum_{\ell=0}^{\min(\gamma_1,\gamma_2)} \frac{(\lambda+\ell)_{\min(\gamma_1,\gamma_2)-\ell}}{(\gamma_1-\ell)!(\gamma_2-\ell)!\ell!} \times a_1^{\gamma_1-\ell} a_2^{\gamma_2-\ell} b_{1,2}^\ell. \quad (19)$$

Remark 5. The result (A.11) was mentioned in [7] without the factorization leading to (19). If $a_1 \neq 0$ and $a_2 \neq 0$, an equivalent formulation of (19) is

$$c_\gamma(\lambda, P_2) = \frac{a_1^{\gamma_1}a_2^{\gamma_2}}{\gamma_1!\gamma_2!}(\lambda)_{\max(\gamma_1,\gamma_2)} \times \sum_{\ell=0}^{\min(\gamma_1,\gamma_2)} (\lambda+\ell)_{\min(\gamma_1,\gamma_2)-\ell} \times \binom{\gamma_1}{\ell}\binom{\gamma_2}{\ell}\ell!\left(\frac{b_{1,2}}{a_1a_2}\right)^\ell. \quad (20)$$

3.2. Trivariate NMDs

Theorem 6. *Consider the affine polynomial with the three variables $\mathbf{z} = (z_1, z_2, z_3)$ defined by*

$$A_3(\mathbf{z}) = 1 - \sum_{T\in\mathfrak{P}_3^*} a_T z^T. \quad (21)$$

The coefficient of $\mathbf{z}^{\gamma}$ *in the Taylor expansion of* $[A_3(\mathbf{z})]^{-\lambda}$ *are*

$$c_{\gamma}(\lambda, P_3) = \sum_{\beta_1=0}^{\gamma_1}\sum_{\beta_2=0}^{\gamma_2}\sum_{\beta_3=0}^{\gamma_3}\sum_{v=\|\gamma-\beta\|}^{\lfloor|\gamma-\beta|/2\rfloor}(\lambda)_v \times \frac{b_{2,3}^{v-\gamma_1+\beta_1} b_{1,3}^{v-\gamma_2+\beta_2} b_{1,2}^{v-\gamma_3+\beta_3}}{\prod_{i=1}^{3}(v-\gamma_i+\beta_i)!} \frac{b_{1,2,3}^{|\gamma-\beta|-2v}}{(|\gamma-\beta|-2v)!} \times \frac{(\lambda+\gamma_1-\beta_1)_{\beta_1}}{\beta_1!}\frac{(\lambda+\gamma_2-\beta_2)_{\beta_2}}{\beta_2!} \times \frac{(\lambda+\gamma_3-\beta_3)_{\beta_3}}{\beta_3!} a_1^{\beta_1}a_2^{\beta_2}a_3^{\beta_3}. \quad (22)$$

When $a_i \neq 0, b_{i,j} \neq 0, i \neq j, i = 1,2,3,\ j = 1,2,3,$ *and* $b_{1,2,3} \neq 0$, *an equivalent expression is*

$$c_{\gamma}(\lambda, P_3) = a_1^{\gamma_1}a_2^{\gamma_2}a_3^{\gamma_3}\sum_{\beta_1=0}^{\gamma_1}\sum_{\beta_2=0}^{\gamma_2}\sum_{\beta_3=0}^{\gamma_3}\sum_{v=\|\gamma-\beta\|}^{\lfloor|\gamma-\beta|/2\rfloor}(\lambda)_v \times \frac{\left(b_{2,3}b_{1,3}b_{1,2}/b_{1,2,3}^2\right)^v}{(|\gamma-\beta|-2v)!\prod_{i=1}^{3}(v-\gamma_i+\beta_i)!} \times \left[\prod_{i=1}^{3}\frac{(\lambda+\gamma_i-\beta_i)_{\beta_i}}{\beta_i!}\right]\left(\frac{b_{1,2,3}}{a_1b_{2,3}}\right)^{\gamma_1-\beta_1} \times \left(\frac{b_{1,2,3}}{a_2b_{1,3}}\right)^{\gamma_2-\beta_2}\left(\frac{b_{1,2,3}}{a_3b_{1,2}}\right)^{\gamma_3-\beta_3} \quad (23)$$

$$= (\lambda)_{\|\gamma\|}\frac{\mathbf{a}^{\gamma}}{\gamma!}\sum_{\alpha_1=0}^{\gamma_1}\sum_{\alpha_2=0}^{\gamma_2}\sum_{\alpha_3=0}^{\gamma_3}\sum_{v=\|\alpha\|}^{\lfloor|\alpha|/2\rfloor}\frac{(\lambda+\|\alpha\|)_{v-\|\alpha\|}}{(|\alpha|-2v)!\prod_{i=1}^{3}(v-\alpha_i)!} \times \left(\frac{b_{2,3}b_{1,3}b_{1,2}}{b_{1,2,3}^2}\right)^v \alpha!\left(\frac{b_{1,2,3}}{a_1b_{2,3}}\right)^{\alpha_1}\left(\frac{b_{1,2,3}}{a_2b_{1,3}}\right)^{\alpha_2}\left(\frac{b_{1,2,3}}{a_3b_{1,2}}\right)^{\alpha_3} \times \prod_{i=1}^{3}\left[(\lambda+\alpha_i)_{\gamma_i-\alpha_i}\binom{\gamma_i}{\alpha_i}\right]. \quad (24)$$

4. Estimating the Polarization Degree of Low Flux Polarimetric Images Using Maximum Likelihood Methods

4.1. Low Flux Polarimetric Images. The state of the polarization of the light can be described by the random behavior of a complex vector $\mathbf{A} = (A_X, A_Y)$, called the Jones vector, whose covariance matrix, called the polarization matrix, is

$$\Gamma = \begin{pmatrix} \mathbb{E}[A_X A_X^*] & \mathbb{E}[A_X A_Y^*] \\ \mathbb{E}[A_Y A_X^*] & \mathbb{E}[A_Y A_Y^*] \end{pmatrix} \triangleq \begin{pmatrix} a_1 & a_3 + ia_4 \\ a_3 - ia_4 & a_2 \end{pmatrix}, \quad (25)$$

where $*$ denotes the complex conjugate. The covariance matrix Γ is a nonnegative Hermitian matrix whose diagonal terms are the intensity components in the X and Y directions. The cross terms of Γ are the correlations between the Jones components. If we assume a fully developed speckle, the Jones vector $\mathbf{A}$ is distributed according to a complex Gaussian distribution whose probability density function (pdf) is [15]:

$$p(\mathbf{A}) = \frac{1}{\pi^2|\Gamma|}\exp\left(-\mathbf{A}^{\dagger}\Gamma^{-1}\mathbf{A}\right), \quad (26)$$

where $|\Gamma|$ is the determinant of the matrix Γ and $\dagger$ denotes the conjugate transpose operator. As a consequence, the statistical properties of $\mathbf{A}$ are fully characterized by the covariance matrix Γ. The different components of Γ can be classically estimated by using four intensity images that are related to the components of the Jones vector as follows (see [20], for more details):

$$\begin{aligned} I_1 &= |A_X|^2, \\ I_2 &= |A_Y|^2, \\ I_3 &= \frac{1}{2}|A_X|^2 + \frac{1}{2}|A_Y|^2 + \mathrm{Re}(A_X A_Y^*), \\ I_4 &= \frac{1}{2}|A_X|^2 + \frac{1}{2}|A_Y|^2 + \mathrm{Im}(A_X A_Y^*). \end{aligned} \quad (27)$$

The state of the polarization of the light is classically characterized by the square DoP defined by [15, pages 134–136]

$$P^2 = 1 - 4\frac{|\Gamma|}{[\mathrm{trace}(\Gamma)]^2} = 1 - \frac{4\left[a_1a_2 - (a_3^2 + a_4^2)\right]}{(a_1 + a_2)^2}, \quad (28)$$

where trace(Γ) is the trace of the matrix Γ. The light is totally depolarized for $P = 0$, totally polarized for $P = 1$, and partially polarized when $P \in]0,1[$. As a consequence, estimating the square DoP of a polarimetric image is important in many practical applications. Different estimation methods of P^2 using several combinations of intensity images were studied in [20]. Since only one realization of the random vector $\mathbf{I} = (I_1, \ldots, I_4)^T$ was available for a given pixel of a polarimetric image, the image was supposed to be locally stationary and ergodic. These assumptions were used to derive square DoP estimators using several neighbor pixels belonging to a so-called estimation window.

This paper considers practical applications where the intensity level of the reflected light is very low (low flux assumption), which leads to an additional source of fluctuations on the detected signal. Under the low flux assumption, the quantum nature of the light leads to a Poisson-distributed noise which can become very important relatively to the mean value of the signal at a low photon level. As a consequence, the observed pixels of the low flux polarimetric image are discrete random variables contained in the vector $\mathbf{N} = (N_1, \ldots, N_4)$ such that the conditional distributions of the random variables $N_l \mid I_l$, for $l = 1, \ldots, 4$ are independent and distributed according to Poisson distributions with means I_l, for $l = 1,\dots,4$. The resulting joint distribution of $\mathbf{N}$ is a multivariate mixed Poisson distribution [22]:

$$P(\mathbf{N}=\mathbf{k}) = \int \underset{(\mathbb{R}^+)^4}{\cdots} \int \prod_{l=1}^{4} \frac{I_l^{k_l}}{k_l!} \exp(-I_l) f(\mathbf{I})\, d\mathbf{I}, \quad (29)$$

where $\mathbf{k} = (k_1,\dots,k_4)$, $k_i \in \mathbb{N}$, and $f(\mathbf{I})$ is the joint pdf of the intensity vector. This section studies estimators of the square DoP P^2 defined in (28) based on several vectors $\mathbf{N}^1,\dots,\mathbf{N}^n$ belonging to the estimation window. These estimators are constructed from estimates of the covariance matrix elements a_i, $i = 1,\dots,4$. As explained in the introduction, several studies have been recently devoted to the estimation of the square DoP using less than four polarimetric images. This paper goes into this direction by deriving estimators based on the observation of 2 or 3 polarimetric images.

The joint distribution of the intensity vector $\mathbf{I}$ is known to be a multivariate gamma distribution whose Laplace transform is [20]

$$E\left[\exp\left(\sum_{k=1}^{4} z_k I_k\right)\right] = \frac{1}{P_4(\mathbf{z})}, \quad (30)$$

where the affine polynomial P_4 is

$$P_4(\mathbf{z}) = 1 + \mathbf{z}\boldsymbol{\mu} + k_a\left[2z_1z_2 + z_3z_4 + (z_1+z_2)(z_3+z_4)\right] \quad (31)$$

with $\mathbf{z} = (z_1,\dots,z_4)$ and

$$k_a = \frac{1}{2}\left(a_1a_2 - a_3^2 - a_4^2\right),$$
$$\boldsymbol{\mu} = \left(a_1, a_2, a_3 + \frac{a_1+a_2}{2}, a_4 + \frac{a_1+a_2}{2}\right)^T. \quad (32)$$

As a consequence, the distribution of $\mathbf{N}$ is an NMD whose PGF can be written as (the interested reader is invited to consult [22] for more details)

$$G_{\mathbf{N}}(\mathbf{z}) = \frac{1}{P_4(z_1-1, z_2-1, z_3-1, z_4-1)}. \quad (33)$$

The results of Section 2 allow us to compute the masses of $\mathbf{N}$ that will be useful for studying the maximum likelihood estimator (MLE) of the square DoP.

4.2. MLE Using Three Polarimetric Images. The PGF of $\widetilde{\mathbf{N}} = (N_1, N_2, N_3)$ can be computed from (33) by setting $z_4 = 1$. The following result can be obtained:

$$G_{\widetilde{\mathbf{N}}}(\mathbf{z}) = \frac{1}{P_3(\mathbf{z})} \quad (34)$$

with

$$P_3(\mathbf{z}) = P_3(\mathbf{0}) + z_1(\mu_1 - 3k_a) + z_2(\mu_2 - 3k_a) + z_3(\mu_3 - 2k_a) + k_a(2z_1z_2 + z_1z_3 + z_2z_3), \quad (35)$$

$\mathbf{z} = (z_1, z_2, z_3)$, and $P_3(\mathbf{0}) = 1 - \sum_{i=1}^{3}\mu_i + 4k_a$. The results of Section 3.2 can then be used to express the masses of $\widetilde{\mathbf{N}}$ as a function of $\boldsymbol{\theta} = (a_1, a_2, a_3, a_4^2)^T$.

The ML estimator of $\boldsymbol{\theta}$ based on several vectors $\widetilde{\mathbf{N}}^k$ belonging to the estimation window (where $k = 1,\dots,K$ and K is the number of pixels of the observation window) is obtained by maximizing the log-likelihood

$$l_3\left(\widetilde{\mathbf{N}}^1,\dots,\widetilde{\mathbf{N}}^K \mid \boldsymbol{\theta}\right) = \sum_{k=1}^{K} \log\left[P\left(\widetilde{\mathbf{N}}^k\right)\right] \quad (36)$$

with respect to $\boldsymbol{\theta}$ (note again that $P(\widetilde{\mathbf{N}}^k)$ is the mass of a trivariate NMD that has been computed in Section 3.2). The practical determination of the ML estimator of $\boldsymbol{\theta}$ is achieved by using a Newton-Raphson procedure. The ML estimators of the vector $\boldsymbol{\theta}$, denoted as $\widetilde{\boldsymbol{\theta}} = \left(\widetilde{a}_1, \widetilde{a}_2, \widetilde{a}_3, \widetilde{a_4^2}\right)^T$, are then plugged into (28) to provide the ML estimator of the square DoP based on three polarimetric images:

$$\widetilde{P}^2 = 1 - \frac{4\left[\widetilde{a}_1\widetilde{a}_2 - \left(\widetilde{a}_3^2 + \widetilde{a_4^2}\right)\right]}{\left(\widetilde{a}_1 + \widetilde{a}_2\right)^2}. \quad (37)$$

4.3. MLE Using Two Polarimetric Images. The PGF of $\underline{\mathbf{N}} = (N_1, N_2)$ can be computed from (34) by setting $z_3 = 1$. The following result can be obtained:

$$G_{\underline{\mathbf{N}}}(\mathbf{z}) = \frac{1}{P_2(\mathbf{z})} \quad (38)$$

with

$$P_2(\mathbf{z}) = P_2(\mathbf{0}) + z_1(\mu_1 - 2k_a) + z_2(\mu_2 - 2k_a) + 2k_a z_1 z_2, \quad (39)$$

$\mathbf{z} = (z_1, z_2)$ and $P_2(\mathbf{0}) = 1 - \sum_{i=1}^{2}\mu_i + 2k_a$. The results of Section 3.1 can then be used to express the masses of $\underline{\mathbf{N}}$ as a function of $\boldsymbol{\theta} = (a_1, a_2, k_a)^T$.

The MLE of $\boldsymbol{\theta}$ based on several vectors $\underline{\mathbf{N}}^k$ belonging to the estimation window is obtained by maximizing the log-likelihood

$$l_2\left(\underline{\mathbf{N}}^1,\dots,\underline{\mathbf{N}}^K \mid \boldsymbol{\theta}\right) = \sum_{k=1}^{K} \log\left[P\left(\underline{\mathbf{N}}^k\right)\right] \quad (40)$$

with respect to $\boldsymbol{\theta}$ (note that $P(\underline{\mathbf{N}}^k)$ is the mass of a bivariate NMD that has been computed in Section 3.1). The practical determination of the ML estimator of $\boldsymbol{\theta}$ is achieved by using a Newton-Raphson procedure. The ML estimators of the vector $\boldsymbol{\theta}$ elements, denoted as $\underline{\boldsymbol{\theta}} = (\underline{a}_1, \underline{a}_2, \underline{k}_a)^T$, are then plugged into (28) to provide the MLE of the square DoP based on two polarimetric images:

$$\underline{P}^2 = 1 - \frac{8\underline{k}_a}{\left(\underline{a}_1 + \underline{a}_2\right)^2}. \quad (41)$$

TABLE 1: Covariance matrix elements and square DoP values for the Jones vector.

	Γ_0	Γ_1	Γ_2	Γ_3	Γ_4	Γ_5	Γ_6	Γ_7	Γ_8	Γ_9	Γ_{10}
a_1	2.00	1.71	2.40	1.82	2.00	2.67	2.40	2.24	2.74	2.00	2.00
a_2	2.00	2.29	1.60	2.18	2.00	1.33	1.60	1.76	1.26	2.00	2.00
a_3	0.00	0.40	0.48	0.91	0.89	0.96	1.07	1.05	1.46	0.60	1.41
a_4	0.00	0.40	0.64	0.58	0.89	0.80	1.05	1.28	0.73	1.80	1.41
P^2	0	0.1	0.2	0.3	0.4	0.5	0.6	0.7	0.8	0.9	1.0

TABLE 2: Simulation results for the estimation of P^2 using 2 or 3 images, obtained from 1000 Monte-Carlo runs ($n = 51 \times 51$).

	2 images MLE				3 images MLE			
	Bias	std	MSE	avar	Bias	std	MSE	avar
Γ_0	$3.57e-03$	$1.31e-02$	$1.85e-04$	—	$4.12e-03$	$9.08e-03$	$9.93e-05$	—
Γ_1	$4.28e-04$	$3.14e-02$	$9.83e-04$	$9.78e-04$	$1.71e-03$	$2.57e-02$	$6.62e-04$	$7.25e-04$
Γ_2	$9.49e-04$	$3.23e-02$	$1.04e-03$	$9.82e-04$	$1.20e-03$	$2.86e-02$	$8.18e-04$	$7.83e-04$
Γ_3	$-2.15e-04$	$3.13e-02$	$9.81e-04$	$9.67e-04$	$4.84e-04$	$2.88e-02$	$8.28e-04$	$8.07e-04$
Γ_4	$2.49e-03$	$3.02e-02$	$9.19e-04$	$8.91e-04$	$1.71e-03$	$2.81e-02$	$7.91e-04$	$7.67e-04$
Γ_5	$2.86e-04$	$2.85e-02$	$8.10e-04$	$7.54e-04$	$4.80e-04$	$2.65e-02$	$7.03e-04$	$6.69e-04$
Γ_6	$5.25e-04$	$2.59e-02$	$6.72e-04$	$6.58e-04$	$6.32e-04$	$2.47e-02$	$6.11e-04$	$5.95e-04$
Γ_7	$1.05e-03$	$2.37e-02$	$5.60e-04$	$5.43e-04$	$1.04e-03$	$2.26e-02$	$5.11e-04$	$4.95e-04$
Γ_8	$1.59e-04$	$2.01e-02$	$4.03e-04$	$4.12e-04$	$3.17e-04$	$1.90e-02$	$3.60e-04$	$3.73e-04$
Γ_9	$-2.88e-04$	$1.76e-02$	$3.10e-04$	$3.20e-04$	$-4.46e-04$	$1.65e-02$	$2.73e-04$	$2.86e-04$
Γ_{10}	$-5.98e-03$	$8.98e-03$	$1.16e-04$	—	$-5.73e-03$	$8.58e-03$	$1.06e-04$	—

5. Simulation Results

5.1. Estimation Performance. The performance of the ML estimators of the square DoP based on two or three polarimetric images has been evaluated via several experiments. The first simulations compare the log Mean Square Errors (MSEs) of the square DoP estimators constructed from two or three images. Eleven different covariance matrices of the Jones vector have been considered in order to define typical values of the DoP. The values of a_i, for $i = 1, \ldots, 4$, (defining the covariance matrix elements of the Jones vector) and the corresponding values of the square DoPs are reported in Table 1. Note that all the covariance matrices Γ_j, $j = 0, \ldots, 10$, have been normalized so that the mean of the total intensity $\mathbf{E}[I_1 + I_2] = \mathbf{E}[N_1 + N_2] = a_1 + a_2$ is equal to 4. Thus, the average number of photons collected on each pixel equals 4 for each matrix of the considered set. This point is in agreement with the low flux assumption.

Figure 1(a) display the empirical log MSEs of the square DoP estimators for the set of covariance matrices defined in Table 1 as a function of the true square DoP value. The red plus markers + correspond to the estimators obtained for two polarimetric images (2D MLE), whereas the blue cross markers × correspond to the MLE obtained for three polarimetric images (3D MLE). Note that these empirical MSEs have been computed for a square observation window of size $n = 51 \times 51$ pixels, based on 1000 Monte-Carlo runs. The theoretical asymptotic log MSEs of the MLE are also displayed in Figure 1(a) with continuous lines. These asymptotic values correspond to the Cramer-Rao Lower Bound (CRLB) for the parameter P^2. The MLE is known to be asymptotically unbiased and efficient under mild regularity conditions (that are satisfied for P^2). Thus, the MSE of the estimates can be approximated a for large sample by the CRLB. More details about the way of computing the square DoP CLRBs can be found in [23]. Figure 1(a) indicates that the empirical MSEs are in good agreement with the corresponding CRLBs, except for the matrices Γ_0 and Γ_{10}. Indeed, the CRLBs for these two matrices cannot be computed since the true value of the parameters belongs to the boundary of its definition domain. The empirical bias, standard deviations ("std"), MSEs, and asymptotic variances ("avar") of the estimators of P^2 are also reported in Table 2. It is interesting to note that the MLE obtained using 3 images is slightly more biased than the one obtained using 2 images. However, the MLE based on 3 images provides lower MSEs than the estimator based on 2 images, as expected.

In order to appreciate the influence of the Poisson noise due to the low flux assumption, experiments have been conducted using the high flux assumption. In this case, the intensity vector $\mathbf{I}$ is assumed to be known. Thus, the high flux MLEs using two and three images can be derived from the intensity vectors $\underline{\mathbf{I}}^k = (I_1^k, I_2^k)^T$ and $\widetilde{\mathbf{I}}^k = (I_1^k, I_2^k, I_3^k)^T$ [20]. The results are depicted in Figure 1(b). A comparison between Figures 1(a) and 1(b) allows one to appreciate a similar global behaviour for all the estimators, with a maximum MSE near $P^2 = 1/3$ and decreasing MSEs as P^2 goes to 0 or 1. The degradation of the estimation performance due to the presence of Poisson noise (due to the low flux assumption) can also be clearly noticed.

The next set of simulations studies the performance of the different estimators as a function of the sample size n. Figures 2(a) and 2(b) show the log MSEs of the square DoP

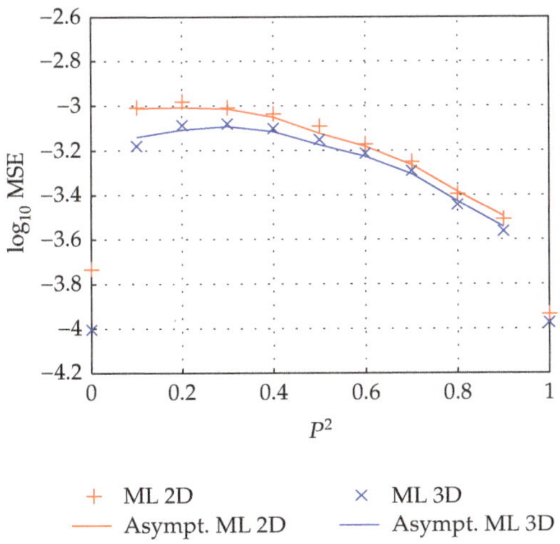

(a) Low flux

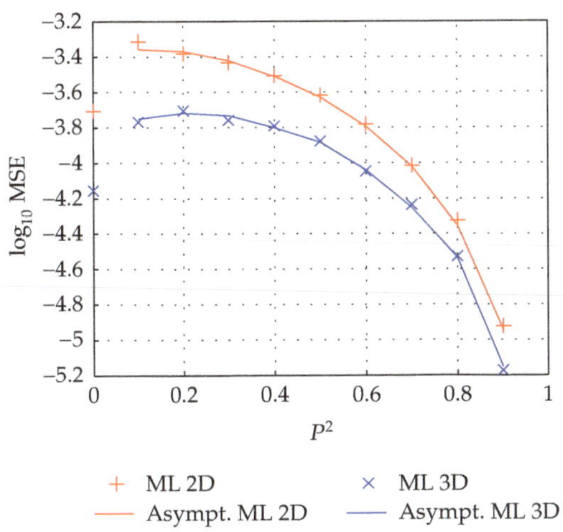

(b) High flux

FIGURE 1: log MSEs of the square DoP estimates using 2 and 3 images versus P^2 for the set of polarization matrices defined in Table 1 under (a) low flux and (b) high flux assumptions ($n = 51 \times 51$, ML: maximum likelihood estimators, and Asympt.: theoretical asymptotic value of the log MSE for a given estimator).

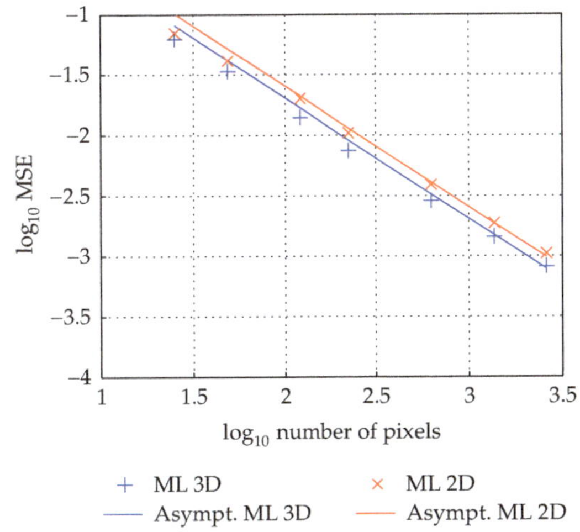

(a) Matrix Γ_2

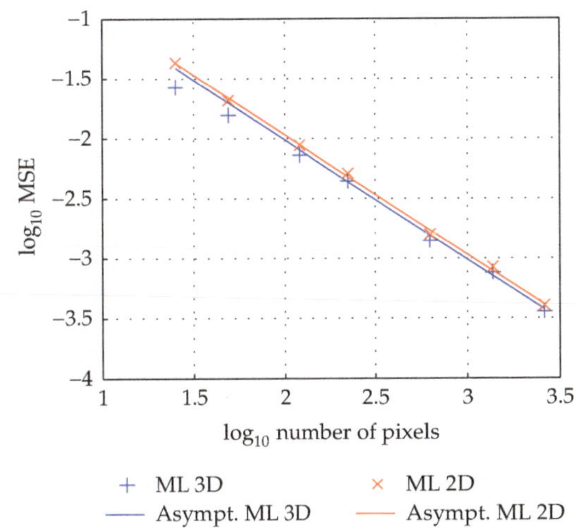

(b) Matrix Γ_8

FIGURE 2: log MSE of the estimated square DoP P^2 using 2 or 3 intensity images versus the logarithm of the sample size for the matrices (a) Γ_2 and (b) Γ_8 (ML: maximum likelihood estimators, and Asympt. theoretical asymptotic value of the log MSE for a given estimator).

estimates obtained for 2 and 3 images (for the two particular matrices Γ_2 and Γ_8). The empirical bias, standard deviations ("std"), MSEs, and asymptotic variances ("avar") are also reported in Tables 3 and 4. One can see that the empirical MSEs are in good agreement with their theoretical asymptotic values for a large enough sample size, that is, $n > 25 \times 25$. Moreover, the gain of a performance using 3 images instead of 2 is more significant for small values of P^2 (indeed, the difference between the different curves is more pronounced in the left figure corresponding to $P^2 = 0.2$ than in the right figure corresponding to $P^2 = 0.8$.)

Figures 3(a) and 3(b) display the MSEs of the MLE under the high flux assumption. By comparing Figures 2 and 3, one can observe that the gain of performance using 3 images instead of 2 is more important in the high flux scenario than under a low flux assumption.

5.2. Application to Synthetic Polarimetric Images. In order to appreciate the estimation performance on polarimetric images, we consider a synthetic polarimetric image of size 512×512 composed of three distinct objects located on a homogeneous background depicted in Figure 4 (see also [20]). The polarimetric properties of these objects and background (i.e., the covariance matrix of the Jones vector and the square DoPs) are reported in Table 5. The polarimetric low flux images generated according to this model are also represented in Figure 5 in negative colors (bright pixels correspond to a small number of photons, dark ones correspond to a large number of photons). Note that these images represent the values of the vector $\mathbf{N} = (N_1, N_2, N_3, N_4)^T$ for each pixel. The square DoP of each pixel $\mathbf{x}^{(i,j)}$ (for $i, j = 1, \ldots, 512$) has been estimated from vectors belonging to windows of size $n = 15 \times 15$ centered around the pixel of coordinates (i, j) in

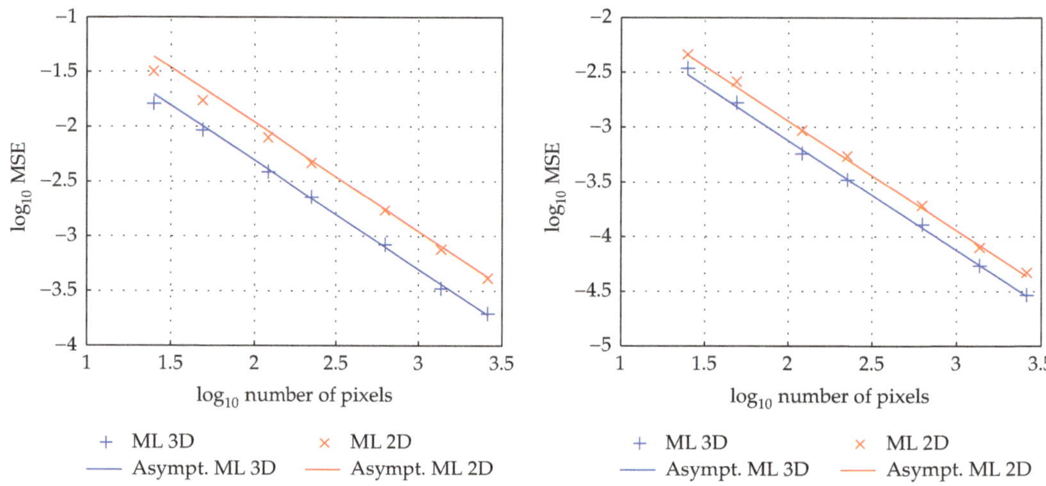

(a) Matrix Γ_2 (b) Matrix Γ_8

FIGURE 3: log MSE of the estimated square DoP P^2 using 2 or 3 intensity images versus the logarithm of the sample size for the matrices (a) Γ_2 and (b) Γ_8 under high flux assumption (ML: maximum likelihood estimators, Asympt.: theoretical asymptotic value of the log MSE for a given estimator).

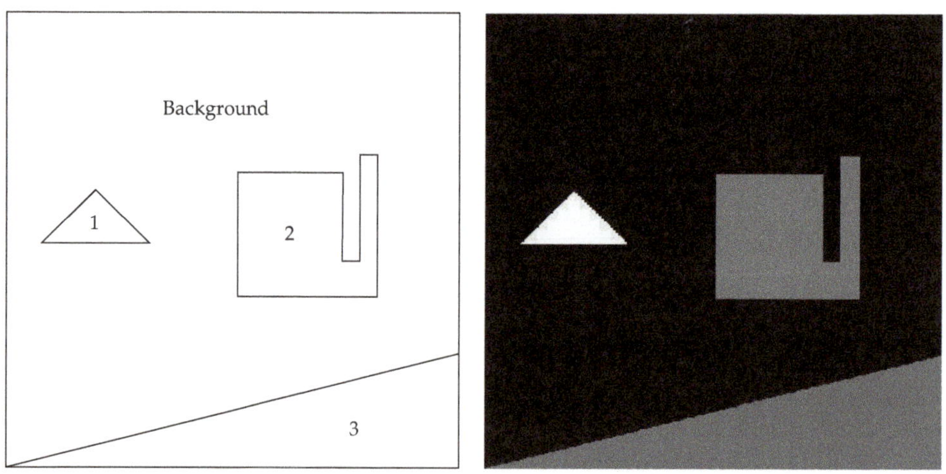

(a) Scene (b) Theoretical squared DoP

FIGURE 4: Composition of the scene used to generate synthetic polarimetric low flux images and associated theoretical squared DoP.

the analyzed image. The estimated square DoPs are depicted in Figures 6(a) and 6(b) for the MLEs using 2 and 3 images, respectively. One can see that the ML method for 3 images gives more homogeneous results on each object than the ML method derived for 2 images. This result confirms that the MLE using 3 images performs better than the MLE using only 2 images. However, the polarimetric properties of a scene can be clearly recovered with 2 or 3 low flux images.

5.3. Estimation Results on Real-Word Polarimetric Images. Finally, the ML estimator based on three images is applied on real polarimetric data. These images are acquired by using a laser as a coherent illumination source. The scene consists of two disks. The first one, intended to provide low DoP, is a grey diffuse material (left object in Figure 7). The second one is made of sand blasted aluminium providing high DoP (right object in Figure 7). Due to the experimental conditions, the measured intensities are quite low. As a consequence, these intensities are assumed to be distributed according to an NMD. The intensity images corresponding to N_1, N_2, N_3, and N_4 are depicted in Figure 7. The interested reader is invited to read [20] for more details on these data. It is important to note that the two disks exhibit the similar level of total reflectivity $N_1 + N_2$ and can hardly be distinguished without a polarimetric processing.

Figure 8 shows the ML estimates of the square DoPs P^2 for 3 images and an estimation window of size $n = 9\times9$ pixels. As expected, the values of the estimates are quite different on each disk: quite higher on the metal than on the plastic disk. This result is in good agreement with the theoretical

TABLE 3: Simulation results for the estimation of P^2 using 2 or 3 images, obtained from 1000 Monte-Carlo runs (matrix Γ_2).

n	2 images MLE				3 images MLE			
	Bias	std	MSE	avar	Bias	std	MSE	avar
5^2	$9.03e-02$	$2.49e-01$	$7.01e-02$	$1.02e-01$	$1.20e-01$	$2.20e-01$	$6.27e-02$	$8.15e-02$
7^2	$5.10e-02$	$1.97e-01$	$4.13e-02$	$5.21e-02$	$7.04e-02$	$1.69e-01$	$3.36e-02$	$4.16e-02$
11^2	$2.60e-02$	$1.39e-01$	$2.01e-02$	$2.11e-02$	$3.22e-02$	$1.14e-01$	$1.39e-02$	$1.68e-02$
15^2	$7.68e-03$	$1.02e-01$	$1.04e-02$	$1.14e-02$	$1.59e-02$	$8.48e-02$	$7.43e-03$	$9.05e-03$
25^2	$-7.42e-04$	$6.25e-02$	$3.90e-03$	$4.09e-03$	$2.16e-03$	$5.36e-02$	$2.87e-03$	$3.26e-03$
37^2	$-5.19e-04$	$4.32e-02$	$1.87e-03$	$1.87e-03$	$6.87e-05$	$3.79e-02$	$1.44e-03$	$1.49e-03$
51^2	$9.49e-04$	$3.23e-02$	$1.04e-03$	$9.82e-04$	$1.20e-03$	$2.86e-02$	$8.18e-04$	$7.83e-04$

TABLE 4: Simulation results for the estimation of P^2 using 2 or 3 images, obtained from 1000 Monte-Carlo runs (matrix Γ_8).

n	2 images MLE				3 images MLE			
	Bias	std	MSE	avar	Bias	std	MSE	avar
5^2	$-3.91e-03$	$2.08e-01$	$4.31e-02$	$4.29e-02$	$2.80e-02$	$1.61e-01$	$2.67e-02$	$3.88e-02$
7^2	$4.04e-03$	$1.44e-01$	$2.08e-02$	$2.19e-02$	$1.89e-02$	$1.24e-01$	$1.56e-02$	$1.98e-02$
11^2	$2.14e-03$	$9.38e-02$	$8.80e-03$	$8.86e-03$	$8.70e-03$	$8.47e-02$	$7.24e-03$	$8.01e-03$
15^2	$-1.27e-03$	$7.14e-02$	$5.10e-03$	$4.76e-03$	$2.70e-03$	$6.63e-02$	$4.39e-03$	$4.31e-03$
25^2	$5.23e-04$	$3.98e-02$	$1.58e-03$	$1.71e-03$	$6.87e-04$	$3.75e-02$	$1.40e-03$	$1.55e-03$
37^2	$2.24e-03$	$2.89e-02$	$8.42e-04$	$7.83e-04$	$2.54e-03$	$2.75e-02$	$7.61e-04$	$7.08e-04$
51^2	$1.59e-04$	$2.01e-02$	$4.03e-04$	$4.12e-04$	$3.17e-04$	$1.90e-02$	$3.60e-04$	$3.73e-04$

properties of the considered material. This emphasizes the interest in considering efficient estimators based on NMDs for polarimetric images.

Appendix

Proofs of the Theorems

Proof of Theorem 1. The set of affine polynomials with real coefficients and variables $(z_1,\ldots,z_n)$ is a vector space of dimension 2^n spanned by the basis $(\mathbf{z}^T)_{T\in\mathfrak{P}_n}$. The set of polynomials $(\mathbf{z}^T\prod_{t\in[n]\setminus T}(1-a_tz_t))_{T\in\mathfrak{P}_n}$ is another basis of this vector space. The proof of the theorem is obtained by expressing the coefficients of A_n in this latter basis. Considering the expansion

$$A_n(z_1,\ldots,z_n) = d_\emptyset^n\prod_{i\in[n]}(1-a_iz_i) - \sum_{T\in\mathfrak{P}_n^*} d_T^n\mathbf{z}^T\prod_{i\in[n]\setminus T}(1-a_iz_i), \tag{A.1}$$

the following results can be obtained.

(1) Substituting $\mathbf{z}=\mathbf{0}$ in (A.1) leads to $d_\emptyset^n=1$.

(2) Substituting $\mathbf{z}_i = (\delta_i(1),\ldots,\delta_i(n))$ in (A.1), where $\delta_i(j)=1$ if $i=j$ and $\delta_i(j)=0$ else leads to

$$1-a_i = 1-a_i-d_{\{i\}}^n, \tag{A.2}$$

or equivalently

$$d_{\{i\}}^n = 0, \quad i\in[n]. \tag{A.3}$$

(3) Without loss of generality, if $|T| = k > 1$, the coefficients d_T^n can be computed for $T=\{1,\ldots,k\}$. Indeed, consider a permutation σ such that $T = \{\sigma(1),\ldots,\sigma(k)\}$. If $d_{\{1,\ldots,k\}}^n = f(\{a_S\}_{S\in\mathfrak{P}_k^*})$, we have $d_T^n = f(\{a_{\sigma(S)}\}_{S\in\mathfrak{P}_k^*})$. Using the relation

$$\begin{aligned}1-A_n(z_1,\ldots,z_{n-1},0) &= \sum_{T\in\mathfrak{P}_{n-1}^*} a_Tz^T\\ &= 1-A_{n-1}(z_1,\ldots,z_{n-1}),\end{aligned} \tag{A.4}$$

the expansion

$$A_n(z_1,\ldots,z_n) = \prod_{i\in[n]}(1-a_iz_i) - \sum_{T\in\mathfrak{P}_n^*,|T|\geqslant 2} d_T^n\mathbf{z}^T\prod_{i\in[n]\setminus T}(1-a_iz_i), \tag{A.5}$$

yields for any $k<n$,

$$d_{[k]}^n = d_{[k]}^{n-1}. \tag{A.6}$$

In order to determine $d_{[n]}^n$, we can assume $a_i\neq 0$, $i\in[n]$ and substitute $z_i = 1/a_i$, $i\in[n]$ in (A.5). The following result is obtained

$$1-n-\sum_{T\in\mathfrak{P}_n^*,|T|\geqslant 2}\frac{a_T}{a^T} = -\frac{d_T^n}{a^{[n]}}, \tag{A.7}$$

TABLE 5: Polarimetric properties of elements that compose the scene displayed in Figure 4.

Object	Polarization matrix Γ	P^2	Remarks
Background	$\begin{pmatrix} 0.79 & 0 \\ 0 & 0.98 \end{pmatrix}$	0.0115	Very depolarizing and dark background
1	$\begin{pmatrix} 3.6 & 0 \\ 0 & 0.22 \end{pmatrix}$	0.783	Very bright and weakly depolarizing object (typically steel)
2	$\begin{pmatrix} 3 & 0.1 \\ 0.1 & 0.6 \end{pmatrix}$	0.447	Bright object quite depolarizing
3	$\begin{pmatrix} 0.7 & 0.5 + 0.2\, i \\ 0.5 - 0.2\, i & 1.07 \end{pmatrix}$	0.414	Dark object whose mean total intensity is the same as the background

(a) Low flux intensity N_1 (b) Low flux intensity N_2

(c) Low flux intensity N_3 (d) Low flux intensity N_4

FIGURE 5: Synthetic intensity images (negative colors) for the scene depicted in Figure 4 and described in Table 5.

hence

$$\begin{aligned} d^n_{[n]} &= \left[(n-1) + \sum_{T \in \mathfrak{P}^*_n, |T| \geq 2} \frac{a_T}{a^T} \right] a^{[n]} \\ &= \sum_{T \in \mathfrak{P}^*_n, |T| \geq 2} a_T a^{[n] \setminus T} + (n-1)\, a^{[n]}. \end{aligned} \tag{A.8}$$

□

Proof of Theorem 3. The relation (10) leads to

$$\begin{aligned} [A_n(\mathbf{z})]^{-\lambda} &= \left(1 - Q_n\left(\frac{z_1}{1 - a_1 z_1}, \ldots, \frac{z_n}{1 - a_n z_n}\right)\right)^{-\lambda} \\ &\quad \times \left[\prod_{i=1}^{n} (1 - a_i z_i)\right]^{-\lambda} \\ &= \left(\sum_{\alpha \in \mathbb{N}^n} c_\alpha (\lambda, 1 - Q_n) \left(\frac{\mathbf{z}}{\mathbf{1} - \mathbf{az}}\right)^{\alpha}\right) \end{aligned}$$

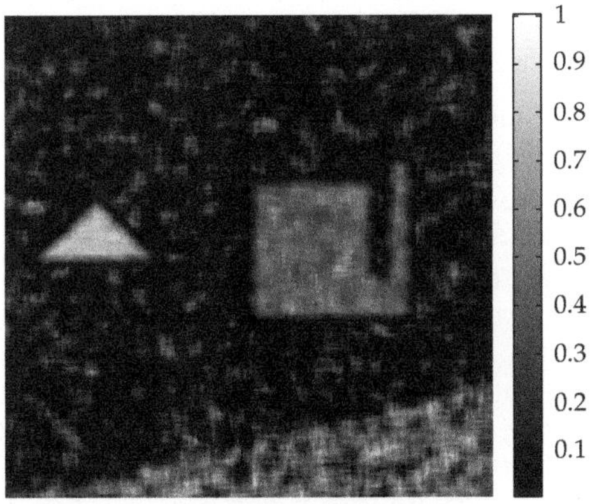

(a) MLE 2 images

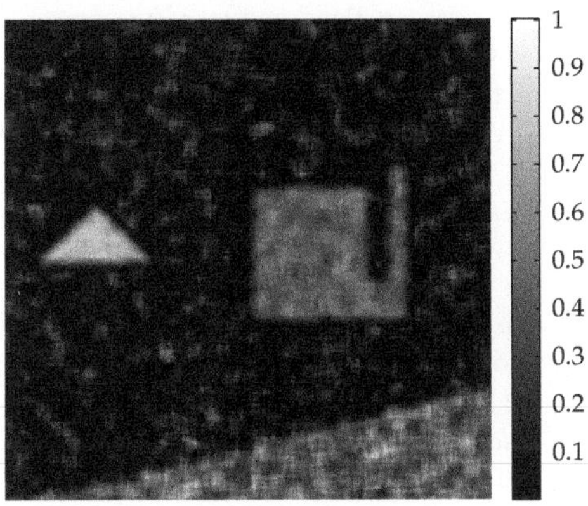

(b) MLE 3 images

FIGURE 6: Estimates of P^2 using 2 or 3 low flux intensity images for the synthetic polarimetric images for an estimation window of size $n = 15 \times 15$. MLE: maximum likelihood estimator.

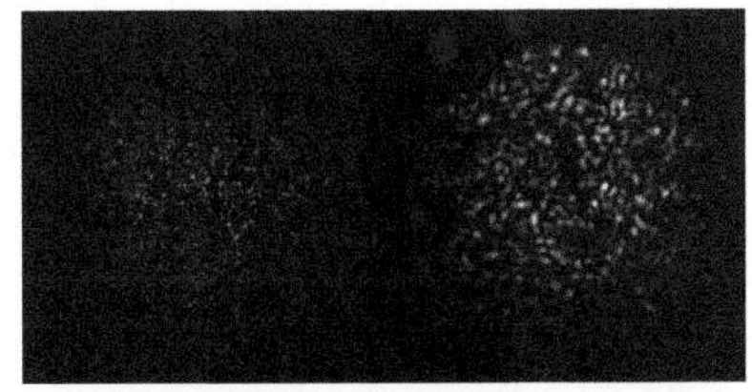

(a) Low flux intensity N_1

(b) Low flux intensity N_2

(c) Low flux intensity N_3

(d) Low flux intensity N_4

FIGURE 7: Real-world polarimetric intensity images of a scene composed of a plastic disk (left) and a steel disk (right).

$$
\begin{aligned}
&\times \left[\prod_{i=1}^{n} (1 - a_i z_i) \right]^{-\lambda} \\
&= \left(\sum_{\alpha \in \mathbb{N}^n} c_\alpha (\lambda, 1 - Q_n)\, \mathbf{z}^\alpha (\mathbf{1} - \mathbf{az})^{-(\lambda \mathbf{1} + \alpha)} \right) \\
&= \sum_{\alpha \in \mathbb{N}^n} c_\alpha (\lambda, 1 - Q_n)\, \mathbf{z}^\alpha \left(\sum_{\beta \in \mathbb{N}^n} (\lambda \mathbf{1} + \alpha)_\beta \frac{\mathbf{a}^\beta \mathbf{z}^\beta}{\beta!} \right) \\
&= \sum_{\alpha \in \mathbb{N}^n} \sum_{\beta \in \mathbb{N}^n} c_\alpha (\lambda, 1 - Q_n) \left((\lambda \mathbf{1} + \alpha)_\beta \frac{\mathbf{a}^\beta}{\beta!} \right) \mathbf{z}^{\alpha + \beta} \\
&= \sum_{\gamma \in \mathbb{N}^n} \left(\sum_{\alpha + \beta = \gamma} c_\alpha (\lambda, 1 - Q_n) (\lambda \mathbf{1} + \alpha)_\beta \frac{\mathbf{a}^\beta}{\beta!} \right) \mathbf{z}^\gamma
\end{aligned}
\tag{A.9}
$$

which proves (15). Straightforward computations allow us to obtain the equalities (16) and (17) from (15). □

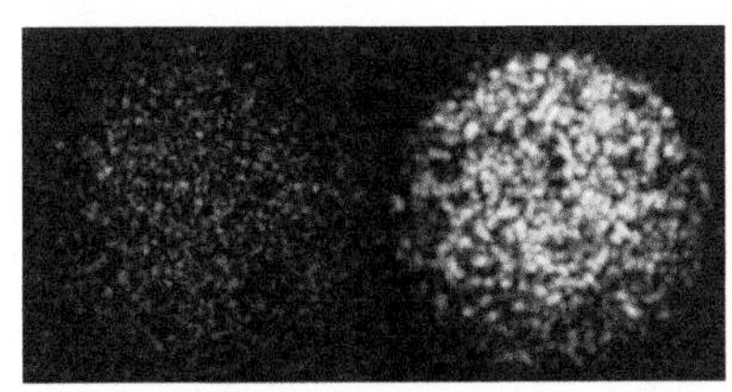

FIGURE 8: Estimates of P^2 using the 3 low flux intensity images N_1, N_2, and N_3 for the real polarimetric images for an estimation window of size $n = 9 \times 9$. "MLE": maximum likelihood estimator.

Proof of Theorem 4. Denote $\|\alpha\| = \max_{i=1,\dots,n}(\alpha_i)$, $|\alpha| = \sum_{i=1}^{n} \alpha_i$ and introduce the notation of [6]

$$
c_\alpha (\lambda, P_n) = \sum_{k \in K_\alpha} (\lambda)_{|k|} \frac{\mathbf{a}^k}{k!}, \tag{A.10}
$$

where $(\lambda)_k = \lambda(\lambda+1)\cdots(\lambda+k-1) = \Gamma(\lambda+k)/\Gamma(\lambda)$. By using Theorem 3, the following results can be obtained

$$\begin{aligned}
c_\gamma(\lambda, P_2) &= \sum_{\alpha+\beta=\gamma} c_\alpha(\lambda, 1-Q_2)(\lambda\mathbf{1}+\alpha)_\beta \frac{\mathbf{a}^\beta}{\beta!} \\
&= \sum_{\ell=0}^{\min(\gamma_1,\gamma_2)} (\lambda)_\ell (\lambda+\ell)_{\gamma_1-\ell}(\lambda+\ell)_{\gamma_2-\ell} \\
&\quad \times \frac{a_1^{\gamma_1-\ell} a_2^{\gamma_2-\ell} b_{1,2}^\ell}{(\gamma_1-\ell)!(\gamma_2-\ell)!\ell!} \\
&= \sum_{\ell=0}^{\min(\gamma_1,\gamma_2)} \frac{\Gamma(\lambda+\ell)}{\Gamma(\lambda)} \frac{\Gamma(\lambda+\gamma_1)}{\Gamma(\lambda+\ell)} \frac{\Gamma(\lambda+\gamma_2)}{\Gamma(\lambda+\ell)} \\
&\quad \times \frac{a_1^{\gamma_1-\ell} a_2^{\gamma_2-\ell} b_{1,2}^\ell}{(\gamma_1-\ell)!(\gamma_2-\ell)!\ell!} \\
&= \frac{\Gamma(\lambda+\max(\gamma_1,\gamma_2))}{\Gamma(\lambda)} \sum_{\ell=0}^{\min(\gamma_1,\gamma_2)} \frac{\Gamma(\lambda+\min(\gamma_1,\gamma_2))}{\Gamma(\lambda)} \\
&\quad \times \frac{\Gamma(\lambda)}{\Gamma(\lambda+\ell)} \frac{b_{1,2}^\ell}{l!} \left(\prod_{i=1}^{2} \frac{a_i^{\gamma_i-\ell}}{(\gamma_i-\ell)!} \right) \\
&= (\lambda)_{\max(\gamma_1,\gamma_2)} \sum_{\ell=0}^{\min(\gamma_1,\gamma_2)} \frac{(\lambda+\ell)_{\min(\gamma_1,\gamma_2)-\ell}}{(\gamma_1-\ell)!(\gamma_2-\ell)!\ell!} \\
&\quad \times a_1^{\gamma_1-\ell} a_2^{\gamma_2-\ell} b_{1,2}^\ell.
\end{aligned} \tag{A.11}$$

□

Proof of Theorem 6. The relation (16) with $Q_3(\mathbf{z}) = b_{1,2}z_1z_2 + b_{1,3}z_1z_3 + b_{2,3}z_2z_3 + b_{1,2,3}z_1z_2z_3$ leads to (22). By using the trivial equality

$$\begin{aligned}
& b_{2,3}^{\upsilon-\gamma_1+\beta_1} b_{1,3}^{\upsilon-\gamma_2+\beta_2} b_{1,2}^{\upsilon-\gamma_3+\beta_3} b_{1,2,3}^{|\gamma-\beta|-2\upsilon} a_1^{\beta_1} a_2^{\beta_2} a_3^{\beta_3} \\
&= \left(\frac{b_{2,3}b_{1,3}b_{1,2}}{b_{1,2,3}^2} \right)^\upsilon \left(\prod_{i=1}^{3} a_i^{\gamma_i} \right) \left(\frac{b_{1,2,3}}{a_1 b_{2,3}} \right)^{\gamma_1-\beta_1} \\
&\quad \times \left(\frac{b_{1,2,3}}{a_2 b_{1,3}} \right)^{\gamma_2-\beta_2} \left(\frac{b_{1,2,3}}{a_3 b_{1,2}} \right)^{\gamma_3-\beta_3}
\end{aligned} \tag{A.12}$$

we easily obtain (23). Note that for $\|\alpha\| \leqslant \upsilon \leqslant \lfloor|\alpha|/2\rfloor$, we have $(\lambda)_\upsilon = (\lambda)_{\|\alpha\|}(\lambda+\|\alpha\|)_{\upsilon-\|\alpha\|}$. By substituting $\alpha_i = \gamma_i - \beta_i$, $i = 1, 2, 3$ in (23), the last result (24) can be obtained. □

Acknowledgments

The authors would like to thank Gérard Letac for fruitful discussions regarding multivariate gamma distributions and negative multinomial distributions. The Authors are also grateful to Mehdi Alouini for acquiring and providing them the polarimetric images.

References

[1] D. C. Doss, "Definition and characterization of multivariate negative binomial distribution," *Journal of Multivariate Analysis*, vol. 9, no. 3, pp. 460–464, 1979.

[2] N. L. Johnson, S. Kotz, and N. Balakrishnan, *Discrete Multivariate Distributions*, John Wiley & Sons, New York, NY, USA, 1997.

[3] F. Le Gall, "The modes of a negative multinomial distribution," *Statistics and Probability Letters*, vol. 76, no. 6, pp. 619–624, 2006.

[4] R. C. Griffiths and R. K. Milne, "A class of infinitely divisible multivariate negative binomial distributions," *Journal of Multivariate Analysis*, vol. 22, no. 1, pp. 13–23, 1987.

[5] S. K. Bar-Lev, D. Bshouty, P. Enis, G. Letac, I. L. Lu, and D. Richards, "The diagnonal multivariate natural exponential families and their classification," *Journal of Theoretical Probability*, vol. 7, no. 4, pp. 883–929, 1994.

[6] P. Bernardoff, "Which negative multinomial distributions are infinitely divisible?" *Bernoulli*, vol. 9, no. 5, pp. 877–893, 2003.

[7] F. Chatelain, S. Lambert-Lacroix, and J. Y. Tourneret, "Pairwise likelihood estimation for multivariate mixed Poisson models generated by gamma intensities," *Statistics and Computing*, vol. 19, no. 3, pp. 283–301, 2009.

[8] C. Brosseau, *Fundamentals of Polarized Light—A Statistical Approach*, John Wiley & Sons, New York, NY, USA, 1998.

[9] J. S. Lee and E. Pottier, *Polarimetric Radar Imaging: From Basics to Applications*, Optical Science and Engineering, Taylor & Francis, London, UK, 2009.

[10] S. Huard, *Polarization of LightStatistical Optics*, John Wiley & Sons, New York, NY, USA, 1997.

[11] H. Mott, *Remote Sensing With Polarimetric Radar*, John Wiley & Sons, New York, NY, USA, 2007.

[12] R. Shirvani, M. Chabert, and J. Y. Tourneret, "Ship and oil-spill detection using the degree of polarization in linear and hybrid/compact dual-pol SAR," *IEEE Journal of Selected Topics in Applied Earth Observations and Remote Sensing*, vol. 5, no. 3, pp. 885–892, 2012.

[13] R. C. Jones, "A new calculus for the treatment of optical systems V. A more general formulation and description of another calculus," vol. 37, no. 2, pp. 107–110, 1947.

[14] E. Wolf, "Coherence properties of partially polarized electromagnetic radiation," *Il Nuovo Cimento*, vol. 13, no. 6, pp. 1165–1181, 1959.

[15] J. Goodman, *Statistical Optics*, John Wiley & Sons, New York, NY, USA, 1985.

[16] J. Fade, M. Roche, and P. Réfrégier, "Precision of moment-based estimation of the degree of polarization in coherent imagery without polarization device," *Journal of the Optical Society of America A*, vol. 25, no. 2, pp. 483–492, 2008.

[17] J. Fade, P. Réfrégier, and M. Roche, "Estimation of the degree of polarization from a single speckle intensity image with photon noise," *Journal of Optics A*, vol. 10, no. 11, Article ID 115301, 2008.

[18] P. Réfrégier, M. Roche, and F. Goudail, "Cramer-Rao lower bound for the estimation of the degree of polarization in active coherent imagery at low photon levels," *Optics Letters*, vol. 31, no. 24, pp. 3565–3567, 2006.

[19] M. Roche, J. Fade, and P. Réfrégier, "Parametric estimation of the square degree of polarization from two intensity images degraded by fully developed speckle noise," *Journal of the Optical Society of America A*, vol. 24, no. 9, pp. 2719–2727, 2007.

[20] F. Chatelain, J. Y. Tourneret, M. Roche, and M. Alouini, "Estimating the polarization degree of polarimetric images in coherent illumination using maximum likelihood methods," *Journal of the Optical Society of America A*, vol. 26, no. 6, pp. 1348–1359, 2009.

[21] W. B. Sparks and D. J. Axon, "Panoramic polarimetry data analysis," *Publications of the Astronomical Society of the Pacific*, vol. 111, no. 764, pp. 1298–1315, 1999.

[22] A. Ferrari, G. Letac, and J. Y. Tourneret, "Multivariate mixed poisson distributions," in *Proceedings of the EUSIPCO-04*, F. Hlawatsch, G. Matz, M. Rupp, and B. Wistawel, Eds., pp. 1067–1070, Vienna, Austria, September 2004.

[23] F. Chatelain, *Lois statistiques multivariees pour le traitement de l'image [Ph.D. thesis]*, Institut National Polytechnique de Toulouse, Toulouse, France, 2007.

Note on Qualitative Robustness of Multivariate Sample Mean and Median

Evgueni Gordienko, Andrey Novikov, and J. Ruiz de Chávez

Department of Mathematics, Autonomous Metropolitan University, Iztapalapa, San Rafael Atlixco 186, Col. Vicentina, C.P. 09340, Mexico City, DF, Mexico

Correspondence should be addressed to J. Ruiz de Chávez; jrch@xanum.uam.mx

Academic Editor: Shein-chung Chow

It is known that the robustness properties of estimators depend on the choice of a metric in the space of distributions. We introduce a version of Hampel's qualitative robustness that takes into account the $\sqrt{n}$-asymptotic normality of estimators in R^k, and examine such robustness of two standard location estimators in $\mathbb{R}^k$. For this purpose, we use certain combination of the Kantorovich and Zolotarev metrics rather than the usual Prokhorov type metric. This choice of the metric is explained by an intention to expose a (theoretical) situation where the robustness properties of sample mean and L_1-sample median are in reverse to the usual ones. Using the mentioned probability metrics we show the qualitative robustness of the sample multivariate mean and prove the inequality which provides a quantitative measure of robustness. On the other hand, we show that L_1-sample median could not be "qualitatively robust" with respect to the same distance between the distributions.

1. Introduction

The following Hampel's definition (originally given for the one-dimensional case) of *qualitative robustness* [1, 2] deals with π-balls in the space of distributions rather than with standard "contamination neighborhoods" (see for the latter, e.g., [3, 4]).

The sequence Q_n, $n \geq 1$, of estimators is qualitatively robust at the distribution $\mathscr{L}$ if for every $\epsilon > 0$ there exists $\delta > 0$ such that $\pi(\mathscr{L}, \widetilde{\mathscr{L}}) < \delta$ entails

$$\sup_{n\geq 1} \pi\left(Q_n\left(X_1, \ldots, X_n\right), Q_n\left(\widetilde{X}_1, \ldots, \widetilde{X}_n\right)\right) < \epsilon. \quad (1)$$

Here, and throughout, π denotes the Prokhorov metric, and $X_1, X_2, \ldots; \widetilde{X}_1, \widetilde{X}_2, \ldots,$ are i.i.d. random vectors distributed, respectively, as $\mathscr{L}$ and $\widetilde{\mathscr{L}}$.

For a metric μ on the space of distributions and random vectors X, Y we will write $\mu(X, Y)$ (as in (1)) having in mind the μ-distance between the distributions of X and Y.

By all means, the use of the Prokhorov metric is only an option. For instance, in [5] other probability metrics in the definition of qualitative robustness were used. (See also [6–9] for using different probability metrics or pseudo-metrics related to the estimation of robustness).

As noted in [1, 2] in $\mathbb{R}^1$, sample means are not qualitatively robust at any $\mathscr{L}$, while sample medians are qualitatively robust at any $\mathscr{L}$ having a unique median. (See also [10] for lack of qualitative robustness of sample means in certain Banach spaces).

Moreover, in [11] it was shown that for symmetric distributions the median is, in certain sense, the "most robust" estimator of a center location when using the pseudo-metric corresponding to neighborhoods of contamination type. (See more with this respect in [8]).

At the same time, it is known from the literature that under different circumstances, in particular, using distinct probability metrics (as, e.g., in (1) or in other definitions) the robustness properties of estimators can change considerably (see, for instance, the discussion in [12]).

The first aim of the present paper is to consider a modified version of Hampel's definition of qualitative robustness taking into account the $\sqrt{n}$-asymptotic normality of the sequence of estimators. The formal definition is given in the next section but, basically, we replace (1) with the following condition:

$$\sup_{n\geq 1} \sqrt{n} l\left(Q_n\left(X_1,\ldots,X_n\right), Q_n\left(\widetilde{X}_1,\ldots,\widetilde{X}_n\right)+\gamma\right) \leq \varepsilon, \quad (2)$$

where γ is some constant, and l is a probability metric (different from the Prokhorov metric in our case).

The second goal of the paper is to present an example of two probability metrics l and μ (on the space of distributions in $\mathbb{R}^k$) for which the following holds:

(i) when $Q_n = T_n$, $n \geq 1$, are multivariate sample means, the left-hand side of inequality (2) (with some γ) is bounded by const $\cdot\mu(\mathscr{L},\widetilde{\mathscr{L}})$;

(ii) when $Q_n = M_n$, $n \geq 1$, are sample medians, we give an example of symmetric smooth distributions $\mathscr{L}$ and $\widetilde{\mathscr{L}}_\alpha$, $\alpha > 0$, in $\mathbb{R}$, such that $\mu(\mathscr{L},\widetilde{\mathscr{L}}_\alpha) \to 0$ as $\alpha \to 0$, while there is a positive constant r such that the left-hand side of inequality (2) (with $\gamma = 0$) is greater than r for all sufficiently small $\alpha > 0$. Therefore, sample medians are not qualitative robust (in our sense) with respect to these metrics.

The metrics l and μ are the following (the complete definitions are given in Section 2). l is the Kantorovich metric (see, e.g., [13]) and μ is certain combination of l and of the Zolotarev metric of order 2 (see [14]).

We should stress that this choice is not determined by any advantages for statistical applications. Moreover, the closeness of distributions in the Kantorovich metric implies the closeness in the Prokhorov metric, but also reduces the probability of "large-valued outliers" (but not the rounding errors). Therefore, our selection of metric is not quite consistent with the standard approach to qualitative robustness where the Prokhorov metric (or its invariant versions) is used.

Nevertheless, our choice allows to unveil the possible unusual robustness properties of sample means and medians and to assess the certain quantitative robustness of the multivariate sample mean (with respect to the considered metrics!). The obtained "robustness inequality" does not work in the "gross error model" but (jointly with inequalities (23)) it could be useful for quantitative assessment of the robustness of sample means, under perturbation of data of "rounding" type.

2. Basic Definitions

Let $(H,|\cdot|)$ be a separable Hilbert space with the norm $|\cdot|$ generated by an inner product $\langle\cdot,\cdot\rangle$, and let $\mathscr{B}(H)$ be the Borel σ-algebra of subsets of H. Let also $X; X_1,\ldots,X_n,\ldots$ and $\widetilde{X}; \widetilde{X}_1,\ldots,\widetilde{X}_n,\ldots$ be two sequences of i.i.d. random vectors in H with their respective distributions denoted by $\mathscr{L}$ and $\widetilde{\mathscr{L}}$.

Under the assumption

$$E|X| < \infty, \qquad E\left|\widetilde{X}\right| < \infty, \quad (3)$$

the *means* EX and $E\widetilde{X}$ are defined as the corresponding Bochner integrals.

The definition of the *median* of X is less standard (consult, e.g., [15] for different definitions in $H = \mathbb{R}^k$). We will use the definition of median given in general setting in [16] (sometimes called $\mathbb{L}_1$-median):

$$M := \arg\min_{x\in H} E\left(|X-x|-|x|\right), \quad (4)$$

$$\widetilde{M} := \arg\min_{x\in H} E\left(\left|\widetilde{X}-x\right|-|x|\right). \quad (5)$$

In [16] it was shown that M exists, and it is unique unless $\mathscr{L}$ is concentrated on a one-dimensional subspace of H. In the last case the set of minimizers in (4) is $\{tx_0 : t \in [a,b]\}$ for some $x_0 \in H$, and one can set $M := ((a+b)/2)x_0$.

In what follows we denote ($n = 1,2,\ldots$):

$$S_n := X_1+\cdots+X_n, \qquad \widetilde{S}_n := \widetilde{X}_1+\cdots+\widetilde{X}_n,$$
$$T_n = \frac{S_n}{n}, \qquad \widetilde{T}_n := \frac{\widetilde{S}_n}{n} \quad \text{(sample means)}. \quad (6)$$

On the other hand, let ($n = 1,2,\ldots$),

$$M_n = M_n\left(X_1,\ldots,X_n\right), \qquad \widetilde{M}_n = \widetilde{M}_n\left(\widetilde{X}_1,\ldots,\widetilde{X}_n\right) \quad (7)$$

be *sample medians* defined by (4) and (5) replacing $\mathscr{L}$ and $\widetilde{\mathscr{L}}$ by the corresponding empirical distributions $\mathscr{L}_n$ and $\widetilde{\mathscr{L}}_n$ (obtained from $X_1,\ldots,X_n$ and $\widetilde{X}_1,\ldots,\widetilde{X}_n$, resp.). Robustness (in terms of ϵ-contamination neighborhoods) and asymptotic normality of M_n, $n \geq 1$, were proved, for instance, in [17] (see also [18]). The qualitative robustness properties of sample means $\{T_n\}$ and sample medians $\{M_n\}$ when the Prokhorov metric is used were discussed in Introduction.

Let us first see what happens with qualitative robustness of $\{T_n\}$ and $\{M_n\}$ if in the above definition we replace π with the Kantorovich metric:

$$l\left(X,\widetilde{X}\right) \equiv l\left(\mathscr{L},\widetilde{\mathscr{L}}\right) := \sup_{\phi\in\mathrm{Lip}} \left|E\phi(X) - E\phi\left(\widetilde{X}\right)\right|, \quad (8)$$

where

$$\mathrm{Lip} := \{\phi : H \longmapsto \mathbb{R} : \phi \text{ is bounded and}$$
$$\left|\phi(x)-\phi(y)\right| \leq |x-y|;\ x,y\in H\}. \quad (9)$$

Under condition (3) $l(X,\widetilde{X}) < \infty$, and it is well known (see, e.g., [13]) that $l(Y_n,Y) \to 0$ if and only if $\pi(Y_n,Y) \to 0$ and $E|Y_n| \to E|Y|$.

For $H = \mathbb{R}$,

$$l\left(X,\widetilde{X}\right) = \int_{-\infty}^{\infty} \left|F_X(x) - F_{\widetilde{X}}(x)\right| dx. \quad (10)$$

Now, applying the regularity properties of l (see, e.g., [13]):

$$l\left(\frac{S_n}{n},\frac{\widetilde{S}_n}{n}\right) \leq \frac{1}{n}\sum_{i=1}^{n} l\left(X_i,\widetilde{X}_i\right) = l\left(\mathscr{L},\widetilde{\mathscr{L}}\right), \quad (11)$$

we see that the sequence of sample means $\{T_n, n \geq 1\}$ is "qualitatively robust" using l instead of the Prokhorov metric.

It seems straightforward (using the approach similar to one given in [10, 19] and results in [16]) to show that the sequence of sample medians $\{M_n, n \geq 1\}$ is also "qualitatively robust" with respect to l. However, this is out of the scope of this paper.

3. l–μ–$\sqrt{n}$-Robustness and Main Results

Now and in what follows we suppose that $H = \mathbb{R}^k$ with the Euclidean norm $|\cdot|$. The results presented in this section are an extension to the multidimensional case of the similar findings for $H = \mathbb{R}$, published in the hardly accessible proceedings [20]. Moreover, we improve the results of [20] even in the one-dimensional case.

In order to simplify calculations in the proof of Theorem 5 below we will use in the definitions of Kantorovich's and Zolotarev's metrics (see below) the following norm $\|x\| := \sum_{i=1}^{k} |x_i|$ in the space $\mathbb{R}^k$.

Thus, in what follows the Kantorovich metric l is defined by relationships (8), (9), where in (9) in place of the norm $|\cdot|$ the norm $\|\cdot\|$ is used.

Let μ be some fixed simple probability metric on the set of all probability distributions on $(\mathbb{R}^k, \mathscr{B}(\mathbb{R}^k))$.

We will consider a sequence Q_n, $n \geq 1$, of estimators of some parameter $\theta \in \mathbb{R}^k$ of the distribution $\mathscr{L}$ ($\widetilde{\theta}$ of the distribution $\widetilde{\mathscr{L}}$, resp.).

Definition 1. We say that a sequence Q_n, $n \geq 1$, is $(l - \mu - \sqrt{n})$-robust at $\mathscr{L}$ if there is some fixed vector $\gamma = \gamma(\mathscr{L}, \widetilde{\mathscr{L}}) \in \mathbb{R}^k$ such that for every $\epsilon > 0$ there exists $\delta > 0$ such that $\mu(\mathscr{L}, \widetilde{\mathscr{L}}) \leq \delta$ entails

$$\sup_{n\geq 1} \sqrt{n} l \left(Q_n (X_1, \ldots, X_n), Q_n (\widetilde{X}_1, \ldots, \widetilde{X}_n) + \gamma\right) \leq \epsilon. \quad (12)$$

Remark 2. Taking into account that $l(aX + b, aY + b) = al(X, Y)$, $a \geq 0$, $b \in \mathbb{R}^k$, we see that (12) can be related with the $\sqrt{n}$-asymptotic normality of estimators Q_n, $n \geq 1$. The "scaling parameter" γ (in (12)) is necessary to ensure the equality of means of the corresponding limit normal distributions (when they exist). Only in case $EX = E\widetilde{X}$ in (12) $\gamma = 0$.

Zolotarev's probability metric of order 2 ζ_2 is defined as follows (see, e.g., [14, 21]):

$$\zeta_2 (X, \widetilde{X}) \equiv \zeta_2 (\mathscr{L}, \widetilde{\mathscr{L}}) := \sup_{\phi \in \mathscr{D}_2} \left| E\phi (X) - E\phi (\widetilde{X}) \right|, \quad (13)$$

where

$$\mathscr{D}_2 := \left\{\phi : \mathbb{R}^k \longmapsto \mathbb{R} : \|D\phi (x) - D\phi (y)\| \leq \|x - y\| ; \; x, y \in \mathbb{R}^k\right\}, \quad (14)$$

and $D\phi = (D_1\phi, \ldots, D_k\phi)$ is the gradient of ϕ.

Remark 3. The distance $\zeta_2(X, \widetilde{X})$ can take infinite value. Particularly, from (13), (14) we see that $\zeta_2(X, \widetilde{X}) = \infty$ if $EX \neq E\widetilde{X}$. On the other hand, if there exist second moments, $EX = E\widetilde{X}$ and $EX^2 < \infty$, $E\widetilde{X}^2 < \infty$, then $\zeta_2(X, \widetilde{X}) < \infty$ (see, e.g., [14]). The function $\phi(x) := (1/2)|x|^2 \in \mathscr{D}_2$ in (14). Therefore, by (13),

$$\left|EX^2 - E\widetilde{X}^2\right| \leq \zeta_2 (X, \widetilde{X}). \quad (15)$$

Thus, if $EX^2 < \infty$ and $\zeta_2(X, \widetilde{X}) < \infty$ then $E\widetilde{X}^2 < \infty$. Consequently, if $EX = E\widetilde{X}$ and $E|X|^2 < \infty$, then $\zeta_2(X, \widetilde{X}) < \infty$ if and only if $E|\widetilde{X}|^2 < \infty$.

We now define the metric μ to work with:

$$\mu (X, \widetilde{X}) := \max \left\{2l (X, \widetilde{X}), \inf_{b\in\mathbb{R}^k} \zeta_2 (X, \widetilde{X} + b)\right\}. \quad (16)$$

3.1. $(l$–μ–$\sqrt{n})$-Robustness of Sample Means. To prove the inequality in Theorem 5 below we need to impose the following restriction on the distribution $\mathscr{L}$.

Assumption 4. *(i) $E|X|^{k+2} < \infty$, and the covariance matrix $\mathscr{M}$ of X is positive definite.*

The distribution $\mathscr{L}$ of X has a density f such that for some $s \geq 1$:

(ii) the density f_s of $S_s := X_1 + \cdots + X_s$ is bounded and differentiable;

(iii) the gradient Df_s is bounded and $|Df_s|$ belongs to $\mathbb{L}_1(\mathbb{R}^k)$;

(iv) for some $\alpha > 0$

$$\int_{|x|>\alpha n} |Df_s| \, dx = O(n^{-1/2}), \quad \text{as } n \longrightarrow \infty. \quad (17)$$

Note that in view of Remark 3 under the condition (i) $\mu(X, \widetilde{X}) < \infty$ if and only if

$$E|\widetilde{X}|^2 < \infty, \quad (18)$$

(see (15)).

Theorem 5. *Under Assumption 4 and supposing* (18) *it holds*

$$\sup_{n\geq 1} \sqrt{n} l (T_n, \widetilde{T}_n + \gamma) \leq c\mu (\mathscr{L}, \widetilde{\mathscr{L}}), \quad (19)$$

where $\gamma = EX - E\widetilde{X}$,

$$c = \max \left\{(10s - 1)^{1/2}, 5.4\, dk\right\}, \quad (20)$$

$$d = \sup_{n\geq s} \max_{1\leq i\leq k} \|D_i g_n\|_{\mathbb{L}_1(\mathbb{R}^k)} < \infty, \quad (21)$$

and g_n is the density of $(1/\sqrt{n})(X_1 + \cdots + X_n)$, $n \geq 1$.

Remark 6. (i) The constant c in (20), (21) is entirely determined by the distribution $\mathscr{L}$ of X. For various particular densities of X the constant d in (21) can be bounded by means of computer calculations. For this one can use the fact (true under wide conditions) that the sequence $D_i g_n$, $n \geq s$, converges in $\mathbb{L}_1$-norm to the corresponding partial derivative of the limit normal density with covariance matrix $\mathscr{M}$ (and zero mean since $||D_i g_n||_{\mathbb{L}_1(\mathbb{R}^k)}$ is invariant under translations).

For example, let $k = 2$ and $X = (X', X'')$, where X' and X'' are independent random variables; X' has the gamma density with $\alpha = 2.1$ and arbitrary $\lambda' > 0$, while X'' has the gamma density with $\alpha = 3$ and arbitrary λ''. Simple computer calculations show that in (21) $d < \max\{0.7065\lambda', 0.5414\lambda''\}$, and since we can take $s = 1$, we obtain in (20) that

$$c < \max\left\{3, 10.8 \max\left\{0.7065\lambda', 0.5414\lambda''\right\}\right\}. \quad (22)$$

For instance, $c < 7.6302$ for $\lambda' = \lambda'' = 1$. (For these values of λ', λ'' we can take $s = 3$ in (20) and obtain $c < 6.3829$.)

(ii) Since under the above assumption $\mu(\mathscr{L}, \widetilde{\mathscr{L}}) < \infty$ entails (18), inequality (19) ensures *$(l - \mu - \sqrt{n})$-robustness of the sequence of sample means T_n, $n \geq 1$.*

(iii) For $k = 1$, in [20] an example is given showing that in general the sequence of sample means T_n, $n \geq 1$ is not $(l - l - \sqrt{n})$-robust (even if (18) holds and $EX = E\widetilde{X}$). It is also almost evident that the sample means T_n, $n \geq 1$, are not $(l - \mu - \sqrt{n})$-robust, for example, if μ is the total variation metric V (or, if $\mu = \max(l, V)$). The appearance of Zolotarev's metric on the right-hand side of (19) is related to closeness of corresponding limit normal distributions.

Corollary 7. *Suppose for a moment that $\theta = \widetilde{\theta}$ and that one evaluates the quality of estimators by mean of absolute errors: $\delta_n := E||T_n - \theta||$, $\widetilde{\delta}_n := E||\widetilde{T}_n - \theta||$. Then from (19) it follows that*

$$\left|\delta_n - \widetilde{\delta}_n\right| \leq n^{-1/2} \text{const}\, \mu\left(\mathscr{L}, \widetilde{\mathscr{L}}\right), \quad n \geq 1. \quad (23)$$

(The simple proof is similar to the one given in [20]).

3.2. About $(l-\mu-\sqrt{n})$-Robustness of Sample Medians. Let again μ be the metric defined in (16). We show that the sequence of sample medians M_n, $n \geq 1$, in general, is not $(l - \mu - \sqrt{n})$-robust even when X and $\widetilde{X}$ have strictly positive, bounded, smooth densities symmetric with respect to the origin, and the sequences of sample medians M_n, $n \geq 1$, $\widetilde{M}_n$, $n \geq 1$, are $\sqrt{n}$-asymptotically normal. We consider a modified version of the corresponding example from [20].

Example 8. Let $H = \mathbb{R}$, $X \sim N(0, 1)$, and for $\epsilon \in (0, 1)$ let $\widetilde{X}_\epsilon$ be a random variable with the density:

$$f_{\widetilde{X}_\epsilon}(x) := \frac{1}{c(\epsilon)}\left(\epsilon + \frac{x^2}{\epsilon^3 + x^2}\right) e^{-x^2/2}, \quad x \in \mathbb{R}, \quad (24)$$

where $c(\epsilon)$ is a normalizing constant. By symmetry of the density, we get

$$M = \widetilde{M} = EX = E\widetilde{X} = 0, \quad (25)$$

and also it is clear that $f_X, f_{\widetilde{X}} \in \mathbb{C}^\infty(\mathbb{R})$, and

$$\sup_{\epsilon \in (0,1), x \in \mathbb{R}} f_{\widetilde{X}_\epsilon}(x) < \infty. \quad (26)$$

First of all let us show that for this example the left-hand side of (12) is infinite for any $\gamma \neq 0$. It is well known (see, e.g., [16]) that $M_n \to M = 0$, $\widetilde{M}_n \to \widetilde{M} = 0$ as $n \to \infty$ with probability 1. Also, from the results of [16] we can obtain that $|M_n| \leq (2/n)\sum_{i=1}^n |X_i|$. Using this inequality it is easy to show that the sequence $|M_n|$, $n \geq 1$, is uniformly integrable. Therefore, $E|M_n| \to 0$ (and also $E|\widetilde{M}_n| \to 0$) as $n \to \infty$, and for this $l(M_n, \widetilde{M}_n + \gamma) \to l(0, \gamma) = \gamma$.

Let now $\gamma = 0$ in (12). By the well-known asymptotic normality (see [22, page 307]),

$$\pi\left(\sqrt{n}M_n, \eta\right) \longrightarrow 0, \qquad \pi\left(\sqrt{n}\widetilde{M}_n, \eta_\epsilon\right) \longrightarrow 0, \quad (27)$$

where

$$\eta \sim N\left(0, \frac{1}{2f_X(0)}\right) = N\left(0, \sqrt{\frac{\pi}{2}}\right),$$
$$\eta_\epsilon \sim N\left(0, \frac{1}{2f_{\widetilde{X}_\epsilon}(0)}\right) = N\left(0, \frac{c(\epsilon)}{\epsilon}\right). \quad (28)$$

From (27) and (28) it follows that there is $r > 0$ such that for all small enough ϵ and all large enough n we get $\pi(\sqrt{n}M_n, \sqrt{n}\widetilde{M}_n) \geq r$. Consequently, for all such ϵ and n,

$$\sqrt{n}l\left(M_n, \widetilde{M}_n\right) = l\left(\sqrt{n}M_n, \sqrt{n}\widetilde{M}_n\right) \geq r^2 > 0, \quad (29)$$

since $l \geq \pi^2$ (see [13, page 86]).

We have obtained that the left-hand side of (12) is positive for all small enough $\epsilon > 0$. On the other hand, $\mu(X, \widetilde{X}_\epsilon) \to 0$ as $\epsilon \to 0$. It follows from the inequality

$$\zeta_2(X, Y) \leq \frac{1}{2}\int_{-\infty}^{\infty} x^2 \left|f_X(x) - f_Y(x)\right| dx \quad (30)$$

(valued when $EX = EY$; see [21, page 376]) and from (10).

Remark 9. The densities as in (24) represent the following somewhat strange type of "contamination." Since $\max_x f_X(x) = f_X(0)$ sample points from f_X tend to concentrate around the origin. But $f_{\widetilde{X}_\epsilon}(0) \to 0$ as $\epsilon \to 0$, and therefore sample points from $f_{\widetilde{X}_\epsilon}$ frequently in some extent are separated from 0.

A natural question is "how to choose the metric $\widetilde{\mu}$ to ensure $(l - \widetilde{\mu} - \sqrt{n})$-robustness of the sequence of sample medians M_n, $n \geq 1$?" Our conjecture is (for $k = 1$, e.g.) to try $\widetilde{\mu}(X, \widetilde{X}) = \max\{l(X, \widetilde{X}), \text{esssup}_{x\in\mathbb{R}}|f_X(x) - f_{\widetilde{X}}(x)|\}$ (supposing the existence of densities).

If $M = \widetilde{M}$ then under certain conditions the closeness in $\widetilde{\mu}$ guarantees the closeness of normal densities which are limiting for $\{\sqrt{n}M_n, n \geq 1\}$ and for $\{\sqrt{n}\widetilde{M}_n, n \geq 1\}$, respectively. To attempt proving $(l - \widetilde{\mu} - \sqrt{n})$-robustness of M_n, $n \geq 1$ (as in (12)) one can show Hampel's qualitative robustness of M_n, $n \geq 1$ with respect to the metric l and then use the property $\sqrt{n}l(M_n, \widetilde{M}_n) = l(\sqrt{n}M_n, \sqrt{n}\widetilde{M}_n)$. A not clear point of this plan is finding conditions under which $E|\sqrt{n}(M_n - M)| \to 0$ as $n \to \infty$.

Example 10. Let us give another (very simple) example of the sequence of estimators which is $(l - \mu - \sqrt{n})$-robust with $\mu := l$ (on the class of distributions described below). For $H = \mathbb{R}$ we consider the class W of all random variables X with bounded supports $\mathrm{Supp}(X) = [0, \theta_X]$, having density f_X such that $f_X(x) \geq \beta > 0$, $x \in [0, \theta_X]$. We suppose that $\theta_X \leq \theta^* < \infty$, $X \in W$, and β is the same for all $X \in W$. Assume that parameter $\theta = \theta_X$ is unknown, and the sequence of estimators $Q_n := \max_{1\leq i\leq n} X_i$ is used to estimate it.

Denoting $\theta = \theta_X$, $\widetilde{\theta} = \widetilde{\theta}_X$ and choosing in (12) $\gamma = \theta - \widetilde{\theta}$, we get

$$\begin{aligned} l\left(Q_n, \widetilde{Q}_n + \gamma\right) &= l\left(\theta - \max_{1\leq i\leq n} X_i, \widetilde{\theta} - \max_{1\leq i\leq n} \widetilde{X}_i\right) \\ &\leq E\left|\left(\theta - \max_{1\leq i\leq n} X_i\right) - \left(\widetilde{\theta} - \max_{1\leq i\leq n} \widetilde{X}_i\right)\right| \\ &\leq E\left(\theta - \max_{1\leq i\leq n} X_i\right) + E\left(\widetilde{\theta} - \max_{1\leq i\leq n} \widetilde{X}_i\right), \end{aligned} \quad (31)$$

because of the metric l is minimal for the compound metric $E|X - Y|$ (see, e.g., [13]).

By elementary calculations we bound the right-hand side of (31) by

$$\frac{2}{\beta(n+1)}. \quad (32)$$

Now for each n fixed, by induction we obtain (see (10))

$$\begin{aligned} l\left(\max_{1\leq i\leq n} X_i, \max_{1\leq i\leq n} \widetilde{X}_i\right) &= \int_0^\infty \left|F_X^n(x) - F_{\widetilde{X}}^n(x)\right| dx \\ &\leq n\int_0^\infty \left|F_X(x) - F_{\widetilde{X}}(x)\right| dx = nl\left(X, \widetilde{X}\right). \end{aligned} \quad (33)$$

On the other hand, let, for example, $\theta \leq \widetilde{\theta}$. Then

$$l\left(X, \widetilde{X}\right) = \int_0^\theta \left|F_X(x) - F_{\widetilde{X}}(x)\right| dx + \int_\theta^{\widetilde{\theta}} \left(1 - F_{\widetilde{X}}(x)\right) dx. \quad (34)$$

But for $x \in [\theta, \widetilde{\theta}]$,

$$1 - F_{\widetilde{X}}(x) = \int_x^{\widetilde{\theta}} f_{\widetilde{X}}(y)\, dy \geq \beta\left(\widetilde{\theta} - x\right). \quad (35)$$

From this relationship it follows that

$$\text{for } X, \widetilde{X} \in W, \quad \widetilde{\theta} \longrightarrow \theta \quad \text{as } l\left(X, \widetilde{X}\right) \longrightarrow 0. \quad (36)$$

Finally, applying (32), we can select N in such a way that

$$\sup_{n\geq N} \sqrt{n} l\left(Q_n, \widetilde{Q}_n + \gamma\right) \leq \frac{\epsilon}{2}. \quad (37)$$

For $n < N$ we can use the inequality $l(X, Y + b) \leq l(X, Y) + |b|$ and (33), (36). Thus, choosing small enough $\delta > 0$ we can ensure inequality (12) for all $X, \widetilde{X} \in W$ with $\mu(X, \widetilde{X}) \equiv l(X, \widetilde{X}) \leq \delta$. In this way we proved $(l - l - \sqrt{n})$-robustness of Q_n on the set W.

Appendix

A. The Proofs

To start with the proof of Theorem 5 of the previous section, first of all we note that in (19)

$$\begin{aligned} l\left(T_n, \widetilde{T}_n + \gamma\right) &= l\left(\frac{S_n}{n}, \frac{\widetilde{S}_n}{n} + EX - E\widetilde{X}\right) \\ &= l\left(\frac{S_n - nEX}{n}, \frac{\widetilde{S}_n - nE\widetilde{X}}{n}\right), \end{aligned} \quad (A.1)$$

because of $l(X + b, Y + b) = l(X, Y)$, $b \in \mathbb{R}^k$. Also it is easy to see that $\zeta_2(X + b, Y + b) = \zeta_2(X, Y)$.

On the other hand, by definition of μ in (16)

$$\begin{aligned} \inf_{b\in\mathbb{R}^k} \zeta_2\left(X, \widetilde{X} + b\right) &= \zeta_2\left(X, \widetilde{X} + EX - E\widetilde{X}\right) \\ &= \zeta_2\left(X - EX, \widetilde{X} - E\widetilde{X}\right) \end{aligned} \quad (A.2)$$

(see Remark 3). Also

$$\begin{aligned} l\left(X - EX, \widetilde{X} - E\widetilde{X}\right) &= l\left(X, \widetilde{X} + EX - E\widetilde{X}\right) \\ &\leq l\left(X, \widetilde{X}\right) + E\left|X - \widetilde{X}\right| \leq 2l\left(X, \widetilde{X}\right), \end{aligned} \quad (A.3)$$

because the metric l is minimal for the metric $E|\cdot|$.

The above arguments show that to establish inequality (19) it suffices to prove the version of (19) with $\gamma = 0$, $EX = E\widetilde{X} = 0$, and the metric $\mu' := \max\{l, \zeta_2\}$ on the right-hand side of (19).

The proof is based on the two following lemmas.

Let $g : \mathbb{R}^k \mapsto \mathbb{R}$ be a differentiable function. We will write $Dg := (D_1 g, \ldots, D_k g)$, where $D_i g(x) := (\partial/\partial x_i) g(x)$, $i = 1, 2, \ldots, k$.

Lemma A.1. *Let X, Y, and ξ be random vectors in $\mathbb{R}^k$ such that*

(a) *ξ is independent of X and Y;*

(b) $E|X|^2 < \infty$, $E|Y|^2 < \infty$; $EX = EY$;

(c) *ξ has a bounded differentiable density f_ξ (with respect to the Lebesgue measure) such that $D_i f_\xi \in \mathbb{L}_1(\mathbb{R}^k)$, $i = 1, 2, \ldots, k$.*

Then

$$l(X + \xi, Y + \xi) \leq k \max_{1\leq i\leq k} \left\|D_i f_\xi\right\|_{\mathbb{L}_1(\mathbb{R}^k)} \zeta_2(X, Y). \quad (A.4)$$

Proof. From the definition (8), (9) of the metric l (with the norm $\|\cdot\|$ instead of $|\cdot|$!) it follows (the proof is simple) that in (8) the class of functions Lip (given in (9)) can be replaced by the class of all bounded differentiable functions $g : \mathbb{R}^k \mapsto \mathbb{R}$ such that

$$\sup_{x\in\mathbb{R}^k} \left|\frac{\partial g(x)}{\partial x_i}\right| \leq 1, \quad i = 1, 2, \ldots, k. \quad (A.5)$$

Let us fix any such function g and arbitrary $i, j \in \{1, 2, \ldots, k\}$. Then (by the Fubini theorem),

$$\begin{aligned} & Eg(X+\xi) - Eg(Y+\xi) \\ & = \int_{\mathbb{R}^k} g(x)\,dx \int_{\mathbb{R}^k} f_\xi(x-t)\,[dF_X(t) - dF_Y(t)] \\ & \quad \times \int_{\mathbb{R}^k} [dF_X(t) - dF_Y(t)] \int_{\mathbb{R}^k} g(x)\, f_\xi(x-t)\,dx. \end{aligned} \tag{A.6}$$

For each $x \in \mathbb{R}$ fixed let

$$\begin{aligned} h(t) &:= \int_{\mathbb{R}^k} g(x)\, f_\xi(x-t)\,dx \\ &= \int_{\mathbb{R}^k} g(y+t)\, f_\xi(y)\,dy, \quad t \in \mathbb{R}^k. \end{aligned} \tag{A.7}$$

Because of boundedness of $\partial g/\partial t_i$ and integrability of $D_i f_\xi$ we can differentiate in (A.7) under the integral sign (see [23, Appendix A]). Thus,

$$\begin{aligned} \frac{\partial h}{\partial t_j} h(t) &= \int_{\mathbb{R}^k} \frac{\partial g}{\partial t_j} g(y+t)\, f_\xi(y)\,dy, \\ \frac{\partial^2 h}{\partial t_i \partial t_j} h(t) &= -\int_{\mathbb{R}^k} \frac{\partial g}{\partial x_j}(x)\, \frac{\partial f_\xi}{\partial t_i}(x-t)\,dx. \end{aligned} \tag{A.8}$$

In view of (A.5) we obtain

$$\left| \frac{\partial^2 h}{\partial t_i \partial t_j} h(t) \right| \le \int_{\mathbb{R}^k} \left| \frac{\partial f_\xi}{\partial t_i}(x-t) \right| dx = \left\| D_i f_\xi \right\|_{\mathbb{L}_1(\mathbb{R}^k)}. \tag{A.9}$$

For every $x, y \in \mathbb{R}^k$ we have

$$\|Dh(x) - Dh(y)\| = \sum_{j=1}^{k} \left| D_j h(x) - D_j h(y) \right|, \tag{A.10}$$

$$\left| D_j h(x) - D_j h(y) \right| \le \sum_{i=1}^{k} \sup_{x \in \mathbb{R}^k} \frac{\partial^2 h}{\partial x_i \partial x_j}(x)\, \|x_i - y_i\|. \tag{A.11}$$

Comparing (A.6)–(A.11) and taking into account the definition of ζ_2 in (13), (14), we obtain inequality (A.4). □

Lemma A.2. *Under the assumption of the previous section the constant d in (21) is finite.*

Remark A.3. A similar assertion was proved in [24] for $\|\cdot\|_{\mathbb{L}_1}$-norm of second derivatives of the densities g_n (under sightly different conditions). For this reason we give only a sketch of the proof of Lemma A.2 indicating only differences in comparison to the proof of Lemma 4.1 in [24] (where the omitted details can be seen).

Proof. As before, let g_n $(n \ge s)$ be the density of $S_n/\sqrt{n}$, and ϕ_n its characteristic function. Let also f_n denote the density of S_n. There is $l \ge s$ such that $|t|\phi_l(t) \in \mathbb{L}_1(\mathbb{R}^k)$ and therefore

$$|t|\,\phi_n(t) \in \mathbb{L}_1\left(\mathbb{R}^k\right), \quad \text{for } n \ge l. \tag{A.12}$$

Indeed, since

$$v_i(x_i) := \frac{\partial}{\partial x_i} f_s(\ldots, x_i, \ldots) \in \mathbb{L}_1(\mathbb{R}) \tag{A.13}$$

(fixing other variables), we can integrate by parts:

$$\begin{aligned} \phi_s(\sqrt{s}t) &= \int_{\mathbb{R}^{k-1}} e^{i\langle t', x' \rangle} dx' \int_{\mathbb{R}} f_s(\ldots, x_i, \ldots)\, e^{it_i x_i} dx \\ &= \int_{\mathbb{R}^{k-1}} e^{i\langle t', x' \rangle} dx' \frac{1}{it_i} \int_{\mathbb{R}} v_i(x_i)\, e^{it_i x_i} dx_i, \end{aligned} \tag{A.14}$$

and (A.12) follows for large enough l.

In view of (A.12) we can write down the inverse Fourier transform for $n \ge l$:

$$g_n(x) = \frac{1}{(2\pi)^k} \int_{\mathbb{R}^k} \phi_n(t)\, e^{-i\langle t, x \rangle} dt, \tag{A.15}$$

and differentiate under the integral sign in (A.15).

The condition $\phi_n \in \mathbb{L}_1(\mathbb{R}^k)$ $(n \ge l)$ and Assumption (i) in Section 2 ensure the hypothesis of in [25, Theorem 19.2, Ch. 4]. By this theorem,

$$g_n(x) = \sum_{m=0}^{k} n^{-m/2} P_m(x)\, \phi_{0,\mathcal{M}}(x) + o\left(\frac{1}{n^{k/2}}\right), \tag{A.16}$$

$x \in \mathbb{R}^k$ (as $n \to \infty$), where P_m, $m = 0, \ldots, k$, are certain polynomials and $\phi_{0,\mathcal{M}}$ is the normal density with zero mean and the covariance matrix $\mathcal{M}$ (see Assumption (i)).

Let q_n denote the Fourier transform of $D_j h_n$, where

$$h_n(x) := g_n(x) - \sum_{m=0}^{k} n^{-m/2} P_m(x)\, \phi_{0,\mathcal{M}}(x). \tag{A.17}$$

Using (A.15), (A.16), and in [25, Lemma 7.2, Ch. 2], we can obtain that for each fixed j

$$q_n(t) = t_j \left[\phi_n(t) - \sum_{m=0}^{k} n^{-m/2} \widehat{P}_m(t) \exp\left(-\frac{1}{2}\langle t, \mathcal{M}t \rangle\right) \right], \tag{A.18}$$

with certain polynomials $\widehat{P}_m$.

By arguments similar to those given in [24], it follows from (A.18) that there exist constants $\beta > 0$, $c < \infty$ such that

$$\int_{|t| \le \beta\sqrt{n}} |q_n(t)|\,dt \le \frac{c}{n^{k/2}} \quad (n \ge l). \tag{A.19}$$

On the other hand, to prove that

$$\int_{|t| > \beta\sqrt{n}} |q_n(t)| = O\left(\frac{1}{n^{k/2}}\right), \quad \text{as } n \longrightarrow \infty, \tag{A.20}$$

it is sufficient to show that

$$\int_{|t| > \gamma\sqrt{n}} |t_j|\, |\phi_n(t)|\,dt = O\left(\frac{1}{n^{k/2}}\right), \quad \text{as } n \longrightarrow \infty. \tag{A.21}$$

But the last equality follows from (A.12) and the fact that

$$\sup_{|t|>\gamma\sqrt{n}} \left|\phi_1\left(\frac{t}{\sqrt{n}}\right)\right| \le \kappa < 1. \tag{A.22}$$

Expressing $D_j h_n$ in terms of q_n (as an inverse Fourier transform), and using (A.16)–(A.20) we can establish that there exist a constant $\overline{c}$ and polynomials $\widetilde{P}_m(x)$, $m = 0, \dots, k$ such that for each $j \in \{1, \dots, k\}$,

$$\sup_{x\in\mathbb{R}^k} \left|D_j g_n(x) - \sum_{m=0}^{k} n^{-m/2} \widetilde{P}_m(x)\,\phi_{0,\mathcal{M}}(x)\right| \le \frac{\overline{c}}{n^{k/2}}, \tag{A.23}$$

$n = l, l+1, \dots$

The next step is to find for each $j \in \{1, \dots, k\}$ an upper bound for $\|D_j g_n\|_{\mathbb{L}_1(\mathbb{R}^k)}$ which does not depend on $n \ge s$.

For $n > s$, $g_n(x) = n^{k/2} f_n(\sqrt{n}x)$ and $f_n(x) = \int_{\mathbb{R}^k} f_s(x - t) f_{n-s(t)} dt$. Thus, using Assumptions (ii), (iii), and the corresponding theorems in [23, Appendix A], we get

$$D_j f_n(x) = \int_{\mathbb{R}^k} D_j f_s(x-t)\, f_{n-s}(t)\, dt. \tag{A.24}$$

We have

$$\left\|D_j g_n\right\|_{\mathbb{L}_1(\mathbb{R}^k)} \le \int_{|x|\le 2\sqrt{n}\alpha} \left|D_j g_n\right| dx + \int_{|x|>2\sqrt{n}\alpha} \left|D_j g_n\right| dx, \tag{A.25}$$

where α is the constant from Assumption (iv). The first summands on the right-hand side of (A.25) are uniformly bounded in $n \ge l$ due to (A.23). To bound the second terms in (A.25) we write (see (A.24))

$$\begin{aligned}\int_{|x|>2\sqrt{n}\alpha} &\left|D_j g_n\right| dx \\ &= n^{1/2} \int_{|z|>2\alpha n} \left|D_j f_n(z)\right| dz \\ &\le n^{1/2} \int_{|z|>2\alpha n} dz \int_{\mathbb{R}^k} \left|D_j f_s(z-t)\right| f_{n-s}(t)\, dt =: n^{1/2} I_n.\end{aligned} \tag{A.26}$$

Now

$$\begin{aligned}I_n = &\int_{|z|>2\alpha n} dz \int_{|t|\le\alpha n} \left|D_j f_s(z-t)\right| f_{n-s}(t)\, dt \\ &+ \int_{|z|>2\alpha n} dz \int_{|t|>\alpha n} \left|D_j f_s(z-t)\right| f_{n-s}(t)\, dt.\end{aligned} \tag{A.27}$$

From $|z| > 2\alpha n$ and $|t| \le \alpha n$ it follows that $|z - t| > \alpha n$. Thus, the first terms on the right-hand side of (A.27) are less than $c'/n^{1/2}$ due to Assumption (iv). Applying the Fubini theorem and using the fact of integrability of $D_j f_s$ we see that the second summand in (A.27) is bounded by const $P(|\sum_{i=1}^{n-s} X_i| > \alpha n)$, which is $O(1/n)$ by the Chebyshev and Rosenthal inequalities. □

Exploiting Lemmas A.1 and A.2, the rest of the proof of the Theorem 5 in Section 3 is carried out exactly as the proof in one-dimensional case given in [20]. The "ideality properties" of the metric ζ_2,

$$\zeta_2\left(a\sum_{i=1}^{n} X_i, a\sum_{i=1}^{n} \widetilde{X}_i\right) \le a^2 \sum_{i=1}^{n} \zeta_2\left(X_i, \widetilde{X}_i\right), \tag{A.28}$$

used there hold true for random vectors (see, e.g., [14]). The general version of inequality (3.15) in [20] which relates the Kantorovich and the total variation metrics is proved in [13, page 89]. Note that the proof presented in [20] uses the so-called convolution approach (see, e.g., [26]) and induction arguments. This method has been widely used to estimate rates of convergence in multidimensional Central Limit Theorems (see, e.g., [13, 14, 21, 26]).

Acknowledgments

The authors thank the National System of Investigators (SNI) of CONACYT, Mexico for the partial support of this work. The authors are grateful to the anonymous referee for his valuable suggestions on the improvement of an earlier version of this paper.

References

[1] F. R. Hampel, "A general qualitative definition of robustness," *Annals of Mathematical Statistics*, vol. 42, pp. 1887–1896, 1971.

[2] F. R. Hampel, E. M. Ronchetti, P. J. Rousseeuw, and W. A. Stahel, *Robust Statistics: The Approach Based on Inuence Functions*, Wiley, New York, NY, USA, 1986.

[3] P. J. Huber, *Robust Statistics*, Wiley, New York, NY, USA, 1981.

[4] J. Jurečková and P. K. Sen, *Robust Statistical Procedures. Asymptotics and Interrelations*, Wiley, New York, NY, USA, 1996.

[5] X. He and G. Wang, "Qualitative robustness of S*-estimators of multivariate location and dispersionl," *Statistica Neerlandica*, vol. 51, no. 3, pp. 257–268, 1997.

[6] L. Davies, "An efficient Fréchet differentiable high breakdown multivariate location and dispersion estimator," *Journal of Multivariate Analysis*, vol. 40, no. 2, pp. 311–327, 1992.

[7] D. L. Donoho and R. C. Liu, "The "automatic" robustness of minimum distance functionals," *Annals of Statistics*, vol. 16, pp. 552–586, 1988.

[8] X. He and D. G. Simpson, "Lower bounds for contamination bies: globally minimax versus locally linear estimation," *The Annals of Statistics*, vol. 21, pp. 314–337, 1993.

[9] A. Y. Kharin and P. A. Shlyk, "Robust multivariate Bayesian forecasting under functional distortions in the χ^2-metric," *Journal of Statistical Planning and Inference*, vol. 139, no. 11, pp. 3842–3846, 2009.

[10] A. Cuevas, "Qualitative robustness in abstract inference," *Journal of Statistical Planning and Inference*, vol. 18, no. 3, pp. 277–289, 1988.

[11] P. J. Huber, "Robust estimation of a location parameter," *Annals of Mathematical Statistics*, vol. 35, pp. 73–101, 1964.

[12] R. Zielinski, "Robustness of sample mean and sample median under restrictions on outliers," *Zastosowania Matematyki-Applicationes Mathematicae*, vol. 19, pp. 239–240, 1987.

[13] S. T. Rachev and L. R. Rüshendorf, *Mass Transportation Problems, vol. II: Applications*, Springer, New York, NY, USA, 1998.

[14] V. Zolotarev, "Probability metrics," *Theory of Probability and Its Applications*, vol. 28, pp. 264–287, 1983.

[15] C. G. Small, "A survey of multidimensional medians," *International Statistical Review*, vol. 58, pp. 263–277, 1990.

[16] J. H. B. Kemperman, "The median of a finite measure on Banach space," in *Statistical Data Analysis Based on the* L_1 *-Norm and Related Methods*, Y. Dodge, Ed., pp. 217–230, Elsevier Science Publishers, North Holland, 1987.

[17] D. Gervini, "Robust functional estimation using the median and spherical principal components," *Biometrika*, vol. 95, no. 3, pp. 587–600, 2008.

[18] P. Chandhuri, "Multivariate location estimation using extension of *R*-estimates through *U*-statistics type approach," *Annals of Statistics*, vol. 30, pp. 897–916, 1992.

[19] I. Mizera, "Qualitative robustness and weak continuity: the extreme function? Nonparametrics and robustness in modern statistical inference and time series analysis," in *A Festschrift in Honor of Professor Jana Jurečková Institute of Mathematical Statistics Collections*, J. Antoch, M. Hušková, and P. K. Sen, Eds., vol. 7, pp. 169–181, Institute of Mathematical Statistics, Beachwood, Ohio, USA, 2010.

[20] E. Gordienko and A. Novikov, "Probability metrics and robustness: is the sample median more robust than the sample mean?" in *Proceedings of the Joint Session of 7th Prague Symposium on Asymptotic Statistics and 15th Prague Conference on Information Theory, Statistics Decision Functions*, pp. 374–386, Charles University in Prague, Prague, Czech Republic, 2006.

[21] S. T. Rachev, *Probability Metrics and the Stability of Stochastic Models*, Wiley, Chichester, UK, 2009.

[22] A. W. Van der Vaart, *Asymptotic Statistics*, Cambridge University Press, Cambridge, UK, 1998.

[23] R. M. Dudley, *Uniform Central Limit Theorem*, Cambridge University Press, Cambridge, UK, 1999.

[24] E. Gordienko, "Comparing the distributions of sums of independent random vectors," *Kybernetika*, vol. 41, no. 4, pp. 519–529, 2005.

[25] R. N. Bhattacharya and R. Ranga Rao, *Normal Approximation and Asymptotic Expansions*, Wiley, New York, NY, USA, 1976.

[26] V. V. Senatov, "Uniform estimates of the rate of convergence in the multidimensional central limit theorem," *Theory of Probability and Its Applications*, vol. 25, pp. 745–759, 1980.

4

Weighted Kappas for 3×3 Tables

Matthijs J. Warrens

Unit of Methodology and Statistics, Institute of Psychology, Leiden University, P.O. Box 9555, 2300 RB Leiden, The Netherlands

Correspondence should be addressed to Matthijs J. Warrens; warrens@fsw.leidenuniv.nl

Academic Editor: Ricardas Zitikis

Weighted kappa is a widely used statistic for summarizing inter-rater agreement on a categorical scale. For rating scales with three categories, there are seven versions of weighted kappa. It is shown analytically how these weighted kappas are related. Several conditional equalities and inequalities between the weighted kappas are derived. The analytical analysis indicates that the weighted kappas are measuring the same thing but to a different extent. One cannot, therefore, use the same magnitude guidelines for all weighted kappas.

1. Introduction

In biomedical, behavioral, and engineering research, it is frequently required that a group of objects is rated on a categorical scale by two observers. Examples are the following: clinicians that classify the extent of disease in patients; pathologists that rate the severity of lesions from scans; and experts that classify production faults. Analysis of the agreement between the two observers can be used to assess the reliability of the rating system. High agreement would indicate consensus in the diagnosis and interchangeability of the observers. Various authors have proposed statistical methodology for analyzing agreement. For example, for modeling patterns of agreement, the loglinear models proposed in Tanner and Young [1] and Agresti [2, 3] can be used. However, in practice researchers are frequently only interested in a single number that quantifies the degree of agreement between the raters [4, 5]. Various statistics have been proposed in the literature [6, 7], but the most popular statistic for summarizing rater agreement is the weighted kappa introduced by Cohen [8].

Weighted kappa allows the use of weighting schemes to describe the closeness of agreement between categories. Each weighting scheme defines a different version or special case of weighted kappa. Different weighting schemes have been proposed for the various scale types. In this paper, we only consider scales of three categories. This is the smallest number of categories for which we can distinguish three types of categorical scales, namely, nominal scales, continuous-ordinal scales, and dichotomous-ordinal scales [9]. A dichotomous-ordinal scale contains a point of "absence" and two points of "presence", for example, no disability, moderate disability, or severe disability. A continuous-ordinal scale does not have a point of "absence". The scale can be described by three categories of "presence", for example, low, moderate, or high. Identity weights are used when the categories are nominal [10]. In this case, weighted kappa becomes the unweighted kappa introduced by Cohen [11], also known as Cohen's kappa. Linear weights [12, 13] or quadratic weights [14, 15] can be used when the categories are continuous ordinal. The modified linear weights introduced in Cicchetti [9] are suitable if the categories are dichotomous ordinal.

Although weighted kappa has been used in thousands of research applications [16], it has also been criticized by various authors [17–19]. Most of the criticism has focused on a particular version of weighted kappa, namely, Cohen's kappa for nominal categories. Weighted kappa and unweighted kappa correct for rater agreement due to chance alone using the marginal distributions. For example, in the context of latent class models, de Mast [18] and de Mast and van Wieringen [6] argued that the premise that chance measurements have the distribution defined by the marginal distributions cannot be defended. It is, therefore, difficult to interpret the value of Cohen's kappa, and it makes the question of how large or how small the value should be arbitrary. Using signal detection theory, Uebersax [19] showed that different agreement studies with different marginal distributions can produce the same value of Cohen's kappa. Again, this makes the value difficult to interpret. Alternative statistics for summarizing inter-rater agreement are discussed in, for example, de Mast [18] and Perreault and Leigh [20].

Although the choice for a specific version of weighted kappa usually depends on the type of categorical scale at hand, it frequently occurs that weighted kappas corresponding to different weighting schemes are applied to the same data. For example, Cohen's kappa for nominal scales [11] is also frequently applied when the categories are continuous ordinal. When different weighted kappas are applied to the same data, they usually produce different values [5, 21]. For understanding the behavior of weighted kappa and its dependence on the weighting scheme, it is useful to compare the different versions of weighted kappa analytically [21]. For example, if the agreement table is tridiagonal, then the value of the quadratically weighted kappa exceeds the value of the linearly weighted kappa, which, in turn, is higher than the value of unweighted kappa [22, 23]. An agreement table is tridiagonal if it has nonzero elements only on the main diagonal and on the two diagonals directly adjacent to the main diagonal. These analytic results explain orderings of the weighted kappas that are observed in practice.

In this paper, we consider scales that consist of three categories and compare the values of seven special cases of weighted kappa. There are several reasons why the case of three categories is an interesting topic of investigation. First of all, various scales that are used in practice consist of three categories only. Examples can be found in Anderson et al. [24] and Martin et al. [25]. Furthermore, the case of three categories is the smallest case where symmetrically weighted kappas in general have different values, since all weighted kappas with symmetric weighting schemes coincide with two categories. Finally, as it turns out, with three categories we may derive several strong analytic results, which do not generalize to the case of four or more categories. The seven weighted kappas belong to two parameter families. For each parameter family, it is shown that there are only two possible orderings of its members. Hence, despite the fact that the paper is limited to weighted kappas for three categories, we present various interesting and useful results that deepen our understanding of the application of weighted kappa.

The paper is organized as follows. In Section 2 we introduce notation and define four versions of weighted kappa. In Section 3, we introduce the three category reliabilities of a 3×3 agreement table as special cases of weighted kappa. The two parameter families are defined in Section 4. In Section 5, we present several results on inequalities between the seven weighted kappas. In Section 6, we consider the case that all special cases of weighted kappa coincide. Section 7 contains a discussion.

2. Weighted Kappas

Suppose that two raters, each, independently classify the same set of objects (individuals, observations) into the same set of three categories that are defined in advance. For a population of n objects, let π_{ij} for $i, j \in \{1, 2, 3\}$ denote the proportion classified into category i by the first observer and into category j by the second observer. Table 1 presents an abstract version of a 3 × 3 population agreement table of proportions. The marginal totals $\pi_{1+}, \pi_{2+}, \pi_{3+}$ and $\pi_{+1}, \pi_{+2}, \pi_{+3}$ indicate how often raters 1 and 2 used the categories 1, 2, and 3. Four examples of 3 × 3 agreement tables from the literature with frequencies are presented in Table 2. The marginal totals of the tables are in bold. For each table, the last column of Table 2 contains the corresponding estimates of seven weighted kappas. Between brackets behind each point estimate is the associated 95% confidence interval. Definitions of the weighted kappas are presented below.

Table 1: Notation for a 3 × 3 agreement table with proportions.

		Rater 2			
		1	2	3	Total
Rater 1	1	π_{11}	π_{12}	π_{13}	π_{1+}
	2	π_{21}	π_{22}	π_{23}	π_{2+}
	3	π_{31}	π_{32}	π_{33}	π_{3+}
	Total	π_{+1}	π_{+2}	π_{+3}	1

Recall that weighted kappa allows the use of weighting schemes to describe the closeness of agreement between categories. For each cell probability π_{ij}, we may specify a weight. A weighting scheme is called symmetric if for all i, j cell probabilities π_{ij} and π_{ji} are assigned the same weight. The weighting schemes can be formulated from either a similarity or a dissimilarity perspective. Definitions of weighted kappa in terms of similarity scaling can be found in Warrens [13, 22]. For notational convenience, we will define the weights in terms of dissimilarity scaling here. For the elements on the agreement diagonal, there is no disagreement. The diagonal elements are, therefore, assigned zero weight [8, page 215]. The other six weights are non-negative real numbers w_i for $i \in \{1, 2, \ldots, 6\}$. The inequality $w_i > 0$ indicates that there is some disagreement between the assignments by the raters. Categories that are more similar are assigned smaller weights. For example, ordinal scale categories that are one unit apart in the natural ordering are assigned smaller weights than categories that are more units apart.

Table 3 presents one general and seven specific weighting schemes from the literature. The identity weighting scheme for nominal categories was introduced in Cohen [11]. The top table in Table 2 is an example of a nominal scale. The quadratic weighting scheme for continuous-ordinal categories was introduced in Cohen [8]. The quadratically weighted kappa is the most popular version of weighted kappa [4, 5, 15]. The linear weighting scheme for continuous-ordinal categories was introduced in Cicchetti and Allison [29] and Cicchetti [30]. The second table in Table 2 is an example of a continuous-ordinal scale. The dichotomous-ordinal weighting scheme was introduced in Cicchetti [9]. The two bottom tables in Table 2 are examples of dichotomous-ordinal scales. All weighting schemes in Table 3, except the general symmetric and the quadratic, are special cases of the weighting scheme with additive weights introduced in Warrens [31].

TABLE 2: Four examples of 3×3 agreement tables from the literature with corresponding values of weighted kappas.

Category labels	3×3 table				Kappas	
					Estimates	95% CI
Psychotic	106	10	4	**120**	$\hat{\kappa} = .429$	(.323–.534)
Neurotic	22	28	10	**60**	$\hat{\kappa}_\ell = .492$	(.393–.592)
Personality disorder	2	12	6	**20**	$\hat{\kappa}_q = .567$	(.458–.676)
	130	**50**	**20**	**200**	$\hat{\kappa}_c = .536$	(.434–.637)
Spitzer et al. [26]					$\hat{\kappa}_1 = .596$	(.481–.710)
Personality types					$\hat{\kappa}_2 = .325$	(.182–.468)
					$\hat{\kappa}_3 = .222$	(.024–.420)
No atopy	136	12	1	**149**	$\hat{\kappa} = .730$	(.645–.815)
Atopy, no neurodermatitis	8	59	4	**71**	$\hat{\kappa}_\ell = .737$	(.652–.822)
Neurodermatitis	2	4	6	**12**	$\hat{\kappa}_q = .748$	(.651–.845)
	146	**75**	**11**	**232**	$\hat{\kappa}_c = .759$	(.678–.840)
Simonoff [27]					$\hat{\kappa}_1 = .786$	(.703–.869)
Stability of atopic disease					$\hat{\kappa}_2 = .720$	(.624–.817)
					$\hat{\kappa}_3 = .497$	(.240–.754)
Negative	1360	63	8	**1431**	$\hat{\kappa} = .675$	(.632–.719)
Low positive	61	66	13	**140**	$\hat{\kappa}_\ell = .761$	(.725–.798)
High positive	10	16	137	**163**	$\hat{\kappa}_q = .830$	(.798–.862)
	1431	**145**	**158**	**1734**	$\hat{\kappa}_c = .744$	(.705–.782)
Castle et al. [28]					$\hat{\kappa}_1 = .716$	(.672–.760)
Results of hybrid capture testing					$\hat{\kappa}_2 = .415$	(.339–.491)
					$\hat{\kappa}_3 = .839$	(.794–.884)
Good recovery	36	4	1	**41**	$\hat{\kappa} = .689$	(.549–.828)
Moderate disability	5	20	4	**29**	$\hat{\kappa}_\ell = .735$	(.610–.861)
Severe disability	0	1	9	**10**	$\hat{\kappa}_q = .788$	(.667–.910)
	41	**25**	**14**	**80**	$\hat{\kappa}_c = .741$	(.614–.868)
Anderson et al. [24]					$\hat{\kappa}_1 = .750$	(.605–.895)
Glasgow outcome scale scores					$\hat{\kappa}_2 = .610$	(.427–.793)
					$\hat{\kappa}_3 = .707$	(.489–.925)

In this paper, we only consider weighted kappas with symmetric weighting schemes. For notational convenience, we define the following six coefficients:

$$\begin{aligned} a_1 &= \pi_{23} + \pi_{32}, & b_1 &= \pi_{2+}\pi_{+3} + \pi_{3+}\pi_{+2}, \\ a_2 &= \pi_{13} + \pi_{31}, & b_2 &= \pi_{1+}\pi_{+3} + \pi_{3+}\pi_{+1}, \\ a_3 &= \pi_{12} + \pi_{21}, & b_3 &= \pi_{1+}\pi_{+2} + \pi_{2+}\pi_{+1}. \end{aligned} \tag{1}$$

To avoid pathological cases, we assume that $b_1, b_2, b_3 > 0$. The coefficients a_1, a_2, and a_3 reflect raw disagreement between the raters, whereas b_1, b_2, and b_3 reflect chance-expected disagreement. The general formula of weighted kappa for 3×3 tables with symmetric weights will be denoted by κ_w. In terms of the coefficients a_1, a_2, a_3 and b_1, b_2, b_3, this weighted kappa is defined as

$$\kappa_w = 1 - \frac{w_1 a_1 + w_2 a_2 + w_3 a_3}{w_1 b_1 + w_2 b_2 + w_3 b_3}. \tag{2}$$

The value of κ_w lies between 1 and $-\infty$. The numerator $w_1 a_1 + w_2 a_2 + w_3 a_3$ of the fraction in (2) reflects raw weighted disagreement. It is a weighted sum of the cell probabilities π_{ij} that are not on the main diagonal of the 3×3 table, and it quantifies the disagreement between the raters. The denominator $w_1 b_1 + w_2 b_2 + w_3 b_3$ of the fraction in (2) reflects weighted disagreement under chance. It is a weighted sum of the products $\pi_{i+}\pi_{+j}$ for $i \neq j$. High values of $w_1 a_1 + w_2 a_2 + w_3 a_3$ correspond to high disagreement. If $w_1 a_1 + w_2 a_2 + w_3 a_3 = 0$, then we have $\kappa_w = 1$, and there is perfect agreement between the observers. Furthermore, we have $\kappa_w = 0$ if the raw weighted disagreement is equal to the weighted disagreement under chance.

Special cases of κ_w are obtained by using the specific weighting schemes in Table 3 in the general formula (2). Unweighted kappa, linearly weighted kappa, quadratically weighted kappa, and Cicchetti's weighted kappa are, respectively, defined as

$$\begin{aligned} \kappa &= 1 - \frac{a_1 + a_2 + a_3}{b_1 + b_2 + b_3}, & \kappa_\ell &= 1 - \frac{a_1 + 2a_2 + a_3}{b_1 + 2b_2 + b_3}, \\ \kappa_q &= 1 - \frac{a_1 + 4a_2 + a_3}{b_1 + 4b_2 + b_3}, & \kappa_c &= 1 - \frac{a_1 + 3a_2 + 2a_3}{b_1 + 3b_2 + 2b_3}. \end{aligned} \tag{3}$$

TABLE 3: Eight weighting schemes for 3 × 3 tables.

Name	Source	Scale type	Symbol	Scheme		
				0	w_3	w_2
General symmetric	[8]		κ_w	w_3	0	w_1
				w_2	w_1	0
				0	1	1
Identity	[11]	Nominal	κ	1	0	1
				1	1	0
				0	1	2
Linear	[29, 30]	Continuous-ordinal	κ_ℓ	1	0	1
				2	1	0
				0	1	4
Quadratic	[8]	Continuous-ordinal	κ_q	1	0	1
				4	1	0
				0	2	3
	[9, 31]	Dichotomous-ordinal	κ_c	2	0	1
				3	1	0
				0	1	1
Reliability category 1		Dichotomous	κ_1	1	0	0
				1	0	0
				0	1	0
Reliability category 2		Dichotomous	κ_2	1	0	1
				0	1	0
				0	0	1
Reliability category 3		Dichotomous	κ_3	0	0	1
				1	1	0

Assuming a multinominal sampling model with the total numbers of objects n fixed, the maximum likelihood estimate of the cell probability π_{ij} for $i, j \in \{1, 2, 3\}$ is given by $\widehat{\pi}_{ij} = n_{ij}/n$, where n_{ij} is the observed frequency. Note that the a_1, a_2, a_3 and b_1, b_2, b_3 are functions of the cell probabilities π_{ij}. The maximum likelihood estimate $\widehat{\kappa}_w$ of κ_w in (2) is obtained by replacing the cell probabilities π_{ij} by $\widehat{\pi}_{ij}$ [32]. The last column of Table 2 contains the estimates of the weighted kappas for each of the four 3 × 3 tables. For example, for the top table of Table 2, we have $\widehat{\kappa} = .429$, $\widehat{\kappa}_\ell = .492$, $\widehat{\kappa}_q = .567$, and $\widehat{\kappa}_c = .536$. Between brackets behind the kappa estimates are the 95% confidence intervals. These were obtained using the asymptotic variance of weighted kappa derived in Fleiss et al. [33].

3. Category Reliabilities

With a categorical scale, it is sometimes desirable to combine some of the categories [34], for example, when two categories are easily confused, and then calculate weighted kappa for the collapsed table. If we combine two of the three categories, the 3 × 3 table collapses into a 2 × 2 table. For a 2 × 2 table, all weighted kappas with symmetric weighting schemes coincide. Since we have three categories, there are three possible ways to combine two categories. The three κ-values of the collapsed 2 × 2 tables are given by

$$\kappa_1 = 1 - \frac{a_2 + a_3}{b_2 + b_3}, \qquad \kappa_2 = 1 - \frac{a_1 + a_3}{b_1 + b_3}, \qquad \kappa_3 = 1 - \frac{a_1 + a_2}{b_1 + b_2}. \tag{4}$$

These three kappas are obtained by using the three bottom weighting schemes in Table 3 in the general formula (2). The last column of Table 2 contains the estimates of these weighted kappas for each of the four 3 × 3 tables.

Weighted kappa κ_i for $i \in \{1, 2, 3\}$ corresponds to the 2 × 2 table that is obtained by combining the two categories other than category i. The 2 × 2 table reflects how often the two raters agreed on the category i and on the category "all others". Weighted kappa κ_i for $i \in \{1, 2, 3\}$, hence, summarizes the agreement or reliability between the raters on the single category i, and it is, therefore, also called the category reliability of i [10]. It quantifies how good category i can be distinguished from the other two categories. For example, for the second table of Table 2, we have $\widehat{\kappa}_1 = .786$, $\widehat{\kappa}_2 = .720$, and $\widehat{\kappa}_3 = .497$. The substantially lower value of κ_3 indicates that the third category is not well distinguished from the other two categories.

Unweighted kappa κ and linearly weighted kappa κ_ℓ are weighted averages of the category reliabilities. Unweighted kappa is a weighted average of κ_1, κ_2, and κ_3, where the weights are the denominators of the category reliabilities [10]:

$$\frac{(b_2+b_3)\kappa_1+(b_1+b_3)\kappa_2+(b_1+b_2)\kappa_3}{(b_2+b_3)+(b_1+b_3)+(b_1+b_2)}=\kappa. \quad (5)$$

Since κ is a weighted average of the category reliabilities, the κ-value always lies between the values of κ_1, κ_2, and κ_3. This property can be verified for all four tables of Table 2. Therefore, when combining two categories, the κ-value can go either up or down, depending on which two categories are combined [34]. The value of κ is a good summary statistic of the category reliabilities if the values of κ_1, κ_2, and κ_3 are (approximately) identical. Table 2 shows that this is not the case in general. With an ordinal scale, it only makes sense to combine categories that are adjacent in the ordering. We should, therefore, ignore κ_2 with ordered categories, since this statistic corresponds to the 2×2 table that is obtained by merging the two categories that are furthest apart. Furthermore, note that for the two bottom 3 × 3 tables of Table 2 the first category is the "absence" category. If the scale is dichotomous ordinal and category 1 is the "absence" category, then κ_1 is the κ-value of the 2 × 2 table that corresponds to "absence" versus "presence" of the characteristic.

The statistic κ_ℓ is a weighted average of κ_1 and κ_3, where the weights are the denominators of the category reliabilities [13, 35]:

$$\frac{(b_2+b_3)\kappa_1+(b_1+b_2)\kappa_3}{(b_2+b_3)+(b_1+b_2)}=\kappa_\ell. \quad (6)$$

Since κ_ℓ is a weighted average of the category reliabilities κ_1 and κ_3, the κ_ℓ-value always lies between the values of κ_1 and κ_3. This property can be verified for all four tables of Table 2. Unlike κ_q, statistic κ_ℓ can be considered an extension of κ to ordinal scales that preserves the "weighted average" property [13, 35]. The value of κ_ℓ is a good summary statistic of κ_1 and κ_3 if the two weighted kappas are (approximately) identical. This is the case for the two bottom tables of Table 2.

The statistic κ_c is also a weighted average of κ_1 and κ_3, where the weights are $2(b_2+b_3)$ and (b_1+b_2):

$$\frac{2(b_2+b_3)\kappa_1+(b_1+b_2)\kappa_3}{2(b_2+b_3)+(b_1+b_2)}=\kappa_c. \quad (7)$$

A proof can be found in Warrens [31].

4. Families of Weighted Kappas

In this section, we show that the seven weighted kappas introduced in Sections 2 and 3 are special cases of two families. Let $r \geq 0$ be a real number. Inspection of the formulas κ_2, κ, κ_ℓ, and κ_q shows that they only differ on how the coefficients a_2 and b_2 are weighted. The first family is, therefore, given by

$$\lambda_r = 1-\frac{a_1+ra_2+a_3}{b_1+rb_2+b_3}. \quad (8)$$

For $r = 0, 1, 2, 4$, we have, respectively, the special cases κ_2, κ, κ_ℓ, and κ_q.

Recall that κ_ℓ and κ_c are weighted averages of the category reliabilities κ_1 and κ_3. This motivates the following definition. Let $s \in [0, 1]$. Then the second family is defined as

$$\mu_s = \frac{(1-s)(b_2+b_3)\kappa_1+s(b_1+b_2)\kappa_3}{(1-s)(b_2+b_3)+s(b_1+b_2)}. \quad (9)$$

The family μ_s consists of the weighted averages of κ_1 and κ_3 where the weights are multiples of (b_2+b_3) and (b_1+b_2). For $s = 0, 1/3, 1/2, 1$, we have, respectively, the special cases κ_1, κ_c, κ_ℓ, and κ_3. Note that κ_ℓ belongs to both λ_r and μ_s.

The following proposition presents a formula for the family in (9) that will be used in Theorem 6 below.

Proposition 1. *The family in* (9) *is equivalent to*

$$\mu_s = 1-\frac{sa_1+a_2+(1-s)a_3}{sb_1+b_2+(1-s)b_3}. \quad (10)$$

Proof. Since κ_1 and κ_3 are equal to, respectively,

$$\kappa_1=\frac{b_2+b_3-(a_2+a_3)}{b_2+b_3}, \qquad \kappa_3=\frac{b_1+b_2-(a_1+a_2)}{b_1+b_2}, \quad (11)$$

we can write (9) as

$$\begin{aligned}\mu_s &= \frac{(1-s)(b_2+b_3-a_2-a_3)+s(b_1+b_2-a_1-a_2)}{(1-s)(b_2+b_3)+s(b_1+b_2)} \\ &= 1-\frac{(1-s)(a_2+a_3)+s(a_1+a_2)}{(1-s)(b_2+b_3)+s(b_1+b_2)},\end{aligned} \quad (12)$$

which is identical to the expression in (10). □

5. Inequalities

In this section, we present inequalities between the seven weighted kappas. We will use the following lemma repeatedly.

Lemma 2. *Let* $u, v \geq 0$ *and* $r, w, z > 0$. *Then one has the following:*

$$\begin{aligned}&(i) \quad \frac{u}{w}<\frac{v}{z} \Longleftrightarrow \frac{u}{w}<\frac{ru+v}{rw+z}; \\ &(ii) \quad \frac{u}{w}=\frac{v}{z} \Longleftrightarrow \frac{u}{w}=\frac{ru+v}{rw+z}; \\ &(iii) \quad \frac{u}{w}>\frac{v}{z} \Longleftrightarrow \frac{u}{w}>\frac{ru+v}{rw+z}.\end{aligned} \quad (13)$$

Proof. Since w and z are positive numbers, we have $u/w < v/z$, or $uz < vw$. Adding ruw to both sides, we obtain $u(rw+z) < w(ru+v)$, or $u/w < (ru+v)/(rw+z)$. □

Theorem 3 classifies the orderings of the special cases of the family λ_r in (8).

Theorem 3. *For $r < r'$ one has the following:*

$$(i) \quad \lambda_r < \lambda_{r'} \Longleftrightarrow \frac{a_1 + a_3}{b_1 + b_3} > \frac{a_2}{b_2};$$
$$(ii) \quad \lambda_r = \lambda_{r'} \Longleftrightarrow \frac{a_1 + a_3}{b_1 + b_3} = \frac{a_2}{b_2}; \qquad (14)$$
$$(iii) \quad \lambda_r > \lambda_{r'} \Longleftrightarrow \frac{a_1 + a_3}{b_1 + b_3} < \frac{a_2}{b_2}.$$

Proof. The inequality $\lambda_r < \lambda_{r'}$ is equivalent to

$$\frac{a_1 + ra_2 + a_3}{b_1 + rb_2 + b_3} > \frac{a_1 + r'a_2 + a_3}{b_1 + r'b_2 + b_3}. \qquad (15)$$

Since $r < r'$, it follows from Lemma 2 that inequality (15) is equivalent to

$$\frac{a_1 + ra_2 + a_3}{b_1 + rb_2 + b_3} > \frac{(r' - r)\,a_2}{(r' - r)\,b_2} = \frac{a_2}{b_2}. \qquad (16)$$

Applying Lemma 2 for a second time, we find that inequality (16) is equivalent to

$$\frac{a_1 + a_3}{b_1 + b_3} > \frac{a_2}{b_2}. \qquad (17)$$

This completes the proof. □

Theorem 3 shows that, in practice, we only observe one of two orderings of κ_2, κ, κ_ℓ, and κ_q. In most cases, we have $\kappa_2 < \kappa < \kappa_\ell < \kappa_q$. For example, in Table 2 all 3 × 3 tables exhibit this ordering. For all these 3 × 3 tables, it holds that $(a_1 + a_3)/(b_1 + b_3) > a_2/b_2$. Furthermore, if the 3 × 3 table would be tridiagonal [22, 23], we would have $a_2 = 0$, and the inequality $(a_1 + a_3)/(b_1 + b_3) > a_2/b_2$ would also hold. The other possibility is that we have $\kappa_2 > \kappa > \kappa_\ell > \kappa_q$. The only example from the literature where we found this ordering is the 3 × 3 table presented in Cohen [11]. The table in Cohen satisfies the condition in (iii) of Theorem 3. We conclude that, with ordinal scales, we almost always have the ordering $\kappa_2 < \kappa < \kappa_\ell < \kappa_q$. The equality condition in Theorem 3 is discussed in Section 6.

Theorem 4 classifies the orderings of the special cases of the family μ_s in (9).

Theorem 4. *For $s < s'$, one has the following:*

$$(i) \quad \mu_s < \mu_{s'} \Longleftrightarrow \frac{a_1 + a_2}{b_1 + b_2} > \frac{a_2 + a_3}{b_2 + b_3};$$
$$(ii) \quad \mu_s = \mu_{s'} \Longleftrightarrow \frac{a_1 + a_2}{b_1 + b_2} = \frac{a_2 + a_3}{b_2 + b_3}; \qquad (18)$$
$$(iii) \quad \mu_s > \mu_{s'} \Longleftrightarrow \frac{a_1 + a_2}{b_1 + b_2} < \frac{a_2 + a_3}{b_2 + b_3}.$$

Proof. The special cases of μ_s are weighted averages of κ_1 and κ_3. For $s < s'$, we have $\mu_s < \mu_{s'}$ if and only if $\kappa_1 < \kappa_3$; that is, a statistic that gives more weight to κ_3 will be higher if the κ_3-value exceeds the κ_1-value. Furthermore, we have $\kappa_1 < \kappa_3 \Leftrightarrow$

$$\frac{a_1 + a_2}{b_1 + b_2} > \frac{a_2 + a_3}{b_2 + b_3}. \qquad (19)$$

This completes the proof. □

Theorem 4 shows that, in practice, we only observe one of two orderings of $\kappa_3, \kappa_\ell, \kappa_c$, and κ_1. We either have the ordering $\kappa_3 < \kappa_\ell < \kappa_c < \kappa_1$, which is the case in the first, second, and fourth 3 × 3 tables of Table 2, or we have $\kappa_3 > \kappa_\ell > \kappa_c > \kappa_1$, which is the case in the third 3 × 3 table in Table 2.

Proposition 5 follows from Theorems 3 and 4 and the fact that κ is a weighted average of κ_1, κ_2, and κ_3 [10].

Proposition 5. *Consider the following:*

$$(i) \quad \kappa < \kappa_3 < \kappa_1 \Longleftrightarrow \kappa_2 < \kappa < \kappa_3 < \kappa_\ell < \kappa_c < \kappa_1;$$
$$(ii) \quad \kappa < \kappa_1 < \kappa_3 \Longleftrightarrow \kappa_2 < \kappa < \kappa_1 < \kappa_c < \kappa_\ell < \kappa_3, \kappa_q;$$
$$(iii) \quad \kappa_3 < \kappa_1 < \kappa \Longleftrightarrow \kappa_3, \kappa_q < \kappa_\ell < \kappa_c < \kappa_1 < \kappa < \kappa_2;$$
$$(iv) \quad \kappa_1 < \kappa_3 < \kappa \Longleftrightarrow \kappa_1 < \kappa_c < \kappa_\ell < \kappa_3 < \kappa < \kappa_2. \qquad (20)$$

Proposition 5 shows that we have an almost complete picture of how the seven weighted kappas are ordered just by comparing the values of κ, κ_1, and κ_3. The double inequality $\kappa < \kappa_3 < \kappa_1$ holds for the fourth 3×3 table of Table 2, whereas the inequality $\kappa < \kappa_1 < \kappa_3$ holds for the third 3 × 3 table of Table 2. Both tables have a dichotomous-ordinal scale. Recall that κ_c corresponds to a weighting scheme specifically formulated for dichotomous-ordinal scales. It turns out that the κ_c-value can be both lower and higher than the κ_ℓ-value with dichotomous-ordinal scales. Which statistic is higher depends on the data. Furthermore, κ tends to be smaller than κ_1 and κ_3. The condition $\kappa < \kappa_1, \kappa_3$ can be interpreted as an increase in the κ-value if we combine the middle category of the 3-category scale with one of the outer categories. This way of merging categories makes sense if the categories are ordered.

6. Equalities

Apart from the equality conditions in (ii) of Theorems 3 and 4, we only considered inequalities between the weighted kappas in the previous section. Unless there is perfect agreement, the values of the weighted kappas are usually different. Table 4 contains three hypothetical agreement tables that we have constructed to illustrate that the three equality conditions in Theorems 3, 4, and 6 (below) are not identical. For the top table in Table 4, we have $(a_1 + a_3)/(b_1 + b_3) = a_2/b_2$, which is equivalent to the equality $\kappa_2 = \kappa = \kappa_\ell = \kappa_q$ (Theorem 3). Although all weighted kappas of the family λ_r coincide, the kappas not belonging to this family produce different values. For the middle table in Table 4 we have $(a_1 + a_2)/(b_1 + b_2) = (a_2 + a_3)/(b_2 + b_3)$, which is equivalent to the equality $\kappa_3 = \kappa_\ell = \kappa_c = \kappa_1$ (Theorem 4). Although all weighted kappas of

TABLE 4: Three hypothetical 3 × 3 agreement tables with corresponding values of weighted kappas.

Categories	3 × 3 table				Kappas	
1	4	1	0	**5**	$\widehat{\kappa} = .617$	$\widehat{\kappa}_1 = .475$
2	1	2	0	**3**	$\widehat{\kappa}_\ell = .617$	$\widehat{\kappa}_2 = .617$
3	3	0	12	**15**	$\widehat{\kappa}_q = .617$	$\widehat{\kappa}_3 = .736$
	8	**3**	**12**	**23**	$\widehat{\kappa}_c = .572$	
1	6	0	1	**7**	$\widehat{\kappa} = .581$	$\widehat{\kappa}_1 = .635$
2	3	6	0	**9**	$\widehat{\kappa}_\ell = .635$	$\widehat{\kappa}_2 = .479$
3	0	3	6	**9**	$\widehat{\kappa}_q = .668$	$\widehat{\kappa}_3 = .635$
	9	**9**	**7**	**25**	$\widehat{\kappa}_c = .635$	
1	11	1	0	**12**	$\widehat{\kappa} = .603$	$\widehat{\kappa}_1 = .603$
2	2	5	0	**7**	$\widehat{\kappa}_\ell = .603$	$\widehat{\kappa}_2 = .603$
3	2	1	3	**6**	$\widehat{\kappa}_q = .603$	$\widehat{\kappa}_3 = .603$
	15	**7**	**3**	**25**	$\widehat{\kappa}_c = .603$	

the family μ_s coincide, the kappas that do not belong to this family produce different values.

For the bottom table in Table 4, we have the stronger condition $a_1/b_1 = a_2/b_2 = a_3/b_3$. Theorem 6 (below) shows that this condition is equivalent to the case that all weighted kappas, that is, all special cases of (2), coincide.

Theorem 6. *The following conditions are equivalent:*

$$\begin{aligned}
&(i) \quad \frac{a_1}{b_1} = \frac{a_2}{b_2} = \frac{a_3}{b_3} = c \geq 0; \\
&(ii) \quad \kappa_w = 1 - c; \\
&(iii) \quad \lambda_r = \lambda_t = \mu_s \quad \text{for } r \neq t,\ s \neq \frac{1}{2}; \\
&(iv) \quad \lambda_r = \mu_s = \mu_t \quad \text{for } r \neq 2,\ s \neq t.
\end{aligned} \tag{21}$$

Proof. In words, (ii) means that all special cases of (2) are identical. Therefore, (ii) ⇒ (iii), (iv). We first show that (i) ⇒ (ii). It then suffices to show that (iii), (iv) ⇒ (i).

If (i) holds, we have

$$\frac{a_2}{b_2} = \frac{c_1 a_1}{c_1 b_1}, \qquad \frac{a_3}{b_3} = \frac{c_2 a_1}{c_2 b_1} \tag{22}$$

for certain $c_1, c_2 > 0$. Hence,

$$\begin{aligned}
\kappa_w &= 1 - \frac{w_1 a_1 + w_2 c_1 a_1 + w_3 c_2 a_1}{w_1 b_1 + w_2 c_1 b_1 + w_3 c_2 b_1} \\
&= 1 - \frac{a_1 (w_1 + w_2 c_1 + w_3 c_2)}{b_1 (w_1 + w_2 c_1 + w_3 c_2)} \\
&= 1 - \frac{a_1}{b_1} = 1 - c.
\end{aligned} \tag{23}$$

Thus, all special cases of weighted kappa in (2) coincide if (i) is valid.

Next, we show that (iii), (iv) ⇒ (i). Consider condition (iii) first. If two special cases of λ_r are identical, it follows from Theorem 3 that all of them are identical. Hence, we have $\kappa_2 = \mu_s$ for a certain $s \in [0, 1]$ with $s \neq 1/2$. Using formula (10), we have $\kappa_2 = \mu_s \Leftrightarrow$

$$\frac{a_1 + a_3}{b_1 + b_3} = \frac{s a_1 + a_2 + (1 - s) a_3}{s b_1 + b_2 + (1 - s) b_3}. \tag{24}$$

Combining (24) with $a_2/b_2 = (a_1 + a_3)/(b_1 + b_3)$ (Theorem 3), we obtain

$$\frac{a_2}{b_2} = \frac{a_1 + a_3}{b_1 + b_3} = \frac{s a_1 + a_2 + (1 - s) a_3}{s b_1 + b_2 + (1 - s) b_3}. \tag{25}$$

Applying Lemma 2 to the outer ratios of (25), we obtain

$$\frac{a_2}{b_2} = \frac{a_1 + a_3}{b_1 + b_3} = \frac{s a_1 + (1 - s) a_3}{s b_1 + (1 - s) b_3}. \tag{26}$$

First, suppose that $s < 1/2$. Applying Lemma 2 to the right-hand side equality of (26), we obtain

$$\frac{a_2}{b_2} = \frac{a_1 + a_3}{b_1 + b_3} = \frac{(1 - 2s) a_3}{(1 - 2s) b_3} = \frac{a_3}{b_3}, \tag{27}$$

or $a_2/b_2 = a_3/b_3$. Applying Lemma 2 to the second and fourth term of the triple equality (27), we obtain $a_1/b_1 = a_3/b_3$. Thus, we have $a_1/b_1 = a_2/b_2 = a_3/b_3$, which completes the proof for $s < 1/2$. Next, suppose that $s > 1/2$. Applying Lemma 2 to the right-hand side equality of (26), we obtain

$$\frac{a_2}{b_2} = \frac{a_1 + a_3}{b_1 + b_3} = \frac{(2s - 1) a_1}{(2s - 1) b_1} = \frac{a_1}{b_1}, \tag{28}$$

or $a_1/b_1 = a_2/b_2$. Applying Lemma 2 to the second and fourth terms of the triple equality (28), we obtain $a_1/b_1 = a_3/b_3$. Thus, we also have $a_1/b_1 = a_2/b_2 = a_3/b_3$ for $s > 1/2$, which completes the proof for condition (iii).

Next, consider condition (iv). If two special cases of μ_s are identical, it follows from Theorem 4 that all of them are identical. Hence, we have $\kappa_1 = \kappa_3 = \lambda_r$ for a certain $r \geq 0$ and $r \neq 2$. We have $\kappa_3 = \kappa_1 = \lambda_r \Leftrightarrow$

$$\frac{a_1 + a_2}{b_1 + b_2} = \frac{a_2 + a_3}{b_2 + b_3} = \frac{a_1 + r a_2 + a_3}{b_1 + r b_2 + b_3}. \tag{29}$$

First, suppose that $r > 2$. Applying Lemma 2 to the outer ratios of (29), we obtain

$$\frac{a_1 + a_2}{b_1 + b_2} = \frac{a_2 + a_3}{b_2 + b_3} = \frac{(r-1)\,a_2 + a_3}{(r-1)\,b_2 + b_3}. \quad (30)$$

Applying Lemma 2 to the right-hand side equality of (30) gives

$$\frac{a_1 + a_2}{b_1 + b_2} = \frac{a_2 + a_3}{b_2 + b_3} = \frac{(r-2)\,a_2}{(r-2)\,b_2} = \frac{a_2}{b_2}. \quad (31)$$

Applying Lemma 2 to the outer ratios of (31), we obtain $a_1/b_1 = a_2/b_2$, while applying Lemma 2 to the second and fourth terms of the triple equality (31), we obtain $a_3/b_3 = a_2/b_2$. Thus, we have $a_1/b_1 = a_2/b_2 = a_3/b_3$.

Finally, if $r < 2$, then consider the equality $\kappa_1 = \kappa_3 = \kappa_\ell = \lambda_r \Leftrightarrow$

$$\begin{aligned}\frac{a_2 + a_3}{b_2 + b_3} = \frac{a_1 + a_2}{b_1 + b_2} &= \frac{a_1 + 2a_2 + a_3}{b_1 + 2b_2 + b_3} \\ &= \frac{(2/r)\,a_1 + 2a_2 + (2/r)\,a_3}{(2/r)\,b_1 + 2b_2 + (2/r)\,b_3}.\end{aligned} \quad (32)$$

Since $2/r > 1$, applying Lemma 2 to the right-hand side equality of (32) gives

$$\begin{aligned}\frac{a_1 + a_2}{b_1 + b_2} = \frac{a_2 + a_3}{b_2 + b_3} &= \frac{a_1 + 2a_2 + a_3}{b_1 + 2b_2 + b_3} \\ &= \frac{(2/r-1)\,a_1 + (2/r-1)\,a_3}{(2/r-1)\,b_1 + (2/r-1)\,b_3} \\ &= \frac{a_1 + a_3}{b_1 + b_3}.\end{aligned} \quad (33)$$

However,

$$\frac{a_1 + a_2}{b_1 + b_2} = \frac{a_2 + a_3}{b_2 + b_3} = \frac{a_1 + a_3}{b_1 + b_3} \quad (34)$$

is equivalent to $\kappa_1 = \kappa_2 = \kappa_3$. Since κ is a weighted average of κ_1, κ_2, and κ_3, we must have $\kappa = \kappa_2$. But then condition (iii) holds, and we have already shown that (iii) ⇒ (i). This completes the proof for condition (iv). □

Theorem 6 shows that all weighted kappas for 3 × 3 tables are identical if we have the double inequality $a_1/b_1 = a_2/b_2 = a_3/b_3$. If this condition holds, the equalities $(a_1+a_3)/(b_1+b_3) = a_2/b_2$ and $(a_1 + a_2)/(b_1 + b_2) = (a_2 + a_3)/(b_2 + b_3)$ also hold. Theorem 6 also shows that if any two special cases of the family λ_r are equal to a member of the family μ_s other than κ_ℓ, then all weighted kappas coincide. Furthermore, if any two special cases of the family μ_s are identical to a member of the family λ_r other than κ_ℓ, then all weighted kappas must be identical.

7. Discussion

Since it frequently happens that different versions of the weighted kappa are applied to the same contingency data, regardless of the scale type of the categories, it is useful to compare the various versions analytically. For rating scales with three categories, we may define seven special cases of weighted kappa. The seven weighted kappas belong to two different parameter families. Only the weighted kappa with linear weights belongs to both families. For both families, it was shown that there are only two possible orderings of its members (Theorems 3 and 4). We conclude that with ordinal scales consisting of three categories, quadratically weighted kappa usually produces higher values than linearly weighted kappa, which in turn has higher values than unweighted kappa.

Since there are only a few possible orderings of the weighted kappas, it appears that the kappas are measuring the same thing, but to a different extent. Various authors have presented magnitude values for evaluating the values of kappa statistics [36–38]. For example, an estimated value of 0.80 generally indicates good or excellent agreement. There is general consensus in the literature that uncritical use of these guidelines leads to questionable decisions in practice. If the weighted kappas are measuring the same thing, but some kappas produce substantially higher values than others, then the same guidelines cannot be applied to all weighted kappas. However, using the same guidelines for different kappas appears to be common practice. If one wants to work with magnitude guidelines, then it seems reasonable to use stricter criteria for the quadratically weighted kappa than for unweighted kappa, since the former statistic generally produces higher values.

The quadratically and linearly weighted kappas were formulated for continuous-ordinal scale data. However, in practice, many scales are dichotomous ordinal (see, e.g., Anderson et al. [24] and Martin et al. [25]). In this case, the application of the weighted kappa proposed by Cicchetti [9] or the additively weighted kappa introduced in Warrens [31] is perhaps more appropriate. Unfortunately, Cicchetti's weighted kappa has been largely ignored in the application of kappa statistics. In most applications, the quadratically weighted kappa is used [4, 5]. The observation that the quadratically weighted kappa tends to produce the highest value for many data may partly explain this popularity. As pointed out by one of the reviewers, to determine whether Cicchetti's weighted kappa has real advantages, the various weighted kappas need to be compared on the quality and efficiency of prediction. This is a possible topic for future work.

Acknowledgments

The author thanks four anonymous reviewers for their helpful comments and valuable suggestions on an earlier version of this paper. This research is part of Project 451-11-026 funded by the Netherlands Organisation for Scientific Research.

References

[1] M. A. Tanner and M. A. Young, "Modeling ordinal scale disagreement," *Psychological Bulletin*, vol. 98, no. 2, pp. 408–415, 1985.

[2] A. Agresti, "A model for agreement between ratings on an ordinal scale," *Biometrics*, vol. 44, no. 2, pp. 539–548, 1988.

[3] A. Agresti, *Analysis of Ordinal Categorical Data*, John Wiley & Sons, Hoboken, NJ, USA, 2nd edition, 2010.

[4] P. Graham and R. Jackson, "The analysis of ordinal agreement data: beyond weighted kappa," *Journal of Clinical Epidemiology*, vol. 46, no. 9, pp. 1055–1062, 1993.

[5] M. Maclure and W. C. Willett, "Misinterpretation and misuse of the Kappa statistic," *American Journal of Epidemiology*, vol. 126, no. 2, pp. 161–169, 1987.

[6] J. de Mast and W. N. van Wieringen, "Measurement system analysis for categorical measurements: agreement and kappa-type indices," *Journal of Quality Technology*, vol. 39, no. 3, pp. 191–202, 2007.

[7] M. J. Warrens, "Inequalities between kappa and kappa-like statistics for $k \times k$ tables," *Psychometrika*, vol. 75, no. 1, pp. 176–185, 2010.

[8] J. Cohen, "Weighted kappa: nominal scale agreement provision for scaled disagreement or partial credit," *Psychological Bulletin*, vol. 70, no. 4, pp. 213–220, 1968.

[9] D. V. Cicchetti, "Assessing inter rater reliability for rating scales: resolving some basic issues," *British Journal of Psychiatry*, vol. 129, no. 11, pp. 452–456, 1976.

[10] M. J. Warrens, "Cohen's kappa is a weighted average," *Statistical Methodology*, vol. 8, no. 6, pp. 473–484, 2011.

[11] J. Cohen, "A coefficient of agreement for nominal scales," *Educational and Psychological Measurement*, vol. 20, pp. 37–46, 1960.

[12] S. Vanbelle and A. Albert, "A note on the linearly weighted kappa coefficient for ordinal scales," *Statistical Methodology*, vol. 6, no. 2, pp. 157–163, 2009.

[13] M. J. Warrens, "Cohen's linearly weighted kappa is a weighted average of 2×2 kappas," *Psychometrika*, vol. 76, no. 3, pp. 471–486, 2011.

[14] J. L. Fleiss and J. Cohen, "The equivalence of weighted kappa and the intraclass correlation coefficient as measures of reliability," *Educational and Psychological Measurement*, vol. 33, pp. 613–619, 1973.

[15] M. J. Warrens, "Some paradoxical results for the quadratically weighted kappa," *Psychometrika*, vol. 77, no. 2, pp. 315–323, 2012.

[16] L. M. Hsu and R. Field, "Interrater agreement measures: comments on kappa$_n$, Cohen's kappa, Scott's π and Aickin's α," *Understanding Statistics*, vol. 2, pp. 205–219, 2003.

[17] E. Bashkansky, T. Gadrich, and D. Knani, "Some metrological aspects of the comparison between two ordinal measuring systems," *Accreditation and Quality Assurance*, vol. 16, no. 2, pp. 63–72, 2011.

[18] J. de Mast, "Agreement and kappa-type indices," *The American Statistician*, vol. 61, no. 2, pp. 148–153, 2007.

[19] J. S. Uebersax, "Diversity of decision-making models and the measurement of interrater agreement," *Psychological Bulletin*, vol. 101, no. 1, pp. 140–146, 1987.

[20] W. D. Perreault and L. E. Leigh, "Reliability of nominal data based on qualitative judgments," *Journal of Marketing Research*, vol. 26, pp. 135–148, 1989.

[21] M. J. Warrens, "Conditional inequalities between Cohen's kappa and weighted kappas," *Statistical Methodology*, vol. 10, pp. 14–22, 2013.

[22] M. J. Warrens, "Weighted kappa is higher than Cohen's kappa for tridiagonal agreement tables," *Statistical Methodology*, vol. 8, no. 2, pp. 268–272, 2011.

[23] M. J. Warrens, "Cohen's quadratically weighted kappa is higher than linearly weighted kappa for tridiagonal agreement tables," *Statistical Methodology*, vol. 9, no. 3, pp. 440–444, 2012.

[24] S. I. Anderson, A. M. Housley, P. A. Jones, J. Slattery, and J. D. Miller, "Glasgow outcome scale: an inter-rater reliability study," *Brain Injury*, vol. 7, no. 4, pp. 309–317, 1993.

[25] C. S. Martin, N. K. Pollock, O. G. Bukstein, and K. G. Lynch, "Inter-rater reliability of the SCID alcohol and substance use disorders section among adolescents," *Drug and Alcohol Dependence*, vol. 59, no. 2, pp. 173–176, 2000.

[26] R. L. Spitzer, J. Cohen, J. L. Fleiss, and J. Endicott, "Quantification of agreement in psychiatric diagnosis. A new approach," *Archives of General Psychiatry*, vol. 17, no. 1, pp. 83–87, 1967.

[27] J. S. Simonoff, *Analyzing Categorical Data*, Springer, New York, NY, USA, 2003.

[28] P. E. Castle, A. T. Lorincz, I. Mielzynska-Lohnas et al., "Results of human papillomavirus DNA testing with the hybrid capture 2 assay are reproducible," *Journal of Clinical Microbiology*, vol. 40, no. 3, pp. 1088–1090, 2002.

[29] D. Cicchetti and T. Allison, "A new procedure for assessing reliability of scoring EEG sleep recordings," *The American Journal of EEG Technology*, vol. 11, pp. 101–110, 1971.

[30] D. V. Cicchetti, "A new measure of agreement between rank ordered variables," in *Proceedings of the Annual Convention of the American Psychological Association*, vol. 7, pp. 17–18, 1972.

[31] M. J. Warrens, "Cohen's weighted kappa with additive weights," *Advances in Data Analysis and Classification*, vol. 7, pp. 41–55, 2013.

[32] Y. M. M. Bishop, S. E. Fienberg, and P. W. Holland, *Discrete Multivariate Analysis: Theory and Practice*, The MIT Press, Cambridge, Mass, USA, 1975.

[33] J. L. Fleiss, J. Cohen, and B. S. Everitt, "Large sample standard errors of kappa and weighted kappa," *Psychological Bulletin*, vol. 72, no. 5, pp. 323–327, 1969.

[34] M. J. Warrens, "Cohen's kappa can always be increased and decreased by combining categories," *Statistical Methodology*, vol. 7, no. 6, pp. 673–677, 2010.

[35] M. J. Warrens, "Cohen's linearly weighted kappa is a weighted average," *Advances in Data Analysis and Classification*, vol. 6, no. 1, pp. 67–79, 2012.

[36] D. V. Cicchetti and S. A. Sparrow, "Developing criteria for establishing interrater reliability of specific items: applications to assessment of adaptive behavior," *American Journal of Mental Deficiency*, vol. 86, no. 2, pp. 127–137, 1981.

[37] P. E. Crewson, "Reader agreement studies," *American Journal of Roentgenology*, vol. 184, no. 5, pp. 1391–1397, 2005.

[38] J. R. Landis and G. G. Koch, "A one-way components of variance model for categorical data," *Biometrics*, vol. 33, no. 4, pp. 159–174, 1977.

Bayesian and Non-Bayesian Inference for Survival Data Using Generalised Exponential Distribution

Chris Bambey Guure and Samuel Bosomprah

Department of Biostatistics, School of Public Health, University of Ghana, Legon Accra, Ghana

Correspondence should be addressed to Chris Bambey Guure; chris@chrisguure.com

Academic Editor: Zhidong Bai

A two-parameter lifetime distribution was introduced by Kundu and Gupta known as generalised exponential distribution. This distribution has been touted to be an alternative to the well-known 2-parameter Weibull and gamma distributions. We seek to determine the parameters and the survival function of this distribution. The survival function determines the probability that a unit under investigation will survive beyond a certain specified time, say, (x). We have employed different data sets to estimate the parameters and see how well the distribution can be used to analyse survival data. A comparison is made about the estimators used in this study. Standard errors of the estimators are determined and used for the comparisons. A simulation study is also carried out, and the mean squared errors and absolute bias are obtained for the purpose of comparison.

1. Introduction

As the need grows for conceptualization, formalization, and abstraction in biology, so too does mathematics' relevance to the field according to Fagerström et al. [1]. Mathematics is particularly important for analysing and characterizing random variation, for example, size and weight of individuals in populations, their sensitivity to chemicals, and time-to-event cases, such as the amount of time an individual needs to recover from illness. The frequency distribution of such data is a major factor determining the type of statistical analysis that can be validly carried out on any data set. Many widely used statistical methods, such as ANOVA (analysis of variance) and regression analysis, require that the data be normally distributed, but only rarely is the frequency distribution of data tested when these techniques are used (Limpert et al.) [2].

Gupta and Kundu [3] recently proposed the two-parameter generalised exponential distribution (*GE*) as an alternative to the lognormal, gamma, and Weibull distributions and did some studies on its properties. Some references on *GE* distribution are Raqab [4], Raqab and Ahsanullah [5], Zheng [6], and Kundu and Gupta [7]. According to Gupta and Kundu [8], the two-parameter $GE(\theta, p)$ can have increasing and decreasing failure rates depending on the shape parameter.

Some research has been done to compare MLE to that of the Bayesian approach in estimating the survival function and the parameters of the Weibull distribution which are similar to the generalised exponential distribution. Amongst others, Sinha [9] determined the Bayes estimates of the reliability function and the hazard rate of the Weibull failure time distribution by employing only squared error loss function; Abdel-Wahid and Winterbottom [10] studied the approximate Bayesian estimates for the Weibull reliability function and hazard rate from censored data by employing a new method that has the potential of reducing the number of terms in Lindley procedure; Guure and Ibrahim [11] did some studies on Bayesian analysis of the survival function and failure rate of Weibull distribution with censored data. Huang and Wu [12] considered Bayesian estimation and prediction for Weibull model with progressive censoring. Similar work can be seen in Guure et al. [13], Zellner [14], Guure et al. [15]. Al-Aboud [16], Al-Athari [17], and Pandey et al. [18].

Maximum likelihood method is one of the most popular estimation techniques for many distributions. According to the classical statistician, maximum likelihood has some outstanding properties. We have in this paper compared the maximum likelihood estimator and that of the Bayesian estimators to determine the best method that can be used to estimate the parameters of the generalised exponential distribution.

The remaining part of the paper is arranged as follows. Section 2 contains materials and methods that deal with the derivative of the parameters under maximum likelihood estimator and that of the Bayesian estimators. Section 3 deals with analysis of some real data sets, followed by a simulation study in Section 4. Results and discussion are in Section 5, and then Section 6 is the conclusion.

2. Materials and Methods

Let $(t_1, \ldots, t_n)$ be the set of n random lifetimes with respect to the generalised exponential distribution, with p and θ as the parameters, where θ is the scale parameter and p is the shape parameter. The cumulative distribution function (cdf), the probability density function (pdf), and the survival function $S(t)$ are given, respectively, as

$$F(t;\theta,p) = \{1-\exp(-\theta t)\}^{p} \quad \theta, p, t > 0, \tag{1}$$

the density function:

$$f(t;\theta,p) = p\theta\{1-\exp(-\theta t)\}^{p-1}\exp(-\theta t), \tag{2}$$

and the survival function:

$$S(t;\theta,p) = 1-\{1-\exp(-\theta t)\}^{p}. \tag{3}$$

Let the GE distribution with the shape parameter p and the scale parameter θ be denoted by $GE(\theta, p)$.

2.1. Maximum Likelihood Estimation. Since $(t_1, \ldots, t_n)$ is the set of n random lifetimes from the generalised exponential distribution with parameters θ and p.

The likelihood function is

$$L(t_i;\theta,p) = \prod_{i=1}^{n}\left[p\theta\{1-\exp(-\theta t)\}^{p-1}\exp(-\theta t)\right]. \tag{4}$$

To obtain the equations for the unknown parameters, we take the log of (4) and partially different with respect to the unknown parameters. By employing an iterative procedure such as Newton-Raphson, the estimates of the scale and shape parameters can easily be obtained. Readers can refer to Kundu and Gupta [7] for details.

Therefore, the survival function can be obtained as

$$\widehat{S}(t) = 1-\left\{1-\exp\left(-\widehat{\theta}t\right)\right\}^{\widehat{p}}, \tag{5}$$

where $(\widehat{\theta})$ and $(\widehat{p})$ are the maximum likelihood estimates of the parameters.

2.2. Bayesian Inference. In Bayesian analysis, the parameter of interest is always considered to be a random variable with a prior distribution. The prior distribution is the distribution of the parameter before any data is observed. The selection of prior distribution is most often than not based on the type of prior information that is available to us. When we have little or no information about the parameter, a noninformative prior should be used else an informative prior. In analysing data from medical, engineering, or biological studies, it is possible to obtain information with respect to similar studies in the past, and if that is even unattainable, information from an expert could be modelled to fit an appropriate prior distribution. This can be referred to as prior elicitation.

We let the two unknown parameters take on the gamma prior distributions by assuming that the hyper parameters are all known and greater than zero, that is, $a, b, c, d > 0$. The gamma prior is assumed for this distribution because both the scale and shape parameters are greater than zero:

$$\begin{aligned} v_1(\theta) &\propto \theta^{b-1}\exp(-\theta a), \quad \theta, a, b > 0, \\ v_2(p) &\propto p^{d-1}\exp(-pc), \quad p, c, d > 0. \end{aligned} \tag{6}$$

Bayesian inference is based on the posterior distribution which is obtained by dividing the joint density function to the marginal distribution function as given below:

$$\pi^*(\theta, pt_i) \propto \frac{v_1(\theta)\, v_2(p)\, L(t_i;\theta,p)}{\int_0^\infty \int_0^\infty v_1(\theta)\, v_2(p)\, L(t_i;\theta,p)\, d\theta dp}. \tag{7}$$

Due to the complex nature of the posterior distribution given in (7), Lindley approximation is employed in order to estimate the unknown parameters.

The Bayes estimator is considered under two loss functions. Since in drawing conclusions about the survival or duration of a living organism, an overestimation could be more detrimental to underestimation or vice versa, we have considered both asymmetric (general entropy) loss function and symmetric (squared error) loss function.

2.2.1. Lindley Approximation. Lindley [19] proposed an approximation for a ratio of integral of the form

$$\frac{\int \omega(\alpha)\exp\{\ell(\alpha)\}\, d\alpha}{\int v(\alpha)\exp\{\ell(\alpha)\}\, d\alpha}, \tag{8}$$

where $\ell(\alpha)$ is the log likelihood and $\omega(\alpha)$, $v(\alpha)$ are arbitrary functions of α, $v(\alpha)$ is the prior distribution for α, and $\omega(\alpha) = u(\alpha)\cdot v(\alpha)$ with $u(\alpha)$ being some function of interest as seen in (8).

The posterior expectation according to Sinha [9] is

$$E\{u(\alpha)\mid x\} = \frac{\int u(\alpha)\exp\{\ell(\alpha)+\rho(\alpha)\}\, d\alpha}{\int \exp\{\ell(\alpha)+\rho(\alpha)\}\, d\alpha}, \tag{9}$$

where $\rho(\alpha) = \log\{v(\alpha)\}$ and $\ell(\alpha)$ represent the log-likelihood function. Considering the Bayesian estimator under the squared error loss function, which is the posterior mean, the posterior expectation can be approximated asymptotically with respect to the two parameters by (10):

$$\hat{u} = u(\hat{\theta}, \hat{p}) + \frac{1}{2}[(u_{11}\sigma_{11}) + (u_{22}\sigma_{22})] + u_1\rho_1\sigma_{11} + u_2\rho_2\sigma_{22} + \frac{1}{2}[(\ell_{30}u_1\sigma_{11}^2) + (\ell_{03}u_2\sigma_{22}^2)], \tag{10}$$

$$u = \theta, \quad u_1 = \frac{\partial u}{\partial \theta} = 1, \quad u_{11} = 0,$$

$$u = p, \quad u_2 = \frac{\partial u}{\partial p} = 1, \quad u_{22} = 0,$$

$$\rho = \ln v_1(\theta) + \ln v_2(p), \tag{11}$$

$$\rho_1 = \frac{b-1}{\theta} - a, \quad \rho_2 = \frac{d-1}{p} - c,$$

$$\sigma_{11} = (-\ell_{20})^{-1}, \quad \sigma_{22} = (-\ell_{02})^{-1}.$$

The second and third derivatives with respect to the scale and shape parameters are

$$\begin{aligned}\ell_{20} &= -\frac{n}{\theta^2} - \sum_{i=1}^{n}\frac{(p-1)[(t_i)^2\exp(-\theta t_i)]}{(1-\exp(-\theta t_i))} - \sum_{i=1}^{n}\frac{(p-1)[(t_i)^2(\exp(-\theta t_i))^2]}{(1-\exp(-\theta t_i))^2},\\ \ell_{30} &= \frac{2n}{\theta^3} + \sum_{i=1}^{n}\frac{(p-1)[(t_i)^3\exp(-\theta t_i)]}{(1-\exp(-\theta t_i))} + \sum_{i=1}^{n}\frac{3(p-1)[(t_i)^3(\exp(-\theta t_i))^2]}{(1-\exp(-\theta t_i))^2} + \sum_{i=1}^{n}\frac{2(p-1)[(t_i)^3(\exp(-\theta t_i))^3]}{(1-\exp(-\theta t_i))^3},\\ \ell_{02} &= -\frac{n}{p^2},\\ \ell_{03} &= \frac{2n}{p^3}.\end{aligned} \tag{12}$$

To estimate the survival function, under the squared error loss function, we let

$$u(S) = 1 - \{1 - \exp(-\theta t)\}^p, \tag{13}$$

where

$$u_1 = \frac{\partial u(S)}{\partial \theta}, \quad u_{11} = \frac{\partial^2 u(S)}{\partial \theta^2}, \quad u_2 = \frac{\partial u(S)}{\partial p}, \quad u_{22} = \frac{\partial^2 u(S)}{\partial p^2}. \tag{14}$$

2.2.2. General Entropy Loss Function. The general entropy loss function is a generalisation of the entropy loss function. The Bayes estimator $\hat{\alpha}_{BG}$ of α under the general entropy loss is

$$\hat{\alpha}_{BG} = [E_\alpha(\alpha^{-k})]^{-1/k}, \tag{15}$$

provided $E_\alpha(\cdot)$ exists and is finite. The Bayes estimator for this loss function with respect to the parameters and the survival function are

$$E\{[(\theta)^{-k}, (p)^{-k}]\} = \frac{\iint u[(\theta)^{-k}, (p)^{-k}]\, v_1(\theta)\, v_2(p)\, L(t_i;\theta,p)\, d\theta\, dp}{\iint v_1(\theta)\, v_2(p)\, L(t_i;\theta,p)\, d\theta\, dp},$$

$$E\{[(S)^{-k}]\} = \frac{\iint u[1-\{1-\exp(-\theta t)\}^p]^{-k} v_1(\theta) v_2(p)\, L(t_i;\theta,p)\, d\theta\, dp}{\iint v_1(\theta)\, v_2(p)\, L(t_i;\theta,p)\, d\theta\, dp}. \tag{16}$$

A similar Lindley approach is used for the general entropy loss function as in the squared error loss function, with

$$\begin{aligned} u &= (\theta)^{-k}, \quad u_1 = \frac{\partial u}{\partial \theta} = -k(\theta)^{-k-1},\\ u_{11} &= \frac{\partial^2 u}{\partial (\theta)^2} = -(-k^2 - k)(\theta)^{-k-2},\\ u &= (p)^{-k}, \quad u_2 = \frac{\partial u}{\partial p} = -k(p)^{-k-1},\\ u_{22} &= \frac{\partial^2 u}{\partial (p)^2} = -(-k^2-k)(p)^{-k-2}.\end{aligned} \tag{17}$$

For the general entropy loss function, the posterior expectation according to Lindley [19] can be approximated by using (18). The survival function is similarly obtained by substituting (14) into (18):

$$\hat{u} = \left\{u(\hat{\theta}, \hat{p}) + \frac{1}{2}[(u_{11}\sigma_{11}) + (u_{22}\sigma_{22})] + u_1\rho_1\sigma_{11} + u_2\rho_2\sigma_{22} + \frac{1}{2}[(\ell_{30}u_1\sigma_{11}^2) + (\ell_{03}u_2\sigma_{22}^2)]\right\}^{-1/k}. \tag{18}$$

3. Real Data Analysis

3.1. Example 1. We analyse two data sets in this section which we have considered to be relatively small and moderate for illustration and comparison purposes. The data is obtained from Lawless [20]. The data represent survival times for two groups of laboratory mice, all of which were exposed to a fixed dose of radiation at an age of 5 to 6. The first group of mice lived in a conventional lab environment, and the second group was kept in a germ-free environment. The cause of death for each mouse was assigned after autopsy to be one of three things: thymic lymphoma (C_1), reticulum cell sarcoma (C_2), or other causes (C_3). The mice all died by the end of the experiment, so there is no censoring. For the purpose of our study, we have considered the first set of data from the control group which seem relatively small with $n = 22$ and the second which seem relatively moderate from the third set of data of the germ-free group with $n = 38$.

Using the iterative procedure suggested from the beginning of this paper, the MLEs of $\widehat{\theta}$ and $\widehat{p}$ for the relatively small samples are 0.015024 and 36.935530, respectively, with their corresponding standard errors as 0.003203 and 8.374682. The relatively moderate samples have $\widehat{\theta} = 0.003735$ and $\widehat{p} = 8.624091$, with their corresponding standard errors as 0.000606 and 1.999012.

Since we do not have any prior information on the hyperparameters, we assume $a = b = c = d = 0.0001$. This makes the priors proper on $\widehat{\theta}$ and $\widehat{p}$ and the corresponding posteriors also proper.

When we compute the Bayes estimators with squared error loss of $\widehat{\theta}$ and $\widehat{p}$ under the relatively small sample sizes, the following parameters estimates and standard errors are obtained, respectively, 0.014605, 36.935530 and 0.003114, 8.374682. For the moderate samples, we have $\widehat{\theta} = 0.003716$ with a standard error of 0.000603. The $\widehat{p} = 8.624091$ with a standard error of 1.999012.

Computing the Bayes estimates of $\widehat{\theta}$ and $\widehat{p}$ and their corresponding standard errors under the general entropy loss functions with the loss parameter being 0.6, we have 0.014270, 35.630480 and 0.003042, 8.096443. With the loss parameter being −0.6, we have 0.014517, 36.600361 and 0.003095, 8.303224, respectively.

Considering the Bayes estimates on the relatively moderate samples of $\widehat{\theta}$ and $\widehat{p}$ and their corresponding standard errors under the general entropy loss functions with the loss parameter being 0.6, we have 0.003646, 8.445543 and 0.000591, 1.970048. With the loss parameter being −0.6, we have 0.003698, 8.578749 and 0.00600, 1.991657, respectively.

Observably, the Bayes estimator under squared error loss for the shape parameter (p) has the same estimate and standard error as compared to that of the classical maximum likelihood estimator but with the scale parameter $(\widehat{\theta})$, and Bayes under squared error has a smaller standard error in juxtaposition to MLE.

3.2. Example 2. In this section, we analyse another real data set which we have considered to be relatively large and complete for illustrative purpose. The data originates from Bjerkedal [21]. The data represents the survival times of guinea pigs injected with different doses of tubercle bacilli. It is known that guinea pigs have high susceptibility to that of human tuberculosis, and this informs our decision to use the data in our study. Here, our primary concern is with the animals in the same cage that were under the same regimen. The regimen number is the common logarithm of the number of bacillary units in 0.5 mL of challenge solution; that is, regimen 6.6 corresponds to 4.4 × 106 bacillary units per 0.5 mL (log 4.4 × 106 = 6.6). Corresponding to regimen 6.6, there were 72 observations.

TABLE 1: Average parameters estimates and their corresponding standard errors ($n = 72$).

par	MLE	BS	BG ($k = 0.6$)	BG ($k = -0.6$)
$\widehat{\theta}$	0.016962 (0.001999)	0.016958 (0.001998)	0.016773 (0.001977)	0.016911 (0.001993)
$\widehat{p}$	2.474104 (0.491576)	2.474104 (0.491576)	2.446856 (0.488365)	2.467235 (0.490766)

TABLE 2: Standard errors of the survival function.

Est	$n = 22$	$n = 38$	$n = 72$
MLE	0.517012	0.467909	0.529970
BS	0.543902	0.471755	0.530072
BG ($k = 0.6$)	0.556815	0.480653	0.531440
BG ($k = -0.6$)	0.547363	0.474025	0.530418

Observing from Table 1, it is evident that the estimator with the smallest parameter estimate and having a corresponding smaller standard error is Bayesian with the generalised entropy loss function. This occurred for both parameters with a positive loss parameter, that is, 0:6.

The importance of the survival function cannot be ignored; therefore, the correctness of its estimate is very crucial to both biological and medical studies. As clearly presented in Table 2, the estimator with the smallest standard error under all the samples is the classical maximum likelihood estimator; therefore, MLE is preferred as a better estimator to the others.

Comparing all the estimators, it is clear from the results that Bayes estimator under general entropy loss function with the loss parameter of +0.6 has the smallest standard error and estimate for both the shape parameter $(\widehat{p})$ and the scale parameter $(\widehat{\theta})$.

4. Simulation Study

Since it is difficult to compare the performance of the estimators theoretically and also to validate the real data employed in this paper, we have performed extensive simulations to compare the estimators through mean squared errors and absolute biases by employing different sample sizes with different parameter values.

We have considered in the simulation study a sample size of $n = 25$, 50, and 100, which is representative of small, moderate, and large data sets. The following steps were employed to generate the data.

The generation of $GE(\theta, p)$ is simple as stated in Gupta and Kundu [8]. If U follows a uniform distribution in the interval $[0, 1]$, then $Y = (-\ln(1 - U^{(1/p)})/\theta)$ follows $GE(\theta, p)$. Consequently, with a very good uniform random number generator, the generation of GE random deviate is immediate.

A lifetime T is generated from the sample sizes indicated above from the GE distribution which represents failure of the product or unit. The values of the assumed actual shape

TABLE 3: Average mean squared errors and absolute biases of $(\hat{\theta})$.

n		$p = 0.8$	$p = 1.6$	$p = 2.2$
25	ML	0.117259 (0.255219)	0.077502 (0.206616)	0.062024 (0.190155)
	BS	0.117108 (0.255095)	0.076553 (0.205513)	0.060763 (0.188471)
	BG ($k = 0.6$)	0.118176 (0.251456)	0.075759 (0.208706)	0.060024 (0.185077)
	BG ($k = -0.6$)	**0.108193 (0.246358)**	**0.075167 (0.203752)**	**0.059216 (0.179142)**
50	ML	0.048628 (0.170036)	0.031950 (0.136181)	0.027519 (0.130563)
	BS	0.048623 (0.170024)	**0.031869 (0.136036)**	0.027383 (0.130287)
	BG ($k = 0.6$)	0.047058 (0.163792)	0.032612 (0.139630)	0.027723 (0.130121)
	BG ($k = -0.6$)	**0.045906 (0.163408)**	0.033207 (0.159536)	**0.027145 (0.127761)**
100	ML	0.023911 (0.119557)	0.015292 (0.097709)	0.012631 (0.088584)
	BS	0.023910 (0.119556)	0.015283 (0.097685)	**0.012616 (0.088533)**
	BG ($k = 0.6$)	0.021514 (0.115652)	0.014627 (0.094909)	0.012650 (0.088935)
	BG ($k = -0.6$)	**0.020102 (0.113236)**	**0.014052 (0.092336)**	0.013054 (0.089552)

TABLE 4: Average mean squared errors and absolute biases of $(\hat{p})$.

n		$p = 0.8$	$p = 1.6$	$p = 2.2$
25	ML	0.076755 (0.191499)	0.490147 (0.476051)	**0.743582 (0.661779)**
	BS	0.076755 (0.191499)	0.490147 (0.476051)	**0.743582 (0.661779)**
	BG ($k = 0.6$)	0.065099 (0.188381)	0.553240 (0.462777)	0.910514 (0.683827)
	BG ($k = -0.6$)	**0.062020 (0.182668)**	**0.481165 (0.451018)**	0.898282 (0.647155)
50	ML	**0.025391 (0.118187)**	**0.147209 (0.283414)**	**0.364344** (0.445543)
	BS	**0.025391 (0.118187)**	**0.147209 (0.283414)**	**0.364344** (0.445543)
	BG ($k = 0.6$)	0.027027 (0.123510)	0.149940 (0.288091)	0.368843 (0.438096)
	BG ($k = -0.6$)	0.026932 (0.126065)	0.159536 (0.294393)	0.364682 **(0.435072)**
100	ML	0.011782 (0.085569)	0.060777 (0.186684)	0.130699 (0.278809)
	BS	0.011782 (0.085569)	0.060777 (0.186684)	0.130699 (0.278809)
	BG ($k = 0.6$)	0.012004 (0.084495)	**0.058056 (0.187923)**	0.141918 (0.286061)
	BG ($k = -0.6$)	**0.011583 (0.084047)**	0.059122 (0.188084)	**0.126457 (0.277587)**

parameter (p) of the *GE* distribution were taken to be 0.8, 1.6, and 2.2. The scale parameter (θ) was considered throughout the paper to be one (4).

We observed that the parameter estimates under the classical maximum likelihood method could not be obtained in close form, and we therefore employed Newton-Raphson iterative approach via the Hessian matrix. This can simply be implemented in the *R* programming language with the package *maxLik*.

To compute the Bayes estimates, an assumption is made such that θ and p take, respectively, Gamma(a, b) and Gamma(c, d) priors. We set the hyperparameters to be 0.0001, that is, $a = b = c = d = 0.0001$, in order to obtain proper priors. This approach was suggested by Press and Tanur [22]. Note that at this point the posterior distribution is also proper.

The values of the loss parameter for the general entropy loss function are $k = \pm 0.6$, which can be extended for other values of the loss parameter. Readers are referred to Calabria and Pulcini [23] for the choice of the loss parameter values. These were iterated $R = 1000$ times. The mean squared error and the absolute bias values are determined and presented below for the purpose of comparison.

5. Results and Discussion

In this study, our objective is to obtain the estimates of the parameters and to observe the performance of the methods used for estimation. To examine the estimates of the parameters, we obtain the absolute biases and mean squared errors of the estimates under different methods of estimation.

From Table 3, it is very clear that the most dominant estimator that had the smallest mean squared errors vis-$\bar{a}$-vis the absolute biases for the scale parameter $(\hat{\theta})$ is Bayesian under general entropy loss function. This is followed closely by Bayes under squared error loss function. What has been observed again is that, as the sample size increases, the mean squared error of all the estimators decrease unswervingly. This is simply an indication of how good and reliable the estimators are.

When we consider Table 4, which contains the mean squared errors and the absolute biases of the estimated shape parameter $(\hat{p})$, we noticed that the mean squared errors and the absolute biases of the two estimators, that is, maximum likelihood and Bayes under squared error loss function, have the same values for the estimated shape parameter.

TABLE 5: Average mean squared errors and absolute biases of the survival function $\widehat{S}(t)$ at $T = 25, 50$, and 100.

	$p = 0.8$	$p = 1.6$	$p = 2.2$
ML	**0.075982** (0.042809)	**0.082151** (0.045894)	**0.078309** (0.044532)
BS	0.075986 (0.042836)	0.082597 (0.046023)	0.078476 (0.044688)
BG ($k = 0.6$)	0.080981 (0.045294)	0.082265 (0.045688)	0.086399 (0.046789)
BG ($k = -0.6$)	0.077210 (**0.042326**)	0.083746 (**0.044032**)	0.083479 (**0.043191**)
ML	0.075706 (**0.031059**)	0.080699 (**0.031094**)	**0.078253** (0.031479)
BS	0.075725 (0.031063)	0.080711 (0.031107)	0.078272 (0.031507)
BG ($k = 0.6$)	0.079284 (0.031347)	0.082140 (0.032464)	0.078831 (**0.031191**)
BG ($k = -0.6$)	**0.075028** (0.031119)	**0.078181** (0.031753)	0.080356 (0.032732)
ML	0.073820 (**0.022000**)	**0.075437** (**0.022445**)	**0.074257** (0.022533)
BS	0.073923 (**0.022000**)	**0.075437** (0.022447)	0.077209 (0.022535)
BG ($k = 0.6$)	0.072898 (0.021704)	0.078784 (0.022582)	0.076435 (**0.022308**)
BG ($k = -0.6$)	**0.070723** (0.022449)	0.075552 (0.022798)	0.076741 (0.022954)

ML: maximum likelihood, BG: general entropy, and BS: squared error.

This is expected in that the priors used for the Bayesian analysis are noninformative. With regards to the survival function, Bayes estimator under the general entropy loss function gives a minimum bias with relatively small samples. Maximum likelihood estimator is slightly ahead of the other estimators with respect to the mean squared error. The bold numbers indicate the smallest MSE and minimum biases of the estimated parameters and their corresponding estimators.

From Table 5, we observed that the classical maximum likelihood estimator (MLE) as compared to Bayes estimators under squared error and general entropy loss function had the smallest mean squared error values as well as minimum absolute bias for the estimated survival function of the generalised exponential distribution. This implies that maximum likelihood estimator may be preferred when estimating the survival function to the others.

6. Conclusion

In this paper, we have considered the Bayes estimation of the unknown parameters of the generalized exponential distribution. We have also assumed a gamma prior on both parameters, and we provide the Bayes estimators under the assumptions of squared error and general entropy loss functions. We observed that the Bayes estimators cannot be obtained in explicit forms, due to the complex nature of the posterior distribution of which Bayes inference is drawn. Therefore, Lindley's numerical approximations procedure is used. It is also observed that the parameter estimates under the classical maximum likelihood method could not be obtained in close form; we therefore employed Newton-Raphson iterative approach via the Hessian matrix.

From the results and discussions above, it is evident that the Bayesian estimator under general entropy loss function performed quiet better than Bayes under squared error loss function and that of maximum likelihood estimator for estimating the scale parameter $\widehat{\theta}$, with both MSE and absolute bias. In the case of the shape parameter ($\widehat{p}$), the Bayesian estimator under the squared error loss function and the maximum likelihood estimator are both almost tantamount in estimating it. For the survival function, maximum likelihood performed better than the other estimators.

References

[1] T. Fagerström, P. Jagers, P. Schuster, and E. Szathmary, "Biologists put on mathematical glasses," *Science*, vol. 274, no. 5295, pp. 2039–2040, 1996.

[2] E. Limpert, W. A. Stahel, and M. Abbt, "Log-normal distributions across the sciences: keys and clues," *BioScience*, vol. 51, no. 5, pp. 341-352, 2001.

[3] R. D. Gupta and D. Kundu, "Generalized exponential distributions," *Australian & New Zealand Journal of Statistics*, vol. 41, no. 2, pp. 173–188, 1999.

[4] M. Z. Raqab, "Inferences for generalized exponential distribution based on record statistics," *Journal of Statistical Planning and Inference*, vol. 104, no. 2, pp. 339–350, 2002.

[5] M. Z. Raqab and M. Ahsanullah, "Estimation of the location and scale parameters of generalized exponential distribution based on order statistics," *Journal of Statistical Computation and Simulation*, vol. 69, no. 2, pp. 109–123, 2001.

[6] G. Zheng, "On the Fisher information matrix in type II censored data from the exponentiated exponential family," *Biometrical Journal*, vol. 44, no. 3, pp. 353–357, 2002.

[7] D. Kundu and R. D. Gupta, "Generalized exponential distribution: Bayesian estimations," *Computational Statistics & Data Analysis*, vol. 52, no. 4, pp. 1873–1883, 2008.

[8] R. D. Gupta and D. Kundu, "Generalized exponential distribution: different method of estimations," *Journal of Statistical Computation and Simulation*, vol. 69, no. 4, pp. 315–337, 2001.

[9] S. K. Sinha, "Bayes estimation of the reliability function and hazard rate of a weibull failure time distribution," *Trabajos de Estadistica*, vol. 1, no. 2, pp. 47–56, 1986.

[10] A. A. Abdel-Wahid and A. Winterbottom, "Approximate Bayesian estimates for the Weibull reliability function and hazard rate from censored data," *Journal of Statistical Planning and Inference*, vol. 16, no. 3, pp. 277–283, 1987.

[11] C. B. Guure and N. A. Ibrahim, "Bayesian analysis of the survival function and failure rate of Weibull distribution with censored data," *Mathematical Problems in Engineering*, vol. 2012, Article ID 329489, 18 pages, 2012.

[12] S. R. Huang and S. J. Wu, "Bayesian estimation and prediction for Weibull model with progressive censoring," *Journal of Statistical Computation and Simulation*, vol. 82, no. 11, pp. 1607–1620, 2012.

[13] C. B. Guure, N. A. Ibrahim, and M. Bakri Adam, "Bayesian inference of the Weibull model based on interval-censored survival data," *Computational and Mathematical Methods in Medicine*, vol. 2013, Article ID 849520, 10 pages, 2013.

[14] A. Zellner, "Bayesian estimation and prediction using asymmetric loss functions," *Journal of the American Statistical Association*, vol. 81, no. 394, pp. 446–451, 1986.

[15] C. B. Guure, N. A. Ibrahim, M. B. Adam, S. Bosomprah, and A. O. M. Ahmed, "Bayesian parameter and reliability estimate of Weibull failure time distribution," *Bulletin of the Malaysian Mathematical Sciences Society*. In press.

[16] F. M. Al-Aboud, "Bayesian estimations for the extreme value distribution using progressive censored data and asymmetric loss," *International Mathematical Forum*, vol. 4, no. 33–36, pp. 1603–1622, 2009.

[17] F. M. Al-Athari, "Parameter estimation for the double-pareto distribution," *Journal of Mathematics and Statistics*, vol. 7, no. 4, pp. 289–294, 2011.

[18] B. N. Pandey, N. Dwividi, and B. Pulastya, "Comparison between Bayesian and maximum likelihood estimation of the scale parameter in weibull distribution with known shape under linex loss function," *Journal of Scientometric Research*, vol. 55, pp. 163–172, 2011.

[19] D. V. Lindley, "Approximate Bayesian methods," *Trabajos Estadist*, vol. 31, pp. 223–245, 1980.

[20] J. F. Lawless, *Statistical Models and Methods for Lifetime Data*, Wiley, New York, NY, USA, 2003.

[21] T. Bjerkedal, "Acquisition of resistance in Guinea pigs infected with different doses of virulent tubercle bacilli," *American Journal of Epidemiology*, vol. 72, no. 1, pp. 130–148, 1960.

[22] S. J. Press and J. M. Tanur, *The Subjectivity of Scientists and the Bayesian Approach*, Wiley, New York, NY, USA, 2001.

[23] R. Calabria and G. Pulcini, "Point estimation under asymmetric loss functions for left-truncated exponential samples," *Communications in Statistics*, vol. 25, no. 3, pp. 585–600, 1996.

6

On the Study of Transience and Recurrence of the Markov Chain Defined by Directed Weighted Circuits Associated with a Random Walk in Fixed Environment

Chrysoula Ganatsiou

Department of Civil Engineering, Section of Mathematics and Statistics, School of Technological Sciences, University of Thessaly, Pedion Areos, 38334 Volos, Greece

Correspondence should be addressed to Chrysoula Ganatsiou; ganatsio@otenet.gr

Academic Editor: Man Lai Tang

By using the cycle representation theory of Markov processes, we investigate proper criterions regarding transience and recurrence of the corresponding Markov chain represented uniquely by directed cycles (especially by directed circuits) and weights of a random walk with jumps in a fixed environment.

1. Introduction

A systematic research has been developed (Kalpazidou [1], MacQueen [2], Minping and Min [3], Zemanian [4], and others) in order to investigate representations of the finite-dimensional distributions of Markov processes (with discrete or continuous parameter) having an invariant measure, as decompositions in terms of the cycle (or circuit) passage functions:

$$J_{\hat{c}}(i,j) = 1, \quad \text{if } i, j \text{ are consecutive states of } \hat{c},$$
$$= 0, \quad \text{otherwise}, \tag{1}$$

for any directed sequence $\hat{c} = (i_1, i_2, \ldots, i_r)$ (or $c = (i_1, i_2, \ldots, i_r, i_1)$) of states called a *cycle (or a circuit)*, $r > 1$, of the corresponding Markov process. The representations are called *cycle (or circuit) representations* while the corresponding discrete parameter Markov processes generated by directed circuits $c = (i_1, i_2, \ldots, i_r, i_1)$, $r > 1$, are called *circuit chains*.

Following the context of the theory of Markov processes' cycle-circuit representation, the present work arises as an attempt to investigate proper criterions regarding the properties of transience and recurrence of the corresponding Markov chain represented uniquely by directed cycles (especially by directed circuits) and weights of a random walk with jumps (having one elastic left barrier) in a fixed ergodic environment (Kalpazidou [1], Derriennic [5]).

The paper is organized as follows. In Section 2, we give a brief account of certain concepts of cycle representation theory of Markov processes that we will need throughout the paper. In Section 3, we present some auxiliary results in order to make the presentation of the paper more comprehensible. In particular, in Section 3, a random walk with jumps (having one elastic left barrier) in a fixed ergodic environment is considered, and the unique representations by directed cycles (especially by directed circuits) and weights of the corresponding Markov chain are investigated. These representations will give us the possibility to study proper criterions regarding transience and recurrence of the abovementioned Markov chain, as it is described in Section 4.

Throughout the paper, we will need the following notations:

$$\aleph = \{0, 1, 2, \ldots\}, \qquad \aleph^* = \{1, 2, \ldots\},$$
$$Z = \{\ldots, -1, 0, 1, \ldots\}. \tag{2}$$

2. Preliminaries

Let us consider a denumerable set S. Then the directed sequence $c = (i_1, i_2, \ldots, i_r, i_1)$ modulo the cyclic permutations, where $i_1, i_2, \ldots, i_r \in S$, $r > 1$, completely defines a *directed circuit* in S. The ordered sequence $\hat{c} = (i_1, i_2, \ldots, i_r)$ associated with the given directed c is called a *directed cycle* in S.

A directed circuit may be considered as $c = (c(r), c(r+1), \ldots, c(r+v-1), c(r+v))$, if there exists an $r \in Z$, such that $i_1 = c(r+0)$, $i_2 = c(r+1), \ldots, i_v = c(r+v-1)$, $i_1 = c(r+v)$, where c is a periodic function from Z to S.

The corresponding directed cycle is defined by the ordered sequence $\hat{c} = (c(r), c(r+1), \ldots, c(r+v-1))$. The values $c(k)$ are the *points* of c, while the directed pairs $(c(k), c(k+1))$, $k \in Z$, are the *directed edges* of c.

The smallest integer $p \equiv p(c) \geq 1$ satisfying the equation $c(r+p) = c(r)$, for all $r \in Z$, is the *period* of c. A directed circuit c such that $p(c) = 1$ is called a *loop*. (In the present work we will use directed circuits with distinct point elements.)

Let us also consider a directed circuit c (or a directed cycle $\hat{c}$) with period $p(c) > 1$. Then we may define by

$$J_c^{(n)}(i,j) = 1, \quad \text{if there exists an } r \in Z \text{ such that } i = c(r),\ j = c(r+n), \qquad (3)$$
$$= 0, \quad \text{otherwise,}$$

the *n-step passage function* associated with the directed circuit c, for any $i, j \in S$, $n \geq 1$.

Furthermore we may define by

$$J_c(i) = 1, \quad \text{if there exists an } r \in Z \text{ such that } i = c(r),$$
$$= 0, \quad \text{otherwise,} \qquad (4)$$

the *passage function* associated with the directed circuit c, for any $i \in S$. The above definitions are due to MacQueen [2] and Kalpazidou [1].

Given a denumerable set S and an infinite denumerable class C of overlapping directed circuits (or directed cycles) with distinct points (except for the terminals) in S such that all the points of S can be reached from one another following paths of circuit edges; that is, for each two distinct points i and j of S there exists a finite sequence $c_1, \ldots, c_k$, $k \geq 1$, of circuits (or cycles) of C such that i lies on c_1 and j lies on c_k, and any pair of consecutive circuits (c_n, c_{n+1}) have at least one point in common. Generally we may assume that the class C contains, among its elements, circuits (or cycles) with period greater or equal to 2.

With each directed circuit (or directed cycle) $c \in C$ let us associate a strictly *positive weight* w_c which must be independent of the choice of the representative of c, that is, it must satisfy the consistency condition $w_{c \circ t_k} = w_c$, $k \in Z$, where t_k is the translation of length k (that is, $t_k(n) \equiv n+k$, $n \in Z$, for any fixed $k \in Z$).

For a given class C of overlapping directed circuits (or cycles) and for a given sequence $(w_c)_{c \in C}$ of weights we may define by

$$p_{ij} = \frac{\sum_{c \in C} w_c \cdot J_c^{(1)}(i,j)}{\sum_{c \in C} w_c \cdot J_c(i)} \qquad (5)$$

the elements of a Markov transition matrix on S, if and only if $\sum_{c \in C} w_c \cdot J_c(i) < \infty$, for any $i \in S$. This means that a given Markov transition matrix $P = (p_{ij})$, $i, j \in S$, can be represented by directed circuits (or cycles) and weights if and only if there exists a class of overlapping directed circuits (or cycles) C and a sequence of positive weights $(w_c)_{c \in C}$ such that the formula (5) holds. In this case, the Markov transition matrix P has a unique stationary distribution p which is a solution of $pP = p$ and is defined by

$$p(i) = \sum_{c \in C} w_c \cdot J_c(i), \quad i \in S. \qquad (6)$$

It is known that the following classes of Markov chains may be represented uniquely by circuits (or cycles) and weights:

(i) the recurrent Markov chains (Minping and Min [3]),

(ii) the reversible Markov chains.

3. Auxiliary Results

Let us consider a Markov chain $(X_n)_{n \geq 0}$ on $\aleph$ with transitions $k \to (k+1)$, $k \to (k-1)$, and $k \to k$ whose the elements of the corresponding Markov transition matrix are defined by

$$P(X_{n+1} = 0 / X_n = 0) = r_0,$$
$$P(X_{n+1} = 1 / X_n = 0) = p_0, \quad p_0 = 1 - r_0,$$
$$P(X_{n+1} = k+1 / X_n = k) = p_k, \quad k \geq 1, \qquad (7)$$
$$P(X_{n+1} = k / X_n = k) = r_k, \quad k \geq 1,$$
$$P(X_{n+1} = k-1 / X_n = k) = q_k, \quad k \geq 1,$$

such that $p_k + q_k + r_k = 1$, $0 \leq p_k, r_k \leq 1$, for every $k \geq 1$, as it is shown in (Figure 1).

Assume that $(p_k)_{k \geq 0}$ and $(r_k)_{k \geq 0}$ are arbitrary fixed sequences with $0 \leq p_0 = 1 - r_0 \leq 1$, $0 \leq p_k, r_k \leq 1$, for every $k > 1$. If we consider the directed circuits $c_k = (k, k+1, k)$, $c_k' = (k, k)$, $k \geq 0$, and the collections of weights $(w_{c_k})_{k \geq 0}$ and $(w_{c_k'})_{k \geq 0}$, respectively, then we may obtain that

$$p_k = \frac{w_{c_k}}{w_{c_{k-1}} + w_{c_k} + w_{c_k'}}, \quad \text{for every } k \geq 1, \qquad (8)$$

with

$$p_0 = \frac{w_{c_0}}{w_{c_0} + w_{c_0'}}. \qquad (9)$$

Here the class $C(k)$ contains the directed circuits $c_k = (k, k+1, k)$, $c_{k-1} = (k-1, k, k-1)$, and $c_k' = (k, k)$. Furthermore we may define

$$q_k = \frac{w_{c_{k-1}}}{w_{c_{k-1}} + w_{c_k} + w_{c_k'}}, \quad \text{for every } k \geq 1 \qquad (10)$$

and $r_k = w_{c_k'} / (w_{c_{k-1}} + w_{c_k} + w_{c_k'})$, such that $p_k + q_k + r_k = 1$, for every $k \geq 1$, with $r_0 = 1 - p_0 = w_{c_0'} / (w_{c_0} + w_{c_0'})$.

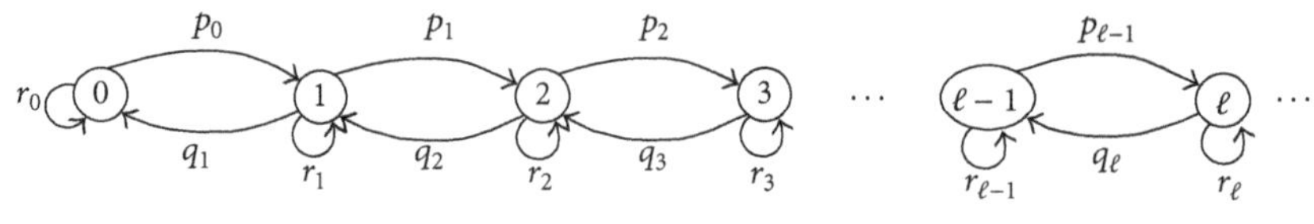

FIGURE 1

The transition matrix $P = (p_{ij})$ with

$$p_{ij} = \frac{\sum_{k=0}^{\infty} w_{c_k} \cdot J_{c_k}^{(1)}(i,j)}{\sum_{k=0}^{\infty} \left[w_{c_k} \cdot J_{c_k}(i) + w_{c'_k} \cdot J_{c'_k}(i) \right]}, \quad \text{for } i \neq j, \tag{11}$$

$$p_{ii} = \frac{\sum_{k=0}^{\infty} w_{c'_k} \cdot J_{c'_k}^{(1)}(i,i)}{\sum_{k=0}^{\infty} \left[w_{c_k} \cdot J_{c_k}(i) + w_{c'_k} \cdot J_{c'_k}(i) \right]}, \tag{12}$$

where $J_{c_k}^{(1)}(i,j) = 1$, if i, j are consecutive points of the circuit c_k, $J_{c_k}(i) = 1$, if i is a point of the circuit c_k, and $J_{c'_k}(i) = 1$, if i is a point of the circuit c'_k, expresses the representation of the Markov chain $(X_n)_{n\geq 0}$ by directed cycles (especially by directed circuits) and weights.

Furthermore we consider also the "*adjoint*" Markov chain $(X'_n)_{n\geq 0}$ on $\aleph$ whose elements of the corresponding Markov transition matrix are defined by

$$\begin{aligned}
&P\left(X'_{n+1} = 0 / X'_n = 0\right) = r'_0, \\
&P\left(X'_{n+1} = 1 / X'_n = 0\right) = q'_0, \quad q'_0 = 1 - r'_0, \\
&P\left(X'_{n+1} = k-1 / X'_n = k\right) = p'_k, \quad k \geq 1, \\
&P\left(X'_{n+1} = k / X'_n = k\right) = r'_k, \quad k \geq 1, \\
&P\left(X'_{n+1} = k+1 / X'_n = k\right) = q'_k, \quad k \geq 1,
\end{aligned} \tag{13}$$

such that $p'_k + q'_k + r'_k = 1$, $0 < p'_k, q'_k, r'_k \leq 1$, for every $k \geq 1$, as it is shown in (Figure 2).

Assume that $(q'_k)_{k\geq 0}$ and $(r'_k)_{k\geq 0}$ are arbitrary fixed sequences with $0 \leq q'_0 = 1 - r'_0 \leq 1$, $0 \leq q'_k, r'_k \leq 1$, for every $k \geq 1$. If we consider the directed circuits $c''_k = (k+1, k, k+1)$, $c'''_k = (k,k)$, $k \geq 0$, and the collections of weights $(w_{c''_k})_{k\geq 0}$, and $(w_{c'''_k})_{k\geq 0}$ respectively, then we may have

$$q'_k = \frac{w_{c''_k}}{w_{c''_{k-1}} + w_{c''_k} + w_{c'''_k}}, \quad \text{for every } k \geq 1, \tag{14}$$

with

$$q'_0 = \frac{w_{c''_0}}{w_{c''_0} + w_{c'''_0}}. \tag{15}$$

Here the class $C'(k)$ contains the directed circuits $c''_k = (k+1, k, k+1)$, $c''_{k-1} = (k, k-1, k)$, and $c'''_k = (k,k)$. Furthermore we may define

$$\begin{aligned}
p'_k &= \frac{w_{c''_{k-1}}}{w''_{c_{k-1}} + w_{c''_k} + w_{c'''_k}}, \quad \text{for every } k \geq 1, \\
r'_k &= \frac{w_{c'''_k}}{w_{c''_{k-1}} + w_{c''_k} + w_{c'''_k}},
\end{aligned} \tag{16}$$

such that $p'_k + q'_k + r'_k = 1$, for every $k \geq 1$, with

$$r'_0 = 1 - q'_0 = \frac{w_{c'''_0}}{w_{c''_0} + w_{c'''_0}}. \tag{17}$$

So the transition matrix $P' = (p'_{ij})$ with elements equivalent to that given by the above-mentioned formulas (11), (12) expresses also the representation of the "adjoint" Markov chain $(X'_n)_{n\geq 0}$ by directed cycles (especially by directed circuits) and weights.

Consequently we have the following.

Proposition 1. *The Markov chain $(X_n)_{n\geq 0}$ defined as above has a unique representation by directed cycles (especially by directed circuits) and weights.*

Proof. Let us consider the set of directed circuits $c_k = (k, k+1, k)$ and $c'_k = (k,k)$, for every $k \geq 0$, since only the transitions from k to $k+1$, k to $k-1$, and k to k are possible. There are three circuits through each point $k \geq 1$, c_{k-1}, c_k, and c'_k and two circuits through 0: c_0, c'_0.

The problem we have to manage is the definition of the weights. We may symbolize by w_k the weight w_{c_k} of the circuit c_k and by w'_k the weight $w_{c'_k}$ of the circuit c'_k, for any $k \geq 0$. The sequences $(w_k)_{k\geq 0}$, $(w'_k)_{k\geq 0}$ must be a solution of

$$\begin{aligned}
p_k &= \frac{w_k}{w_{k-1} + w_k + w'_k}, \quad k \geq 1 \text{ with } p_0 = \frac{w_0}{w_0 + w'_0}, \\
r_k &= \frac{w'_k}{w_{k-1} + w_k + w'_k}, \quad k \geq 1 \text{ with } r_0 = \frac{w'_0}{w_0 + w'_0}, \\
q_k &= 1 - p_k - r_k, \quad k \geq 1.
\end{aligned} \tag{18}$$

Let us take by $b_k = w_k / w_{k-1}$, $\gamma_k = w'_k / w'_{k-1}$, $k \geq 1$.

As a consequence we may have

$$\begin{aligned}
b_k &= \frac{p_k}{q_k} = \frac{p_k}{1 - p_k - r_k}, \\
\gamma_k &= \frac{r_k}{r_{k-1}} \frac{p_{k-1}}{p_k} b_k, \quad \text{for every } k \geq 1.
\end{aligned} \tag{19}$$

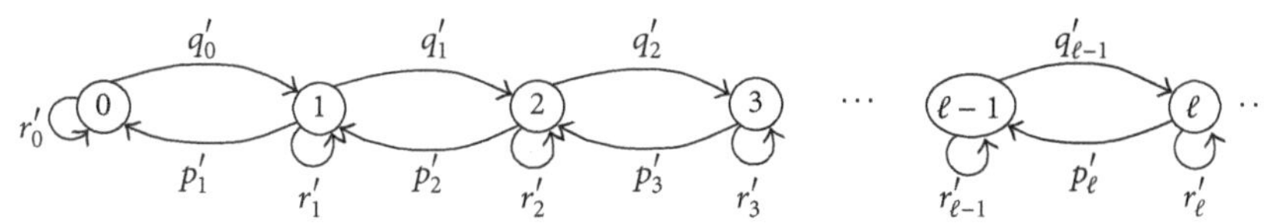

FIGURE 2

Given the sequences $(p_k)_{k\geq 0}$ and $(r_k)_{k\geq 0}$, it is clear that the above sequences $(b_k)_{k\geq 1}$ and $(\gamma_k)_{k\geq 1}$ exist and are unique. This means that the sequences $(w_k)_{k\geq 0}$ and $(w'_k)_{k\geq 0}$ are defined uniquely, up to multiplicative constant factors, by

$$w_k = w_0 \cdot b_1 \cdots b_k, \qquad w'_k = w'_0 \cdot \gamma_1 \cdots \gamma_k. \tag{20}$$

(The unicity is understood up to the constant factors w_0, w'_0.) □

Proposition 2. *The "adjoint" Markov chain $(X'_n)_{n\geq 0}$ defined as above has a unique representation by directed cycles (especially by directed circuits) and weights.*

Proof. Following an analogous way of that given in the proof of Proposition 1 the problem we have also to manage here is the definition of the weights. To this direction we may symbolize by w''_k the weight w''_{c_k} of the circuit c''_k and by w'''_k the weight w'''_{c_k} of the circuit c'''_k, for every $k \geq 0$. The sequences $(w''_k)_{k\geq 0}$ and $(w'''_k)_{k\geq 0}$ must be solutions of

$$q'_k = \frac{w''_k}{w''_{k-1} + w''_k + w'''_k}, \quad k \geq 1 \text{ with } q'_0 = \frac{w''_0}{w''_0 + w'''_0},$$
$$r'_k = \frac{w'''_k}{w''_{k-1} + w''_k + w'''_k}, \quad k \geq 1 \text{ with } r'_0 = \frac{w'''_0}{w''_0 + w'''_0}, \tag{21}$$
$$p'_k = 1 - q'_k - r'_k, \quad k \geq 1.$$

By considering the sequences $(s_k)_k$ and $(t_k)_k$ where $s_k = w''_{k-1}/w''_k$, $t_k = w'''_k/w'''_{k-1}$, $k \geq 1$, we may obtain that

$$s_k = \frac{1 - q'_k - r'_k}{q'_k},$$
$$t_k = \frac{r'_k}{r'_{k-1}} \cdot \frac{q'_k}{q'_{k-1}} \cdot s_k, \quad \text{for every } k \geq 1. \tag{22}$$

For given sequences $(q'_k)_{k\geq 0}$, $(r'_k)_{k\geq 0}$ it is obvious that $(s_k)_{k\geq 1}$, $(t_k)_{k\geq 1}$ exist and are unique for those sequences, that is, the sequences $(w''_k)_{k\geq 0}$, $(w'''_k)_{k\geq 0}$ are defined uniquely, up to multiplicative constant factors, by

$$w''_k = \frac{w''_0}{s_1 \cdot s_2 \cdots s_k}, \tag{23}$$
$$w'''_k = w'''_0 \cdot t_1, t_2 \cdots t_k.$$

(The unicity is based to the constant factors w''_0, w'''_0.) □

4. Recurrence and Transience of the Markov Chains $(X_n)_{n\geq 0}$ and $(X'_n)_{n\geq 0}$

We have that for the Markov chain $(X_n)_{n\geq 0}$, there is a unique invariant measure up to a multiplicative constant factor $\mu_k = w_{k-1} + w_k + w'_k$, $k \geq 1$, $\mu_0 = w_0 + w'_0$, while for the Markov chain $(X'_n)_{n\geq 0}$, $\mu'_k = w''_{k-1} + w''_k + w'''_k$, $k \geq 1$ with $\mu'_0 = w''_0 + w'''_0$. In the case that an irreducible chain is recurrent there is only one invariant measure (finite or not), so we may obtain the following.

Proposition 3. *(i) The Markov chain $(X_n)_{n\geq 0}$ defined as above is positive recurrent if and only if*

$$\sum_{k=1}^{\infty} b_1 \cdot b_2 \cdots b_k < +\infty \quad \left(or\ \frac{1}{w_0} \cdot \sum_{k=1}^{\infty} w_k < +\infty \right),$$
$$\sum_{k=1}^{\infty} \gamma_1 \cdot \gamma_2 \cdots \gamma_k < +\infty \quad \left(or\ \frac{1}{w'_0} \cdot \sum_{k=1}^{\infty} w'_k < +\infty \right). \tag{24}$$

(ii) The Markov chain $(X'_n)_{n\geq 0}$ defined as above is positive recurrent if and only if

$$\sum_{k=1}^{\infty} \frac{1}{s_1 \cdot s_2 \cdots s_k} < +\infty \quad \left(or\ \frac{1}{w''_0} \cdot \sum_{k=1}^{\infty} w''_k < +\infty \right),$$
$$\sum_{k=1}^{\infty} t_1 \cdot t_2 \cdots t_k = +\infty \quad \left(or\ \frac{1}{w'''_0} \cdot \sum_{k=1}^{\infty} w'''_k = +\infty \right). \tag{25}$$

In order to obtain recurrence and transience criterions for the Markov chains $(X_n)_{n\geq 0}$ and $(X'_n)_{n\geq 0}$ we shall need the following proposition (Karlin and Taylor [6]).

Proposition 4. *Let us consider a Markov chain on $\aleph$ which is irreducible. Then if there exists a strictly increasing function that is harmonic on the complement of a finite interval and that is bounded, then the chain is transient. In the case that there exists such a function which is unbounded then the chain is recurrent.*

Following this direction, we shall use a well-known method-theorem based on the Foster-Kendall theorem (Karlin and Taylor [6]) by considering the harmonic function $g = (g_k, k \geq 1)$. For the Markov chain $(X_n)_{n\geq 0}$ this is a solution of

$$p_0 \cdot g_1 + r_0 \cdot g_0 = g_0,$$
$$p_k \cdot g_{k+1} + q_k \cdot g_{k-1} + r_k \cdot g_k = g_k, \quad k \geq 1. \tag{26}$$

Since $\Delta g_k = g_k - g_{k-1}$, for every $k \geq 1$, we obtain that

$$\begin{aligned}
&p_k \cdot g_{k+1} + q_k \cdot g_k \\
&\quad - q_k \cdot g_k + q_k \cdot g_{k-1} + r_k \cdot g_k = g_k, \\
&p_k \cdot (\Delta g_{k+1} + g_k) + q_k \cdot g_k \\
&\quad - q_k \cdot g_k + q_k \cdot g_{k-1} + r_k \cdot g_k = g_k \\
&p_k \cdot \Delta g_{k+1} + (p_k + q_k + r_k) \cdot g_k \\
&\quad - q_k \cdot g_k + q_k \cdot g_{k-1} = g_k \\
&p_k \cdot \Delta g_{k+1} - q_k \cdot (g_k - g_{k-1}) = 0 \\
&p_k \cdot \Delta g_{k+1} = q_k \cdot \Delta g_k.
\end{aligned} \tag{27}$$

If we put $m_k = \Delta g_k / \Delta g_{k+1}$, we get $m_k = p_k / q_k$ (with $p_k = 1 - q_k - r_k$), $k \geq 1$, which is the equation of definition of the sequences $(s_k)_{k\geq 1}$ and $(t_k)_{k\geq 1}$ (as a multiplicative factor of the $(s_k)_{k\geq 1}$) for the Markov chain $(X'_n)_{n\geq 0}$ such that $q'_k = q_k$, $r'_k = r_k$, for every $k \geq 1$. This means that the strictly increasing harmonic functions of the Markov chain $(X_n)_{n\geq 0}$ are in correspondence with the weight representations of the Markov chain $(X'_n)_{n\geq 0}$ such that

$$\begin{aligned}
q'_k &= P\left(X'_{n+1} = k + 1 / X'_n = k\right) \\
&= P\left(X_{n+1} = k - 1 / X_n = k\right) = q_k, \\
r'_k &= P\left(X'_{n+1} = k / X'_n = k\right) \\
&= P\left(X_{n+1} = k / X_n = k\right) = r_k, \\
p'_k &= 1 - q'_k - r'_k \\
&= 1 - q_k - r_k = p_k, \quad \text{for every } k \geq 1.
\end{aligned} \tag{28}$$

To express this kind of duality we will call the Markov chain $(X'_n)_{n\geq 0}$, the *adjoint* of the Markov chain $(X_n)_{n\geq 0}$ and reciprocally in the case that the relation (28) is satisfied.

Equivalently for the Markov chain $(X'_n)_{n\geq 0}$, the harmonic function $g' = (g'_k, k \geq 1)$ satisfies the equation

$$\begin{aligned}
&r'_0 \cdot g'_0 + q'_0 \cdot g'_1 = g'_0, \\
&q'_k \cdot g'_{k+1} + p'_k \cdot g'_{k-1} + r'_k \cdot g'_k = g'_k, \quad k \geq 1.
\end{aligned} \tag{29}$$

Since $\Delta g'_k = g'_k - g'_{k-1}$, for every $k \geq 1$, we have

$$\begin{aligned}
&q'_k \cdot \left(\Delta g'_{k+1} + g'_k\right) + p'_k \cdot g'_k - p'_k \cdot g'_k \\
&\quad + p'_k \cdot g'_{k-1} + r'_k \cdot g'_k = g'_k, \\
&\left(p'_k + q'_k + r'_k\right) \cdot g'_k + q'_k \cdot \Delta g'_{k+1} \\
&\quad - p'_k \cdot g'_k + p'_k \cdot g'_{k-1} = g'_k, \\
&q'_k \cdot \Delta g'_{k+1} = p'_k \cdot \left(g'_k - g'_{k-1}\right) = p'_k \cdot \Delta g'_k.
\end{aligned} \tag{30}$$

If we put $\ell_k = \Delta g'_{k+1} / \Delta g'_k$, we get $\ell_k = p'_k / q'_k$ (with $q'_k = 1 - p'_k - r'_k$), $k \geq 1$, which is the equation of the definition of the sequences $(b_k)_{k\geq 1}$ and $(\gamma_k)_{k\geq 1}$ (as a multiplicative factor of the $(b_k)_{k\geq 1}$) for the Markov chain $(X_n)_{n\geq 0}$ such that $p'_k = p_k$, $r'_k = r_k$, for every $k \geq 1$. By considering a similar approximation of that given before for the Markov chain $(X_n)_{n\geq 0}$, we may say that the strictly increasing harmonic functions of the Markov chain $(X'_n)_{n\geq 0}$ are in correspondence with the weight representations of the Markov chain $(X_n)_{n\geq 0}$ such that equivalent equations of (28) are satisfied.

So we may have the following.

Proposition 5. *The Markov chain $(X_n)_{n\geq 0}$ defined as above is transient if and only if the adjoint Markov chain $(X'_n)_{n\geq 0}$ is positive recurrent and reciprocally. Moreover the adjoint Markov chains $(X_n)_{n\geq 0}$ and $(X'_n)_{n\geq 0}$ are null recurrent simultaneously.*
In particular

(i) *the Markov chain $(X_n)_{n\geq 0}$ defined as above is transient if and only if $(1/w''_0) \cdot \sum_{k=1}^{\infty} w''_k < +\infty$ and $(1/w'''_0) \cdot \sum_{k=1}^{\infty} w'''_k = +\infty$;*

(ii) *the Markov chain $(X'_n)_{n\geq 0}$ defined as above is transient if and only if $(1/w_0) \cdot \sum_{k=1}^{\infty} w_k < +\infty$ and $(1/w'_0) \cdot \sum_{k=1}^{\infty} w'_k < +\infty$;*

(iii) *the adjoint Markov chains $(X_n)_{n\geq 0}$ and $(X'_n)_{n\geq 0}$ are null recurrent if*

$$\begin{aligned}
\frac{1}{w_0} \cdot \sum_{k=1}^{\infty} w_k &= \frac{1}{w''_0} \cdot \sum_{k=1}^{\infty} w''_k = +\infty, \\
\frac{1}{w'_0} \cdot \sum_{k=1}^{\infty} w'_k &= \frac{1}{w'''_0} \cdot \sum_{k=1}^{\infty} w'''_k = +\infty.
\end{aligned} \tag{31}$$

Proof. The proof of Proposition 5 is an application mainly of Proposition 4 as well as of Proposition 3. □

Acknowledgment

C. Ganatsiou is indebted to the referee for the valuable comments, which led to a significant change of the first version.

References

[1] S. Kalpazidou, *Cycle Representations of Markov Processes*, Springer, New York, NY, USA, 1995.

[2] J. MacQueen, "Circuit processes," *Annals of Probability*, vol. 9, pp. 604–610, 1981.

[3] Q. Minping and Q. Min, "Circulation for recurrent markov chains," *Zeitschrift für Wahrscheinlichkeitstheorie und Verwandte Gebiete*, vol. 59, no. 2, pp. 203–210, 1982.

[4] A. H. Zemanian, *Infinite Electrical Networks*, Cambridge University Press, Cambridge, UK, 1991.

[5] Yv. Derriennic, "Random walks with jumps in random environments (examples of cycle and weight representations)," in *Probability Theory and Mathematical Statistics: Proceedings of The 7th Vilnius Conference 1998*, B. Grigelionis, J. Kubilius, V. Paulauskas, V. A. Statulevicius, and H. Pragarauskas, Eds., pp. 199–212, VSP, Vilnius, Lithuania, 1999.

[6] S. Karlin and H. Taylor, *A First Course in Stochastic Processes*, Academic Press, New York, NY, USA, 1975.

On the Generalized Lognormal Distribution

Thomas L. Toulias and Christos P. Kitsos

Technological Educational Institute of Athens, Department of Mathematics, Ag. Spyridonos & Palikaridi Street, 12210 Egaleo, Athens, Greece

Correspondence should be addressed to Thomas L. Toulias; t.toulias@teiath.gr

Academic Editor: Mohammad Fraiwan Al-Saleh

This paper introduces, investigates, and discusses the γ-order generalized lognormal distribution (γ-GLD). Under certain values of the extra shape parameter γ, the usual lognormal, log-Laplace, and log-uniform distribution, are obtained, as well as the degenerate Dirac distribution. The shape of all the members of the γ-GLD family is extensively discussed. The cumulative distribution function is evaluated through the generalized error function, while series expansion forms are derived. Moreover, the moments for the γ-GLD are also studied.

1. Introduction

Lognormal distribution has been widely applied in many different aspects of life sciences, including biology, ecology, geology, and meteorology as well as in economics, finance, and risk analysis, see [1]. Also, it plays an important role in Astrophysics and Cosmology; see [2–4] among others, while for Lognormal expansions see [5].

In principle, the lognormal distribution is defined as the distribution of a random variable whose logarithm is normally distributed, and usually it is formulated with two parameters. Furthermore, log-uniform and log-laplace distributions can be similarly defined with applications in finance; see [6, 7]. Specifically, the power-tail phenomenon of the Log-Laplace distributions [8] attracts attention quite often in environmental sciences, physics, economics, and finance as well as in longitudinal studies [9]. Recently, Log-Laplace distributions have been proposed for modeling growth rates as stock prices [10] and currency exchange rates [7].

In this paper a generalized form of Lognormal distribution is introduced, involving a third shape parameter. With this generalization, a family of distributions is emerged, which combines theoretically all the properties of Lognormal, Log-Uniform, and Log-Laplace distributions, depending on the value of this third parameter.

The generalized γ-order Lognormal distribution (γ-GLD) is the distribution of a random vector whose logarithm follows the γ-order normal distribution, an exponential power generalization of the usual normal distribution, introduced by [11, 12]. This family of p-dimensional generalized normal distributions, denoted by $\mathcal{N}_\gamma^p(\mu, \Sigma)$, is equipped with an extra shape parameter γ and constructed to play the role of normal distribution for the generalized Fisher's entropy type of information; see also [13, 14].

The density function f_X of a p-variate, γ-order, normally distributed random variable $Y \sim \mathcal{N}_\gamma^p(\mu, \Sigma)$, with location vector $\mu \in \mathbb{R}^p$, positive definite scale matrix $\Sigma \in \mathbb{R}^{p\times p}$, and shape parameter $\gamma \in \mathbb{R} \setminus [0, 1]$, is given by [11].

$$f_Y(y) = f_Y(y; \mu, \Sigma, \gamma) = C_\gamma^p |\det \Sigma|^{-1/2} \exp\left\{-\frac{\gamma-1}{\gamma} Q_\theta(y)^{\gamma/2(\gamma-1)}\right\}, \quad y \in \mathbb{R}^p, \tag{1}$$

where Q_θ is the quadratic form $Q_\theta(y) = (y-\mu)^{\mathrm{T}}\Sigma^{-1}(y-\mu)$, $\theta = (\mu, \Sigma)$, while C_γ^p being the normalizing factor

$$C_\gamma^p = \pi^{-p/2} \frac{\Gamma(p/2+1)}{\Gamma(p((\gamma-1)/\gamma))} \left(\frac{\gamma-1}{\gamma}\right)^{p((\gamma-1)/\gamma)-1}. \tag{2}$$

From (1), notice that the second-ordered normal is the known multivariate normal distribution; that is, $\mathcal{N}_2^p(\mu, \Sigma) = \mathcal{N}^p(\mu, \Sigma)$; see also [13, 15].

In Section 2, a generalized form of the Lognormal distribution is introduced, which is derived from the univariate family of $\mathcal{N}_\gamma(\mu, \sigma^2) = \mathcal{N}_\gamma^1(\mu, \sigma^2)$ distributions, denoted by $\mathcal{LN}_\gamma(\mu, \sigma)$, and includes the Log-Laplace distribution as well as the Log-Uniform distribution. The shape of the $\mathcal{LN}_\gamma(\mu, \sigma)$ members is extensively discussed while it is connected to the tailing behavior of $\mathcal{LN}_\gamma$ through the study of the c.d.f. In Section 3, an investigation of the moments of the generalized Lognormal distribution, as well as the special cases of Log-Uniform and Log-Laplace distributions, is presented.

The generalized error function, that is briefly provided here, plays an important role in the development of $\mathcal{LN}_\gamma(\mu, \sigma)$; see Section 2. The generalized error function denoted by Erf_a and the generalized complementary error function $\mathrm{Erfc}_a = 1 - \mathrm{Erf}_a$, $a \geq 0$ [16], are defined, respectively, as

$$\mathrm{Erf}_a(x) := \frac{\Gamma(a+1)}{\sqrt{\pi}} \int_0^x e^{-t^a} dt, \quad x \in \mathbb{R}. \tag{3}$$

The generalized error function can be expressed (changing to t^a variable), through the lower incomplete gamma function $\gamma(a, x)$ or the upper (complementary) incomplete gamma function $\Gamma(a, x) = \Gamma(a) - \gamma(a, x)$, in the form

$$\mathrm{Erf}_a(x) = \frac{\Gamma(a)}{\sqrt{\pi}} \gamma\left(\frac{1}{a}, x^a\right) = \frac{\Gamma(a)}{\sqrt{\pi}} \left[\Gamma\left(\frac{1}{a}\right) - \Gamma\left(\frac{1}{a}, x^a\right)\right], \quad x \in \mathbb{R}; \tag{4}$$

see [16]. Moreover, adopting the series expansion form of the lower incomplete gamma function,

$$\gamma(a, x) := \int_0^x t^{a-1} e^{-t} dt = \sum_{k=0}^{\infty} \frac{(-1)^k}{k!\,(a+k)} x^{a+k}, \quad x, a \in \mathbb{R}_+, \tag{5}$$

a series expansion form of the generalized error function is extracted:

$$\mathrm{Erf}_a(x) = \frac{\Gamma(a+1)}{\sqrt{\pi}} \sum_{k=0}^{\infty} \frac{(-1)^k}{k!\,(ka+1)} x^{ka+1}, \quad x, a \in \mathbb{R}_+. \tag{6}$$

Notice that, Erf_2 is the known error function erf, that is, $\mathrm{Erf}_2(x) = \mathrm{erf}(x)$, while Erf_0 is the function of a straight line through the origin with slope $(e\sqrt{\pi})^{-1}$. Applying $a = 2$, the known incomplete gamma function identities such as $\gamma(1/2, x) = \sqrt{\pi}\,\mathrm{erf}\,\sqrt{x}$ and $\Gamma(1/2, x) = \sqrt{\pi}(1 - \mathrm{erf}\,\sqrt{x}) = \sqrt{\pi}\,\mathrm{erfc}\,\sqrt{x}$, $x \geq 0$ are obtained. Moreover, $\mathrm{Erf}_a\, 0 = 0$ for all $a \in \mathbb{R}_+$ and

$$\lim_{x \to \pm\infty} \mathrm{Erf}_a x = \pm \frac{1}{\sqrt{\pi}} \Gamma(a) \Gamma\left(\frac{1}{a}\right), \quad a \in \mathbb{R}_+, \tag{7}$$

as $\gamma(a, x) \to \Gamma(a)$ when $x \to +\infty$.

2. The γ-Order Lognormal Distribution

The generalized univariate Lognormal distribution is defined, through the univariate generalized γ-order normal distribution, as follows.

Definition 1. When the logarithm of a random variable X follows the univariate γ-order normal distribution, that is, $\log X \sim \mathcal{N}_\gamma(\mu, \sigma^2)$, then X is said to follow the generalized Lognormal distribution, denoted by $\mathcal{LN}_\gamma(\mu, \sigma)$; that is, $X \sim \mathcal{LN}_\gamma(\mu, \sigma)$.

The $\mathcal{LN}_\gamma(\mu, \sigma)$ is referred to as the (generalized) γ-order Lognormal distribution (γ-GLD). Like the usual Lognormal distribution, the parameter $\mu \in \mathbb{R}$ is considered to be log-scaled, while the non log-scaled μ (i.e. e^μ when μ is assumed log-scaled) is referred to as the location parameter of $\mathcal{LN}_\gamma(\mu, \sigma)$. Hence, if $X \sim \mathcal{LN}_\gamma(\mu, \sigma)$, then $\log X$ is a γ-order normally distributed variable; that is, $\log X \sim \mathcal{N}_\gamma(\mu, \sigma^2)$. Therefore, the location parameter $\mu \in \mathbb{R}$ of X is in fact the mean of X's natural logarithm, that is, $\mathrm{E}[\log X] = \mu$, while

$$\mathrm{Var}[\log X] = \left(\frac{\gamma}{\gamma-1}\right)^{2((\gamma-1)/\gamma)} \frac{\Gamma(3((\gamma-1)/\gamma))}{\Gamma(((\gamma-1)/\gamma))} \sigma^2, \tag{8}$$

$$\mathrm{Kurt}[\log X] = \frac{\Gamma(\gamma - 1/\gamma)\,\Gamma(5(\gamma - 1/\gamma))}{\Gamma^2(3((\gamma-1)/\gamma))}; \tag{9}$$

see [15] for details on $\mathcal{N}_\gamma$.

Let $Y := \log X \sim \mathcal{N}_\gamma(\mu, \sigma^2)$ with density function as in (1) and $X = g(Y) = e^Y$. Then, the density function f_X of $X \sim \mathcal{LN}_\gamma(\mu, \sigma)$ can be written, through (1), as

$$\begin{aligned} f_X(x) = f_X(x; \mu, \sigma, \gamma) &= f_Y\left(g^{-1}(x)\right) \left|\frac{d}{dx} g^{-1}(x)\right| \\ &= f_Y(\log x) \frac{1}{x} \\ &= \frac{\exp\left\{-((\gamma-1)/\gamma) \left|(\log x - \mu)/\sigma\right|^{\gamma/(\gamma-1)}\right\}}{2\sigma((\gamma-1)/\gamma)^{1/\gamma} \Gamma((\gamma-1)/\gamma)\, x}. \end{aligned} \tag{10}$$

The probability density function f_X, as in (10), is defined in $\mathbb{R}_+^* = \mathbb{R}_+ \backslash 0$; that is, $\mathcal{LN}_\gamma(\mu, \sigma)$ has zero threshold. Therefore, the following definition extends Definition 1.

Definition 2. When the logarithm of a random variable $X + \vartheta$ follows the univariate γ-order normal distribution, that is, $\log(X + \vartheta) \sim \mathcal{N}_\gamma(\mu, \sigma^2)$, then X is said to follow the generalized Lognormal distribution with threshold $\vartheta \in \mathbb{R}$; that is, $X \sim \mathcal{LN}_\gamma(\mu, \sigma; \vartheta)$.

It is clear that when $X \sim \mathcal{LN}_\gamma(\mu, \sigma; \vartheta,)$, $\log(X - \vartheta)$ is a γ-order normally distributed variable, that is, $\log(X - \vartheta) \sim \mathcal{N}_\gamma(\mu, \sigma^2)$, and thus, μ is the mean of $(X - \vartheta)$'s natural logarithm while $\mathrm{Var}[\log X]$ is the same as in (8).

Let $Y = \log(X + \vartheta) \sim \mathcal{N}_\gamma(\mu, \sigma^2)$. The density function of $X = e^Y - \vartheta \sim \mathcal{LN}_\gamma(\mu, \sigma; \vartheta)$ is given by $f_X(x) = f_X(x - \vartheta)$, $x > 0$.

Let $z = (\log(x - \vartheta) - \mu)/\sigma$. Then, the limiting threshold density value of $f_X(x)$ with $x \to \vartheta^+$ implies that

$$\lim_{x\to\vartheta^+} f_X(x) = \sigma^{-1}C^1_\gamma \lim_{z\to-\infty} \exp\left\{-z\left(\sigma + \frac{\mu}{z} - \frac{\gamma-1}{\gamma}|z|^{1/(\gamma-1)}\right)\right\} \quad (11)$$
$$= \sigma^{-1}C^1_\gamma e^{(\operatorname{sgn}\gamma)(-\infty)},$$

and therefore

$$\lim_{x\to\vartheta^+} f_X(x) = \begin{cases} 0, & \gamma \in (1, +\infty), \\ +\infty, & \gamma \in (-\infty, 0); \end{cases} \quad (12)$$

that is, the f_X's defining domain, for the positive-ordered Lognormal random variable X, can be extended to include threshold point ϑ by letting $f_X(\vartheta) = 0$.

The generalized Lognormal family of distributions $\mathscr{LN}_\gamma$ is a wide range family bridging the Log-Uniform $\mathscr{LU}$, Lognormal $\mathscr{LN}$, and Log-Laplace $\mathscr{LL}$ distributions, as well as the degenerate Dirac $\mathscr{D}$ distributions. We have the following.

Theorem 3. *The generalized Lognormal distribution $\mathscr{LN}_\gamma(\mu, \sigma)$, for order values of $\gamma = 0, 1, 2, \pm\infty$, is reduced to*

$$\mathscr{LN}_\gamma(\mu,\sigma) = \begin{cases} \mathscr{D}(e^\mu), & \gamma = 0, \\ \mathscr{LU}(e^{\mu-\sigma}, e^{\mu+\sigma}), & \gamma = 1, \\ \mathscr{LN}(\mu,\sigma), & \gamma = 2, \\ \mathscr{LL}\left(e^\mu, \frac{1}{\sigma}, \frac{1}{\sigma}\right), & \gamma = \pm\infty. \end{cases} \quad (13)$$

Proof. From definition (1) of $\mathscr{N}_\gamma$ the order γ value is a real number outside the closed interval $[0, 1]$. Let $X_\gamma \sim \mathscr{LN}_\gamma(\mu, \sigma)$ with density function f_{X_γ} as in (10). We consider the following cases.

(i) The limiting case $\gamma = 1$: let $x \in \mathbb{R}^*_+$ such that $|\log x - \mu| \le 1$. Using the gamma function additive identity $\Gamma(z + 1) = z\Gamma(z)$, $z \in \mathbb{R}_+$, in (10), we have $\mathscr{LN}_1(\mu, \sigma) = \lim_{\gamma\to1^+} \mathscr{LN}_\gamma(\mu, \sigma)$ with

$$f_{X_1}(x) := \lim_{\gamma\to1^+} f_{X_\gamma}(x) = \begin{cases} \frac{1}{2\sigma x}, & x \in [e^{\mu-\sigma}, e^{\mu+\sigma}], \\ 0, & x \in (-\infty, e^{\mu-\sigma}) \cup (e^{\mu+\sigma}, +\infty), \end{cases} \quad (14)$$

which is the density function of the Log-Uniform distribution $\mathscr{LU}(a, b)$, $0 < a < b$, with $a = e^{\mu-\sigma}$ and $b = e^{\mu+\sigma}$; that is, $\mu = (1/2)\log(ab)$ and $\sigma = (1/2)\log(b/a)$. Therefore, first-ordered Lognormal distribution is in fact the Log-Uniform distribution, with vanishing threshold density, $f_{X_1}(0) = \lim_{x\to0^+} f_{X_1}(x) = 0$. For the purposes of statistical application the Log-Uniform moments are not the same as the model parameters; that is, although $\mu_X = \mu$, $\sigma_X = \sigma/\sqrt{3}$.

(ii) The "normal" case $\gamma = 2$: it is clear that $\mathscr{LN}_2(\mu, \sigma) = \mathscr{LN}(\mu, \sigma)$, as f_{X_2} coincides with the Lognormal density function, and therefore the second-ordered Lognormal distribution is in fact the usual Lognormal distribution.

(iii) The limiting case $\gamma = \pm\infty$: we have $\mathscr{LN}_{\pm\infty}(\mu, \sigma) := \lim_{\gamma\to\pm\infty} \mathscr{LN}_\gamma(\mu, \sigma)$ with

$$f_{X_{\pm\infty}}(x) := \lim_{\gamma\to\pm\infty} f_{X_\gamma}(x) = \frac{1}{2\sigma x}\exp\left\{-\left|\frac{\log x - \mu}{\sigma}\right|\right\} = \begin{cases} \frac{e^{-\mu/\sigma}}{2\sigma} x^{(1-\sigma)/\sigma}, & x \in (0, e^\mu], \\ \frac{e^{\mu/\sigma}}{2\sigma} x^{-(\sigma+1)/\sigma}, & x > e^\mu, \end{cases} \quad (15)$$

which coincides with the density function of the known Log-Laplace distribution (symmetric log-exponential distribution) $\mathscr{LL}(\mu', \alpha, \beta)$ with $\mu' = e^\mu$ and $\alpha = \beta = 1/\sigma$; see [8]. Therefore, the infinite-ordered log-normal distribution is in fact the Log-Laplace distribution, with threshold density

$$f_{X_{\pm\infty}}(0) = \lim_{x\to0^+} f_{X_{\pm\infty}}(x) = \begin{cases} 0, & \sigma < 1, \\ 1, & \sigma = 1, \\ +\infty, & \sigma > 1. \end{cases} \quad (16)$$

For the purposes of statistical application, the Log-Laplace moments are not the same as the model parameters; that is, although $\mu_X = \mu$, $\sigma_X = \sqrt{2}\sigma$.

(iv) The limiting case $\gamma = 0$: we have

$$\lim_{\gamma\to0^-} f_{X_\gamma}(x) = \lim_{k:=[(\gamma-1)/\gamma]\to\infty} f_{X_\gamma}(x), \quad x \in \mathbb{R}^*_+, \quad (17)$$

where $[a]$ is the integer value of $a \in \mathbb{R}$. For the value $x = e^\mu$, the p.d.f. as in (10) implies

$$\lim_{\gamma\to0^-} f_{X_\gamma}(e^\mu) = \frac{1}{2\sigma e^\mu}\left(\lim_{k\to\infty}\frac{k^k}{k!}\right)\cdot e^0 = +\infty, \quad (18)$$

through Stirling's asymptotic formula $k! \approx \sqrt{2\pi k}(k/e)^k$ as $k \to \infty$. Assuming now $x \ne e^\mu$, (10), through (18), implies

$$f_{X_0}(x) := \lim_{\gamma\to0^-} f_{X_\gamma}(x) = \frac{|\log x - \mu|}{2\sqrt{2\pi\sigma^2}x}\cdot 0 \cdot \frac{1}{e} = 0; \quad (19)$$

that is, $\mathscr{LN}_0(\mu, \sigma) := \lim_{\gamma\to0^-} \mathscr{LN}_\gamma(\mu, \sigma) = \mathscr{D}(e^\mu)$ as f_{X_0} coincides with the Dirac density function, with the (non-log-scaled) location parameter e^μ of $\mathscr{LN}_0(\mu, \sigma)$ being the singular (infinity) point. Therefore, the zero-ordered Lognormal distribution $\mathscr{LN}_0$ is in fact the degenerate Dirac distribution with pole at the location parameter of $\mathscr{LN}_{\gamma\to0^-}$ (with vanishing threshold density $f_{X_0}(0) = \lim_{x\to0^+} f_{X_0}(x) = 0$).

From the above limiting cases (i), (iii), and (iv), the defining domain $\mathbb{R} \setminus [0, 1]$ of the order values γ, used in (1), is safely extended to include the values $\gamma = 0, 1, \pm\infty$; that is, γ can now be defined outside the open interval $(0, 1)$. Eventually, the family of the γ-order normals can include the Log-Uniform, Lognormal, Log-Laplace, and the degenerate Dirac distributions as (13) holds. □

From Theorem 3, (12), and (15), the domain of the density functions $f_{X_\gamma}(x)$, $x > 0$, can also be extended to include the threshold point $x = 0$ by setting $f_{X_\gamma}(0) := 0$ for all non-negative-ordered Lognormals, that is, for all $\gamma \in 0 \cup [1, +\infty)$, while for the Log-Laplace case of $\gamma = +\infty$ with $\sigma = 1$ by setting $f_{X_{+\infty}}(0) := (1/2)e^{-\mu}$.

From the fact that $\mathcal{N}_0(\mu, \cdot) = \mathcal{D}(\mu)$, see [15], one can say that the degenerate log-Dirac distribution, say $\mathcal{LD}(\mu)$, equals $\mathcal{LN}_0(\mu, \cdot)$, and hence, through Theorem 3, we can write $\mathcal{LD}(\mu) = \mathcal{D}(e^\mu)$.

Proposition 4. *The mode of the positive-ordered Lognormal random variable* $X_\gamma \sim \mathcal{LN}_\gamma(\mu, \sigma)$, $X < e^\mu$ $\gamma \in (1, +\infty)$, *is given by*

$$\mathit{Mode}\, X_\gamma = e^{\mu - \sigma^\gamma}, \tag{20}$$

with corresponding maximum density value,

$$\max f_{X_\gamma} = f_{X_\gamma}\left(\mathit{Mode}\, X_\gamma\right) = \frac{\exp\{(\sigma^\gamma)/\gamma - \mu\}}{2\sigma((\gamma-1)/\gamma)^{1/\gamma}\Gamma((\gamma-1)/\gamma)}. \tag{21}$$

Proof. Recall the density function of $X_\gamma \sim \mathcal{LN}_\gamma(\mu, \sigma)$ as in (10), and let $m = \text{Mode}\, X_\gamma > 0$. Then it holds that $(d/dx) f_{X_\gamma}(m; \mu, \sigma, \gamma) = 0$; that is,

$$\left(\frac{1}{\sigma}\left|\log m - \mu\right|\right)^{1/(\gamma-1)} \frac{d}{dx}\left|\log x - \mu\right|_{x=m} = -\frac{\sigma}{m}. \tag{22}$$

From $(d/dx)|x| = \operatorname{sgn} x$, $x \in \mathbb{R}$, we have

$$m = e^{\mu - \sigma^\gamma}, \tag{23}$$

provided that $x < e^\mu$. Otherwise, (23) holds trivially, as (22) implies $\sigma = 0$; that is, $m = e^\mu$. Moreover, $(d/dx) f_{X_\gamma}(x) > 0$ when

$$1 + \operatorname{sgn}(\log x - \mu)\, \sigma^{\gamma/(1-\gamma)} \left|\log x - \mu\right|^{1/(\gamma-1)} > 0, \quad x > 0, \tag{24}$$

and thus, f_{X_γ} is a strictly ascending density function on $(0, e^{\mu-\sigma^\gamma})$ when $\gamma > 1$ and also on $(e^{\mu-\sigma^\gamma}, e^\mu)$ when $\gamma < 0$. Similarly, with $(d/dx) f_{X_\gamma}(x) < 0$, f_X is a strictly descending density function on $(e^{\mu-\sigma^\gamma}, +\infty)$ when $\gamma > 1$ and also on $(0, e^{\mu-\sigma^\gamma}) \cup (e^\mu, +\infty)$ when $\gamma < 0$. Specifically, for $\gamma < 0$, the point e^μ is a nonsmooth point of f_X, as

$$\lim_{x \to e^\mu} \frac{d}{dx} f_{X_\gamma}(x) = -\frac{C_\gamma^1}{(\sigma e^\mu)^2}\left[\sigma + \lim_{x \to e^\mu} \operatorname{sgn}(\log x - \mu) \times \left|\frac{\sigma}{\log x - \mu}\right|^{1/(1-\gamma)}\right] = +\infty. \tag{25}$$

Therefore, the positive-ordered Lognormals are formed by a unimodal density function with mode as in (20) and corresponding maximum density as in (21); see Figures 1(a1), 1(a2), and 1(a3). □

Proposition 5. *The global mode point of the negative-ordered Lognormal random variable* $X_\gamma \sim \mathcal{LN}_\gamma(\mu, \sigma)$, $\gamma \in (-\infty, 0)$, *is (in limit) the threshold 0 which is an infinite (probability) density point. Moreover, the location parameter (i.e.,* e^μ*) of* $\mathcal{LN}_\gamma(\mu, \sigma)$ *is a nonsmooth (local) mode point for all* $X_{\gamma<0}$ *that corresponds to locally maximum density*

$$f_{X_\gamma}(e^\mu) = \frac{1}{2\sigma((\gamma-1)/\gamma)^{1/\gamma}\Gamma((\gamma-1)/\gamma)\, e^\mu}, \tag{26}$$

while $e^{\mu-\sigma^\gamma}$ *is a local minimum (probability) density point with corresponding locally minimum density* $f_{X_\gamma}(e^{\mu-\sigma^\gamma})$.

Proof. The negative-ordered Lognormals are formed by density functions admitting threshold 0 (in limit) for their global mode point (of infinite density), as shown in (12). Moreover, from the previously discussed monotonicity of f_{X_γ} in Proposition 4, all the negative-ordered Lognormals admit also e^μ as a local nonsmooth mode point and $\exp\{\mu - \sigma^\gamma\}$ as a local minimum density point, with densities as in (26) and $f_{X_\gamma}(e^{\mu-\sigma^\gamma})$, respectively; see Figures 1(b1), 1(b2), and 1(b3). □

Furthermore, Proposition 5 holds (in limit) for random variables X_1 and $X_{\pm\infty}$ which provide Log-Uniform and Log-Laplace distributions, respectively. Indeed, for given $a, b \in \mathbb{R}_+^*$, $X_1 \sim \mathcal{LU}(a, b) = \mathcal{LN}_1(\mu, \sigma)$ with $\mu = (1/2)\log(ab)$ and $\sigma = (1/2)\log(b/a)$, we get, through (20) and (21), that $\text{Mode}\, X_1 := \text{Mode}\, X_{\gamma \to 1^+} = e^{\mu-\sigma} = a$ with corresponding maximum density (i.e., the maximum value of the density function),

$$\max f_{X_1} = f_{X_1}(a) = \frac{e^{\sigma-\mu}}{2\sigma} \lim_{\gamma \to 1^+} \frac{((\gamma-1)/\gamma)^{(\gamma-1)/\gamma}}{\Gamma(((\gamma-1)/\gamma)+1)} = \frac{1}{a\log(b/a)}. \tag{27}$$

Moreover, the nonzero minimum density (i.e., the minimum, but not zero, value of the density function) is obtained at $x = b$ with $\min f_{X_1} = f_{X_1}(b) = 1/2\sigma b = 1/(b\log(b/a))$. These results are in accordance with the Log-Uniform density function in (14).

For $X_{\pm\infty} \sim \mathcal{LN}_{\pm\infty}(\log\mu, 1/\sigma) = \mathcal{LL}(\mu, \sigma, \sigma)$, we evaluate, through (20) and (21), that

$$\text{Mode} X_{+\infty} = \begin{cases} 0, & \sigma < 1, \\ \dfrac{\mu}{e} & \sigma = 1, \\ \mu, & \sigma > 1, \end{cases} \tag{28}$$

with the corresponding maximum density value being infinite; that is, $\max f_{X_\gamma} = f_{X_\gamma}(0) = +\infty$, provided $\sigma < 1$, and

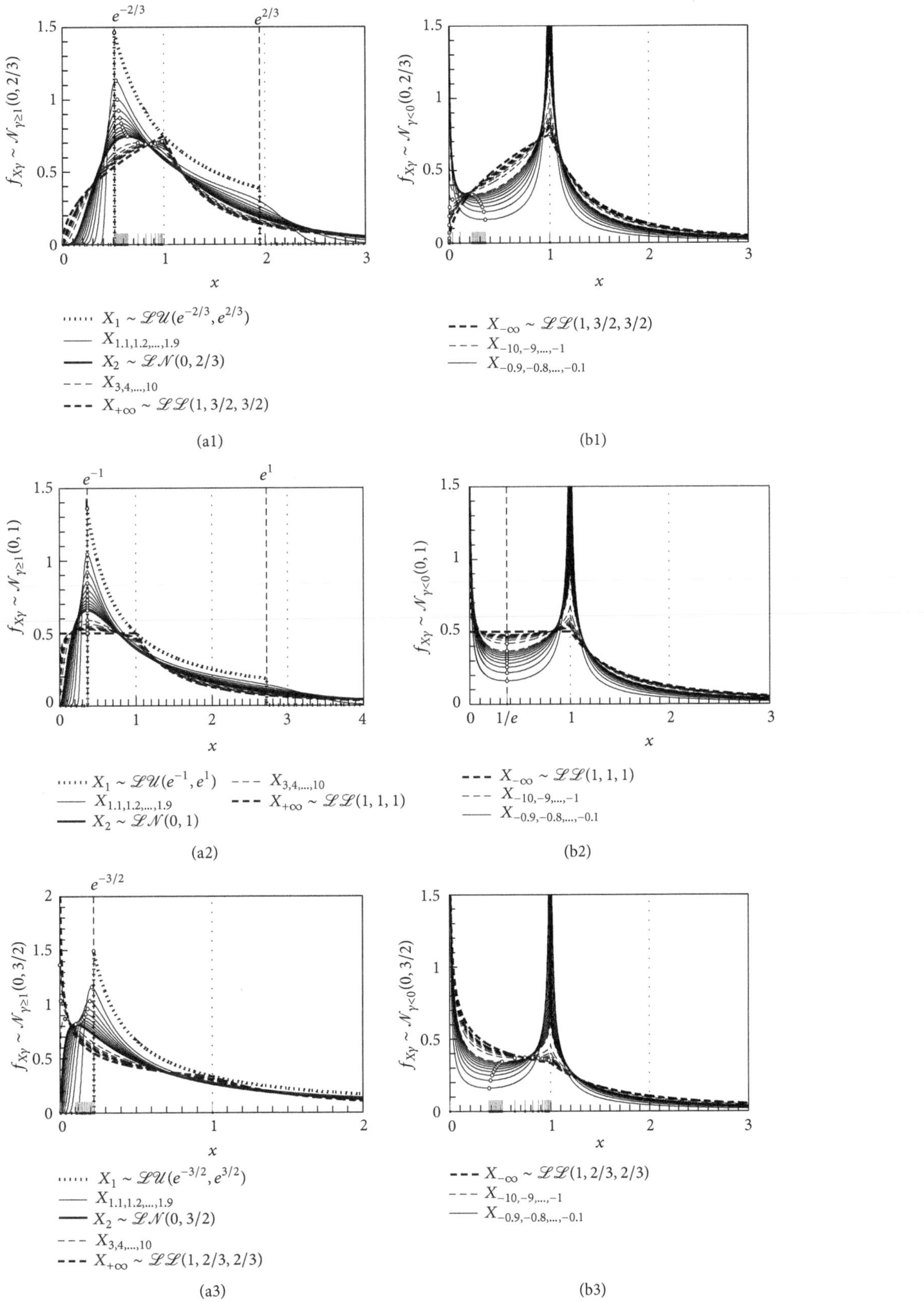

FIGURE 1: Graphs of the density functions f_{X_γ}, $X_\gamma \sim \mathcal{LN}_\gamma(0, \sigma)$, for $\sigma = 2/3, 1, 3/2$, and various positive (left subfigures) and negative (right subfigures) γ values.

$\max f_{X_{+\infty}} = \sigma/(2\mu)$, provided $\sigma \geq 1$. The same result can also be derived through (26) as $\gamma \to -\infty$. These results are in accordance with the Log-Laplace density function in (15), although for $\sigma = 1$, Mode$X_{\pm\infty}$ can be defined, through (15), for any value inside the interval $(0, \mu]$.

The above discussion on behavior of the modes with respect to shape parameter γ is formed in the following propositions.

Proposition 6. *Consider the positive-ordered Lognormal family of distributions $\mathscr{L}\mathscr{N}_\gamma(\mu, \sigma)$ with fixed parameters μ, σ and $\gamma \geq 1$. When γ rises, that is, when one moves from Log-Uniform to Log-Laplace distribution inside the $\mathscr{L}\mathscr{N}_\gamma$ family, the mode points of $\mathscr{L}\mathscr{N}_\gamma$ are*

(i) *strictly increasing from $e^{\mu-\sigma}$ (Log-Uniform case) to e^{μ} (Log-Laplace case) provided that $\sigma < 1$ (with their corresponding maximum density values moving smoothly from $(1/2\sigma)e^{\sigma-\mu}$ to $+\infty$),*

(ii) *fixed at $e^{\mu-1}$ for all $\mathscr{L}\mathscr{N}_{\gamma\geq 1}(\mu, \sigma = 1)$ (with the corresponding maximum density values moving smoothly from $(1/2)e^{1-\mu}$ to $(1/2)e^{-\mu}$),*

(iii) *strictly decreasing from $e^{\mu-\sigma}$ (Log-Uniform case) to threshold 0 (Log-Laplace case) provided that $\sigma > 1$ (with their corresponding maximum density values moving smoothly from $(1/2\sigma)e^{\sigma-\mu}$ to $(1/2\sigma)e^{-\mu}$).*

Proof. Let $X_\gamma \sim \mathscr{L}\mathscr{N}_\gamma(\mu, \sigma)$, Mode X_γ is a smooth monotonous function of $\gamma \in (-\infty, 0) \cup (1, +\infty)$ for positive-and negative-ordered X_γ, as

$$\frac{d}{d\gamma}\text{Mode } X_\gamma = -\sigma^\gamma \log(\sigma)\, e^{\mu-\sigma^\gamma}. \tag{29}$$

For $X_{\pm\infty} \sim \mathscr{L}\mathscr{N}_{\pm\infty}(\mu, \sigma)$ we evaluate, through (20) and (21), that

$$\text{Mode } X_{+\infty} = \begin{cases} e^{\mu}, & \sigma < 1, \\ e^{\mu-1}, & \sigma = 1, \\ 0, & \sigma > 1, \end{cases} \tag{30}$$

with the corresponding maximum density value being infinite; that is, $\max f_{X_\gamma} = f_{X_\gamma}(0) = +\infty$, provided $\sigma > 1$, and $\max f_{X_{+\infty}} = 1/(2\sigma e^{\mu})$, provided $\sigma \leq 1$.

Assume that $\gamma \geq 1$. Considering (27), (30) with (29) and Proposition 4, the results for the positive-ordered Lognormals hold. □

Proposition 7. *For the negative-ordered Lognormal family of distributions $\mathscr{L}\mathscr{N}_\gamma(\mu, \sigma)$ with $\gamma < 0$, when γ rises, that is, when one moves from Log-Laplace to degenerate Dirac distribution inside the $\mathscr{L}\mathscr{N}_\gamma$ family, the local minimum (probability) density points of $\mathscr{L}\mathscr{N}_\gamma$ are*

(i) *strictly increasing from threshold 0 (Log-Laplace case) to $e^{\mu-1}$ (Dirac case) provided that $\sigma < 1$,*

(ii) *fixed at $e^{\mu-1}$ for all $\mathscr{L}\mathscr{N}_\gamma(\mu, \sigma = 1)$,*

(iii) *strictly decreasing from e^{μ} (Log-Laplace case) to $e^{\mu-1}$ (Dirac case) provided that $\sigma > 1$.*

Proof. Assume now that $\gamma < 0$. From (29) we have $(d/d\gamma)(e^{\mu-\sigma^\gamma}) < 0$ when $\sigma > 1$ and $(d/d\gamma)(e^{\mu-\sigma^\gamma}) > 0$ when $\sigma < 1$. Therefore, the local minimum density point $e^{\mu-\sigma^\gamma}$ (see Proposition 5) for $\sigma > 1$ is decreasing from $e^{\mu-\sigma^\gamma}|_{\gamma\to-\infty} = e^{\mu}$ to Mode $X_0 = e^{\mu-1}$ through (20). When $\sigma = 1$, $e^{\mu-\sigma^\gamma} = e^{\mu-1}$ for all $\gamma < 0$, while for $\sigma < 1$, $e^{\mu-\sigma^\gamma}$ increases from $e^{\mu-\sigma^\gamma}|_{\gamma\to-\infty} = 0$ to Mode $X_0 = e^{\mu-1}$ through (20). □

It is easy to see that for the Log-Laplace case $\mathscr{L}\mathscr{L}(\mu, \sigma, \sigma)$, the local minimum density point $e^{\mu-\sigma^\gamma}$ of $X_{\gamma<0}$ with $\sigma > 1$ coincides (in limit) with the local nonsmooth mode point e^{μ} of X_γ; see Figure 1(b3). Also, notice that the local minimum density point $e^{\mu-\sigma^\gamma}$, $\gamma < 0$, for the Dirac case $\mathscr{D}(e^{\mu})$, is the limiting point $e^{\mu-1}$ although the (probability) density in $\mathscr{D}(e^{\mu})$ case vanishes everywhere except at the infinite pole e^{μ}.

Figure 1 illustrates the probability density functions f_{X_γ} curves for scale parameters $\sigma = 2/3, 1, 3/2$ of the positive-ordered lognormally distributed $X_\gamma \sim \mathscr{L}\mathscr{N}_{\gamma\geq 1}(0, \sigma)$ in Figures 1(a1)–1(a3), respectively, while the p.d.f. of negative-ordered lognormally distributed $X_\gamma \sim \mathscr{L}\mathscr{N}_{\gamma<0}(0, \sigma)$ are depicted in Figures 1(b1)–1(b3), respectively. Moreover, the density points $e^{\mu-\sigma^\gamma}$ on f_{X_γ} are also depicted (small circles over p.d.f. curves with their corresponding ticks on x-axis). According to Proposition 7, in Figures 1(a1)–1(a3), that is, for positive-ordered $X_{\gamma\geq 1}$, these density points represent the mode points on f_{X_γ} while in Figures 1(b1)–1(b3), that is, for negative-ordered $X_{\gamma<0}$, represent the local minimum density points on f_{X_γ} curves.

For the evaluation of the cumulative distribution function (c.d.f.) of the generalized Lognormal distribution, the following theorem is stated and proved.

Theorem 8. *The c.d.f. F_{X_γ} of a γ-order Lognormal random variable $X_\gamma \sim \mathscr{L}\mathscr{N}_\gamma(\mu, \sigma)$ is given by*

$$\begin{aligned} F_{X_\gamma}(x) &= \frac{1}{2} + \frac{\sqrt{\pi}}{2\Gamma((\gamma-1)/\gamma)\,\Gamma(\gamma/(\gamma-1))} \\ &\quad \times \operatorname{Erf}_{\gamma/(\gamma-1)}\left\{\left(\frac{\gamma-1}{\gamma}\right)^{(\gamma-1)/\gamma}\frac{\log x-\mu}{\sigma}\right\} \end{aligned} \tag{31}$$

$$= 1 - \frac{1}{2\Gamma((\gamma-1)/\gamma)}\Gamma\left(\frac{\gamma-1}{\gamma}, \frac{\gamma-1}{\gamma}\left(\frac{\log x-\mu}{\sigma}\right)^{\gamma/(\gamma-1)}\right), \quad x \in \mathbb{R}_+^*. \tag{32}$$

Proof. From density function f_{X_γ}, as in (10), we have

$$\begin{aligned} F_{X_\gamma}(x) &= F_{X_\gamma}(x; \mu, \sigma, \gamma) = \int_0^x f_{X_\gamma}(t)\,dt \\ &= \sigma^{-1}C_\gamma^1 \int_0^x t^{-1}\exp\left\{-\frac{\gamma-1}{\gamma}\left|\frac{\log t-\mu}{\sigma}\right|^{\gamma/(\gamma-1)}\right\}dt. \end{aligned} \tag{33}$$

Applying the transformation $w = (\log t - \mu)/\sigma$, $t > 0$, the above c.d.f. is reduced to

$$\begin{aligned} F_{X_\gamma}(x) &= C_\gamma^1 \int_{-\infty}^{(\log x-\mu)/\sigma} \exp\left\{-\frac{\gamma-1}{\gamma}|w|^{\gamma/(\gamma-1)}\right\} dw \\ &= \Phi_{Z_\gamma}\left(\frac{\log x-\mu}{\sigma}\right), \end{aligned} \quad (34)$$

where Φ_{Z_γ} is the c.d.f. of the standardized γ-order normal distribution $Z_\gamma = (1/\sigma)(\log X_\gamma - \mu) \sim \mathcal{N}_\gamma(0,1)$. Moreover, Φ_{Z_γ} can be expressed in terms of the generalized error function. In particular,

$$\begin{aligned} \Phi_{Z_\gamma}(z) &= C_\gamma^1 \int_{-\infty}^{z} \exp\left\{-\frac{\gamma-1}{\gamma}|w|^{\gamma/(\gamma-1)}\right\} dw \\ &= \Phi_{Z_\gamma}(0) + C_\gamma^1 \int_0^z \exp\left\{-\frac{\gamma-1}{\gamma}|w|^{\gamma/(\gamma-1)}\right\} dw, \end{aligned} \quad (35)$$

and as f_{Z_γ} is a symmetric density function around zero, we have

$$\begin{aligned} \Phi_{Z_\gamma}(z) &= \frac{1}{2} + C_\gamma^1 \int_0^z \exp\left\{-\frac{\gamma-1}{\gamma}|w|^{\gamma/(\gamma-1)}\right\} dw \\ &= \frac{1}{2} + C_\gamma^1 \int_0^z \exp\left\{-\left|\left(\frac{\gamma-1}{\gamma}\right)^{(\gamma-1)/\gamma} w\right|^{\gamma/(\gamma-1)}\right\} dw, \end{aligned} \quad (36)$$

and thus

$$\begin{aligned} \Phi_{Z_\gamma}(z) &= \frac{1}{2} + C_\gamma^1 \left(\frac{\gamma}{\gamma-1}\right)^{(\gamma-1)/\gamma} \\ &\quad \times \int_0^{((\gamma-1)/\gamma)^{(\gamma-1)/\gamma} z} \exp\left\{-u^{\gamma/(\gamma-1)}\right\} du. \end{aligned} \quad (37)$$

Substituting the normalizing factor, as in (2), and using (3), we obtain

$$\begin{aligned} \Phi_{Z_\gamma}(z) &= \frac{1}{2} + \frac{\sqrt{\pi}}{2\Gamma\left(((\gamma-1)/\gamma)+1\right)\Gamma\left((2\gamma-1)/(\gamma-1)\right)} \\ &\quad \times \operatorname{Erf}_{\gamma/(\gamma-1)}\left\{\left(\frac{\gamma-1}{\gamma}\right)^{(\gamma-1)/\gamma} z\right\}, \quad z \in \mathbb{R}, \end{aligned} \quad (38)$$

and finally, through (34), we derive (31), which forms (32) through (4). □

It is essential for numeric calculations to express (31) considering positive arguments for Erf. Indeed, through (37), we have

$$\begin{aligned} F_{X_\gamma}(x) &= \frac{1}{2} + \frac{\operatorname{sgn}(\log x-\mu)\sqrt{\pi}}{2\Gamma\left((\gamma-1)/\gamma\right)\Gamma\left(\gamma/(\gamma-1)\right)} \\ &\quad \times \operatorname{Erf}_{\gamma/(\gamma-1)}\left\{\left(\frac{\gamma-1}{\gamma}\right)^{(\gamma-1)/\gamma}\left|\frac{\log x-\mu}{\sigma}\right|\right\}, \end{aligned} \quad (39)$$

while applying (4) into (39) it is obtained that

$$\begin{aligned} F_{X_\gamma}(x) &= \frac{1+\operatorname{sgn}(\log x-\mu)}{2} - \frac{\operatorname{sgn}(\log x-\mu)}{2\Gamma\left((\gamma-1)/\gamma\right)} \\ &\quad \times \Gamma\left(\frac{\gamma-1}{\gamma}, \frac{\gamma-1}{\gamma}\left|\frac{\log x-\mu}{\sigma}\right|^{\gamma/(\gamma-1)}\right). \end{aligned} \quad (40)$$

As the generalized error function Erf_a is defined in (4), through the upper incomplete gamma function $\Gamma(a^{-1}, \cdot)$, series expansions can be used for a more "numerical-oriented" form of (4). Here some expansions of the c.d.f. of the generalized Lognormal distribution are presented.

Corollary 9. *The c.d.f. F_{X_γ} can be expressed in the series expansion form*

$$\begin{aligned} F_{X_\gamma}(x) &= \frac{1}{2} + \frac{((\gamma-1)/\gamma)^{(\gamma-1)/\gamma}}{(2/\gamma)\Gamma\left((\gamma-1)/\gamma\right)}\left(\frac{\log x-\mu}{\sigma}\right) \\ &\quad \times \sum_{k=0}^{\infty} \frac{\left(((1-\gamma)/\gamma)\left|(\log x-\mu)/\sigma\right|^{\gamma/(\gamma-1)}\right)^k}{k!\left[(k+1)\gamma-1\right]}, \\ &\quad x \in \mathbb{R}_+^*. \end{aligned} \quad (41)$$

Proof. Substituting the series expansion form of (6) into (39) and expressing the infinite series using the integer powers k, the series expansion as in (41) is derived. □

Corollary 10. *For the negative-ordered lognormally distributed random variable X_γ with $\gamma = 1/(1-n) \in \mathbb{R}_-$, $n \in \mathbb{N}$, $n \geq 2$, the finite expansion is obtained as*

$$\begin{aligned} F_{X_\gamma}(x) &= \frac{1}{2} + \frac{1}{2}\operatorname{sgn}(\log x-\mu) - \frac{\operatorname{sgn}(\log x-\mu)}{2\exp\left\{n\left|(\log x-\mu)/\sigma\right|^{1/n}\right\}} \\ &\quad \times \sum_{k=0}^{n-1} \frac{n^k}{k!}\left|\frac{\log x-\mu}{\sigma}\right|^{k/n}. \end{aligned} \quad (42)$$

Proof. Applying the following finite expansion form of the upper incomplete gamma function,

$$\Gamma(n,x) = (n-1)!e^{-x}\sum_{k=0}^{n-1}\frac{x^k}{k}, \quad x \in \mathbb{R},\ n \in \mathbb{N}^* = \mathbb{N}\setminus 0, \quad (43)$$

into (40), we readily get (42). □

Example 11. For the (-1)-ordered lognormally distributed X_{-1} (i.e., for $n = 2$), we have

$$\begin{aligned} F_{X_{-1}}(x) &= \frac{1}{2} + \frac{1}{2}\operatorname{sgn}(\log x-\mu) - \operatorname{sgn}(\log x-\mu) \\ &\quad \times \frac{1+2\sqrt{\left|(\log x-\mu)/\sigma\right|}}{2\exp\left\{2\sqrt{\left|(\log x-\mu)/\sigma\right|}\right\}}, \end{aligned} \quad (44)$$

while for the $(-1/2)$-ordered lognormally distributed $X_{-1/2}$ (i.e., for $n=3$), we have

$$F_{X_{-1/2}}(x)=\frac{1}{2}+\frac{1}{2}\operatorname{sgn}(\log x-\mu)-\operatorname{sgn}(\log x-\mu)\times\frac{1+3\sqrt[3]{\left|(\log x-\mu)/\sigma\right|}+9\sqrt[3]{((\log x-\mu)/\sigma)^2}}{2\exp\left\{3\sqrt[3]{\left|(\log x-\mu)/\sigma\right|}\right\}}. \quad (45)$$

Example 12. For the second-ordered Lognormal random variable $X_2\sim\mathscr{LN}_2(\mu,\sigma)$, we immediately derive, from (31), that

$$F_{X_2}(x)=\Phi_{X_2}\left(\frac{\log x-\mu}{\sigma}\right)=\frac{1}{2}+\frac{1}{2}\operatorname{Erf}_2\left(\frac{\log x-\mu}{\sqrt{2}\sigma}\right)=\frac{1}{2}+\frac{1}{2}\operatorname{erf}\left(\frac{\log x-\mu}{\sqrt{2}\sigma}\right); \quad (46)$$

that is, the c.d.f. of the usual Lognormal is derived, as it is expected, due to $\mathscr{LN}_2=\mathscr{LN}$; see Theorem 3.

Example 13. For the infinite-ordered Lognormal $X_{\pm\infty}\sim\mathscr{LN}_{\pm\infty}(\mu,\sigma)$, setting $(\gamma-1)/\gamma=1$, we obtain through (41) and the exponential series expansion that

$$\begin{aligned}F_{X_{\pm\infty}}(x)&=\frac{1}{2}+\frac{1}{2}\sqrt{\pi}\operatorname{Erf}_1\left(\frac{\log x-\mu}{\sigma}\right)\\&=\frac{1}{2}-\frac{1}{2}\operatorname{sgn}(\log x-\mu)\sum_{k=0}^{\infty}\frac{1}{(k+1)!}\left(-\left|\frac{\log x-\mu}{\sigma}\right|\right)^{k+1}\\&=\frac{1}{2}+\frac{1}{2}\operatorname{sgn}(\log x-\mu)-\frac{1}{2}\operatorname{sgn}(\log x-\mu)\times\exp\left\{-\left|\frac{\log x-\mu}{\sigma}\right|\right\},\end{aligned} \quad (47)$$

and hence

$$F_{X_{\pm\infty}}(x)=\begin{cases}\frac{1}{2}e^{\mu/\sigma}x^{1/\sigma}, & x\in(0,e^{\mu}],\\ 1-\frac{e^{\mu/\sigma}}{2x^{1/\sigma}}, & x\in(e^{\mu},+\infty),\end{cases} \quad (48)$$

which is the c.d.f. of the Log-Laplace distribution as in (15). This is expected as $\mathscr{LN}_{\pm\infty}(\mu,\sigma)=\mathscr{LL}(e^{\mu},1/\sigma,1/\sigma)$; see Theorem 3.

It is interesting to mention here that the same result can also be derived through (42), as this finite expansion can be extended for $n=1$, which provides (in limit) the c.d.f. of the infinite-ordered Lognormal distribution.

Example 14. Similarly, for the first-ordered random variable $X_1\sim\mathscr{LN}_1(\mu,\sigma)$, the expansion (41) can be written as

$$\begin{aligned}F_{X_\gamma}(x)&=\frac{1}{2}+\frac{((\gamma-1)/\gamma)^{(\gamma-1)/\gamma}}{2\Gamma(((\gamma-1)/\gamma)+1)}\left(\frac{\log x-\mu}{\sigma}\right)\\&\quad\times\left[1+(\gamma-1)\sum_{k=1}^{\infty}\frac{\left(((1-\gamma)/\gamma)\left|(\log x-\mu)/\sigma\right|^{\gamma/(\gamma-1)}\right)^k}{k!\,[(k+1)\gamma-1]}\right],\end{aligned} \quad (49)$$

and provided that $(\log x-\mu)/\sigma\le 1$, we obtain

$$F_{X_1}(x)=\lim_{\gamma\to 1^+}F_{X_\gamma}(x)=\frac{1}{2}+\frac{\log x-\mu}{2\sigma}(1+0)=\frac{\log x-\mu+\sigma}{2\sigma}, \quad (50)$$

with $F_{X_1}(e^{\mu-\sigma})=0$ and $F_{X_1}(e^{\mu+\sigma})=1$. Therefore,

$$F_{X_1}(x)=\begin{cases}0, & x\in(0,e^{\mu-\sigma}),\\ \frac{1}{2\sigma}(\log x-\mu+\sigma), & x\in[e^{\mu-\sigma},e^{\mu+\sigma}],\\ 1, & x\in(e^{\mu+\sigma},+\infty),\end{cases} \quad (51)$$

coincides with the c.d.f. of the Log-Uniform distribution $\mathscr{LU}(a=e^{\mu-\sigma},b=e^{\mu+\sigma})$ as in (15). This is expected as $\mathscr{LN}_1(\mu,\sigma)=\mathscr{LU}(e^{\mu-\sigma},e^{\mu+\sigma})$; see Theorem 3.

Table 1 provides the probability values $P_{\gamma;1}=\Pr\{X_\gamma\le i\}$, $i=1/2,1,2,\ldots,5$, for various $X_\gamma\sim\mathscr{LN}_\gamma(0,1)$. Notice that $P_{\gamma;1}=1/2$ for all γ values due to the fact that $1=e^{\mu}|_{\mu=0}=\operatorname{Med}X_\gamma$ (see Theorem 8); that is, the point 1 coincides with the γ-invariant median of the $\mathscr{LN}_\gamma(0,1)$ family discussed previously. Moreover, the last two columns provide also the 1st and 3rd quartile points $q_{1;\gamma}$ and $q_{3;\gamma}$ of X_γ; that is, $\Pr\{X_\gamma\le q_{k;\gamma}\}=k/4$, $k=1,3$, for various γ values. These quartiles are evaluated using the quantile function Q_{X_γ} of r.v. X_γ; that is,

$$\begin{aligned}Q_{X_\gamma}(P)&:=\inf\left\{x\in\mathbb{R}_+^*\mid F_{X_\gamma}(x)\ge P\right\}\\&=\exp\left\{\operatorname{sgn}(2P-1)\sigma\times\left[\frac{\gamma}{\gamma-1}\Gamma^{-1}\left(\frac{\gamma-1}{\gamma},|2P-1|\right)\right]^{(\gamma-1)/\gamma}\right\},\\&\qquad P\in(0,1),\end{aligned} \quad (52)$$

for $P=1/4,3/4$, that is derived through (40). The values of the inverse upper incomplete gamma function $\Gamma^{-1}((\gamma-1)/\gamma,\cdot)$ were numerically calculated.

Figure 2 illustrates the c.d.f. F_{X_γ} curves, as in (39), for certain r.v. $X_\gamma\sim\mathscr{LN}_\gamma(0,\sigma)$ and for scale parameters $\sigma=2/3$,

TABLE 1: Probability mass values $P_{\gamma;i} = \Pr\{X_\gamma \le i\}, i = 1/2, 1, 2, \ldots, 5$, and the 1st and 3rd quartiles $q_{1;\gamma}, q_{3;\gamma}$ for various generalized lognormally distributed $X_\gamma \sim \mathscr{LN}_\gamma(0,1)$.

γ	$P_{\gamma;1/2}$	$P_{\gamma;1}$	$P_{\gamma;2}$	$P_{\gamma;3}$	$P_{\gamma;4}$	$P_{\gamma;5}$	$q_{1;\gamma}$	$q_{3;\gamma}$
−50	0.2501	0.5000	0.7499	0.8326	0.8739	0.8987	0.4998	2.0008
−10	0.2505	0.5000	0.7495	0.8297	0.8698	0.8940	0.4990	2.0038
−5	0.2508	0.5000	0.7492	0.8264	0.8652	0.8887	0.4982	2.0071
−2	0.2515	0.5000	0.7485	0.8187	0.8539	0.8756	0.4964	2.0145
−1	0.2521	0.5000	0.7479	0.8097	0.8408	0.8601	0.4945	2.0223
−1/2	0.2524	0.5000	0.7476	0.7989	0.8248	0.8410	0.4925	2.0303
−1/10	0.2528	0.5000	0.7482	0.7757	0.7895	0.7984	0.4986	2.0426
1	0.1534	0.5000	0.8466	1.0000	1.0000	1.0000	0.6065	1.6487
3/2	0.2381	0.5000	0.7619	0.8848	0.9437	0.9721	0.5172	1.9334
2	0.2441	0.5000	0.7559	0.8640	0.9172	0.9462	0.5094	1.9630
3	0.2472	0.5000	0.7528	0.8505	0.8989	0.9267	0.5049	1.9804
4	0.2481	0.5000	0.7519	0.8452	0.8917	0.9188	0.5034	1.9867
5	0.2486	0.5000	0.7514	0.8425	0.8878	0.9145	0.5025	1.9899
10	0.2494	0.5000	0.7506	0.8375	0.8810	0.9068	0.5011	1.9954
50	0.2499	0.5000	0.7501	0.8341	0.8761	0.9013	0.5002	1.9992
$\pm\infty$	0.2500	0.5000	0.7500	0.8333	0.8750	0.9000	0.5000	2.0000

1, 3/2 in the 3 subfigures, respectively. Moreover, the 1st and 3rd quartile points $Q_{X_\gamma}(1/4)$ and $Q_{X_\gamma}(3/4)$ are also depicted (small circles over c.d.f. curves with their corresponding ticks on x-axis).

Theorem 15. *The (non-log-scaled) location parameter e^μ is in fact the geometric mean as well as the median for all generalized lognormally distributed $X_\gamma \sim \mathscr{LN}_\gamma(\mu,\sigma)$. Moreover, this median is also characterized by vanishing median absolute deviation.*

Proof. Considering (39) and the fact that $\mathrm{Erf}_a 0 = 0$, $a \in \mathbb{R}_+^*$, it holds that $\mathrm{Med}\, X_\gamma = F_{X_\gamma}^{-1}(1/2) = e^\mu$. For the geometric mean $(\mu_g)_{X_\gamma} = e^{\mathrm{E}[\log X_\gamma]}$, we readily obtain $(\mu_g)_{X_\gamma} = e^\mu$ as $\log X_\gamma \sim \mathscr{N}_\gamma(\mu,\sigma^2)$ with $\mathrm{E}[X_\gamma] = \mu$. A dispersion measure for the median is the so-called median absolute deviation or MAD, defined by $\mathrm{MAD}(X_\gamma) = \mathrm{Med} \mid X_\gamma - \mathrm{Med}\, X_\gamma \mid$. For $X_\gamma \sim \mathscr{LN}_\gamma(\mu,\sigma)$, we have $X_\gamma - \mathrm{Med} X_\gamma = X_\gamma - e^\mu \sim \mathscr{LN}_\gamma(\mu,\sigma; -e^\mu)$; that is, $X_\gamma - e^\mu$ follows the generalized Lognormal distribution with threshold $-e^\mu$. Furthermore, $|Y|$ is the "folded distribution" case of $Y := X_\gamma - e^\mu$ which is distributed through p.d.f. of the form

$$f_{|Y|}(x) = f_Y(-x) + f_Y(x), \quad x \in \mathbb{R}_+, \tag{53}$$

where f_Y is the p.d.f. of Y. For example, see [17] on the folded normal distribution. However, the density function f_Y is defined in $(-e^\mu, +\infty)$ due to threshold $-e^\mu$, while it vanishes elsewhere; that is,

$$f_{|Y|}(x) = \begin{cases} f_Y(-x) + f_Y(x), & 0 \le x \le e^\mu, \\ f_Y(x), & x > e^\mu. \end{cases} \tag{54}$$

Therefore, the c.d.f. of $|Y|$ is given by

$$F_{|Y|}(x) = \int_0^x f_{|Y|}(t)\,dt = \int_0^{e^\mu} f_Y(-t)\,dt + \int_0^{e^\mu} f_Y(t)\,dt + \int_{e^\mu}^x f_Y(t)\,dt, \quad x \in \mathbb{R}_+. \tag{55}$$

Applying the transformation $w = (1/\sigma)\,[\log(e^\mu - t) - \mu]$, $t < e^\mu$, into the first integral of (55) and $z = (1/\sigma)\,[\log(t+e^\mu)-\mu]$, $t > -e^\mu$, into the other two integrals, we obtain

$$\begin{aligned} F_{|Y|}(x) &= C_\gamma^1 \int_{-\infty}^0 \exp\left\{-\frac{\gamma-1}{\gamma}|w|^{\gamma/(\gamma-1)}\right\} dw + C_\gamma^1 \\ &\quad \times \int_0^{(1/\sigma)\log 2} \exp\left\{-\frac{\gamma-1}{\gamma}|z|^{\gamma/(\gamma-1)}\right\} dz \\ &\quad + C_\gamma^1 \int_{(1/\sigma)\log 2}^{g(x)} \exp\left\{-\frac{\gamma-1}{\gamma}|z|^{\gamma/(\gamma-1)}\right\} dz, \\ g(x) &:= \frac{\log(x+e^\mu)-\mu}{\sigma}, \quad x \in \mathbb{R}_+, \end{aligned} \tag{56}$$

and hence

$$\begin{aligned} F_{|Y|}(x) &= \Phi_Z(0) + \left[\Phi_Z\left(\frac{\log 2}{\sigma}\right) - \Phi_Z(0)\right] \\ &\quad + \left[\Phi_Z(g(x)) - \Phi_Z\left(\frac{\log 2}{\sigma}\right)\right] \\ &= \Phi_Z\left(\frac{\log(x+e^\mu)-\mu}{\sigma}\right), \end{aligned} \tag{57}$$

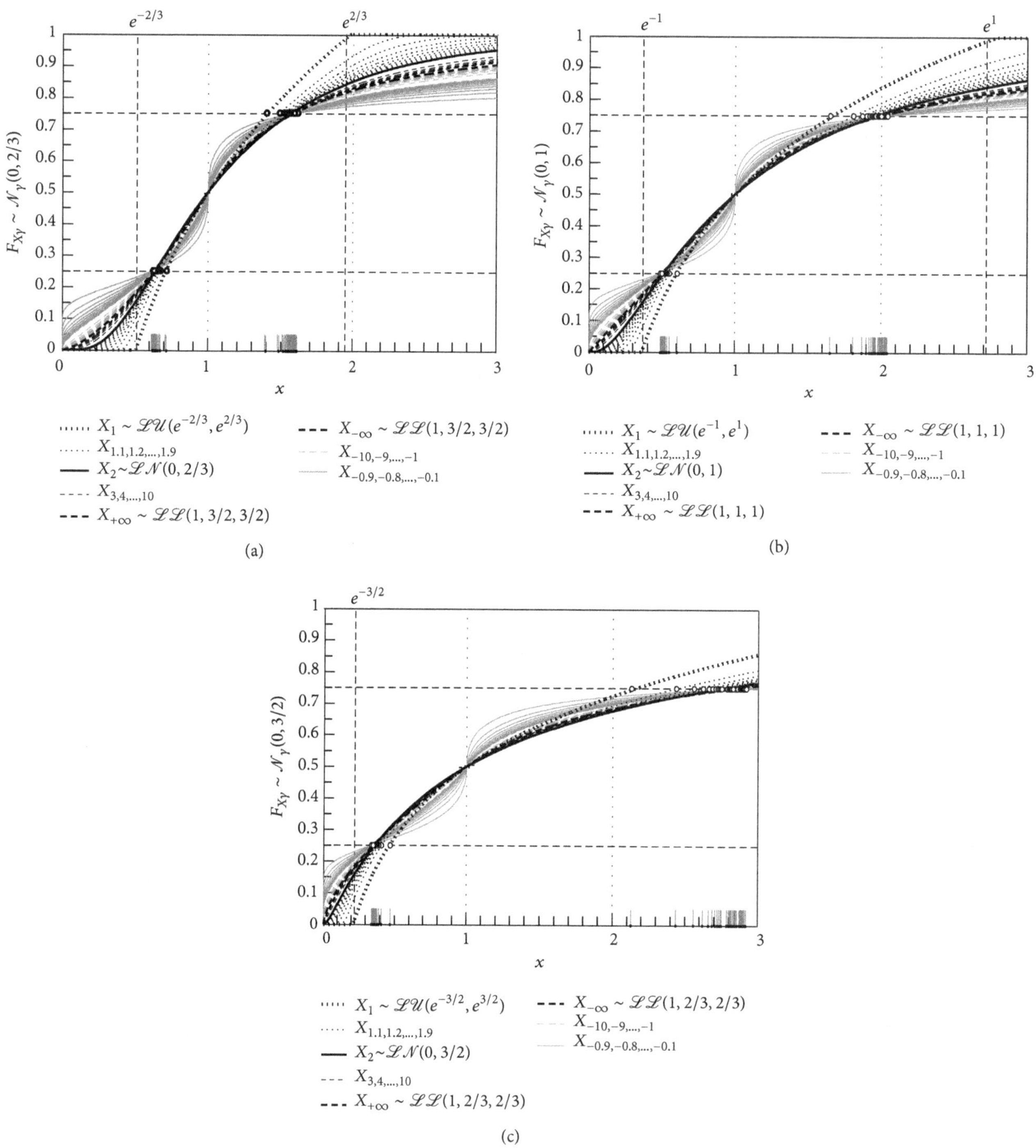

FIGURE 2: Graphs of the c.d.f. F_{X_γ}, $X_\gamma \sim \mathcal{LN}_\gamma(0, \sigma)$, for $\sigma = 2/3, 1, 3/2$, and various γ values.

with Φ_Z being the c.d.f. of the standardized r.v. $Z \sim \mathcal{N}_\gamma(0, 1)$. From (38) and the fact that $\mathrm{Erf}_a 0 = 0$, $a \in \mathbb{R}_+^*$, it is clear that (57) implies $\mathrm{MAD}X_\gamma = \mathrm{Med}|X_\gamma - e^\mu| = F^{-1}_{|X_\gamma - e^\mu|}(1/2) = 0$, for every $X_\gamma \sim \mathcal{LN}_\gamma(\mu, \sigma)$, and the theorem has been proved. □

3. Moments of the γ-Order Lognormal Distribution

For the evaluation of the moments of the generalized Lognormal distribution, the following holds.

Proposition 16. *The tth raw moment $\widetilde{\mu}_X^{(t)}$ of a generalized lognormally distributed random variable $X \sim \mathscr{L}\mathscr{N}_\gamma(\mu, \sigma)$ is given by*

$$\widetilde{\mu}_X^{(t)} = \frac{e^{t\mu}}{\Gamma((\gamma-1)/\gamma)} \sum_{n=0}^{\infty} \frac{(t\sigma)^{2n}}{(2n)!} \left(\frac{\gamma}{\gamma-1}\right)^{2n((\gamma-1)/\gamma)} \times \Gamma\left((2n+1)\frac{\gamma-1}{\gamma}\right) \tag{58}$$

and coincides with the moment generating function of the γ-order normally distributed $\log X$; that is, $M_{\log X}(t) = \widetilde{\mu}_X^{(t)}$.

Proof. From the definition of the tth raw moment $\widetilde{\mu}_X^{(t)}$, we have

$$\widetilde{\mu}_X^{(t)} = \mathrm{E}\left[X^t\right] = \int_{\mathbb{R}_+} x^t f_X(x)\, dx = \frac{1}{\sigma} C_\gamma^1 \int_{\mathbb{R}_+} x^{t-1} \exp\left\{-\frac{\gamma-1}{\gamma}\left|\frac{\log x - \mu}{\sigma}\right|^{\gamma/(\gamma-1)}\right\} dx, \tag{59}$$

and applying the transformation $z = ((\gamma-1)/\gamma)^{(\gamma-1)/\gamma}(1/\sigma)(\log x - \mu)$, $x > 0$, we get

$$\widetilde{\mu}_X^{(t)} = C_\gamma^1 \left(\frac{\gamma}{\gamma-1}\right)^{(\gamma-1)/\gamma} \int_{\mathbb{R}} \exp\left\{t\mu + n\left(\frac{\gamma}{\gamma-1}\right)^{(\gamma-1)/\gamma} \sigma z\right\} \times \exp\left\{-|z|^{\gamma/(\gamma-1)}\right\} dz. \tag{60}$$

Through the exponential series expansion

$$\exp\left\{t\left(\frac{\gamma}{\gamma-1}\right)^{(\gamma-1)/\gamma} \sigma z\right\} = \sum_{n=0}^{\infty} \frac{(t\sigma)^n}{n!} \left(\frac{\gamma}{\gamma-1}\right)^{n((\gamma-1)/\gamma)} z^n, \tag{61}$$

it is obtained that

$$\widetilde{\mu}_X^{(t)} = 2C_\gamma^1 \left(\frac{\gamma}{\gamma-1}\right)^{(\gamma-1)/\gamma} e^{t\mu} \sum_{n=0}^{\infty} \frac{(t\sigma)^{2n}}{(2n)!} \left(\frac{\gamma}{\gamma-1}\right)^{2n((\gamma-1)/\gamma)} \times \int_{\mathbb{R}_+} z^{2n} \exp\left\{-z^{\gamma/(\gamma-1)}\right\} dz. \tag{62}$$

Finally, substituting the normalizing factor C_γ^1 as in (2) into (62) and utilizing the known integral [16],

$$\int_{\mathbb{R}+} x^m e^{-bx^n} dx = \frac{\Gamma((m+1)/n)}{nb^{(m+1)/n}}, \quad n, m, b \in \mathbb{R}_+^*, \tag{63}$$

we obtain (58).

Moreover, for $Y := \log X \sim \mathscr{N}_\gamma(\mu, \sigma^2)$ we have $M_Y(t) = \mathrm{E}[e^{tY}] = \mathrm{E}[X^t] = \widetilde{\mu}_X^{(t)}$, and the proposition has been proved. □

Example 17. For the second-ordered lognormally distributed $X \sim \mathscr{L}\mathscr{N}_2(\mu, \sigma)$, (58) implies

$$\widetilde{\mu}_X^{(t)} = \frac{e^{t\mu}}{\sqrt{\pi}} \sum_{n=0}^{\infty} \frac{\left(2t^2\sigma^2\right)^n}{(2n)!} \Gamma\left(n + \frac{1}{2}\right), \quad t \in \mathbb{R}_+, \tag{64}$$

and through the gamma function identity

$$\Gamma\left(n + \frac{1}{2}\right) = \frac{(2n)!}{2^{2n} n!} \sqrt{\pi}, \quad n \in \mathbb{N}, \tag{65}$$

we have

$$\widetilde{\mu}_X^{(t)} = e^{t\mu} \sum_{n=0}^{\infty} \frac{(t\sigma)^{2n}}{2^n n!} = e^{t\mu} \sum_{n=0}^{\infty} \frac{1}{n!}\left(\frac{1}{2} t^2 \sigma^2\right)^n = e^{t\mu + (1/2)(t\sigma)^2}, \quad t \in \mathbb{R}_+, \tag{66}$$

which is the tth raw moment of the usual lognormally distributed $X \sim \mathscr{L}\mathscr{N}(\mu, \sigma)$, with mean $\mu_X := \widetilde{\mu}_X^{(1)} = \mathrm{E}[X] = \exp\{\mu + (1/2)\sigma^2\}$. This is true as $M_{\log X}(t) = \widetilde{\mu}_X^{(t)} = \exp\{t\mu + (1/2)(t\sigma)^2\}$ is the known moment-generating function of the normally distributed $\log X \sim \mathscr{N}(\mu, \sigma^2)$.

Theorem 18. *The kth central moment (about the mean) $\mu_X^{(t)}$ of a generalized lognormally distributed random variable $X \sim \mathscr{L}\mathscr{N}_\gamma(\mu, \sigma)$ is given by*

$$\mu_X^{(k)} = \frac{e^{k\mu}}{\Gamma((\gamma-1)/\gamma)} \sum_{n=0}^{k} \binom{k}{n} \left(-\frac{\mu_X}{e^\mu}\right)^n S_{k-n}, \quad k \in \mathbb{N}, \tag{67}$$

where

$$S_k = \sum_{m=0}^{\infty} \frac{(k\sigma)^{2m}}{(2m)!} \left(\frac{\gamma}{\gamma-1}\right)^{2m((\gamma-1)/\gamma)} \Gamma\left((2m+1)\frac{\gamma-1}{\gamma}\right), \quad k \in \mathbb{N}. \tag{68}$$

Proof. From the definition of the kth central moment $\mu_X^{(k)}$ we have

$$\mu_X^{(k)} := \mathrm{E}\left[(X - \mu_X)^k\right] = \int_{\mathbb{R}_+} (x - \mu_X)^k f_X(x; \mu, \sigma, \gamma)\, dx, \tag{69}$$

while using the binomial identity we get

$$\mu_X^{(k)} = \sum_{n=0}^{k} \binom{k}{n} (-\mu_X)^n \int_{\mathbb{R}_+} x^{k-n} f_X(x)\, dx = \sum_{n=0}^{k} \binom{k}{n} (-\mu_X)^n \widetilde{\mu}_X^{(k-n)}. \tag{70}$$

Applying Proposition 16, (70) implies that

$$\mu_X^{(k)} = \frac{e^{k\mu}}{\Gamma((\gamma-1)/\gamma)} \sum_{n=0}^{k} \binom{k}{n} \left(-\frac{\mu_X}{e^\mu}\right)^n \sum_{m=0}^{\infty} \frac{[(k-n)\sigma]^{2m}}{(2m)!} \times \left(\frac{\gamma}{\gamma-1}\right)^{2m((\gamma-1)/\gamma)} \Gamma((2m+1)((\gamma-1)/\gamma)), \tag{71}$$

while taking the summation index n until $k-1$, we finally obtain (67), and the theorem has been proved. □

Example 19. Recall Example 17. Substituting (66) and the mean $\mu_X = e^{\mu+(1/2)\sigma^2}$ into (70), the second-ordered lognormally distributed $X \sim \mathscr{LN}_2(\mu,\sigma)$ provides

$$\mu_X^{(k)} = \sum_{n=0}^{k} \binom{k}{n} (-1)^n e^{k\mu+(1/2)[n+(k-n)^2]\sigma^2}, \quad k \in \mathbb{N}, \tag{72}$$

while

$$\sigma_X^2 := \operatorname{Var}[X] = \mu_X^{(2)} = e^{2\mu+\sigma^2}\left(e^{\sigma^2}-1\right), \tag{73}$$

which are the kth central moment and the variance, respectively, of the usual lognormally distributed $X \sim \mathscr{LN}(\mu,\sigma)$. The same result can be derived directly through (67) for $\gamma = 2$ and the use of the known gamma function identity, as in (65).

Theorem 20. *The mean $\mu_X := \mathrm{E}[X]$, variance $\sigma_X^2 := \operatorname{Var}[X]$, coefficient of variation CV_X, skewness λ_X and kurtosis κ_X of the generalized lognormally distributed $X \sim \mathscr{LN}_\gamma(\mu,\sigma)$ are, respectively, given by*

$$\mu_X = \frac{e^{\mu}}{\Gamma((\gamma-1)/\gamma)} S_1, \tag{74}$$

$$\sigma_X^2 = -\mu_X^2 + \frac{e^{2\mu}}{\Gamma((\gamma-1)/\gamma)} S_2, \tag{75}$$

$$CV_X^2 = \Gamma\left(\frac{\gamma-1}{\gamma}\right)\frac{S_2}{S_1^2} - 1, \tag{76}$$

$$\lambda_X = -CV_X^{-3} - CV_X^{-1} + \frac{e^{3\mu}}{\sigma_X^3\Gamma((\gamma-1)/\gamma)} S_3, \tag{77}$$

$$\kappa_X = -CV_X^{-4} - 6CV_X^{-2} - 4\frac{\lambda_X}{CV_X} + \frac{e^{4\mu}}{\sigma_X^4\Gamma((\gamma-1)/\gamma)} S_4, \tag{78}$$

where the sums S_i, $i = 1,\ldots,4$, are given by (68).

Proof. From Proposition 16 we easily obtain (74), as $\mu_X := \tilde{\mu}_X^{(1)}$. From Theorem 18 we have

$$\sigma_X^2 := \mu_X^{(2)} = \mu_X^2 + \left[\Gamma\left(\frac{\gamma-1}{\gamma}\right)\right]^{-1}\left(e^{2\mu}S_2 - 2e^{\mu}\mu_X S_1\right). \tag{79}$$

Hence, substituting S_1 from (74), (75) holds. Moreover, the squared coefficient of variation is readily obtained via (75) and (74). By definition, skewness λ_X is the standardized third (central) moment; that is, $\lambda_X := \operatorname{Skew}[X] = \mu_X^{(3)}/\sigma_X^3$. Theorem 18 provides that

$$\lambda_X = -CV_X^{-3} + \left[\sigma_X^3\Gamma\left(\frac{\gamma-1}{\gamma}\right)\right]^{-1} \times \left(e^{3\mu}S_3 - 3e^{2\mu}\mu_X S_2 + 3e^{\mu}\mu_X S_1\right). \tag{80}$$

Substituting S_1 and S_2 from (74) and (75), we obtain (77). Finally, kurtosis κ_X is (by definition) the standardized fourth (central) moment; that is, $\kappa_X := \operatorname{Kurt}[X] = \mu_X^{(4)}/\sigma_X^4$, which provides, through Theorem 18, that

$$\kappa_X = CV_X^{-4} + \left[\sigma_X^4\Gamma\left(\frac{\gamma-1}{\gamma}\right)\right]^{-1} \times \left(e^{4\mu}S_4 - 4e^{3\mu}\mu_X S_3 + 6e^{2\mu}\mu_X^2 S_2 - 4e^{\mu}\mu_X^3 S_1\right). \tag{81}$$

Substituting S_i, $i = 1,2,3$, from (74), (75), and (77), we obtain (78). □

Example 21. For the second-ordered lognormally distributed $X \sim \mathscr{LN}_2(\mu,\sigma)$, utilizing (65) into (68) we get $S_n = \sqrt{\pi}e^{(n^2\sigma^2)/2}$, $n \in \mathbb{N}^*$. Applying this to Theorem 20 we derive (after some algebra)

$$\mu_X = e^{\mu+(1/2)\sigma^2}, \qquad \sigma_X^2 = e^{2\mu+\sigma^2}\left(e^{\sigma^2}-1\right),$$
$$CV_X = \sqrt{e^{\sigma^2}-1},$$
$$\lambda_X = \left(e^{\sigma^2}+2\right)\sqrt{e^{\sigma^2}-1}, \qquad \kappa_X = e^{4\sigma^2} + 2e^{3\sigma^2} + 3e^{2\sigma^2} - 3, \tag{82}$$

which are the mean, variance, coefficient of variation, skewness, and kurtosis, respectively, of usual lognormally distributed $X \sim \mathscr{LN}(\mu,\sigma)$.

For the usual lognormally distributed random variable $X \sim \mathscr{LN}$, it is known that $\operatorname{Mode} X < \operatorname{Med} X < \mu_X$. The following corollary examines this inequality for the $\mathscr{LN}_\gamma$ family of distributions.

Corollary 22. *For the γ-ordered lognormally distributed $X_\gamma \sim \mathscr{LN}_\gamma(\mu,\sigma)$, it is true that $\operatorname{Mode} X_\gamma \le \operatorname{Med} X_\gamma = (\mu_g)_{X_\gamma} \le \mu_{X_\gamma}$. The first equality holds for the Log-Laplace distributed $X_{+\infty}$ with $\sigma < 1$ as well as for all the negative-ordered $X_{\gamma<0}$ where $\operatorname{Mode} X_\gamma$ is considered to be the local (nonsmooth) mode point of X_γ. The second equality holds for the degenerate Dirac case of X_0.*

Proof. From (74) and Theorem 15 we have

$$\operatorname{Med} X_\gamma = \left(\mu_g\right)_{X_\gamma} = e^{\mu} < \mu_{X_\gamma}, \tag{83}$$

for every $X_\gamma \sim \mathscr{LN}_\gamma(\mu,\sigma)$. The above inequality becomes equality for the limiting Dirac case of X_0. For the relation between the mode and the median of X_γ, the following cases are considered.

(i) The positive-ordered Lognormal case $\gamma > 1$: from (20) we have

$$\operatorname{Mode} X_\gamma = e^{\mu-\sigma^\gamma} < e^{\mu} = \operatorname{Med} X_\gamma. \tag{84}$$

For the Log-Laplace case of $X_{+\infty}$, it holds

$$\operatorname{Mode} X_\gamma = e^{\mu} = \operatorname{Med} X_\gamma, \tag{85}$$

provided that $\sigma < 1$, while for $\sigma \geq 1$ we have

$$\text{Mode}\, X_\gamma = 0 < e^{\mu} = \text{Med}\, X_\gamma. \quad (86)$$

For $\sigma = 1$, the inequality (84) clearly holds.

(ii) The negative-ordered Lognormal case $\gamma < 0$: from Proposition 4 the inequality as in (86) holds. Moreover, if Mode X_γ is considered as the nonsmooth local mode point of the negative-ordered X_γ then the equality as in (85) holds.

From the above cases and (83), the corollary holds true. □

Corollary 23. *The raw and central moments of a Log-Uniformly distributed random variable $X \sim \mathscr{LU}(a,b)$, $0 < a < b$, are given by*

$$\widetilde{\mu}_X^{(t)} = \frac{b^t - a^t}{t\log(b/a)}, \quad t \in \mathbb{R}, \quad (87)$$

$$\mu_X^{(k)} = \frac{(a-b)^k}{\log^k(b/a)} + \frac{1}{\log(b/a)} \times \sum_{n=0}^{k-1}\binom{k}{n}\frac{(a-b)^n\left(b^{k-n} - a^{k-n}\right)}{(k-n)\log^n(b/a)}, \quad k \in \mathbb{N}, \quad (88)$$

respectively, while the mean, variance, coefficient of variation, skewness, and kurtosis of X are given, respectively, by

$$\mu_X = \frac{b-a}{\log(b/a)}, \quad (89)$$

$$\sigma_X^2 = \frac{(b-a)^2}{\log^2(b/a)} + \frac{(b-a)(b+a)}{2\log(b/a)}, \quad (90)$$

$$CV_X = \sqrt{1 + \frac{b+a}{2(b-a)}\log\frac{b}{a}}, \quad (91)$$

$$\lambda_X = \frac{1}{\sigma_X^3}\left[\frac{b^3 - a^3}{3\log(b/a)} - 3\frac{(b-a)^2(b+a)}{2\log^2(b/a)} + 2\frac{(b-a)^3}{\log^3(b/a)}\right], \quad (92)$$

$$\kappa_X = \frac{1}{\sigma_X^4}\left[\frac{b^4 - a^4}{4\log(b/a)} - 4\frac{(b-a)\left(b^3 - a^3\right)}{3\log^2(b/a)} + 3\frac{(b-a)^3(b+a)}{\log^3(b/a)} - 3\frac{(b-a)^4}{\log^4(b/a)}\right]. \quad (93)$$

Proof. Recall Proposition 16 with $X_\gamma \sim \mathscr{LN}_\gamma(\mu,\sigma)$. Through the gamma function additive identity (58) can be written as

$$\widetilde{\mu}_{X_\gamma}^{(t)} = \frac{e^{t\mu}}{t\sigma\Gamma((\gamma-1)/\gamma+1)}\sum_{m=0}^{\infty}\frac{(t\sigma)^{2m+1}}{(2m+1)!}\left(\frac{\gamma}{\gamma-1}\right)^{2m((\gamma-1)/\gamma)} \times\Gamma\left((2m+1)\frac{\gamma-1}{\gamma}+1\right). \quad (94)$$

Thus, letting $X := X_1 \sim \mathscr{LN}_1(\mu,\sigma) = \mathscr{LU}(a,b)$ with $\mu = (1/2)\log(ab)$ and $\sigma = (1/2)\log(b/a)$, it holds (recall the exponential odd series expansion) that

$$\widetilde{\mu}_X^{(t)} = \lim_{\gamma\to1^+}\mu_{X_\gamma}^{(t)} = \frac{e^{t\mu}}{t\sigma}\sum_{m=0}^{\infty}\frac{(t\sigma)^{2m+1}}{(2m+1)!} = \frac{e^{t(\mu+\sigma)} - e^{t(\mu-\sigma)}}{2t\sigma}, \quad t \in \mathbb{R}, \quad (95)$$

and hence (87) holds. Moreover, $\mu_X := \mu_X^{(1)} = \mathrm{E}[X] = (1/2\sigma)(e^{\mu+\sigma} - e^{\mu-\sigma})$, and therefore (89) holds.

Working similarly, (67) implies

$$\mu_X^{(k)} = \lim_{\gamma\to1^+}\mu_{X_\gamma}^{(k)} = e^{k\mu}\sum_{n=0}^{k}\binom{k}{n}\left(-\frac{\mu_X}{e^{\mu}}\right)^n\sum_{m=0}^{\infty}\frac{[(k-n)\sigma]^{2m}}{(2m+1)!}, \quad k \in \mathbb{N}. \quad (96)$$

Using the exponential odd series expansion, the above expansion becomes

$$\mu_X^{(k)} = \frac{e^{k\mu}}{2\sigma}\sum_{n=0}^{k}\binom{k}{n}\left(-\frac{\mu_X}{e^{\mu}}\right)^n\frac{e^{(k-n)\sigma} - e^{-(k-n)\sigma}}{(k-n)}, \quad k \in \mathbb{N}, \quad (97)$$

and, through (89), we obtain (88). Moreover, for $k = 2$, $\sigma_X^2 := \mathrm{Var}[X] = \mu_X^{(2)} - \mu_X^2$ implies (90), and hence (91) also holds. For $k = 3$ and $k = 4$, through $\mu_X^{(3)}$ and $\mu_X^{(4)}$, we obtain (92) and (93), respectively. □

Corollary 24. *The raw and central moments of a Log-Laplace distributed random variable $X \sim \mathscr{LL}(\mu,\sigma,\sigma)$ are given by*

$$\widetilde{\mu}_X^{(t)} = \frac{\mu^t\sigma^2}{\sigma^2 - t^2} > \mu^t, \quad \sigma > t,\ t \in \mathbb{R}, \quad (98)$$

$$\mu_X^{(k)} = \mu^k\sum_{n=0}^{k}\binom{k}{n}\frac{\sigma^{2(n+1)}}{(1-\sigma^2)^n\left[\sigma^2 - (k-n)^2\right]}, \quad \sigma > k,\ k \in \mathbb{N}. \quad (99)$$

The mean, variance, coefficient of variation, skewness, and kurtosis of X are given, respectively, by

$$\mu_X = \frac{\mu\sigma^2}{\sigma^2 - 1} > \mu, \quad \sigma > 1, \quad (100)$$

$$\sigma_X^2 = \frac{\mu^2\sigma^2\left(2\sigma^2 + 1\right)}{(\sigma^2 - 4)(\sigma^2 - 1)^2}, \quad \sigma > 2, \quad (101)$$

$$CV_X = \frac{1}{\sigma}\sqrt{\frac{2\sigma^2 + 1}{\sigma^2 - 4}}, \quad \sigma > 2, \quad (102)$$

$$\lambda_X = \frac{2\left(15\sigma^4 + 7\sigma^2 + 2\right)}{\sigma(\sigma^2 - 9)}\sqrt{\frac{\sigma^2 - 4}{(2\sigma^2 + 1)^3}}, \quad \sigma > 3, \quad (103)$$

$$\kappa_X = \frac{3\left(8\sigma^8 + 212\sigma^6 + 95\sigma^4 + 33\sigma^2 + 12\right)\left(\sigma^2 - 4\right)}{\left(\sigma^2 - 16\right)\left(\sigma^2 - 9\right)\left(2\sigma^2 + 1\right)^2}, \quad \sigma > 4. \tag{104}$$

Proof. Let $X_\gamma \sim \mathscr{LL}_\gamma(\mu,\sigma,\sigma) = \mathscr{LN}_\gamma(\log\mu, 1/\sigma, 1/\sigma)$. For $\gamma = \pm\infty$, that is, $\gamma/(\gamma-1) = 1$, the raw moments as in (58) provide

$$\widetilde{\mu}_X^{(t)} = \widetilde{\mu}_{X_{\pm\infty}}^{(t)} = \mu^t \sum_{k=0}^{\infty}\left(\frac{t}{\sigma}\right)^{2k}, \quad t \in \mathbb{R}, \tag{105}$$

as $X = X_{\pm\infty}$, while through the even geometric series expansion, it is

$$\widetilde{\mu}_{X_{\pm\infty}}^{(t)} = \frac{1}{2}\mu^t\left[\sum_{k=0}^{\infty}\left(\frac{t}{\sigma}\right)^k + \sum_{k=0}^{\infty}\left(-\frac{t}{\sigma}\right)^k\right] = \frac{1}{2}\mu^t\left(\frac{\sigma}{\sigma-t} + \frac{\sigma}{\sigma+t}\right), \tag{106}$$

provided that $\sigma > t$, and hence (98) holds. Moreover, $\mu_X := \widetilde{\mu}_X^{(1)} = \mathrm{E}[X]$, and hence (100) holds.

Working similarly, (67) implies

$$\mu_X^{(k)} = \mu^k\sigma^2 \sum_{n=0}^{k}\binom{k}{n}\frac{\left(-\mu_X/\mu\right)^n}{\sigma^2 - (k-n)^2}, \quad k \in \mathbb{N}, \tag{107}$$

provided $\sigma > k$, and hence, through (100), the central moments (99) are obtained.

Moreover, for $k = 2$ and due to $\sigma_X^2 := \mathrm{Var}[X] = \widetilde{\mu}_X^{(2)} - \mu_X^2$, (101) holds true, while for $k = 3$ and $k = 4$ we derive, through $\mu_X^{(3)}$ and $\mu_X^{(4)}$, (103) and (104), respectively. □

Example 25. For a uniformly distributed r.v. $U \sim \mathscr{U}(a,b) = \mathscr{N}_1(\mu,\sigma)$ with $a = \mu - \sigma$ and $b = \mu + \sigma$, it holds that $\mathrm{LU} := e^U \sim \mathscr{LU}(e^{\mu-\sigma}, e^{\mu+\sigma})$ due to Theorem 3, and therefore LU is a Log-Uniform distributed r.v. as $\mathrm{LU} \sim \mathscr{LU}(e^a, e^b)$. Applying (87), the known moment-generating function of the uniformly distributed $U \sim \mathscr{U}(a,b)$ is derived; that is, $M_U(t) := \mathrm{E}[e^{tU}] = \widetilde{\mu}_{\mathrm{LU}}^{(t)} = (e^{tb} - e^{ta})(1/t(b-a))$.

Similarly, for a Laplace distributed r.v. $L \sim \mathscr{L}(\mu,\sigma) = \mathscr{N}_{\pm\infty}(\mu,\sigma)$, it holds that $\mathrm{LL} := e^L \sim \mathscr{LL}(e^\mu, 1/\sigma, 1/\sigma)$ due to Theorem 3, and therefore LL is a Log-Laplace distributed random variable. Applying (98), we derive the known moment-generating function of the Laplace distributed $L \sim \mathscr{L}(\mu,\sigma)$; that is, $M_L(t) := \mathrm{E}[e^{tL}] = \widetilde{\mu}_{\mathrm{LL}}^{(t)} = e^{t\sigma}(1 - t^2\sigma^2)^{-1}$.

4. Conclusion

The family of the γ-order Lognormal distributions was introduced, which under certain values of γ includes the Log-Uniform, Lognormal, and Log-Laplace distributions as well as the degenerate Dirac distribution. The shape of these distributions for positive and negative shape parameters γ as well as the cumulative distribution functions, was extensively discussed and evaluated through corresponding tables and figures. Moreover, a thorough study of moments was carried out, in which nonclosed forms as well as approximations were obtained and investigated in various examples. This generalized family of distributions derived through the family of the γ-order normal distribution is based on a strong theoretical background as the logarithmic Sobolev inequalities provide. Further examinations and calculations can be produced while an application to real data is upcoming.

Acknowledgment

The authors would like to thank the referee for his valuable comments that helped improve the quality of this paper.

References

[1] E. L. Crow and K. Shimizu, *Lognormal Distributions*, Marcel Dekker, New York, NY, USA, 1988.

[2] A. Parravano, N. Sanchez, and E. J. Alfaro, "The Dependence of prestellar core mass distributions on the structure of the parental cloud," *The Astrophysical Journal*, vol. 754, no. 2, article 150, 2012.

[3] F. Bernardeau and L. Kofman, "Properties of the cosmological density distribution function," *Monthly Notices of the Royal Astronomical Society*, vol. 443, pp. 479–498, 1995.

[4] P. Blasi, S. Burles, and A. V. Olinto, "Cosmological magnetic field limits in an inhomogeneous Universe," *The Astrophysical Journal Letters*, vol. 514, no. 2, pp. L79–L82, 1999.

[5] F. S. Kitaura, "Non-Gaussian gravitational clustering field statistics," *Monthly Notices of the Royal Astronomical Society*, vol. 420, no. 4, pp. 2737–2755, 2012.

[6] G. Yan and F. B. Hanson, "Option pricing for a stochastic-volatility jump-diffusion model with log-uniform jump-amplitudes," in *Proceedings of the American Control Conference*, 2006.

[7] T. J. Kozubowski and K. Podgórski, "Asymmetric Laplace distributions," *The Mathematical Scientist*, vol. 25, no. 1, pp. 37–46, 2000.

[8] T. J. Kozubowski and K. Podgórski, "Asymmetric Laplace laws and modeling financial data," *Mathematical and Computer Modelling*, vol. 34, no. 9–11, pp. 1003–1021, 2001.

[9] M. Geraci and M. Bottai, "Quantile regression for longitudinal data using the asymmetric Laplace distribution," *Biostatistics*, vol. 8, no. 1, pp. 140–154, 2007.

[10] D. B. Madan, "The variance gamma process and option pricing," *The European Financial Review*, vol. 2, pp. 79–105, 1998.

[11] C. P. Kitsos and N. K. Tavoularis, "Logarithmic Sobolev inequalities for information measures," *IEEE Transactions on Information Theory*, vol. 55, no. 6, pp. 2554–2561, 2009.

[12] C. P. Kitsos and N. K. Tavoularis, "New entropy type information measures," in *Proceedings of the Information Technology Interfaces (ITI '09)*, Cavtat, Croatia, June 2009.

[13] C. P. Kitsos and T. L. Toulias, "New information measures for the generalized normal distribution," *Information*, vol. 1, no. 1, pp. 13–27, 2010.

[14] C. P. Kitsos and T. L. Toulias, "Evaluating information measures for the -order Multivariate Gaussian," in *Proceedings by IEEE of the 14th Panhellenic Conference on Informatics (PCI '10)*, pp. 153–157, Tripoli, Greece, September 2010.

[15] C. P. Kitsos, T. L. Toulias, and P. C. Trandafir, "On the multivariate γ-ordered normal distribution," *Far East Journal of Theoretical Statistics*, vol. 38, no. 1, pp. 49–73, 2012.

[16] I. S. Gradshteyn and I. M. Ryzhik, *Table of Integrals, Series, and Products*, Elsevier, 2007.

[17] F. C. Leone, L. S. Nelson, and R. B. Nottingham, "The folded normal distribution," *Technometrics*, vol. 3, pp. 543–550, 1961.

The Complementary Exponentiated Exponential Geometric Lifetime Distribution

Francisco Louzada,[1] Vitor Marchi,[2] and James Carpenter[3]

[1] *Department of Applied Mathematics and Statistics, ICMC, University of São Paulo, 13560-970 São Carlos, SP, Brazil*
[2] *Department of Statistics, Federal University of São Paulo, 13565-905 São Carlos, SP, Brazil*
[3] *London School of Hygiene and Tropical Medicine, University of London, Keppel Street, London WC1E 7HT, UK*

Correspondence should be addressed to Francisco Louzada; louzada@icmc.usp.br

Academic Editor: Gauss M. Cordeiro

We proposed a new family of lifetime distributions, namely, complementary exponentiated exponential geometric distribution. This new family arises on a latent competing risk scenario, where the lifetime associated with a particular risk is not observable but only the maximum lifetime value among all risks. The properties of the proposed distribution are discussed, including a formal proof of its probability density function and explicit algebraic formulas for its survival and hazard functions, moments, rth moment of the ith order statistic, mean residual lifetime, and modal value. Inference is implemented via a straightforwardly maximum likelihood procedure. The practical importance of the new distribution was demonstrated in three applications where our distribution outperforms several former lifetime distributions, such as the exponential, the exponential-geometric, the Weibull, the modified Weibull, and the generalized exponential-Poisson distribution.

1. Introduction

Several new classes of models have been introduced in recent years grounded in the simple exponential distribution. The main idea is to propose lifetime distributions which can accommodate practical applications where the underlying hazard functions are nonconstant, presenting monotone shapes, since the exponential distribution does not provide a reasonable fit in such situations. For instance, we can cite [1], which proposed a variation of the exponential distribution, the exponential geometric (EG) distribution, with decreasing hazard function, [2], which introduced the exponentiated exponential distribution as a generalization of the usual exponential distribution, which can accommodate data with increasing and decreasing hazard functions, [3], which proposed a generalized exponential distribution, which can accommodate data with increasing and decreasing hazard functions, [4], which proposed the exponentiated type distributions extending the Fréchet, gamma, Gumbel, and Weibull distributions, [5], which proposed another modification of the exponential distribution with decreasing hazard function, [6], which generalizes the distribution proposed by [5] by including a power parameter in this distribution, which can accommodate increasing, decreasing, and unimodal hazard functions, [7], which proposed the Poisson-exponential distribution, and [8], which proposed the complementary exponential geometric distribution, which is complementary to the exponential geometric distribution proposed by [1]. The last two proposed distributions accommodate increasing hazard functions.

In this paper, following [8], we propose a new distribution family by extending the exponentiated exponential distribution [2] by compounding it with a geometric distribution, hereafter the complementary exponentiated exponential geometric distribution or simplistically the CE2G distribution. The new distribution genesis is stated on a complementary risk problem base [9] in presence of latent risks, in the sense that there is no information about which factor was responsible for the component failure and only the maximum lifetime value among all risks is observed. This family have one shape and two scale parameters accommodating increasing, decreasing, and bathtub failure rates.

The paper is organized as follows. In Section 2 we introduce the new CE2G distribution, derive the expressions for the probability density, survival, and hazard functions and the pth quantile, and present its genesis. In Section 3 we present some of its properties, such as its characteristic function, rth raw moment, mean and variance, order statistics, rth moment of the ith order statistic, mean residual lifetime, and modal value. In Section 8 we present the inferential procedure. In Section 10 the practical importance of the new distribution was demonstrated in three applications where our distribution outperforms several former lifetime distributions, such as the exponential, the exponential-geometric, the Weibull, the modified Weibull, and the generalized exponential Poisson distribution. Some final comments in Section 11 conclude the paper.

2. The CE2G Model

Let Y be a nonnegative random variable denoting the lifetime of a component in some population. The random variable Y is said to have a CE2G distribution with parameters $\lambda > 0$, $\alpha > 0$, and $0 < \theta < 1$ if its probability density function (pdf) is given by

$$f(y) = \frac{\alpha\lambda\theta e^{-\lambda y}\left(1 - e^{-\lambda y}\right)^{\alpha-1}}{\left[1 - (1-\theta)\left(1 - e^{-\lambda y}\right)^{\alpha}\right]^{2}}, \quad y > 0, \qquad (1)$$

where λ is a scale parameter of the distribution, and α and θ are shape parameters. Figure 1(a) shows the CE2G probability density function for $\lambda = 1$, $\theta = 0.05, 0.5, 0.95$, and $\alpha = 0.3, 1.0, 3$ and we can see that the function can be decreasing or unimodal.

The survival function of a CE2G distributed random variable is given by

$$S(y) = \frac{1 - \left(1 - e^{-\lambda y}\right)^{\alpha}}{1 - (1-\theta)\left(1 - e^{-\lambda y}\right)^{\alpha}}, \quad y > 0, \qquad (2)$$

where, $\alpha > 0, \theta \in (0,1)$, and $\lambda > 0$.

From (2) and (1), the failure rate function, according to the relationship $h(y) = f(y)/S(y)$, is given by

$$h(y) = \frac{\alpha\lambda\theta e^{-\lambda y}\left(1 - e^{-\lambda y}\right)^{\alpha-1}}{\left[1 - \left(1 - e^{-\lambda y}\right)^{\alpha}\right]\left[1 - (1-\theta)\left(1 - e^{-\lambda y}\right)^{\alpha}\right]}. \qquad (3)$$

The initial value is not finite if $\alpha < 1$ and otherwise is given by $h(0) = \lambda\theta$ if $\alpha = 1$ or $h(0) = 0$ if $\alpha > 1$ and the long-term hazard function value is $h(\infty) = \lambda$. The failure rate (3) can be increasing, decreasing, or bathtub as shown in Figure 1(b), which shows some failure rate function shapes to $\lambda = 1, \theta = 0.05, 0.5, 0.95$, and $\alpha = 0.3, 1.0, 3$.

The pth quantile of the CE2G distribution is given by

$$Q(u) = F^{-1}(u) = -\frac{\ln\left(1 - \left(u/\left(\theta(1-u) + u\right)\right)^{1/\alpha}\right)}{\lambda}, \qquad (4)$$

where u has the uniform $U(0,1)$ distribution and $F(y) = 1 - S(y)$ is the distribution function of Y.

Consider that in the study of reliability we can observe only the maximum component lifetime for each component among all risks. On many occasions, the information about what risk produces the dead of the component in analysis is not available or it is impossible that the true cause of failure is specified. Complementary risks (CR) problems arise in several areas and an extensive literature is available. Interested readers can see [10–12].

Then, in this context, our model can be derived as follows. Let M be a random variable denoting the number of failure causes, $m = 1, 2, \ldots$ and considering M with geometrical probability distribution given by

$$P(M = m) = \theta(1-\theta)^{m-1}, \qquad (5)$$

where $0 < \theta < 1$ and $M = 1, 2, \ldots$.

Also consider t_i, $i = 1, 2, 3, \ldots$ realizations of a random variable denoting the failure times, that is, the time-to-event due to the ith CR and, from [2], T_i has an exponentiated exponential probability distribution with parameters λ and α, given by

$$\begin{aligned} f(t_i; \lambda, \alpha) &= \alpha g(t_i; \lambda) G(t_i; \lambda) \\ &= \alpha\lambda \exp\{-\lambda t_i\}\left(1 - \exp\{-\lambda t_i\}\right)^{\alpha-1}, \end{aligned} \qquad (6)$$

where $g(\cdot)$ and $G(\cdot)$ are the pdf and df, respectively, of the exponential distribution with parameter λ.

In the latent complementary risks scenario, the number of causes M and the lifetime t_j associated with a particular cause are not observable (latent variables), but only the maximum lifetime Y among all causes is usually observed. So, we only observe the random variable given by

$$Y = \max\{T_1, T_2, \ldots, T_M\}. \qquad (7)$$

The following result shows that the random variable Y has probability density function given by (1).

Proposition 1. *If the random variable Y is defined as (7), then, considering (5) and (6), Y is distributed according to a CE2G distribution, with probability density function given by (1).*

Proof. The conditional density function of (7) given $M = m$ is given by

$$\begin{aligned} f(y \mid M = m, \lambda, \alpha) &= m\alpha\lambda e^{-\lambda y}\left(1 - e^{-\lambda y}\right)^{\alpha-1}\left[\left(1 - e^{-\lambda y}\right)^{\alpha}\right]^{m-1}; \\ &\qquad t > 0,\ m = 1, \ldots \end{aligned} \qquad (8)$$

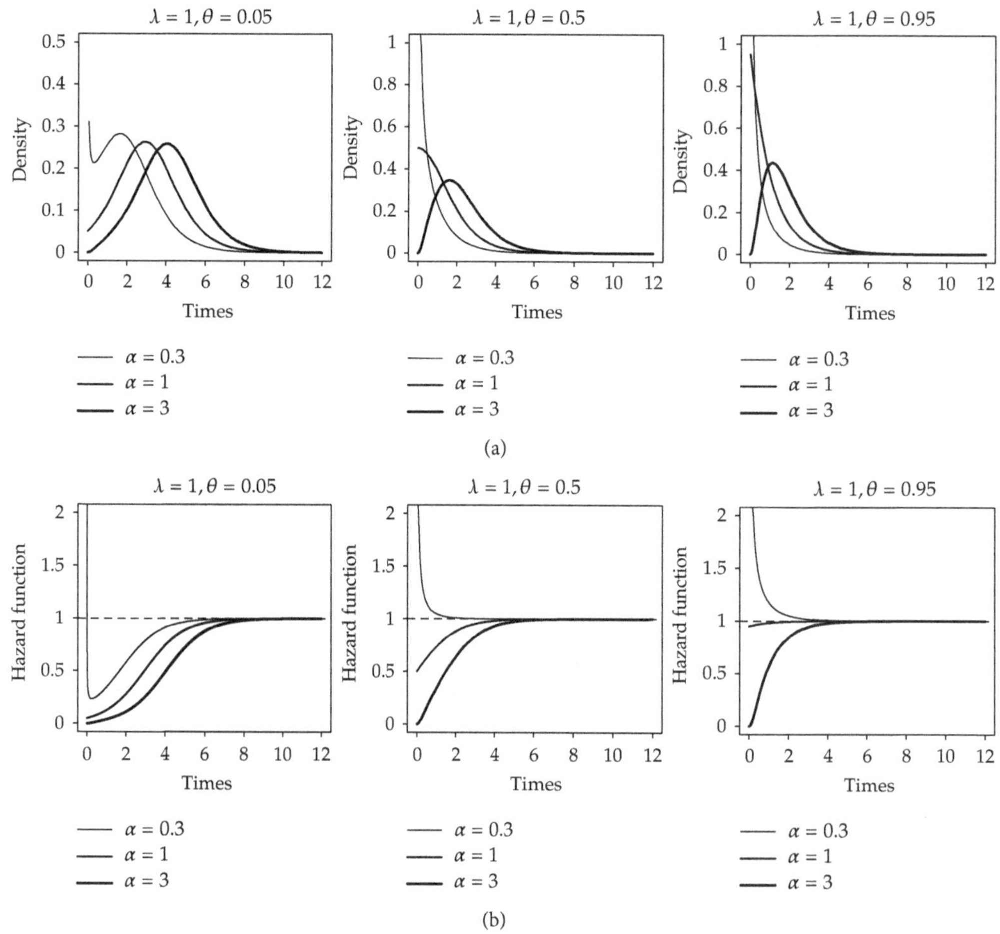

FIGURE 1: (a) Probability density function of the CE2G distribution. (b) Failure rate function of the CE2G distribution. We fixed $\lambda = 1$.

Then, the marginal probability density function of Y is given by

$$f(y) = \sum_{m=1}^{\infty} m\alpha\lambda e^{-\lambda y}\left(1-e^{-\lambda y}\right)^{\alpha-1}\left[\left(1-e^{-\lambda y}\right)^{\alpha}\right]^{m-1} \times \theta(1-\theta)^{m-1}$$

$$= \theta\alpha\lambda e^{-\lambda y}\left(1-e^{-\lambda y}\right)^{\alpha-1}\sum_{m=1}^{\infty} m\left[\left(1-e^{-\lambda y}\right)^{\alpha}(1-\theta)\right]^{m-1}$$

$$= \theta\alpha\lambda e^{-\lambda y}\left(1-e^{-\lambda y}\right)^{\alpha-1}\sum_{m=1}^{\infty}\frac{\left[\left(1-e^{-\lambda y}\right)^{\alpha}(1-\theta)\right]^{m-1}}{1-\left(1-e^{-\lambda y}\right)^{\alpha}(1-\theta)}$$

$$= \theta\alpha\lambda e^{-\lambda y}\left(1-e^{-\lambda y}\right)^{\alpha-1}\left[\frac{1}{1-(1-\theta)\left(1-e^{-\lambda y}\right)^{\alpha}}\right]^{2}. \quad (9)$$

This completes the proof. □

3. Some Properties

Many of the most important features and characteristics of a distribution can be studied through its moments, such as mean and variance. A general expression for rth ordinary moment $\mu_r' = E(Y^r)$ of the CE2G distribution is hard to be obtained and we resume the mean and variance as follows.

The moment generating function of the Y variable with density function given by (1) can be obtained analytically, if we consider the expression, given in [13, page 329, Equation (1.6)].

$$\int_0^1 z^{p-1}(1-z)^{n-1}(1+bz^m)^l dz = \Gamma(n)\sum_{k=0}^{\infty}\binom{l}{k}\frac{(b)^k\Gamma(p+km)}{\Gamma(p+n+km)}. \quad (10)$$

For any real number t, let $\Phi_Y(t)$ be the characteristic function of Y, that is, $\Phi_Y(t) = E[e^{itY}]$, where i denotes the imaginary unit. With the preceding notations, we state the following.

Proposition 2. *For the random variable Y with CE2G distribution, we have that its characteristic function is given by*

$$\Phi(t) = \alpha\theta\Gamma\left(1-\frac{it}{\lambda}\right)\sum_{k=0}^{\infty}\binom{-2}{k}\frac{\Gamma(\alpha[k+1])(\theta-1)^k}{\Gamma(\alpha[k+1]+1-it/\lambda)}, \quad (11)$$

where $i = \sqrt{-1}$.

Proof. Consider the following:

$$\Phi_Y(t) = \int_0^\infty e^{ity} f(y)\,dy$$
$$= \int_0^\infty e^{ity} \frac{\alpha\lambda\theta e^{-\lambda y}\left(1-e^{-\lambda y}\right)^{\alpha-1}}{\left[1-(1-\theta)\left(1-e^{-\lambda y}\right)^{\alpha}\right]^2} dy \quad (12)$$
$$= \alpha\theta \int_0^1 \frac{z^{\alpha-1}(1-z)^{-it/\lambda}}{\left(1-(1-\theta)z^{\alpha}\right)^2} dz,$$

where the last equality follows from the change of variable $z = 1 - e^{-\lambda y}$.

Comparing the last integral with (10), obtaining $n = 1 - it/\lambda$, $b = \theta - 1$, $m = \alpha = p$, and $l = -2$, and making the appropriate substitutions completed the proof. □

Proposition 3. *A random variable Y with density given by* (1) *has mean and variance given, respectively, by*

$$E(Y) = \frac{\theta}{\lambda}\sum_{k=0}^{\infty}\binom{-2}{k}\frac{(\theta-1)^k}{(k+1)}\left[\Psi(0,\alpha[k+1]+1) - \Psi(0,1)\right],$$

$$\begin{aligned}\operatorname{Var}(Y) = \frac{\theta}{\lambda^2}\Bigg\{\sum_{k=0}^{\infty}\Bigg[\binom{-2}{k}\frac{(\theta-1)^k}{(k+1)} &\\ -\Bigg(\Psi(0,1)^2 + \frac{\pi^2}{6} + \Psi(0,\alpha[k+1]+1) &\\ \times\left[\Psi(0,\alpha[k+1]+1) - 2\Psi(0,1)\right] &\\ -\Psi(1,\alpha[k+1]+1)\Bigg)\Bigg] &\\ -\theta\Bigg[\sum_{k=0}^{\infty}\binom{-2}{k}\frac{(\theta-1)^k}{(k+1)} &\\ \times\left(\Psi(0,\alpha[k+1]+1) - \Psi(0,1)\right)\Bigg]^2\Bigg\}, &\end{aligned} \quad (13)$$

where $\Psi(n,z) = (d^{n+1}/dz^{n+1})\ln(\Gamma(z))$ *is known as PsiGamma function.*

Proof. The first result follows from the relationship $\Phi'_Y(t)/i|_{t=0} = E(Y)$. From the literature, $\Phi''_Y(t)/i^2|_{t=0} = E(Y^2)$ and $\operatorname{Var}(Y) = E(Y^2) - [E(Y)]^2$, and with a little algebra follow the results. □

Skewness is a measure of the asymmetry of the probability distribution. The skewness value can be positive or negative, or even undefined. Qualitatively, a negative skew indicates that the tail on the left side of the probability density function is longer than the right side and the bulk of the values lie to the right of the mean. A positive skew indicates that the tail on the right side is longer than the left side and the bulk of the values lie to the left of the mean. The skewness of a random variable Y, say γ_1, is given by the third standardized moment

$$\gamma_1 = \frac{E\left[(Y-\mu)^3\right]}{\left(E\left[(Y-\mu)^2\right]\right)^{3/2}} = \frac{E\left(Y^3\right) - 3E\left(Y^2\right)E(Y) + 3E^2(Y)E(Y) - E^3(Y)}{\left[E\left(Y^2\right) - E^2(Y)\right]^{3/2}}. \quad (14)$$

Kurtosis is any measure of the "peakedness" of the probability distribution of a real-valued random variable. In a similar way to the concept of skewness, kurtosis is a descriptor of the shape of a probability distribution. It is common practice to use the kurtosis to provide a comparison of the shape of a given distribution to that of the normal distribution. One common measure of kurtosis, originating with Karl Pearson, say γ_2, is based on a scaled version of the fourth moment, given by

$$\gamma_2 = \frac{E\left[(Y-\mu)^4\right]}{\left(E\left[(Y-\mu)^2\right]\right)^2} = \frac{E\left(Y^4\right) - 4E\left(Y^3\right)E(Y) + 6E\left(Y^2\right)E^2(Y) - 3E^4(Y)}{\left[E\left(Y^2\right) - E^2(Y)\right]^2}. \quad (15)$$

Algebraic expressions of kurtosis and skewness are extensive to show, due to the fact that is necessary the algebraic moment expressions up order four. This moment can be obtained by algebraic manipulation to determine $E(Y)$, $E(Y^2)$, $E(Y^3)$, and $E(Y^4)$ in (14) and (15) through the Equation (11). Figure 2 shows the kurtosis (γ_2) and skewness (γ_1) of the CE2G distribution for α with $\lambda = 1$, $\theta = 0.1, 0.5, 0.9$ and for θ with $\lambda = 1$, $\alpha = 0.3, 1.0, 3$.

4. Order Statistics

Order statistics are among the most fundamental tools in nonparametric statistics and inference. Let $Y_1, \ldots, Y_n$ be a random sample taken from the CE2G distribution and $Y_{1:n}, \ldots, Y_{n:n}$ denote the corresponding order statistics. Then, the pdf $f_{i:n}(y)$ of the ith order statistics $Y_{i:n}$ is given by

$$f_{i:n}(x) = \frac{n!}{(k-1)!(n-k)!}F(y)^{k-1}\left(1-F(y)\right)^{n-k}f(y). \quad (16)$$

The rth moment of the ith order statistic $Y_{i:n}$ can be obtained from the following result due to [14]:

$$E\left[Y_{i:n}^r\right] = r\sum_{p=n-i+1}^{n}(-1)^{p-n+i-1}\binom{p-1}{n-i}\binom{n}{p}\int_0^\infty y^{r-1}[S(y)]^p dy. \quad (17)$$

Consider the binomial series expansion given by

$$(1-x)^{-r} = \sum_{k=0}^{\infty}\frac{(r)_k}{k!}x^k, \quad (18)$$

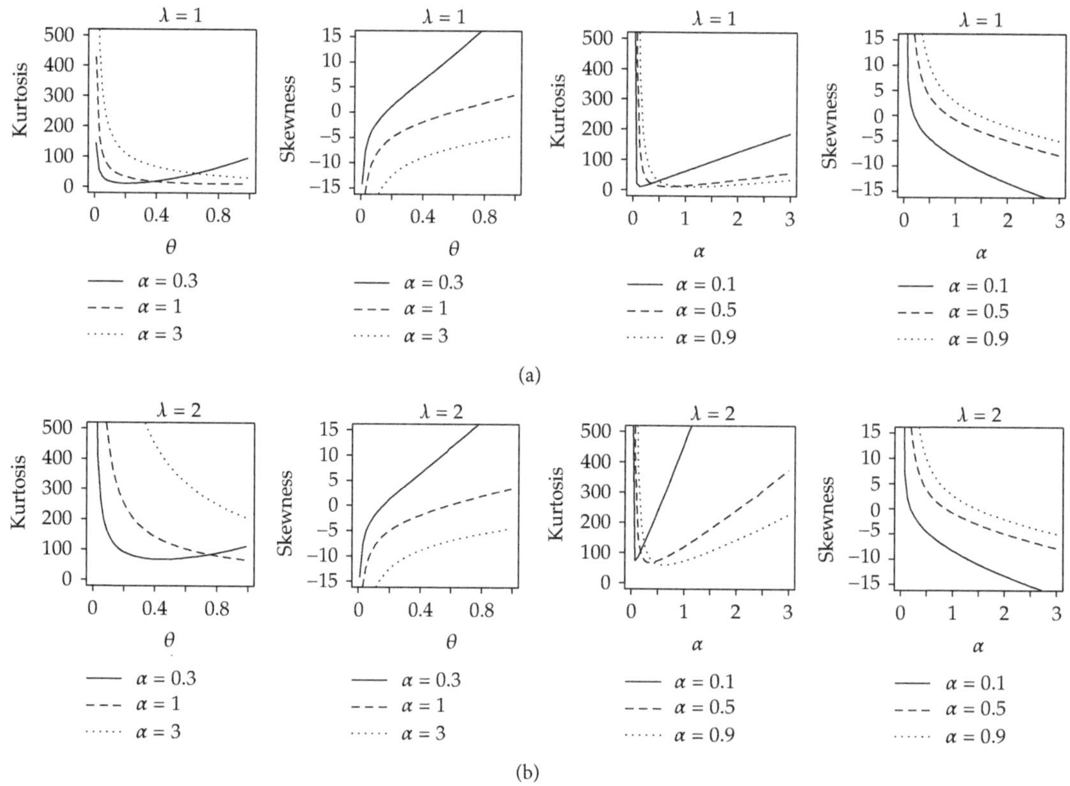

FIGURE 2: (a) Kurtosis and skewness of CE2G distribution for fixed $\lambda = 1$. (b) Kurtosis and skewness of CE2G distribution for fixed $\lambda = 2$.

where $(r)_k$ is a Pochhammer symbol, given $(r)_k = r(r+1)\cdots(r+k-1)$ and if $|x| < 1$ the series converge, and

$$(-r)_k = (-1)^k (r-k+1)_k. \qquad (19)$$

Proposition 4. *For the random variable Y with CE2G distribution, we have that rth moment of the ith order statistic is given by*

$$E\left[Y_{i:n}^r\right] = \frac{r!}{\lambda^r} \sum_{p=n-i+1}^{n} \sum_{j=0}^{\infty} \sum_{k=0}^{\infty} \sum_{l=0}^{p} \sum_{m=0}^{\infty} (-1)^{p-n+i+r+m+l-2} \binom{p-1}{n-i} \binom{n}{p} \times \frac{(1-\theta)^j (p)_j (p-l+1)_l (\alpha(j+l)+k-m+1)_m}{j!\,l!\,m!\,(m+1)^r}. \qquad (20)$$

Proof. From (2) and (18), we have that

$$\begin{aligned}
&\int_0^\infty y^{r-1} [S(y)]^p \, dy \\
&= \int_0^\infty y^{r-1} \left(\frac{1-\left(1-e^{-\lambda y}\right)^\alpha}{1-(1-\theta)\left(1-e^{-\lambda y}\right)^\alpha} \right)^p dy \\
&= \frac{(-1)^{r-1}}{\lambda^r} \int_0^1 \frac{\ln^{r-1}(1-x)}{(1-x)} \left(\frac{1-x^\alpha}{1-(1-\theta)x^\alpha} \right)^p dx \\
&= \frac{(-1)^{r-1}}{\lambda^r} \sum_{j=0}^{\infty} \sum_{k=0}^{\infty} \sum_{l=0}^{p} \frac{(1-\theta)^j (p)_j (-p)_l}{j!\,l!} \\
&\quad \times \int_0^1 x^{\alpha(j+l)+k} \ln^{r-1}(1-x) \, dx.
\end{aligned} \qquad (21)$$

Using the change of variable $\ln(1-x) = -u$ and the expansion (18) results in the kernel of the gamma distribution function as

$$\begin{aligned}
&\int_0^\infty y^{r-1} [S(y)]^p \, dy \\
&= \frac{(-1)^{r-1}}{\lambda^r} \sum_{j=0}^{\infty} \sum_{k=0}^{\infty} \sum_{l=0}^{p} \sum_{m=0}^{\infty} \frac{(1-\theta)^j (p)_j (-p)_l}{j!\,l!} \\
&\quad \times \frac{(-[\alpha(j+l)+k])_m}{m!} \frac{(r-1)!}{(m+1)^r}.
\end{aligned} \qquad (22)$$

Now considering (22) in (17) and the property (19), the result follows. □

5. Entropy

An entropy of a random variable Y is a measure of variation of the uncertainty. A popular entropy measure is Rényi entropy [15].

If Y has the probability density function (1) then Rényi entropy is defined by

$$\gamma(\rho) = \frac{1}{1-\rho}\log\left(\int f^{\rho}(y)\,dy\right), \qquad (23)$$

where $\rho > 0$ and $\rho \neq 1$.

Proposition 5. *If the random variable Y is defined as (7), then, the Rényi entropy, is given by*

$$\gamma(\rho) = \frac{1}{1-\rho} \times \log\left((\alpha\theta)^{\rho}\lambda^{\rho-1}\sum_{k=0}^{\infty}\left[(1-\theta)^{k}(2\rho)_{k}\Gamma(\rho(\alpha-1)+k\alpha+1) \times \Gamma(\rho)\left(k!\Gamma(\alpha(\rho+k)+1)\right)^{-1}\right]\right). \qquad (24)$$

Proof. From (23), we can calculate

$$\begin{aligned}
\int f^{\rho}(y)\,dy &= \int_0^{\infty}\frac{(\alpha\lambda\theta)^{\rho}e^{-\lambda\rho y}\left(1-e^{-\lambda y}\right)^{\rho(\alpha-1)}}{\left[1-(1-\theta)(1-e^{-\lambda y})^{\alpha}\right]^{2\rho}}dy \\
&= (\alpha\lambda\theta)^{\rho}\int_0^{\infty}\sum_{k=0}^{\infty}\left[e^{-\lambda\rho y}\left(1-e^{-\lambda y}\right)^{\rho(\alpha-1)+k\alpha} \times (1-\theta)^{k}\frac{(2\rho)_k}{k!}\right]dy \\
&= (\alpha\theta)^{\rho}\int_0^{\infty}\sum_{k=0}^{\infty}\left[\left(1-e^{-\lambda y}\right)^{\rho(\alpha-1)+k\alpha}(1-\theta)^{k} \times \frac{(2\rho)_k}{k!}\left(\lambda e^{-\lambda y}\right)^{\rho-1}\right]\lambda e^{-\lambda y}dy \\
&= (\alpha\theta)^{\rho}\lambda^{\rho-1}\sum_{k=0}^{\infty}\left[(1-\theta)^{k}\frac{(2\rho)_k}{k!}\int_0^{\infty}u^{\rho(\alpha-1)+k\alpha} \times (1-u)^{\rho-1}du\right] \\
&= (\alpha\theta)^{\rho}\lambda^{\rho-1}\sum_{k=0}^{\infty}\left[(1-\theta)^{k}\frac{(2\rho)_k}{k!} \times \frac{\Gamma(\rho(\alpha-1)+k\alpha+1)\Gamma(\rho)}{\Gamma(\alpha(\rho+k)+1)}\right].
\end{aligned} \qquad (25)$$

So, using the (25) in $\gamma(\rho)$, the result follows. □

6. Reliability

In the context of reliability, the stress-strength model describes the life of a component which has a random strength Y that is subjected to a random stress X. The component fails at the instant hat, the stress applied to it exceeds the strength, and the component will function satisfactorily whenever $Y > X$. So, $R = \Pr(X < Y)$ is a measure of component reliability. In the area of stress-strength models there has been a large amount of work as regards estimation of the reliability R when Y and X are independent random variables belonging to the same univariate family of distributions.

Proposition 6. *If the random variable Y is defined as (7), then, the reliability $R = P(X, Y)$ for X and Y i.i.d is given by*

$$\theta^2\sum_{k=0}^{\infty}\frac{(1-\theta)^k(3)_k}{k!(k+2)}. \qquad (26)$$

Proof. For X and Y i.i.d. CE2G r.v.'s where X is the stress and Y is the strength, the reliability $R = P(X < Y)$ is given by

$$\begin{aligned}
R &= \int_0^{\infty}\int_0^{y}\frac{\alpha\lambda\theta e^{-\lambda x}\left(1-e^{-\lambda x}\right)^{\alpha-1}}{\left[1-(1-\theta)(1-e^{-\lambda x})^{\alpha}\right]^2} \times \frac{\alpha\lambda\theta e^{-\lambda y}\left(1-e^{-\lambda y}\right)^{\alpha-1}}{\left[1-(1-\theta)(1-e^{-\lambda y})^{\alpha}\right]^2}dx\,dy \\
&= \int_0^{\infty}\frac{\theta\left(1-e^{-\lambda y}\right)^{\alpha}}{\left[1-(1-\theta)(1-e^{-\lambda y})^{\alpha}\right]} \times \frac{\alpha\lambda\theta e^{-\lambda y}\left(1-e^{-\lambda y}\right)^{\alpha-1}}{\left[1-(1-\theta)(1-e^{-\lambda y})^{\alpha}\right]^2}dy \\
&= \sum_{k=0}^{\infty}\theta^2\alpha\lambda\frac{(3)_k}{k!}(1-\theta)^k \times \int_0^{\infty}\left(1-e^{-\lambda y}\right)^{\alpha(k+2)-1}e^{-\lambda y}dy \\
&= \sum_{k=0}^{\infty}\sum_{j=0}^{\infty}\theta^2\alpha\lambda\frac{(3)_k(1-\alpha(k+2))_j}{k!j!}(1-\theta)^k \times \int_0^{\infty}e^{-\lambda(j+1)y}dy \\
&= \sum_{k=0}^{\infty}\sum_{j=0}^{\infty}\theta^2\alpha\frac{(3)_k(1-\alpha(k+2))_j}{k!j!(j+1)}(1-\theta)^k \\
&= \sum_{k=0}^{\infty}\theta^2\frac{(3)_k}{k!(k+2)}(1-\theta)^k.
\end{aligned} \qquad (27)$$

This completes the proof. □

7. Residual Lifetime Distribution

Given that there was no failure prior to time t, the residual lifetime distribution of a random variable X, distributed as CE2G distribution, has the survival function given by

$$S_t(x) = \Pr\left[X > x + t \mid X > t\right] = \left(\frac{1-\left(1-e^{-\lambda(x+t)}\right)^{\alpha}}{1-\left(1-e^{-\lambda t}\right)^{\alpha}}\right) \times \left(\frac{1-(1-\theta)\left(1-e^{-\lambda t}\right)^{\alpha}}{1-(1-\theta)\left(1-e^{-\lambda(x+t)}\right)^{\alpha}}\right). \quad (28)$$

The mean residual lifetime of a continuous distribution with survival function $\overline{F}(x)$ is given by

$$\mu(t) = E(X-t \mid X>t) = \frac{1}{S(t)}\int_t^{\infty} S(u)\,du. \quad (29)$$

Proposition 7. *For the random variable Y with CE2G distribution, we have that the mean residual lifetime is given by*

$$\mu(t) = \frac{1}{\lambda}\left(\frac{1-(1-\theta)\left(1-e^{-\lambda t}\right)^{\alpha}}{1-\left(1-e^{-\lambda t}\right)^{\alpha}}\right) \times \sum_{k=0}^{\infty}\sum_{i=0}^{\infty}\sum_{j=0}^{1}\frac{(1-\theta)^{i}(-1)_{j}}{j!} \times \left(\frac{1-\left(1-e^{\lambda t}\right)^{\alpha(i+j)+k+1}}{\alpha(i+j)+k+1}\right). \quad (30)$$

Proof. From (29) and using $S(y)$ given by (2), we have that

$$\frac{1}{S(t)}\int_t^{\infty} S(u)\,du = \frac{1-(1-\theta)\left(1-e^{-\lambda t}\right)^{\alpha}}{1-\left(1-e^{-\lambda t}\right)^{\alpha}} \times \int_t^{\infty}\frac{1-\left(1-e^{-\lambda u}\right)^{\alpha}}{1-(1-\theta)\left(1-e^{-\lambda u}\right)^{\alpha}}du = \frac{1}{\lambda}\frac{1-(1-\theta)\left(1-e^{-\lambda t}\right)}{1-\left(1-e^{-\lambda t}\right)^{\alpha}} \times \int_{1-e^{-\lambda t}}^{1}\frac{1-x^{\alpha}}{\left(1-x^{\alpha}(1-\theta)\right)(1-x)}dx. \quad (31)$$

Now using (18) and making a binomial expansion in a similar way of the proof of Proposition 4 on (22), the result follows. □

8. Inference

Assuming the lifetimes are independently distributed and are independent from the censoring mechanism, the maximum likelihood estimates (MLEs) of the parameters are obtained by direct maximization of the log-likelihood function given by

$$\ell(\theta,\lambda,\alpha) = \ln(\alpha\theta\lambda)\sum_{i=1}^{n}c_i - \lambda\sum_{i=1}^{n}c_i y_i + (\alpha-1)\sum_{i=1}^{n}c_i\ln\left(1-e^{-\lambda y_i}\right) + \sum_{i=1}^{n}(1-c_i)\ln\left(1-\left(1-e^{-\lambda y_i}\right)^{\alpha}\right) - \sum_{i=1}^{n}(1+c_i)\ln\left(1-(1-\theta)\left(1-e^{-\lambda y_i}\right)^{\alpha}\right), \quad (32)$$

where c_i is a censoring indicator, which is equal to 0 or 1, respectively, if the data is censored or observed. The advantage of this procedure is that it runs immediately using existing statistical packages. We have considered the *optim* routine of the R [16].

Large-sample inference for the parameters are based on the MLEs and their estimated standard errors. For (α,θ,λ), we consider the observed Fisher information matrix given by

$$I_F(\alpha,\theta,\lambda) = \left.\begin{pmatrix} I_{\alpha\alpha} & I_{\alpha\theta} & I_{\alpha\lambda} \\ I_{\theta\alpha} & I_{\theta\theta} & I_{\theta\lambda} \\ I_{\lambda\alpha} & I_{\lambda\theta} & I_{\lambda\lambda}\end{pmatrix}\right|_{(\alpha,\theta,\lambda)=(\widehat{\alpha},\widehat{\theta},\widehat{\lambda})}, \quad (33)$$

where the elements of the matrix $I_F(\alpha,\theta,\lambda)$ are given in the appendix.

Under conditions that are fulfilled for the parameters α, θ, and λ in the interior of the parameter space, the asymptotic distribution of $(\widehat{\alpha},\widehat{\theta},\widehat{\lambda})$, as $n \to \infty$, is a normal 3-variate with zero mean and variance covariance matrix $I_F^{-1}(\alpha,\theta,\lambda)$.

In order to compare different distributions, we relied upon several authors in the literature, for example, [6, 17–19], which use the Akaike information criterion (AIC) and Bayesian information criterion (BIC) values, which are defined, respectively, by $-2\ell(\cdot) + 2q$ and $-2\ell(\cdot) + q\log(n)$, where $\ell(\cdot)$ is the LogLikehood evaluated in the MLE vector on respective distribution, q is the number of parameters estimated, and n is the sample size. The best distribution corresponds to a lower AIC and BIC values.

9. Simulation Study

Regarding the performance of the MLEs in the process of estimation, a study was performed based on one hundred generated dataset from the CE2G with six different sets of parameters for n = 20, 50, 100, 200, 500, and 1000. In order to have unbounded parameters, we consider the following restrictions on the parameters in estimation process. For

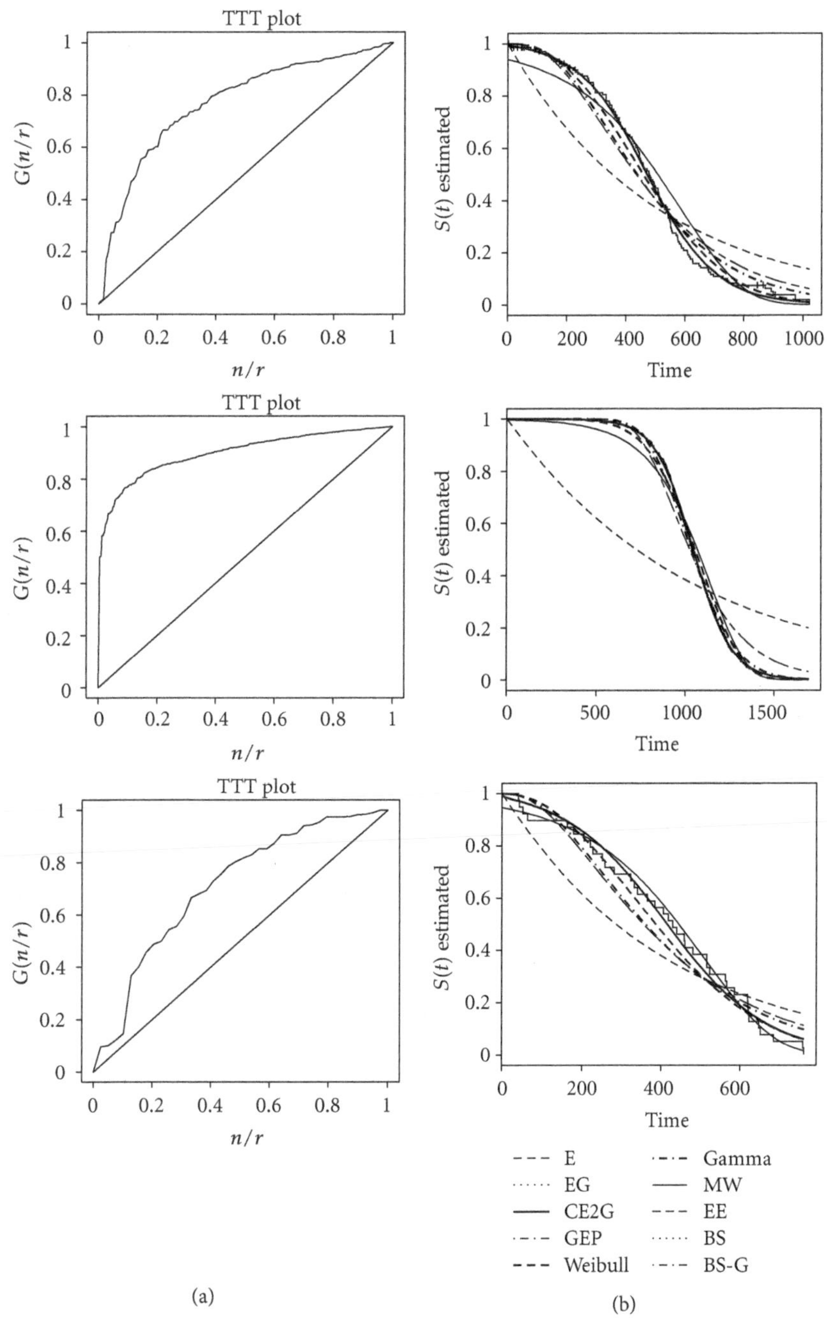

FIGURE 3: (a) Empirical TTT plot for the dataset $T1$, $T2$, and $T3$, respectively. (b) Models fitting for the dataset $T1$, $T2$, and $T3$, respectively.

the parameter θ, we considered the transformation $\theta = e^{\theta^*}/(1 + e^{\theta^*})$, where $\theta^* \in \mathcal{R}$, and for α and λ consider an exponential transformation. Based on the literature of the MLEs, we can return on the original parameters thought of the transformations. For the calculus of their variances, we use the delta method. The values $(\alpha, \lambda, \theta) = (1, 1, 0.5)$ were used as the initial values for all numerics simulations since $\lambda > 0$, $\alpha > 0$, and $0 < \theta < 1$.

The results are condensated in Table 1, which shows the averages of the MLEs, $\text{Av}(\widehat{\alpha}, \widehat{\lambda}, \widehat{\theta})$, together with coverage probability of the 95% confidence intervals for parameters of the CE2G, $C(\alpha, \lambda, \theta)$, the bias, the mean squarer error, MSE, and their deviance, $\text{Sd}(\widehat{\alpha}, \widehat{\lambda}, \widehat{\theta})$. These results suggest that the MLEs estimates have performed adequately. The deviance of the MLEs decrease when sample size increases. The empirical coverage probabilities are close to the nominal coverage level, particularly, as sample size increases.

10. Applications

In this section, we compare the CE2G distribution fit with several usual lifetime distributions on three datasets extracted

Table 1: Mean of the MLEs, their deviances, coverages, bias and MSE.

	n	$\text{Av}(\hat{\alpha}, \hat{\lambda}, \hat{\theta})$	$\text{Sd}(\hat{\alpha}, \hat{\lambda}, \hat{\theta})$	Bias	MSE	$C(\alpha, \lambda, \theta)$
$(\alpha, \lambda, \theta) = (1.48, 3.10, 0.75)$	20	(1.5716, 3.4497, 0.7522)	(0.7890, 1.1204, 0.3327)	(0.0916, 0.3497, 0.0022)	(0.6247, 1.3651, 0.1096)	(0.99, 0.99, 0.80)
	50	(1.4902, 3.4026, 0.7145)	(0.4478, 0.7103, 0.3066)	(0.0102, 0.3026, −0.0355)	(0.1987, 0.5911, 0.0943)	(0.99, 0.99, 0.86)
	100	(1.4765, 3.2589, 0.7233)	(0.2683, 0.4964, 0.2494)	(−0.0035, 0.1589, −0.0267)	(0.0713, 0.2692, 0.0623)	(0.99, 0.99, 0.91)
	200	(1.4798, 3.1846, 0.7379)	(0.2090, 0.3846, 0.2176)	(−0.0002, 0.0846, −0.0121)	(0.0433, 0.1536, 0.0470)	(0.99, 0.99, 0.97)
	500	(1.4725, 3.1617, 0.7361)	(0.1584, 0.2977, 0.1811)	(−0.0075, 0.0617, −0.0139)	(0.0249, 0.0916, 0.0326)	(0.99, 0.99, 0.99)
	1000	(1.5020, 3.1116, 0.7697)	(0.1061, 0.1832, 0.1321)	(0.0220, 0.0116, 0.0197)	(0.0116, 0.0334, 0.0177)	(0.99, 0.99, 0.92)
$(\alpha, \lambda, \theta) = (1.25, 2.63, 0.24)$	20	(1.6389, 2.7783, 0.4016)	(1.0305, 0.8411, 0.3342)	(0.3889, 0.1483, 0.1616)	(1.2026, 0.7224, 0.1367)	(0.99, 0.99, 0.99)
	50	(1.4826, 2.7004, 0.3459)	(0.7378, 0.5976, 0.2589)	(0.2326, 0.0704, 0.1059)	(0.5930, 0.3586, 0.0776)	(0.99, 0.99, 0.99)
	100	(1.3892, 2.6563, 0.3046)	(0.5549, 0.3699, 0.1893)	(0.1392, 0.0263, 0.0646)	(0.3242, 0.1362, 0.0396)	(0.99, 0.99, 0.99)
	200	(1.2869, 2.6143, 0.2729)	(0.3339, 0.2520, 0.1229)	(0.0369, −0.0157, 0.0329)	(0.1117, 0.0631, 0.0160)	(0.99, 0.99, 0.99)
	500	(1.2609, 2.6029, 0.2497)	(0.1980, 0.1444, 0.0632)	(0.0109, −0.0271, 0.0097)	(0.0389, 0.0214, 0.0041)	(0.99, 0.99, 0.99)
	1000	(1.2696, 2.6243, 0.2479)	(0.1621, 0.1123, 0.0517)	(0.0196, −0.0057, 0.0079)	(0.0264, 0.0125, 0.0027)	(0.99, 0.99, 0.99)
$(\alpha, \lambda, \theta) = (0.25, 0.63, 0.20)$	20	(0.3852, 0.6554, 0.4163)	(0.2658, 0.2378, 0.3376)	(0.1352, 0.0254, 0.2163)	(0.0882, 0.0566, 0.1596)	(0.92, 0.99, 0.99)
	50	(0.2809, 0.6400, 0.2641)	(0.1264, 0.1368, 0.1973)	(0.0309, 0.0100, 0.0641)	(0.0168, 0.0186, 0.0427)	(0.99, 0.99, 0.99)
	100	(0.2935, 0.6064, 0.2841)	(0.1162, 0.0931, 0.1732)	(0.0435, −0.0236, 0.0841)	(0.0152, 0.0091, 0.0368)	(0.99, 0.99, 0.99)
	200	(0.2657, 0.6354, 0.2246)	(0.0810, 0.0744, 0.1009)	(0.0157, 0.0054, 0.0246)	(0.0067, 0.0055, 0.0107)	(0.99, 0.99, 0.99)
	500	(0.2569, 0.6388, 0.2078)	(0.0429, 0.0492, 0.0537)	(0.0069, 0.0088, 0.0078)	(0.0019, 0.0025, 0.0029)	(0.99, 0.99, 0.99)
	1000	(0.2536, 0.6313, 0.2044)	(0.0307, 0.0303, 0.0339)	(0.0036, 0.0013, 0.0044)	(0.0009, 0.0009, 0.0012)	(0.99, 0.99, 0.99)
$(\alpha, \lambda, \theta) = (0.30, 0.60, 0.90)$	20	(0.3258, 0.7817, 0.8033)	(0.1165, 0.3750, 0.2751)	(0.0258, 0.1817, −0.0967)	(0.0141, 0.1723, 0.0843)	(0.99, 0.99, 0.80)
	50	(0.2813, 0.6879, 0.7639)	(0.0658, 0.2013, 0.2639)	(−0.0187, 0.0879, −0.1361)	(0.0046, 0.0479, 0.0875)	(0.99, 0.99, 0.85)
	100	(0.2869, 0.6535, 0.8123)	(0.0489, 0.1406, 0.2222)	(−0.0131, 0.0535, −0.0877)	(0.0025, 0.0224, 0.0566)	(0.99, 0.99, 0.93)
	200	(0.2905, 0.6325, 0.8364)	(0.0343, 0.0921, 0.1553)	(−0.0095, 0.0325, −0.0636)	(0.0013, 0.0095, 0.0279)	(0.99, 0.99, 0.97)
	500	(0.3007, 0.6117, 0.8884)	(0.0219, 0.0647, 0.1214)	(0.0007, 0.0117, −0.0116)	(0.0005, 0.0043, 0.0147)	(0.99, 0.99, 0.97)
	1000	(0.2970, 0.6053, 0.8821)	(0.0184, 0.0455, 0.1003)	(−0.0030, 0.0053, −0.0179)	(0.0003, 0.0021, 0.0103)	(0.99, 0.99, 0.98)
$(\alpha, \lambda, \theta) = (0.50, 2.00, 0.40)$	20	(0.5748, 2.3413, 0.4948)	(0.2790, 0.8066, 0.3586)	(0.0748, 0.3413, 0.0948)	(0.0826, 0.7606, 0.1363)	(0.99, 0.99, 0.99)
	50	(0.6019, 2.0303, 0.5348)	(0.2218, 0.4461, 0.2941)	(0.1019, 0.0303, 0.1348)	(0.0591, 0.1979, 0.1038)	(0.99, 0.99, 0.99)
	100	(0.5100, 2.0592, 0.4423)	(0.1622, 0.3178, 0.2465)	(0.0100, 0.0592, 0.0423)	(0.0262, 0.1035, 0.0620)	(0.99, 0.99, 0.99)
	200	(0.5307, 2.0009, 0.4503)	(0.1091, 0.2491, 0.1864)	(0.0307, 0.0009, 0.0503)	(0.0127, 0.0614, 0.0369)	(0.99, 0.99, 0.99)
	500	(0.5045, 1.9954, 0.4194)	(0.0727, 0.1594, 0.1154)	(0.0045, −0.0046, 0.0194)	(0.0053, 0.0252, 0.0136)	(0.99, 0.99, 0.99)
	1000	(0.5051, 2.0072, 0.4034)	(0.0493, 0.1002, 0.0598)	(0.0051, 0.0072, 0.0034)	(0.0024, 0.0100, 0.0036)	(0.99, 0.99, 0.98)
$(\alpha, \lambda, \theta) = (2.00, 0.25, 0.80)$	20	(2.1599, 0.3199, 0.6131)	(1.0176, 0.1112, 0.3449)	(0.1599, 0.0699, −0.1869)	(1.0508, 0.0171, 0.1527)	(0.99, 0.99, 0.79)
	50	(2.0826, 0.2743, 0.7193)	(0.5220, 0.0528, 0.2874)	(0.0826, 0.0243, −0.0807)	(0.2766, 0.0033, 0.0883)	(0.99, 0.99, 0.88)
	100	(1.9984, 0.2629, 0.7519)	(0.4419, 0.0418, 0.2711)	(−0.0016, 0.0129, −0.0481)	(0.1933, 0.0019, 0.0751)	(0.99, 0.99, 0.87)
	200	(2.0322, 0.2569, 0.7808)	(0.3046, 0.0272, 0.2050)	(0.0322, 0.0069, −0.0192)	(0.0929, 0.0008, 0.0420)	(0.99, 0.99, 0.97)
	500	(1.9945, 0.2552, 0.7849)	(0.1613, 0.0218, 0.1783)	(−0.0055, 0.0052, −0.0151)	(0.0258, 0.0005, 0.0317)	(0.99, 0.99, 0.92)
	1000	(1.9659, 0.2526, 0.7774)	(0.1358, 0.0160, 0.1496)	(−0.0341, 0.0026, −0.0226)	(0.0194, 0.0003, 0.0227)	(0.99, 0.99, 0.96)

TABLE 2: Values of the—max $\ell(\cdot)$ and AIC for all fitted distributions.

	E	EE	EG	Weibull	Gamma	CE2G	MW	GEP	BS	BS-G
$T1$										
AIC	1723.7	1657.2	1725.8	1630.5	1649.4	1616.0	1660.0	1659.3	1919.7	1708.5
BIC	1726.7	1663.2	1731.7	1636.5	1655.3	1624.9	1668.9	1668.2	1925.6	1717.3
$T2$										
AIC	6649.8	5703.2	6651.8	5599.0	5605.9	5571.0	5664.7	5705.3	5648.3	5601.3
BIC	6653.9	5711.3	6659.9	5607.1	5613.8	5583.1	5676.8	5717.4	5656.3	5613.4
$T3$										
AIC	549.8	538.2	551.8	530.3	536.5	530.6	530.7	540.3	550.8	534.0
BIC	551.5	541.6	555.2	533.7	539.8	535.6	535.7	545.3	554.1	539.0

from the literature. The first dataset, $T1$, refers to the serum-reversal time (days) of 143 children contaminated with HIV from vertical transmission at the university hospital of the Ribeirão Preto Scholl of Medicine (Hospital das Clínicas da Faculdade de Medicina de Ribeirão Preto) from 1986 to 2001 [20]. Serum reversal can occur in children born from mothers infected with HIV.

The second dataset, $T2$, is lifetimes in hours of 417 forty-watt, 110-volt internally frosted incandescent lamps taken from 42 weekly quality control [21]. Survival times, in days, are given for the control group of lamps on original dataset.

The third dataset, $T3$, gives the survival times for laboratory mice, which were exposed to a fixed dose of radiation at an age of 5 to 6 weeks. The cause of death for each mouse was determined after autopsy to be one of three possibilities: thymic lymphoma (C1), reticulum cell sarcoma (C2), or other causes (C3) [22]. Consider here the minces of C3 in the control group.

Firstly, in order to identify the shape of a lifetime data failure rate function, we will consider a graphical method based on the TTT plot [23]. In its empirical version, the TTT plot is given by $G(r/n) = [(\sum_{i=1}^{r} Y_{i:n}) + (n-r)Y_{r:n}]/(\sum_{i=1}^{n} Y_{i:n})$, where $r = 1,\ldots,n$ and $Y_{i:n}$, $i = 1,\ldots,n$ represent the order statistics of the sample. It has been shown that the failure rate function is increasing (decreasing) if the TTT plot is concave (convex). Figure 3(a) shows concave TTT plots for the $T1$, $T2$, and $T3$ datasets, indicating increasing failure rate functions.

We compare the CE2G distribution fits with the exponential distribution with probability density function given by $f(x) = \lambda e^{-\lambda x}$, the exponentiated exponential distribution, EE, with probability density function given by $f(x) = \alpha * \lambda e^{-\lambda x}(1 - e^{-\lambda x})^{\alpha-1}$, the EG distribution [1] with probability density function given by $f(x) = \lambda(1 - (1 - \theta)e^{-\lambda x})^{-1}$, the Weibull distribution with probability density function given by $f(x) = (\theta/\lambda)(x/\lambda)^{\theta-1}e^{-(x/\lambda)^{\theta}}$, where the shape parameter is θ and scale parameter is λ, the gamma distribution with probability density function given by $f(x) = (1/\lambda^{\theta}\Gamma(\theta))x^{\theta-1}e^{-x/\lambda}$, with shape parameter θ and scale parameter λ, the modified Weibull (MW) distribution with probability density function given by $f(x) = \alpha x^{\theta-1}(\theta + \lambda x)e^{\lambda x}e^{-\alpha x^{\theta}\exp\{\lambda x\}}$, where $\alpha, \theta \geq 0$ and $\lambda > 0$, the generalized exponential Poisson (GEP) distribution [6] with probability density function given by $f(x) = (\alpha\beta\lambda/(1 - e^{-\lambda})^{\alpha})(1 - e^{-\lambda+\lambda\exp(-\beta x)})^{\alpha-1}e^{-\lambda-\beta x+\lambda\exp(-\beta x)}$, the generalized Birnbaum-Saunders (BS-G) distribution [24] with probability density function given by $f(y) = ((\sqrt{(y-\mu)/\beta} + \sqrt{\beta/(x-\mu)})/2\alpha(x-\mu))\phi([\sqrt{(y-\mu)/\beta} - \sqrt{\beta/(x-\mu)}]/\alpha)$, where $\phi(\cdot)$ is the probability density distribution of the standard normal distribution, and the Birnbaum-Saunders (BS) distribution. The BS distribution is obtained considering $\mu = 0$ in the BS-G probability density function.

Table 2 provides the AIC and BIC criterion values for each distribution. They provide evidence in favor of our CE2G distribution for the datasets $T1$ and $T2$ in all of the three comparison criterion. For the dataset $T3$, the CE2G distribution provides similar fitting to the Weibull and MW distributions, implying that the CE2G distribution is a competitor to the usual survival distributions. These results are corroborated by the empirical Kaplan-Meier survival functions and the fitted survival functions shown in Figure 3(b). The MLEs (and their corresponding standard errors in parentheses) of the parameters α, $\theta(\times 1000)$, and $\lambda(\times 10000)$ of the CE2G distribution are given, respectively, by 3.7469 (0.5688), 41.4860 (9.7659), and 17536.46 (7.1814) for $T1$, by 5.1765 (19.4159), 0.2625 (0.9915), and 94.6676 (3.8720) for $T2$, and by 0.0018180 (0.9818), 0.0698 (0.3770), and 78.7704 (11.5084) for $T3$.

11. Concluding Remarks

In this paper, a new lifetime distribution is provided and discussed. The CE2G distribution accommodates increasing, decreasing, and bathtub failure rate functions and arises in a latent complementary risks scenario, where the lifetime associated with a particular risk is not observable but only the maximum lifetime value among all risks. The properties of the proposed distribution are discussed, including a formal proof of its probability density function and explicit algebraic formulas for its survival and hazard functions, moments, rth moment of the ith order statistic, mean residual lifetime, modal value, and the observed Fisher information matrix. Maximum likelihood inference is implemented straightforwardly. The practical importance of the new distribution was demonstrated in three applications where the CE2G distribution provided the best fit in comparison with several other former lifetime distributions.

Appendix

In this appendix, we show the values of the elements of the observed Fisher information matrix in (33). From (32), we obtain

$$\begin{aligned}
I_{\alpha\alpha} &= \sum_{i=1}^{n}\left(\frac{c_i}{\alpha^2}+\frac{(1-c_i)L_i^{\alpha}\ln^2(L_i)}{R_i}+\frac{(1-c_i)L_i^{2\alpha}\ln^2(L_i)}{R_i^2}\right.\\
&\quad-\frac{(1+c_i)(1-\theta)L_i^{\alpha}\ln^2(L_i)}{T_i}\\
&\quad\left.-\frac{(1+c_i)(1-\theta)^2L_i^{2\alpha}\ln^2(L_i)}{T_i^2}\right),\\
I_{\alpha\theta} = I_{\theta\alpha} &= \sum_{i=1}^{n}\left(\frac{(1+c_i)L_i^{\alpha}\ln(L_i)}{T_i}\right.\\
&\quad\left.+\frac{(1+c_i)(1-\theta)L_i^{2\alpha}\ln(L_i)}{T_i^2}\right),\\
I_{\alpha\lambda} = I_{\lambda\alpha} &= \sum_{i=1}^{n}\left(\frac{-c_iX_i}{L_i}+\frac{\alpha(1-c_i)L_i^{\alpha}\ln(L_i)X_i}{L_iR_i}\right.\\
&\quad+\frac{(1-c_i)L_i^{\alpha}X_i}{L_iR_i}+\frac{\alpha(1-c_i)L_i^{2\alpha}\ln(L_i)X_i}{L_iR_i^2}\\
&\quad-\frac{\alpha(1+c_i)(1-\theta)L_i^{\alpha}\ln(L_i)X_i}{L_iT_i}\\
&\quad-\frac{(1+c_i)(1-\theta)L_i^{\alpha}X_i}{L_iT_i}\\
&\quad\left.-\frac{\alpha(1+c_i)(1-\theta)^2L_i^{2\alpha}\ln(L_i)X_i}{L_iT_i^2}\right),\\
I_{\theta\theta} &= \sum_{i=1}^{n}\left(\frac{c_i}{\theta^2}-\frac{(1+c_i)L_i^{2\alpha}}{T_i^2}\right),\\
I_{\theta\lambda} = I_{\lambda\theta} &= \sum_{i=1}^{n}\left(\frac{\alpha(1+c_i)L_i^{\alpha}X_i}{L_iT_i}+\frac{\alpha(1+c_i)(1-\theta)L_i^{2\alpha}X_i}{L_iT_i^2}\right),\\
I_{\lambda\lambda} &= \sum_{i=1}^{n}\left(\frac{c_i}{\lambda^2}+\frac{(\alpha-1)c_iy_iX_i}{L_i}+\frac{(\alpha-1)c_iX_i^2}{L_i^2}\right.\\
&\quad-\frac{\alpha(1-c_i)L_i^{\alpha}y_iX_i}{L_iR_i}-\frac{\alpha(1-c_i)L_i^{\alpha}X_i^2(1-\alpha)}{L_i^2R_i}\\
&\quad+\frac{\alpha^2(1-c_i)L_i^{2\alpha}X_i^2}{L_i^2R_i^2}+\frac{\alpha(1+c_i)(1-\theta)L_i^{\alpha}y_iX_i}{L_iT_i}\\
&\quad+\frac{\alpha(1+c_i)(1-\theta)L_i^{\alpha}X_i^2(1-\alpha)}{L_i^2T_i}\\
&\quad\left.-\frac{\alpha^2(1+c_i)(1-\theta)^2L_i^{2\alpha}X_i^2}{L_i^2T_i^2}\right),
\end{aligned}\tag{A.1}$$

where $L_i = 1 - e^{-\lambda y_i}$, $R_i = 1 - L_i^{\alpha}$, $T_i = 1 - (1-\theta)L_i^{\alpha}$, and $X_i = y_ie^{-\lambda y_i}$.

Acknowledgments

V. Marchi and F. Louzada are supported by the Brazilian organizations CAPES and CNPq, respectively. The authors are grateful to Dr. Gauss Cordeiro, Editor of this special issue in, as well as to the anonymous Referees for their comments, criticisms, and suggestions, which lead to important improvements.

References

[1] K. Adamidis and S. Loukas, "A lifetime distribution with decreasing failure rate," *Statistics & Probability Letters*, vol. 39, no. 1, pp. 35–42, 1998.

[2] R. D. Gupta and D. Kundu, "Exponentiated exponential family: an alternative to gamma and Weibull distributions," *Biometrical Journal*, vol. 43, no. 1, pp. 117–130, 2001.

[3] R. D. Gupta and D. Kundu, "Generalized exponential distributions," *Australian & New Zealand Journal of Statistics*, vol. 41, no. 2, pp. 173–188, 1999.

[4] S. Nadarajah and S. Kotz, "The exponentiated type distributions," *Acta Applicandae Mathematicae*, vol. 92, no. 2, pp. 97–111, 2006.

[5] C. Kuş, "A new lifetime distribution," *Computational Statistics & Data Analysis*, vol. 51, no. 9, pp. 4497–4509, 2007.

[6] W. Barreto-Souza and F. Cribari-Neto, "A generalization of the exponential-Poisson distribution," *Statistics & Probability Letters*, vol. 79, no. 24, pp. 2493–2500, 2009.

[7] F. Louzada-Neto, V. G. Cancho, and G. D. C. Barriga, "The Poisson-exponential distribution: a Bayesian approach," *Journal of Applied Statistics*, vol. 38, no. 6, pp. 1239–1248, 2011.

[8] F. Louzada, M. Roman, and V. G. Cancho, "The complementary exponential geometric distribution: model, properties, and a comparison with its counterpart," *Computational Statistics & Data Analysis*, vol. 55, no. 8, pp. 2516–2524, 2011.

[9] F. Louzada-Neto, "Poly-hazard regression models for lifetime data," *Biometrics*, vol. 55, no. 4, pp. 1121–1125, 1999.

[10] J. F. Lawless, *Statistical Models and Methods for Lifetime Data*, Wiley Series in Probability and Statistics, John Wiley & Sons, New York, NY, USA, 2nd edition, 2003.

[11] M. J. Crowder, A. C. Kimber, R. L. Smith, and T. J. Sweeting, *Statistical Analysis of Reliability Data*, Chapman & Hall, London, UK, 1991.

[12] D. R. Cox and D. Oakes, *Analysis of Survival Data*, Monographs on Statistics and Applied Probability, Chapman & Hall, London, UK, 1984.

[13] I. S. Gradshteyn and I. M. Ryzhik, *Table of Integrals, Series, and Products*, Elsevier, New York, NY, USA, 7th edition, 2007.

[14] H. M. Barakat and Y. H. Abdelkader, "Computing the moments of order statistics from nonidentical random variables," *Statistical Methods & Applications*, vol. 13, no. 1, pp. 15–26, 2004.

[15] A. Rényi, "On measures of entropy and information," in *Proceedings of the Fourth Berkeley Symposium on Mathematical Statistics and Probability*, vol. 1 of *Contributions to the Theory of Statistics*, pp. 547–561, University of California Press, Berkeley, Calif, USA, 1961.

[16] R Development Core Team, R Foundation for Statistical Computing, Vienna, Austria, 2010, http://www.R-project.org.

[17] W. Barreto-Souza, A. L. de Morais, and G. M. Cordeiro, "The weibull-geometric distribution," *Journal of Statistical Computation and Simulation*, vol. 81, no. 5, pp. 645–657, 2011.

[18] K. Adamidis, T. Dimitrakopoulou, and S. Loukas, "On an extension of the exponential-geometric distribution," *Statistics & Probability Letters*, vol. 73, no. 3, pp. 259–269, 2005.

[19] G. S. Mudholkar and D. K. Srivasta, "Exponentiated weibull family: a reanalysis of the bus-motor-failure data," *Technometrics*, vol. 37, no. 4, pp. 436–445, 1995.

[20] G. S. C. Perdoná and F. Louzada-Neto, "A general hazard model for lifetime data in the presence of cure rate," *Journal of Applied Statistics*, vol. 38, no. 7, pp. 1395–1405, 2011.

[21] D. Davis, "An analysis of some failure data," *Journal of the American Statistical Association*, vol. 47, no. 258, pp. 113–150, 1952.

[22] D. G. Hoel, "A representation of mortality data by competing risks," *Biometrics*, vol. 28, no. 2, pp. 475–488, 1972.

[23] M. V. Aarset, "How to identify a bathtub hazard rate," *IEEE Transactions on Reliability*, vol. R-36, no. 1, pp. 106–108, 1987.

[24] Z. W. Birnbaum and S. C. Saunders, "A new family of life distributions," *Journal of Applied Probability*, vol. 6, pp. 319–327, 1969.

9

Using Time Deformation to Filter Nonstationary Time Series with Multiple Time-Frequency Structures

Mengyuan Xu,[1] Wayne A. Woodward,[2] and Henry L. Gray[2]

[1] *Statistics Genomics Unit, NIH/NIMH (National Institutes of Health/National Institute of Mental Health), Bethesda, MD 20892, USA*
[2] *Department of Statistical Science, Southern Methodist University, Dallas, TX 75205, USA*

Correspondence should be addressed to Mengyuan Xu; mengyuantracy@hotmail.com

Academic Editor: Zhidong Bai

For nonstationary time series consisting of multiple time-varying frequency (TVF) components where the frequency of components overlaps in time, classical linear filters fail to extract components. The G-filter based on time deformation has been developed to extract components of multicomponent G-stationary processes. In this paper, we explore the wide application of the G-filter for filtering different types of nonstationary processes with multiple time-frequency structure. Simulation examples illustrate that the G-filter can be applied to filter a broad range of multicomponent nonstationary process where TVF components may in fact overlap in time.

1. Introduction and Background

The traditional linear filter is defined as $y(t) = x * a(t) = \int_{-\infty}^{+\infty} x(t-u)a(u)du$, where $t \in (-\infty, \infty)$ and $\int_{-\infty}^{\infty} |a(t)|dt < \infty$ and where $x(t)$ and $y(t)$ are the input and output processes. Papoulis [1] has shown that $P_Y(f) = |A(f)|^2 P_X(f)$, where $P_X(f)$ and $P_Y(f)$ denote the power spectra of the stationary input and output processes, $x(t)$ and $y(t)$, and $A(f) = \int_{-\infty}^{\infty} a(t)e^{-2\pi i f t}dt$ is the frequency response function. Based on this, certain filters (e.g., the Butterworth filter [2]) have been designed to filter or pass low frequency or high frequency components of the input process to the output process. Given a cutoff frequency, a low-pass (high-pass) Butterworth filter can extract components whose frequency content is below (above) this cutoff frequency.

In general, traditional linear filters, such as the Butterworth filter, are time invariant. They are designed to extract components from stationary processes where the frequency behavior of the signal does not change with time. However, for time series data with time-varying frequency behavior (TVF), these filters can produce very poor results because the time-invariant nature of the filters does not properly account for the time-varying frequency behavior of the data. That is, these filters do not properly adjust the cutoff frequency with time according to the frequency behavior of the data. See discussion and examples in Xu et al. [3].

In order to address the filtering problem for nonstationary data with TVF, Xu et al. [3] developed the G-filter utilizing the time-deformation method by deforming the index set (time axis) of the time series data. The use of time-deformation or time-warping methods to define or analyze nonstationary data has been investigated by several researchers. For example, self-similar processes are obtained by applying the Lamperti operator on the time scale of a stationary process [4]. Gray and Zhang [5] base time deformation on a log transformation of the time axis. They refer to processes that are stationary on the log scale as M-stationary processes and show that the resulting spectral representation is based on the Mellin transform. Flandrin et al. [4] point out that the log transformation of Gray and Zhang [5] is a special case of the Lamperti transform. Gray et al. [6] extend the M-stationary model to analyze data collected at discrete time points. Jiang et al. [7] define a more general family of nonstationary processes called G-stationary processes whose frequencies monotonically change with time. See also Woodward et al. [8, Chapter 13].

In Section 2, we give a brief review of G-stationary processes and the G-filter along with an illustrative example.

In Section 3 we extend the applicability of the G-filter to filter nonstationary time varying signals with varieties of time-frequency structures. We give two examples illustrating the techniques.

2. G-Stationary Processes and the G-Filter

In this section, we define and discuss G-stationary processes, we define the G-filtering procedure given by Xu et al. [3], and, provide an example.

Definition 1 (Dual Process). A process $\{Y(u), u \in (-\infty, \infty)\}$ is called a dual process of the process $\{X(t), t \in (a, b)\}$ if there exists a mapping $u = g(t), t \in (a, b)$ of which a well-defined inverse function, $t = g^{-1}(u)$, $u \in (g(a), g(b))$, exists, so that $X(t) = Y(g(t))$ for $t \in (a, b)$. The mapping $u = g(t)$, $t \in (a, b)$ is called the time-deformation function.

Definition 2 (G-stationary Process). A process $\{X(t), t \in (a, b)\}$ is a G-stationary process if and only if a stationary dual process $\{Y(u), u \in (-\infty, \infty)\}$ exists under some time-deformation function $u = g(t)$, $t \in (a, b)$.

Some special cases of G-stationary processes are given in the following:

(1) when $g(t) = at + b$, $t \in (-\infty, \infty)$, the G-stationary process is simply the traditional weakly stationary process;

(2) when $g(t) = \ln(t)$, $t \in (0, \infty)$, $X(t)$ is called an M-stationary process ((Gray et al. (1988)), [6]);

(3) when $g(t) = (t^\lambda - 1)/\lambda$, $t \in (0, \infty)$ (a Box-Cox transformation), $X(t)$ is called a $G(\lambda)$-stationary process [7];

(4) when $g(t) = at^2 + bt$, $t \in (0, \infty)$ with $a > 0$ and $b \geq 0$, $X(t)$ is called the generalized linear chirp process [9].

Jiang et al. [7] define the generalized instantaneous period (GIP) for the G-stationary process to be $P(t; g, \tau) = g^{-1}(g(t)+\tau)-t$, where τ is the period of the dual process, and the generalized instantaneous frequency (GIF) is $1/P(t; g, \tau)$. They showed that the GIF is approximately proportional to $g'(t)$. So, for the M-stationary process where $g(t) = \ln(t)$, it follows that GIF $\propto 1/t$, and GIP $\propto t$; that is, the generalized instantaneous period is changing linearly with time. For the $G(\lambda)$-stationary process where $g(t) = (t^\lambda - 1)/\lambda$, the GIF changes like a polynomial function, that is, like $t^{\lambda-1}$. For the linear chirp stationary process where $g(t) = at^2 + bt$, the GIF changes linearly with time, that is, like $2at + b$.

Given a realization, the frequency change measured by the GIF can be visually represented using time-frequency plots. In this paper we will use the nonparametric Wigner-Ville plots for this purpose. These plots display the time-frequency behavior in the data by computing inverse Fourier transforms of windowed versions of the sample autocovariance function (see [8, 10, 11]).

Xu et al. [3] define the G-filter mathematically and give a straightforward procedure for G-filtering a set of G-stationary data with one or more time-varying frequency components.

2.1. G–Filtering Procedure

Step 1. Fit a G-stationary model to the data and transform the time axis to obtain a stationary dual process.

Step 2. Apply traditional time-invariant filtering methods to the dual data to extract components.

Step 3. Convert filtered dual components back into the original scale.

See Xu et al. [3] who also point out that the G-filtering procedure described previously is an application of the principle of unitary equivalence systems [12].

In Example 3 we will show how to apply the G-filter to filter linear chirp data. G-stationary processes have an origin, and a given realization will begin at some offsets from the process origin. For a realization of length n, our procedure is to obtain a dual realization (of length n) associated with a transformation $u = g(t)$. The n values, t_k, needed for obtaining the stationary dual based on the discrete Box-Cox time transformation, $k = (t_k^\lambda - 1)/\Delta_n\lambda - \xi$, are $t_k = ((k + \xi)\Delta_n\lambda + 1)^{1/\lambda}$ where $(\xi\Delta_n\lambda + 1)^{1/\lambda}$ is called the *realization offset* and Δ_n is the sampling increment. Jiang et al. [7] employ a search routine to determine the values of λ and offset that produce the "most stationary" dual. For each set of λ and offset values considered in the search, the data, $X(t_k)$, $k = 1, 2, \ldots$, are approximated at the t_k's using interpolation. By then indexing on k, the dual realization associated with the given λ and offset is obtained, and for each combination of λ and offset, the resulting dual realization is checked for stationarity. Jiang et al. [7] suggest measuring stationarity by examining characteristics (e.g., sample autocorrelations) of the first and second halves of the transformed data. They employ a Q-statistic for measuring the difference between the sample autocorrelations of the two halves. This measure is based on the fact that the correlation structure under stationarity stays constant across the realization. The software, GWS, written in S+, can be used to perform this search procedure, and it is available from the authors at the website http://www.texasoft.com/atsa/. This software program can be used to fit a $G(\lambda)$ model to a set of data, and it provides methods for spectral analysis, forecasting, and so forth. In the examples here involving analysis of TVF data, we use the GWS software package.

Example 3. We consider the linear chirp process

$$Y(t) = \cos\left(\frac{0.0002\pi t^2}{2}\right) + \cos\left(\frac{0.0008\pi t^2}{2}\right), \quad (1)$$

where $t \in [0, 400]$. This is a linear chirp stationary process consisting of two chirp components. A realization of the process is shown in Figure 1(a) where it can be seen that frequency behavior is increasing (periods are shortening) in time. It also can be seen that there appear to be more than one TVF component. The Wigner-Ville plot in Figure 1(b) clearly shows two monotonically increasing frequency components in the data, and these frequency components visually appear to be increasing linearly with time. In Figures 1(c) and 1(e),

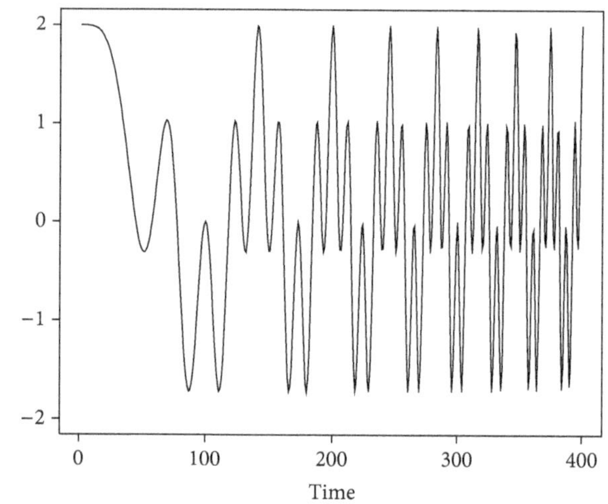

(a) TVF signal

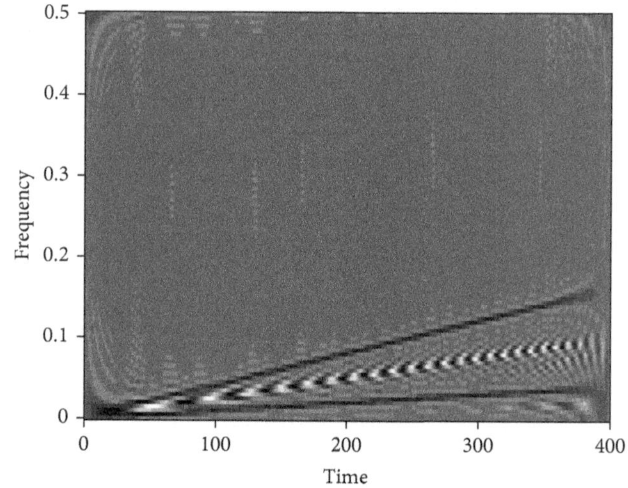

(b) Wigner-Ville plot for (a)

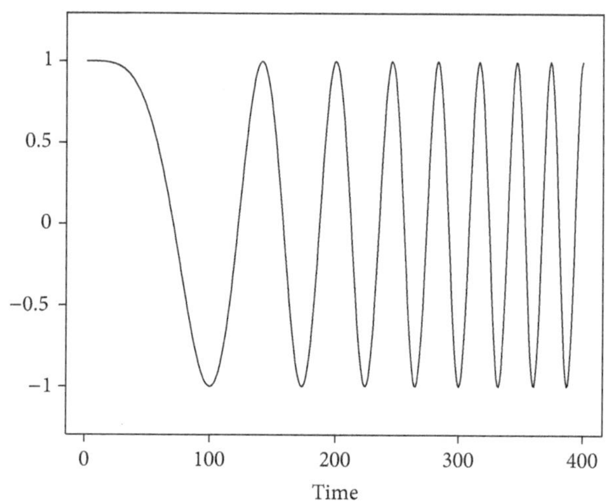

(c) Low frequency component

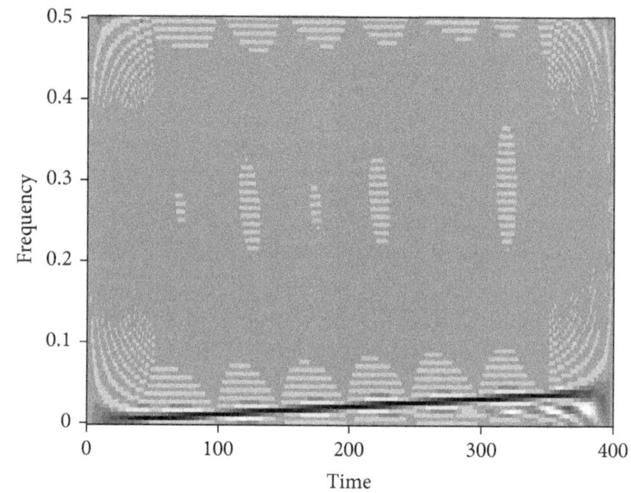

(d) Wigner-Ville plot for (c)

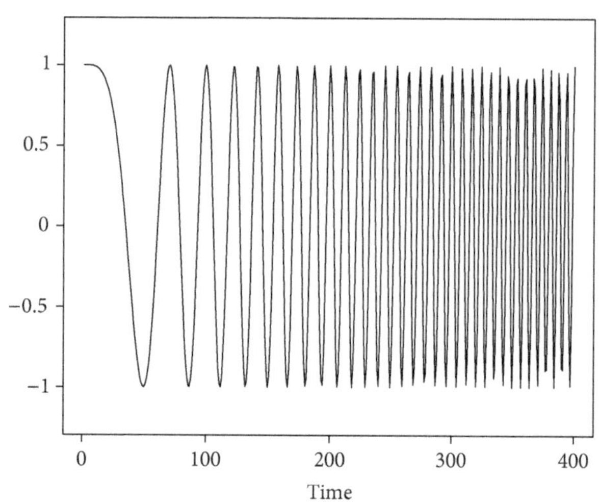

(e) High frequency component

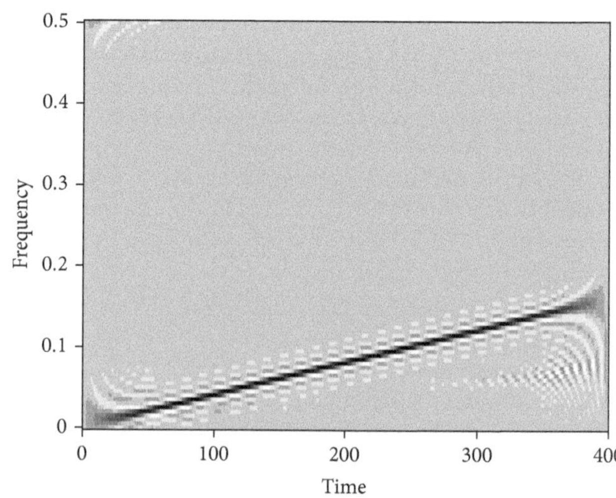

(f) Wigner-Ville plot for (e)

FIGURE 1: (a) A realization from the linear chirp model in Example 3; (b) the Wigner-Ville time-frequency plot for the data in (a); (c) lower frequency TVF component, $Y_1(t)$; (d) Wigner-Ville plot for the signal in (c); (e) higher frequency TVF component, $Y_2(t)$; (f) Wigner-Ville plot for the signal in (e).

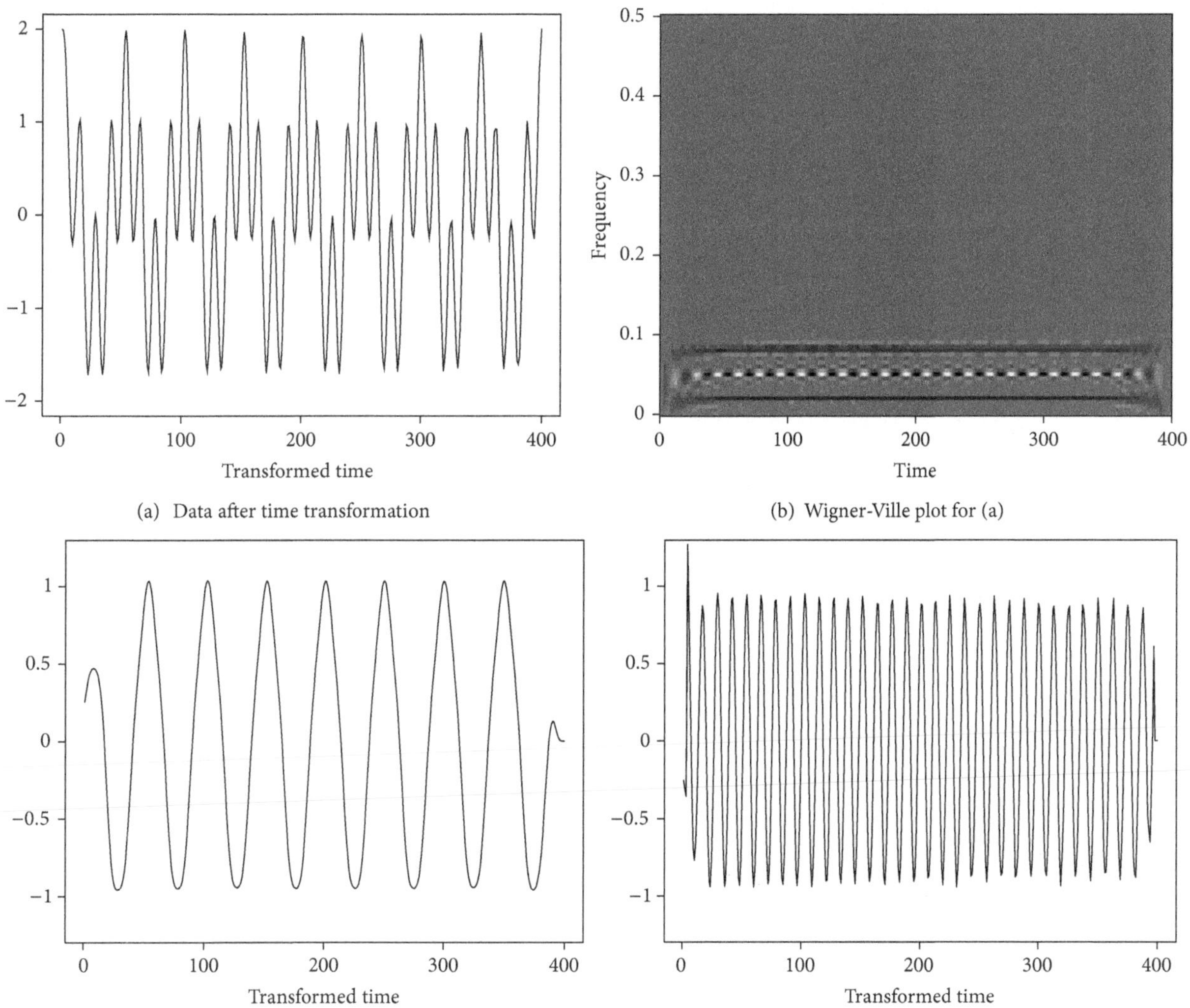

(a) Data after time transformation

(b) Wigner-Ville plot for (a)

(c) Low-pass filtered data

(d) Recovered high frequency TVF signal

FIGURE 2: (a) The dual data obtained from Figure 1(a) using GWS software; (b) Wigner-Ville time-frequency plot for (a); (c) recovered low frequency component from (a); (d) recovered high frequency component from (a).

we show the components $Y_1(t)$ and $Y_2(t)$ where $Y_1(t) = \cos(0.0002\pi t^2/2)$ and $Y_2(t) = \cos(0.0008\pi t^2/2)$. The Wigner-Ville plots associated with these two component signals are shown in Figures 1(d) and 1(f) where it can be seen that $Y_1(t)$ and $Y_2(t)$ display the lower frequency and higher frequency TVF behavior (resp.) shown in Figure 1(b).

To perform Step 1 of the G-filtering procedure, we use the method of Jiang et al. [7] (using GWS software) and find that a $G(\lambda)$ model with $\lambda = 2.1$ and offset 19 is chosen as the transformation which most nearly transform the data in Figure 1(a) into a stationary dual. The resulting dual data are shown in Figure 2(a). Visual inspection suggests that the dual data have stable (nonchanging) frequency behavior with more than one frequency component. The Wigner-Ville plot for the dual data is shown in Figure 2(b) where it is seen that the data contain two stable frequency components at about $f = 0.025$ and $f = 0.075$. Step 2 involves filtering the dual data, which we accomplish by using low-pass and high-pass Butterworth filters with a cutoff frequency $f_c = 0.05$. (The R-function `butterworthT` is available at the website listed above.)

The low and high frequency dual components are shown in Figures 2(c) and 2(d), respectively. The Wigner-Ville plot (not shown) for the data in Figure 2(c) would have the appearance of a single horizontal line at about $f = 0.025$, while the Wigner-Ville plot for Figure 2(d) would show a single horizontal line at about $f = 0.075$. Step 3 in the G-filtering procedure is accomplished by transforming the dual components in Figures 2(c) and 2(d) back to the original time scale. (This is accomplished using the R-function `reinterpol` which is available at the website.) The recovered low and high frequency TVF components of the original data are shown in Figures 3(a) and 3(b). These closely resemble the actual components in Figures 1(c) and 1(e), respectively. The Wigner-Ville plots of the two filtered components are shown

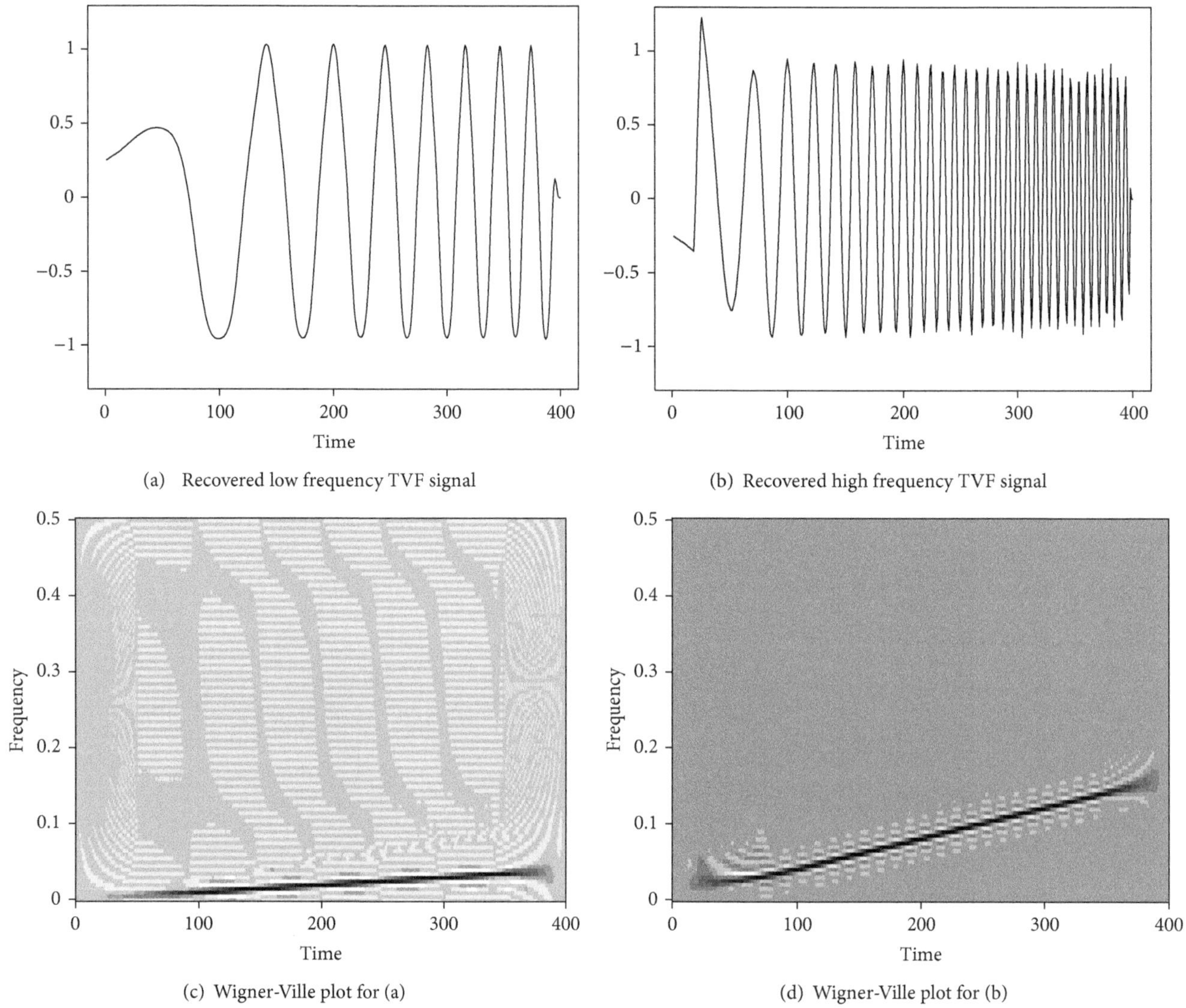

FIGURE 3: (a) Recovered low frequency TVF component; (b) recovered high frequency TVF component; (c) Wigner-Ville plot for (a); (d) Wigner-Ville plot for (b).

in Figures 3(c) and 3(d) where it can be seen that these plots indicate linearly decreasing frequency behavior very similar to that shown in Figures 1(d) and 1(f), respectively, for the true TVF components.

It is important to note that we could not have recovered the TVF components by application of the Butterworth filter directly to the data in Figure 1(a). To see this we notice from Figure 1(b) that there is no single cutoff frequency that can be used to separate the signals across the entire time span ($t = 0$ to $t = 400$). For example, $f_c = 0.1$ could be used to separate the signals from $t = 0$ to about $t = 150$, but beyond $t = 300$ both TVF components are associated with frequencies below $f_c = 0.1$.

For more details concerning the filtering method used in Example 3, see Xu et al. [3]. In the next section we illustrate the use of the time-transformation method to filter more general types of nonstationary processes with multiple time-varying frequency structures.

3. Filtering More General Nonstationary Processes Using the G-Filter

In Example 3, we showed how to use the G-filter to filter a nonstationary processes, called a G-stationary process, by first transforming the realization from a G-stationary process into a realization from a stationary process using time deformation (i.e., transforming the time scale). Specifically, in Example 3 we used the $G(\lambda)$ time transformation $g(t) = (t^{2.1} - 1)/2.1 = u$. In Figures 2(a) and 2(b), it is clear that this transformation approximately stabilized the frequency structure essentially eliminating frequency change across the realization. In many cases, however, a single stationarizing time transformation does not exist. In this section, we will illustrate the extension of the techniques used in Example 3 to such cases.

First, we discuss how the time-deformation technique will affect the spectrum of the data. Suppose that a component

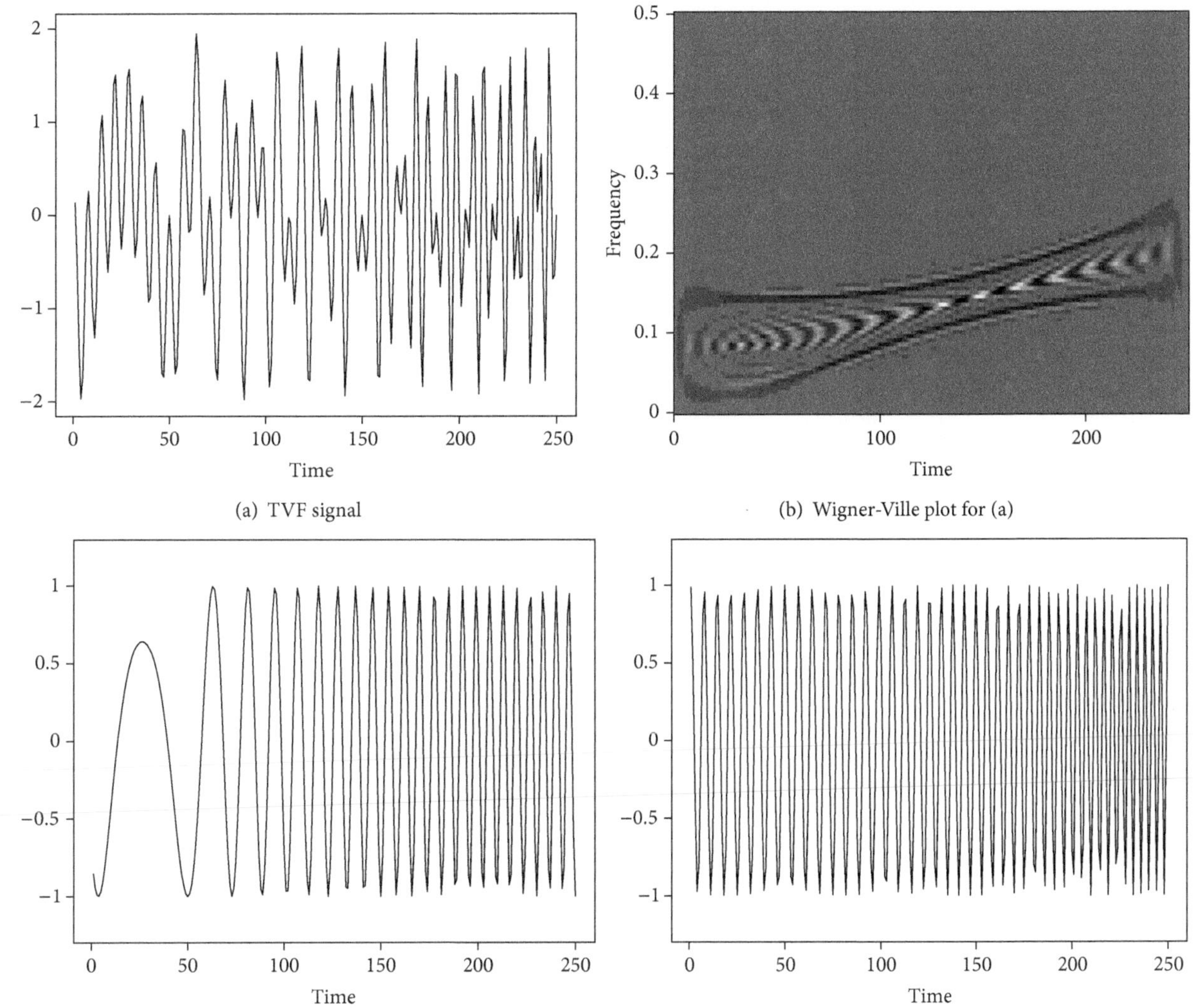

(a) TVF signal

(b) Wigner-Ville plot for (a)

(c) Low frequency TVF component

(d) High frequency TVF component

FIGURE 4: (a) A realization of the process in Example 4; (b) the Wigner-Ville time-frequency distribution of the data; (c), (d) the two components.

has a phase function $2\pi\varphi(t)$; that is, the instantaneous frequency is $\varphi'(t)$ [11]. Let $u = g(t)$ be a monotonic time-deformation function. Then in the dual space, the phase function becomes $2\pi\varphi(g^{-1}(u))$, and the instantaneous frequency is $d\varphi(g^{-1}(u))/du = \varphi'(g^{-1}(u))(dg^{-1}(u)/du) = \varphi'(g^{-1}(u))(1/g'(t)) = \varphi'(t)/g'(t) = \varphi'(g^{-1}(u))/g'(g^{-1}(u))$. In order to stationarize a signal with one TVF, the procedure is to find a time transformation, $g(t)$, such that the instantaneous frequency, $\varphi'(t)/g'(t)$, is constant.

For a process consisting of two components, one with a phase function $2\pi\varphi_1(t)$ and the other $2\pi\varphi_2(t)$, the instantaneous frequencies of these two components are $\varphi_1'(t)$ and $\varphi_2'(t)$. In the dual space, they are $\varphi_1'(t)/g'(t) = \varphi_1'(g^{-1}(u))/g'(g^{-1}(u))$ and $\varphi_2'(t)/g'(t) = \varphi_2'(g^{-1}(u))/g'(g^{-1}(u))$. In Example 3, we obtained a single time transformation, $g(t) = (t^{2.1} - 1)/2.1 = u$, for which $\varphi_1'(t)/g'(t)$ and $\varphi_2'(t)/g'(t)$ were both approximately constant. Visually, from Figure 2(b) we see that $\varphi_1'(t)/g'(t) \approx 0.09$ and $\varphi_2'(t)/g'(t) \approx 0.02$ are represented by the upper and lower horizontal lines, respectively, in the Wigner-Ville plot. This allowed us to select a cutoff frequency between 0.02 and 0.09 for purposes of separating the two components in the dual process.

More generally, it is important to note that regardless of whether $\varphi_1'(t)/g'(t)$ and $\varphi_2'(t)/g'(t)$ are constants, it follows that if $\varphi_1'(t) \geq \varphi_2'(t)$ and $g'(t) > 0$ for all $t \in (a, b)$, then

$$\frac{\varphi_1'\left(g^{-1}(u)\right)}{g'\left(g^{-1}(u)\right)} \geq \frac{\varphi_2'\left(g^{-1}(u)\right)}{g'\left(g^{-1}(u)\right)}. \tag{2}$$

In other words, the component with a higher instantaneous frequency in the original time space still has a higher instantaneous frequency in the dual space. Suppose there exists a constant frequency, f_c, such that

$$\frac{\varphi_1'(t)}{g'(t)} \leq f_c \leq \frac{\varphi_2'(t)}{g'(t)} \tag{3}$$

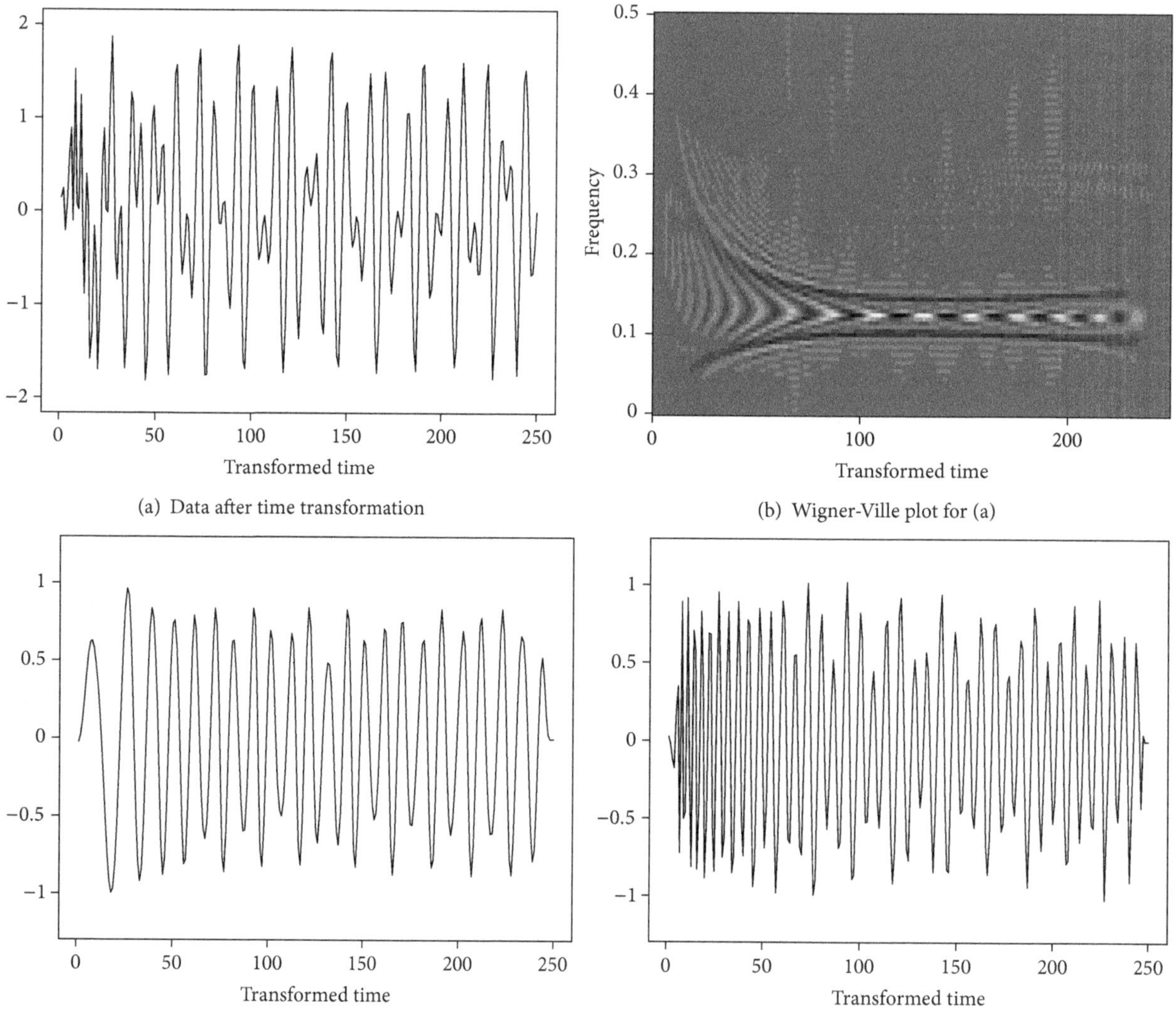

(a) Data after time transformation

(b) Wigner-Ville plot for (a)

(c) Low-pass filtered data from (a)

(d) High pass filtered data from (a)

FIGURE 5: (a) Data on transformed time axis; (b) the Wigner-Ville time-frequency distribution of (a); (c) data in (a) after low-pass filter; (d) data in (a) after high-pass filter.

or

$$\varphi_1'(t) \le f_c g'(t) \le \varphi_2'(t) \quad (4)$$

for all $t \in (a,b)$ (assuming $g'(t) > 0$). Note that (4) says that $f_c g'(t)$ stays between the upper and lower curves shown in the Wigner-Ville plot. From (3) we see that $\varphi_1'(t)/g'(t)$ and $\varphi_2'(t)/g'(t)$ are depicted by the upper and lower Wigner-Ville curves in the dual space, and in this case a Butterworth filter with cutoff frequency f_c should be able to separate these two components in the dual space. The important point is that this separation is possible even if the time transformation did not produce a stationary dual. Thus, if we can find a well-defined time-deformation function $u = g(t)$ and a constant f_c, such that $\varphi_1'(t) \le f_c g'(t) \le \varphi_2'(t)$, for $t \in (a,b)$, then these two components can be extracted from the original data. Example 4 illustrates these comments.

Example 4. Here, we consider a simulated data set which is the signal consisting of the sum of two quadratic chirps,

$$\begin{aligned} X(t) &= \cos\left(2\pi\left(-\left(\frac{t+50}{100}-3\right)^3+0.15t\right)\right) \\ &\quad + \cos\left(2\pi\left(\left(\frac{t+50}{100}-1\right)^3+0.14t\right)\right) \\ &= X_L(t) + X_H(t) \end{aligned} \quad (5)$$

for $0 < t < 250$. The data and the Wigner-Ville time-frequency distribution are displayed in Figures 4(a) and 4(b). The lower frequency TVF component, $X_L(t)$, is shown in Figure 4(c), while the higher frequency component, $X_H(t)$, is shown in Figure 4(d). By examining the Wigner-Ville plot in Figure 4(b) and the component plots in Figures 4(c) and 4(d), it can be seen that the frequency behavior in $X_L(t)$ for

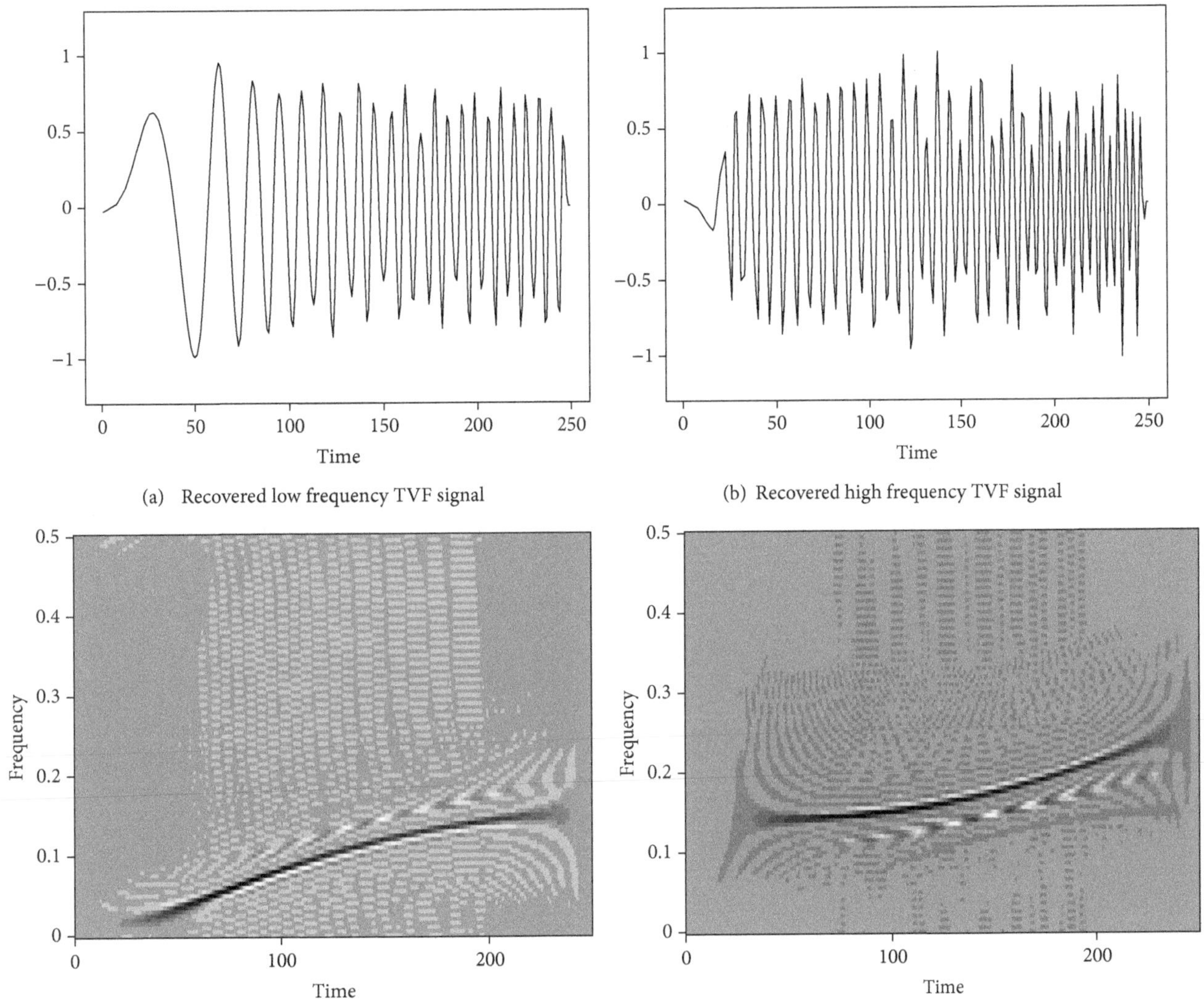

(a) Recovered low frequency TVF signal

(b) Recovered high frequency TVF signal

(c) Wigner-Ville plot for (a)

(d) Wigner-Ville plot for (b)

FIGURE 6: (a) Figure 5(c) plotted on original time scale; (b) Figure 5(d) plotted on original time scale; (c) Wigner-Ville plot for data in (a); (d) Wigner-Ville plot for data in (b).

$200 < t < 250$ is similar to the frequency behavior in $X_H(t)$ for $0 < t < 50$. Consequently, a simple Butterworth filter will not be able to separate the two components.

We apply the G-filtering procedure in an attempt to recover the two TVF components. Step 1 involves using the procedure of Jiang et al. [7] in an attempt to stationarize the signal in Figure 4(a). However, the nature of the two quadratic chirp components is such that there is not a single time transformation, $g(t)$, that will stationarize the signal $X(t)$. Using the GWS software, a $G(\lambda)$ time transformation with $\lambda = 1.7$ and offset $= 3$ is selected as the transformation that most nearly stationarizes the data. Figures 5(a) and 5(b) show the "dual" data and the associated Wigner-Ville plot where it can be seen that the dual does not appear to be stationary. The key point, however, is that even though the transformation has not stationarized the data (the frequency lines in the Wigner-Ville plot are not straight horizontal lines), the upper and lower frequency components can be separated with a simple Butterworth filter at, for example, a cutoff frequency $f = 0.13$. Step 2 in the G-filtering procedure can thus be performed, and the resulting low and high frequency components after filtering the dual data with a Butterworth filter (e.g., using R-function `butterworthT`) are shown in Figures 5(c) and 5(d), respectively. We complete the example by applying Step 3 to plot the filtered dual components back on the original time scale. In Figures 6(a) and 6(b), we show the filtered data plotted on the original time scale (using R-function `reinterpol`), and in Figures 6(c) and 6(d) we show the Wigner-Ville plots for these components. The Wigner-Ville plots in Figures 6(c) and 6(d) are good approximations to the lower and upper curves, respectively, in Figure 4(b) associated with the lower and upper components of the original signal. We comment that the poor behavior at the endpoints is a well-known property of filters such as the Butterworth, and no

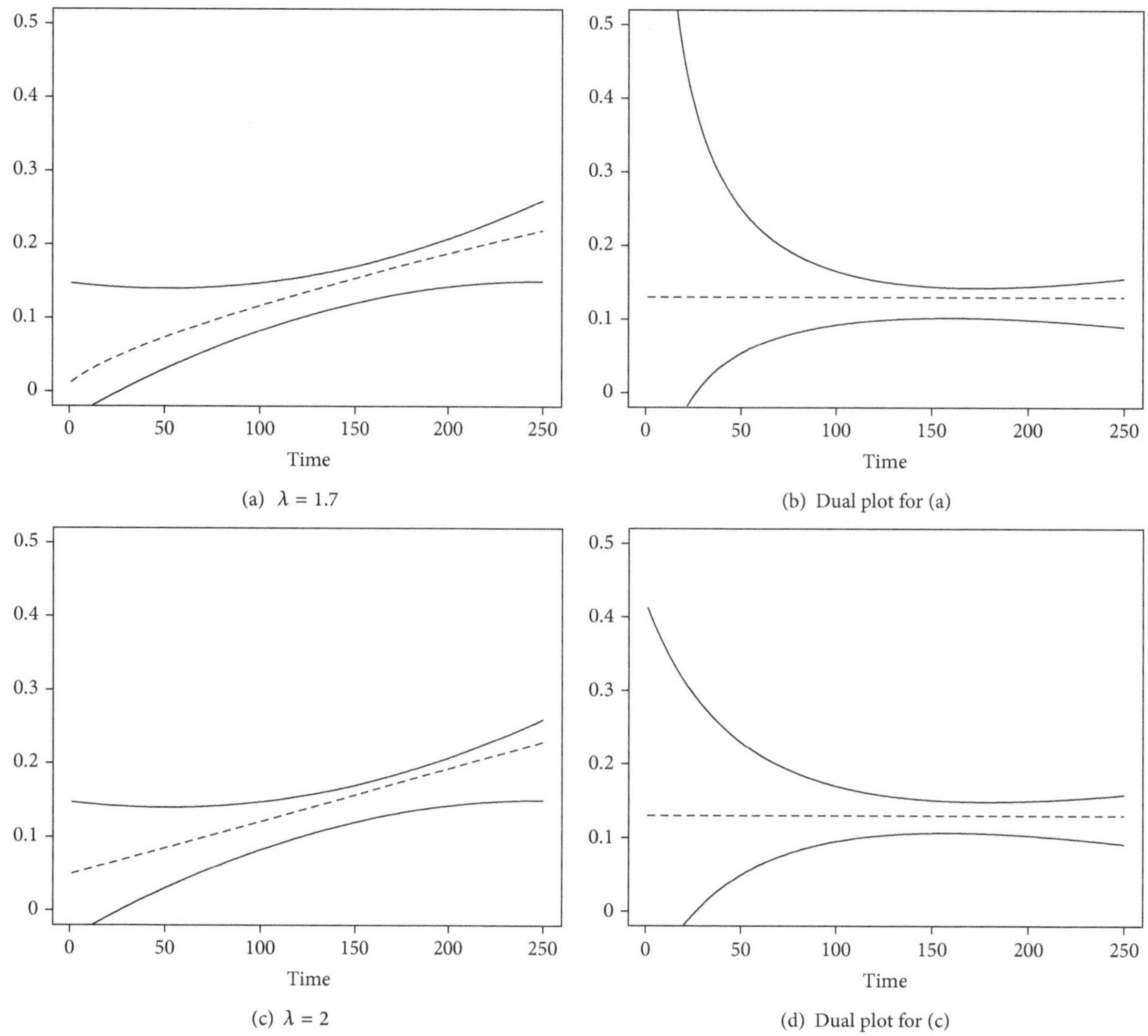

FIGURE 7: (a) Sketch of upper and lower Wigner-Ville curves for Example 4 showing $f_c g(t)$ for $f_c = 0.13$ and $g(t)$ given in (6); (b) sketch of Wigner-Ville curves in dual space; (c) sketch of upper and lower Wigner-Ville curves for Example 4 showing $f_c g(t)$ for $f_c = 0.13$ and $g(t)$ given in (7); (d) sketch of Wigner-Ville curves in dual space.

endpoint adjustments have been made here. Also, the loss of amplitude in the recovered higher frequency regions (see, e.g., Figure 6(d) compared to Figure 4(b)) is at least partially due to the linear interpolation that is used. Haney [13] has shown that the use of Fourier-based interpolation may be able to adjust for some of the signal loss.

Note that Example 4 illustrates the fact that if we can find a time transformation, $g(t)$, and frequency, f_c, such that (4) holds, then the two time-varying frequencies can be separated. This is illustrated in Figure 7(a) where the dotted curve represents $f_c g'(t)$ for $f_c = 0.13$ and where $g(t)$ is the $G(\lambda)$ transformation used in Example 4 with $\lambda = 1.7$ and offset 3; that is,

$$g(t) = \frac{(t+3)^{1.7} - 1}{1.7}. \quad (6)$$

The solid curves are the upper and lower Wigner-Ville curves shown in Figure 4(b) which represent $\varphi_1'(t)$ and $\varphi_2'(t)$. It can be seen that for the time transformation in (6), the curve $f_c g'(t)$ stays between the upper and lower Wigner-Ville curves. Figure 7(b) shows upper and lower Wigner-Ville curves in the associated dual space; that is, from (3) these are $\varphi_1'(t)/g'(t)$ and $\varphi_2'(t)/g'(t)$, respectively. The straight dashed line corresponds to $f_c = 0.13$.

In Figures 7(c) and 7(d) we show the corresponding plots associated with making a time transformation

$$g(t) = \frac{(t+68)^{2} - 1}{2}. \quad (7)$$

It can be seen that either of the time transformation in (6) or (7) provides separation in the dual space sufficient for application of a filter such as the Butterworth.

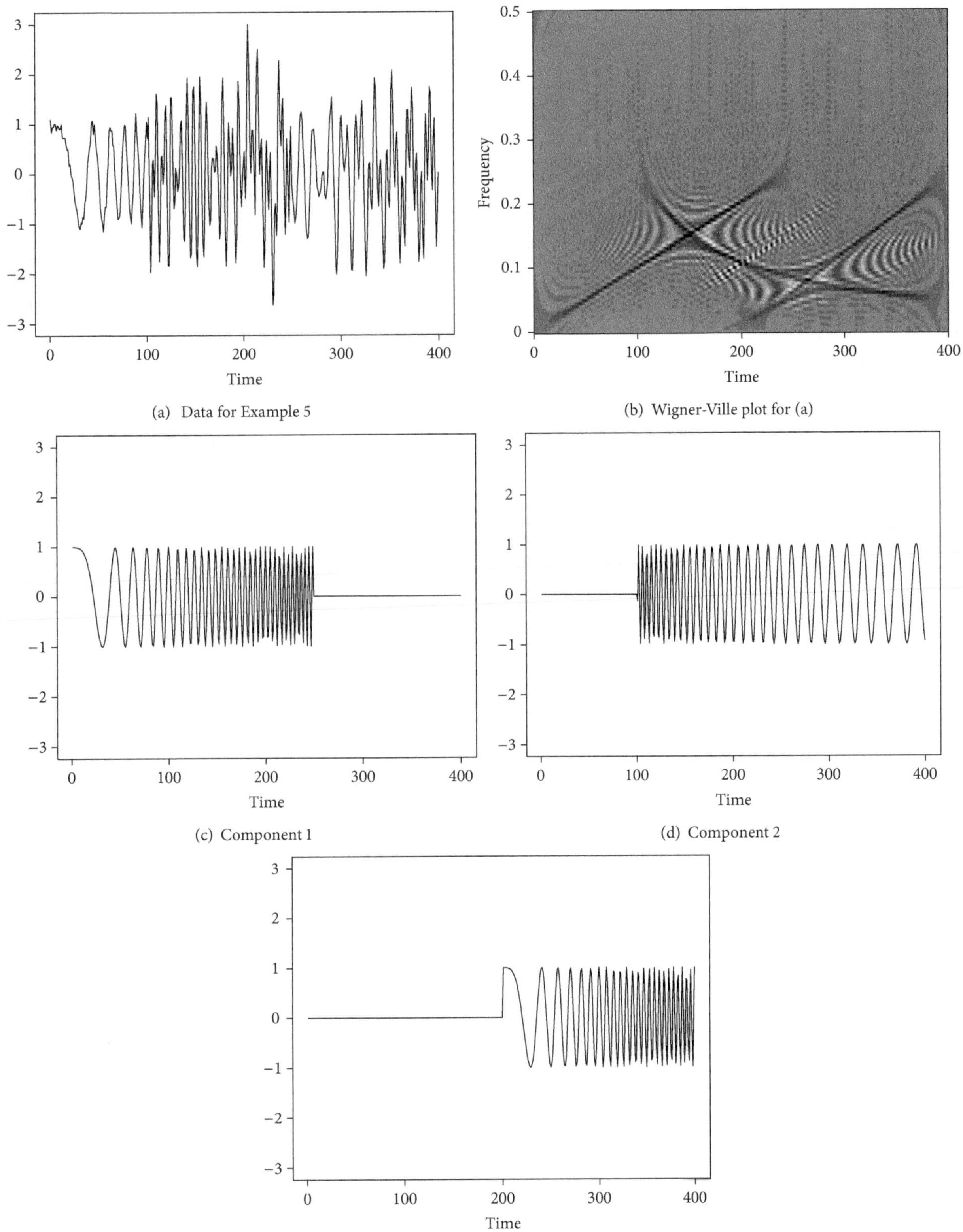

(a) Data for Example 5

(b) Wigner-Ville plot for (a)

(c) Component 1

(d) Component 2

(e) Component 3

FIGURE 8: (a) A realization of the process in Example 5; (b) the Wigner-Ville time-frequency distribution of the data; (c), (d), and (e) the three components.

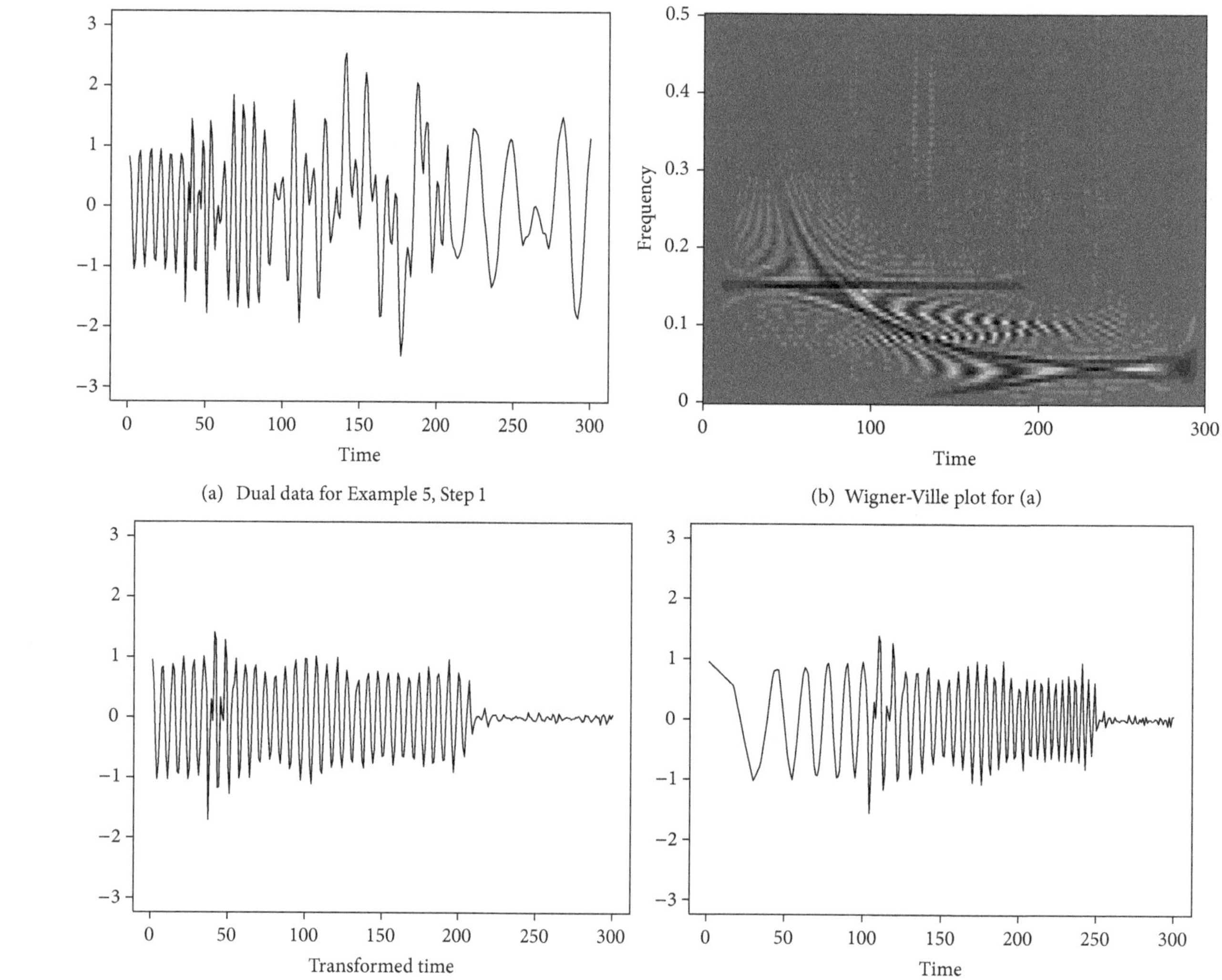

(a) Dual data for Example 5, Step 1

(b) Wigner-Ville plot for (a)

(c) Extracted dual component 1

(d) Recovered chirp component 1

FIGURE 9: (a) The dual data (1 : 300) after the first time deformation; (b) the Wigner-Ville time-frequency distribution of the dual; (c) the extracted stationary dual component; (d) the recovered first linear chirp component.

4. Extracting TVF Components Entering the Signal at Different Times

In this section we discuss methods for extracting components in more complicated TVF settings in which components may enter and/or exit the overall signal at different times. The procedure will be illustrated in Example 5. Suppose the high (low) frequency component in a TVF signal is G-stationary over a certain time span, and we apply the appropriate time transformation to stationarize this G-stationary component over that time span. Then by (2), on the transformed version of the time span being considered, this component will become stationary with a constant frequency that is still higher (lower) than the instantaneous frequency of the other component in the dual space. The key idea addressed in Example 5 is that in order to extract signals, it may be useful to segment the overall realization length into sections over which straightforward signal separations can be obtained using the techniques illustrated in Examples 3 and 4. We use these ideas in the following example.

Example 5. In this example, we simulate a situation similar to that studied by Papandreou-Suppappola and Suppappola [14]. The synthetic signal is shown in Figure 8(a), and the Wigner-Ville plot of this realization is shown in Figure 8(b). From Figure 8(b) we observe the following.

(1) The overall signal consists of three TVF component signals. We will refer to these as Components 1, 2, and 3. Based on the Wigner-Ville plot we conclude that

   (a) component 1 exists for approximately $1 \leq t \leq 250$;

   (b) component 2 exists for approximately $t \geq 100$;

   (c) component 3 exists for approximately $t \geq 200$.

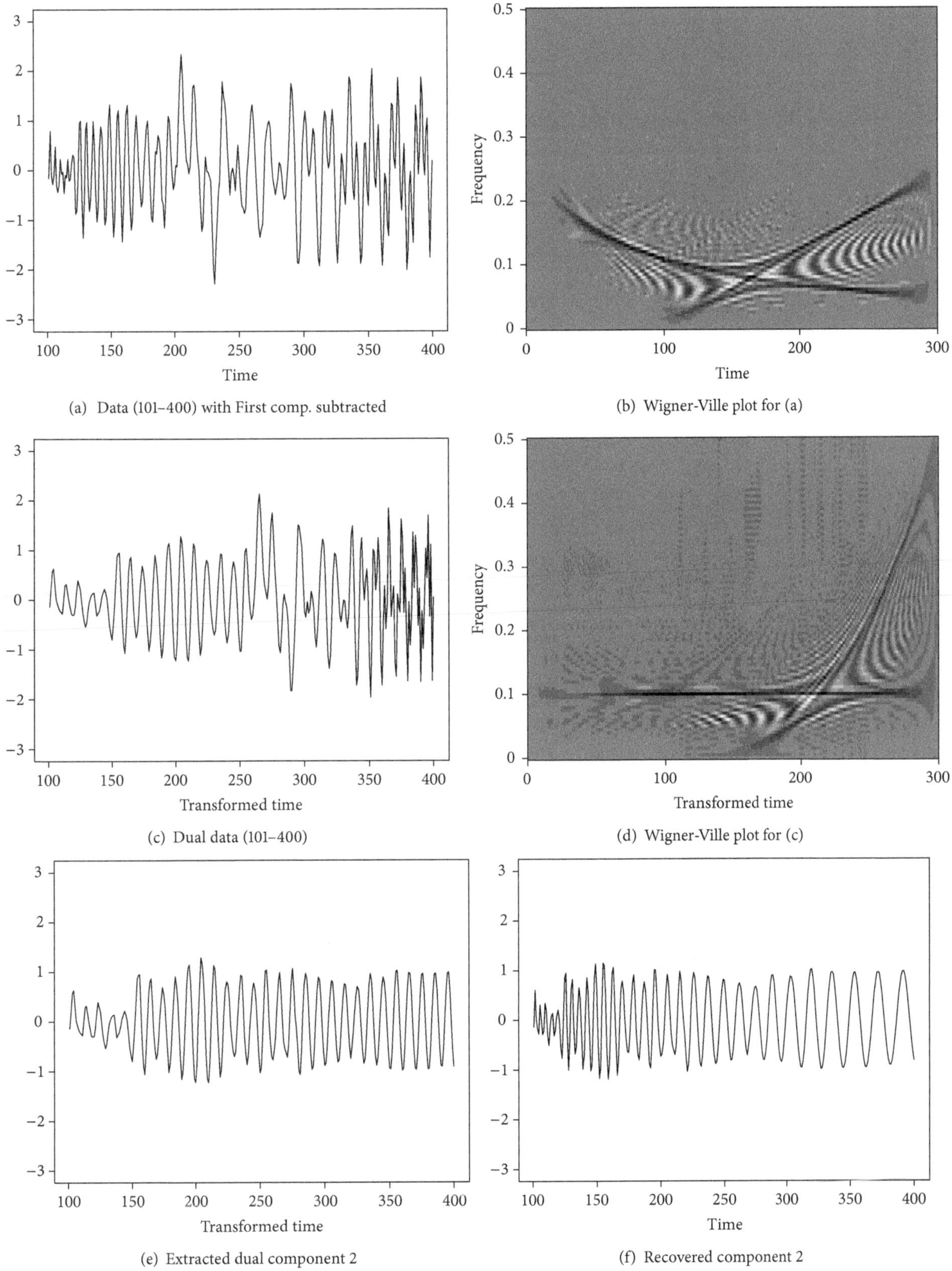

FIGURE 10: (a) The data (101 : 400) with the first filtered component subtracted; (b) the Wigner-Ville time-frequency distribution of the data; (c) the dual data; (d) the Wigner-Ville time-frequency distribution of the dual; (e) the extracted stationary dual component; (f) the recovered M-stationary component.

(2) Components 1 and 3 have Wigner-Ville curves that appear to be straight lines, suggesting that each of these is a linear chirp signal.

(3) Component 1 exists in the signal by itself for about $t \le 100$.

(4) The frequencies associated with the components intersect at two places:

(a) frequencies associated with components 1 and 2 intersect at about $t = 150$;

(b) frequencies associated with components 2 and 3 intersect at about $t = 275$.

Figures 8(b)-8(c) show the component signals actually used to create the actual signal, X_t, which is the sum of the components; that is, $X_t = X_t^{(1)} + X_t^{(2)} + X_t^{(3)}$, $t = 1, \ldots, 400$. We see that our observations in conclusion 1 mentioned above are correct. Also, it should be noted that components 1 and 3 were generated as linear chirps (as was speculated in conclusion 2 mentioned above) and component 2 is M-stationary, but this information will not be assumed to be known in the following analysis.

Step 1 (Extract Component 1). Since component 1 exists by itself for $t \le 100$, we use the method of Jiang et al. [7] (using the GWS software) to find a $G(\lambda)$ model for stationarizing the signal composed of the first 100 points. A $G(2)$ model is selected with offset 0. We apply this time transformation to the first 300 data values, and the resulting dual process is shown in Figure 9(a). The associated Wigner-Ville plot is shown in Figure 9(b) where it is seen that component 1 has now been stationarized (the associated frequency behavior has been stabilized, and it appears as a horizontal line at about $f = 0.15$ in the Wigner-Ville plot). Note that although the segment of the original data to which we applied the time deformation ($1 \le t \le 300$) and the resulting dual data set both have 300 time values, the original time axis has been deformed. Consequently, the first 100 data values (involving only component 1) in the original time scale correspond to about the first 50 values in transformed time. By examining Figure 9(b) we see that the dual data are now split into three segments in the deformed time axis as follows:

(i) the first segment (1–50) which contains only the transformed component 1 signal;

(ii) the middle segment (51–100) where the frequencies intersect;

(iii) the last segment (101–300) where the frequencies are well separated. In this segment the transformed component 1 signal is consistently higher frequency than the transformed component 2 and component 3 signals. In this range, a constant cutoff frequency (e.g., 0.12) can be used to separate component 1 from the other two components.

In order to extract the stationarized dual signal associated with component 1 we note that no extraction is necessary for the first 50 dual time values (the only signal present is the one associated with the first linear chirp component). As suggested in note (iii) above, extraction of component 1 in the range 101–300 can be accomplished using a high-pass Butterworth filter.

Extraction of the first linear chirp component in the range 51–100 is more difficult. The procedure we have selected is to fit AR(p) models to the two extracted signals (i.e., 1–50 and 101–300) and then to use these models to forecast and backcast on the middle segment (51–100). The resulting extracted dual signal associated with component 1 is shown in Figure 9(c). Transforming this signal back to the original time scale gives Figure 9(d) which is similar to the true component shown in Figure 8(c). Note again that although the first dual component terminates at about a transformed time value of 210, the first component in the original time scale terminates at about $t = 250$ as expected.

Step 2 (Extract Component 2). In the second step, we will extract component 2. We first subtract the recovered component 1 (shown in Figure 9(d)) from the original signal leaving only components 2 and 3. The resulting data ($101 \le t \le 400$) and corresponding Wigner-Ville plot are shown in Figures 10(a) and 10(b). We note the following:

(i) only TVF components 2 and 3 remain;

(ii) component 2 exists by itself for $101 \le t \le 200$.

Based on note (ii) above, we apply the method of Jiang et al. [7] to find a stationarizing transformation for the data in time segment 101–200 in Figure 10(a). Using this method, a $G(\lambda)$ model with $\lambda = 0$ is selected with offset 69. Applying this time transformation to the data for $101 \le t \le 400$, we obtain the dual data shown in Figure 10(c) and associated Wigner-Ville plot in Figure 10(d). From the Wigner-Ville plot in Figure 10(d) we note the following:

(i) the second component seems to have been stationarized using this procedure (frequency behavior is represented by a horizontal line at about $f = 0.1$ in Wigner-Ville plot);

(ii) the first segment (101–250) contains only the stationarized component 2;

(iii) transformed component 2 is consistently lower frequency than transformed component 3 for the range 251–280 and the signals could be separated by a constant cutoff frequency of about 0.075;

(iv) the frequencies intersect in the range 281–340;

(v) the transformed component 2 is consistently lower frequency than transformed component 3 for the range 341–400. A constant cutoff frequency of about 0.175 could be used to separate the signals in this range.

Based on these comments we apply a high-pass Butterworth filter to extract component 2 in the range 101–280 and a low-pass Butterworth filter to extract the dual for component 2 for the range 341–400. Forecasting and backcasting are used as in Step 1 for recovering the component 2 dual for the range 281–340. The resulting extracted signal associated with the dual

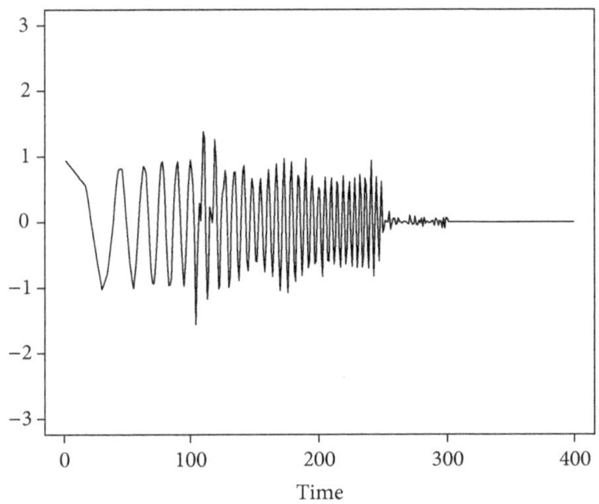

(a) Recovered first component (linear chirp)

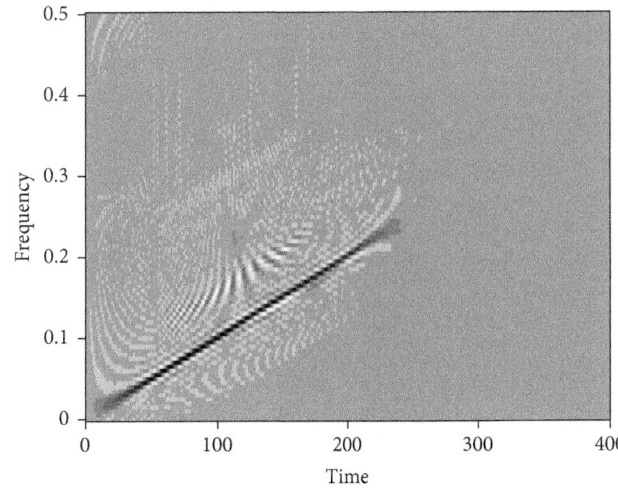

(b) Wigner-Ville plot for (a)

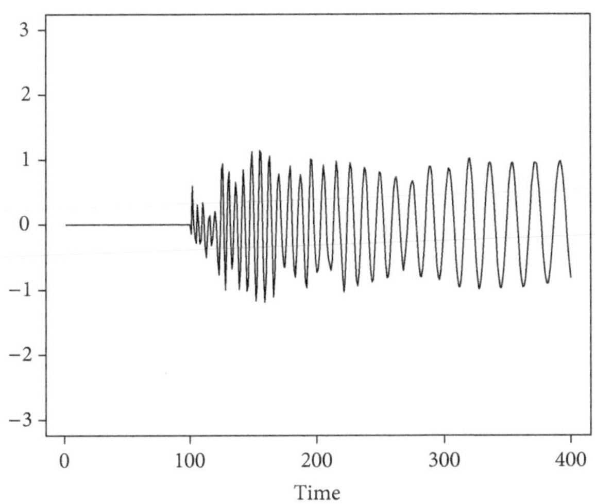

(c) Recovered second component (M-stationary)

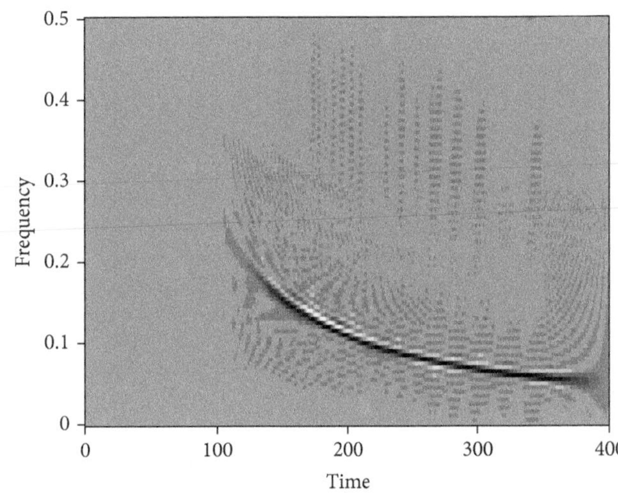

(d) Wigner-Ville plot for (c)

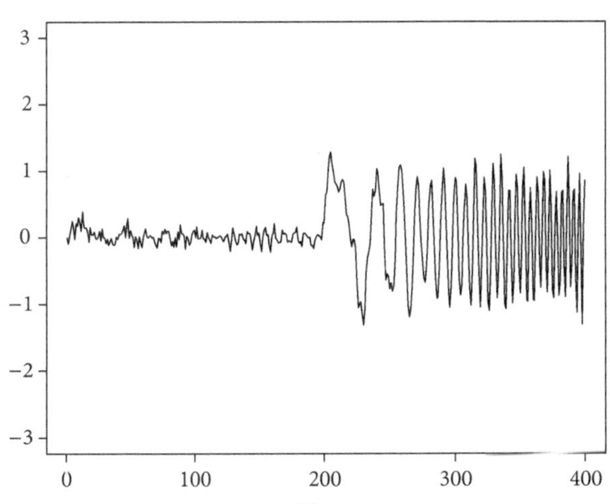

(e) Recovered third component (linear chirp)

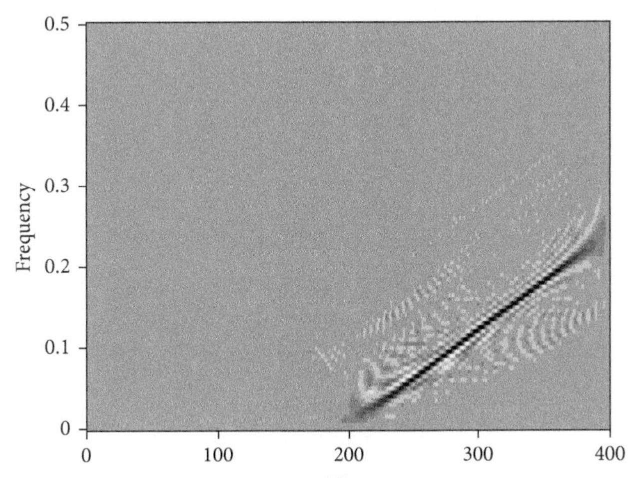

(f) Wigner-Ville plot for (e)

FIGURE 11: (a), (c), and (e): recovered components for Example 5 and (b), (d), and (e): the corresponding Wigner-Ville plots.

for component 2 is shown in Figure 10(e), and Figure 10(f) shows this component transformed back to the original time scale.

Step 3 (Extract Component 3). The third component is obtained by subtracting the two recovered components from the original data; that is, $\widehat{X}_t^{(3)} = X_t - \widehat{X}_t^{(1)} - \widehat{X}_t^{(2)}$, for $t = 1, \ldots, 400$ where $\widehat{X}_t^{(k)}$ represents the recovered kth component.

Figure 11 summarizes the results showing the three recovered components and associated Wigner-Ville plots which correspond well with those shown Figure 8.

5. Conclusion

For nonstationary time series with multiple time-varying frequency structure, especially where the frequencies of components overlap over time, traditional linear filters are not able to successfully extract components. In order to address this problem, the G-filter, a time-variant filter based on the time-deformation method, was studied by Xu et al. [3]. In that paper the G-filter was applied to filter components from certain types of G-stationary processes by transforming them to stationarity using an appropriate time transformation of the time scale.

In practical situations, most nonstationary data with multiple time-varying frequency structure cannot be transformed to stationary by applying any time deformation. In this paper, we showed that the G-filtering procedure can be extended to cases in which the individual component signals may or may not be able to be fully stationarized using a time transformation. We first discussed the case (see Example 4) in which the frequency behaviors of the components of a process can be separated by a time transformation even though a stationarizing transformation is not available. This significantly extends the class of TVF processes that can be filtered using the G-filter. Secondly, we show in Example 5 that the G-filter can be used to filter signals with multiple G-stationary components having varying points of entry into and exit from the signal.

References

[1] A. Papoulis, *Probability, Random Variables, and Stochastic Processes*, McGraw-Hill, New York, NY, USA, 1984.

[2] S. Butterworth, "On the theory of tilter amplifiers," *Experimental Wireless and the Wireless Engineer*, vol. 7, pp. 536–541, 1930.

[3] M. Xu, K. B. Cohlmia, W. A. Woodward, and H. L. Gray, "G-filtering nonstationary time series," *Journal of Probability and Statistics*, vol. 2012, Article ID 738636, 15 pages, 2012.

[4] P. Flandrin, P. Borgnat, and P. O. Amblard, "From stationarity to self-similarity, and back: variations on the Lamperti transformation," in *Processes With Long-Range Correlations*, G. Raganjaran and M. Ding, Eds., pp. 88–117, Springer, New York, NY, USA, 2003.

[5] H. L. Gray and N. F. Zhang, "On a class of nonstationary processes," *Journal of Time Series Analysis*, vol. 9, no. 2, pp. 133–154, 1988.

[6] H. L. Gray, Chu-P.C. Vijverberg, and W. A. Woodward, "Nonstationary data analysis by time deformation," *Communications in Statistics*, vol. 34, no. 1, pp. 163–192, 2005.

[7] H. Jiang, H. L. Gray, and W. A. Woodward, "Time-frequency analysis—G(λ)-stationary processes," *Computational Statistics & Data Analysis*, vol. 51, no. 3, pp. 1997–2028, 2006.

[8] W. A. Woodward, H. L. Gray, and A. C. Elliott, *Applied Time Series Analysis*, Chapman and Hall/CRC, Boca Raton, Fla, USA, 2012.

[9] S. D. Robertson, H. L. Gray, and W. A. Woodward, "The generalized linear chirp process," *Journal of Statistical Planning and Inference*, vol. 140, no. 12, pp. 3676–3687, 2010.

[10] W. Martin and P. Flandrin, "Wigner-Ville spectral analysis of nonstationary processes," *IEEE Transactions on Acoustics, Speech, and Signal Processing*, vol. 33, no. 6, pp. 1461–1470, 1985.

[11] B. Boashash, *Time Frequency Analysis*, Elsevier, Oxford, UK, 2003.

[12] R. G. Baraniuk and D. L. Jones, "Unitary equivalence: a new twist on signal processing," *IEEE Transactions on Signal Processing*, vol. 43, no. 10, pp. 2269–2282, 1995.

[13] J. R. Haney, *Analyzing time series with time-varying frequency behavior and conditional heteroskedasticity [Ph.D. thesis]*, Southern Methodist University, Department of Statistical Science, Dallas, Texas, USA, 2010.

[14] A. Papandreou-Suppappola and S. B. Suppappola, "Analysis and classification of time-varying signals with multiple time-frequency structures," *IEEE Signal Processing Letters*, vol. 9, no. 3, pp. 92–95, 2002.

10

Growth Estimators and Confidence Intervals for the Mean of Negative Binomial Random Variables with Unknown Dispersion

David Shilane[1] and Derek Bean[2]

[1] *Department of Health Research and Policy, Stanford University, Stanford, CA 94305, USA*
[2] *Department of Statistics, University of California, Berkeley, CA 94720, USA*

Correspondence should be addressed to David Shilane; dshilane@gmail.com

Academic Editor: Peter van der Heijden

The negative binomial distribution becomes highly skewed under extreme dispersion. Even at moderately large sample sizes, the sample mean exhibits a heavy right tail. The standard normal approximation often does not provide adequate inferences about the data's expected value in this setting. In previous work, we have examined alternative methods of generating confidence intervals for the expected value. These methods were based upon Gamma and Chi Square approximations or tail probability bounds such as Bernstein's inequality. We now propose growth estimators of the negative binomial mean. Under high dispersion, zero values are likely to be overrepresented in the data. A growth estimator constructs a normal-style confidence interval by effectively removing a small, predetermined number of zeros from the data. We propose growth estimators based upon multiplicative adjustments of the sample mean and direct removal of zeros from the sample. These methods do not require estimating the nuisance dispersion parameter. We will demonstrate that the growth estimators' confidence intervals provide improved coverage over a wide range of parameter values and asymptotically converge to the sample mean. Interestingly, the proposed methods succeed despite adding both bias and variance to the normal approximation.

1. Introduction

Confidence intervals are routinely applied to limited samples of data based upon their asymptotic properties. For instance, the central limit theorem states that the sample mean $\overline{X}$ will have an approximately Normal distribution for large sample sizes provided that the data's second moment is finite. This Normal approximation is a fundamental tool for inferences about the data's expected value in a wide variety of settings. Confidence intervals for the mean are often based upon Normal quantiles even when the sample size is very moderate (e.g. 30 or 50). However, the Normal approximation's quality cannot be ensured for highly skewed distributions [1]. In this setting, the sample mean may converge to the Normal in distribution at a much slower rate. The Negative Binomial distribution is known to have an extremely heavy right tail, especially under high dispersion.

In previous work [2], we established that the Normal confidence interval significantly undercovers the mean at moderate sample sizes. We also suggested alternatives based upon Gamma and Chi Square approximations along with tail probability bounds such as Bernstein's inequality. We now propose growth estimators for the mean. These estimators seek to account for the relative overrepresentation of zero values in highly dispersed negative binomial data. This may be accomplished by imposing a correction factor as an adjustment to the mean or by directly removing a small number of zero values from the sample. We will demonstrate that these alternative procedures provide a confidence interval with improved coverage. Estimators based on the growth method can also be shown to asymptotically converge in sample size to the Normal approximation.

Section 2 reviews the negative binomial distribution and provides some discussion of parameter estimation under high dispersion. Section 3 reviews existing methods of constructing confidence intervals for the mean of negative binomial random variables. In Section 4, we introduce the growth method and propose two new confidence intervals for the negative binomial mean. Section 5 conducts a simulation experiment to compare these methods to existing procedures in terms of coverage probability. Finally, we will conclude the paper with a discussion in Section 6.

2. The Negative Binomial Distribution

The Negative Binomial distribution models the probability that a total of $k \in \mathbb{Z}^+$ failures will result before $\theta \in \mathbb{R}^+$ successes are observed. Each count X is constructed from (possibly unobserved) independent trials that each result in success with a fixed probability $p \in (0, 1)$. The expected value of X is $\mu = \theta(1/p - 1)$. The Negative Binomial distribution may be alternatively parameterized in terms of the mean μ and *dispersion* θ directly [3]. For any $\mu \in \mathbb{R}^+$ and $\theta \in \mathbb{R}^+$, a Negative Binomial random variable $X \sim \text{NB}(\mu, \theta)$ has the probability mass function

$$P(X = x) = \frac{\mu^x}{x!} \frac{\Gamma(\theta + x)}{\Gamma(\theta)[\mu + \theta]^x} \frac{1}{(1 + \mu/\theta)^\theta}, \quad x \in \mathbb{Z}^+. \quad (1)$$

The variance of X is given by $\sigma^2 = \mu + \mu^2/\theta$. As θ grows large, (1) converges to the probability mass function of a Poisson random variable. Smaller values of θ lead to larger variances, so the selection of θ controls the degree of dispersion in the data. At very small values of θ, the dispersion becomes extreme, and the data may be relatively sparse. Because the Negative Binomial distribution has a heavy right tail, the small number of nonzero values may be spread over an extremely wide range. In light of these concerns, it is not surprising that the normal approximation exhibits a slow convergence as a function of sample size.

For all methods, we assume that the data consist of $n \in \mathbb{Z}^+$ independent, identically distributed (i.i.d.) $\text{NB}(\mu, \theta)$ random variables, where μ and θ are unknown. We seek to generate accurate and reliable inferences about μ. The dispersion θ may be considered a nuisance parameter. In the previous work of Shilane et al. [2], some methods relied upon estimates of θ while others directly estimated the variance σ^2 with the unbiased estimator s^2. In general, we prefer to estimate the variance directly where possible. A variety of research suggests that estimating small values of θ is especially difficult in small sample sizes. Some existing procedures include the method of moments estimator $\hat{\theta} = \overline{X}/((s^2/\overline{X}) - 1)$ and an iterative maximum likelihood estimator (MLE) [4, 5]. Aragón et al. [6] and Ferreri [7] provide conditions for the existence and uniqueness of the MLE. Meanwhile, Pieters et al. [8] compare an MLE procedure to the method of moments at small sample sizes. These procedures encounter difficulties when the variance estimate s^2 is less than the sample mean $\overline{X}$. The method of moments estimator will provide an implausible negative number, while maximum likelihood procedures will produce highly variable results by constraining s^2 to be at least as large as $\overline{X}$. (Additionally, the **glm.nb** method in the **R** Statistical programming language will often produce computational errors in this setting rather than return an MLE for θ.)

For any $\alpha \in (0, 1)$, we seek to construct high-quality $1 - \alpha$ confidence intervals for the mean μ based upon a sample $X_1, \ldots, X_n$ of i.i.d. $\text{NB}(\mu, \theta)$ random variables. A method's *coverage probability* is the chance that the interval will contain the parameter of interest μ as a function of the sample size n and the parameters μ and θ. We will primarily judge the quality of a confidence interval in terms of its coverage probability, which ideally would be exactly $1-\alpha$ across all sample sizes and parameter values. However, there are many secondary factors that can impact the selection of methods. Shorter intervals provide greater precision and insight about the underlying scientific problem. The variability of this length should also be minimized. When this variability is too great, the resulting interval may significantly understate or overstate the degree of certainty about the parameter range. Where possible, we prefer methods that can be assured of producing plausible parameter ranges. Since the Negative Binomial distribution draws from a nonnegative sample space, we therefore prefer methods that will result in a nonnegative confidence interval.

3. Prior Methods

The normal approximation and the bootstrap Bias Correct and accelerated (BCA) method [9] are considered standard techniques for the construction of $1 - \alpha$ confidence intervals for the mean. Shilane et al. [2] found that these procedures perform similarly over a wide variety of sample sizes and parameter values in negative binomial models. They also proposed several alternative methods, including Gamma and Chi square approximations along with tail probability bounds such as Bernstein's inequality. These methods improved upon the standard techniques in terms of coverage probabilities over complimentary subspaces of parameter values. We will briefly review these techniques in the following subsections.

3.1. The Gamma Approximation. Shilane et al. [2] proved a limit theorem stating that $\overline{X}$ converges to a Gamma distribution as the sample size n grows large and the dispersion parameter θ approaches zero. The shape parameter is θn, and the rate parameter is given by $\theta n/\mu$. The Gamma approximation therefore requires estimates of μ and θ. The sample mean $\overline{X}$ may be plugged in for μ directly. Due to the difficulties previously discussed with MLE methods under high dispersion, we will rely upon a method of moments estimator of θ as a default. (In practice, an MLE may be substituted where possible). To account for the possibility of negative values, we recommend truncating all values below a small number (such as 10^{-5} by default) to ensure positive estimates. Once the shape and rate parameters are estimated, a $1 - \alpha$ confidence interval for μ is given by the $(\alpha/2)$nd and $(1 - \alpha/2)$th quantiles of the Gamma distribution.

The Gamma approximation generally performs best when n is large and θ is small. In this setting, the Gamma method improves upon the Normal approximation by a coverage probability of about 1-2%. At more moderate sample sizes and dispersions, the Gamma approximation is not especially accurate and often performs worse than the Normal approximation.

3.2. The Chi Square Approximation. A special case of the Gamma approximation of Section 3.1 occurs when $\mu = 2n\theta$. In this setting, the Gamma parameters correspond to a Chi Square distribution with μ degrees of freedom. This Chi Square approximation works especially well when the parameter relationship is approximately equivalent. The parameter relationship may also be phrased in terms of a *ratio statistic* $\mu/(2n\theta)$. Simulation studies conducted by Shilane et al. [2] demonstrate that the Chi Square approximation will provide reasonably good coverage when the ratio statistic is reasonably close to 1, such as values between 2/3 and 1.5. Moreover, the ratio statistic provides information about whether the Chi Square interval is too narrow or too wide. When the ratio statistic exceeds 1, the interval is too wide, and when the ratio is less than 1, it is too narrow. This information may be used to compare the results of other procedures even when the Chi Square method performs poorly. However, it should be emphasized that the Chi Square approximation should be limited in its applications, and asymptotically it will severely overcover the mean.

3.3. Bernstein's Inequality. At small sample sizes and high dispersion, parametric methods that construct confidence intervals by inverting hypothesis testing procedures [10–13] may be inadequate. Tail probability bounds provide an alternative methodology that typically relies upon more mild assumptions about the data. Bounds such as Bernstein's inequality [14], Bennett's inequality [15, 16], or methods based on the work of Hoeffding [17] and Berry-Esseen [18–21] may be employed. Rosenblum and van der laan [22] apply these bounds to produce confidence intervals based on the estimators' empirical influence curves.

In the Negative Binomial setting, Shilane et al. [2] adapted Bernstein's inequality to provide an improvement over a naive confidence interval. The method requires only independent data with finite variance and imposes a heuristic assumption of boundedness in a range $(a, b) \in \mathbb{R}$. Under these conditions, a variant [23] of Bernstein's inequality may be applied to generate the following confidence interval for μ:

$$\overline{X} \pm \left(\frac{-2}{3} (b-a) \log(\alpha/2) + \sqrt{\frac{4}{9}(b-a)^2 [\log(\alpha/2)]^2 - 8n\sigma^2 \log(\alpha/2)} \right) \times (2n)^{-1}. \quad (2)$$

In addition to estimating the variance σ^2 with s^2, the Bernstein confidence interval requires the selection of a bounding range (a, b). Since Negative Binomial variables are nonnegative, $a = 0$ is a natural choice. However these data are unbounded above, so any finite selection of b will impose a heuristic bound on the data. As a default, one may select the sample's maximum value or some multiple thereof. Shilane et al. [2] also considered a variant of Bernstein's Inequality for unbounded data. However, they found the method to be impractical due to the extremely conservative nature of the tail bound in this setting.

The bounded variant of Bernstein's Inequality improved upon the alternative methods in coverage for small sample sizes and high dispersions. However, its confidence intervals were far wider and more variable than the other candidates' results. While this is an improvement over a naive interval for the data's mean (e.g., all values between zero and the sample's maximum), the Bernstein method lacks the interpretative quality provided by parametric approximations. Furthermore, Bernstein's Inequality is a conservative bound, so asymptotically the method will significantly overcover the mean.

4. Growth Estimators for μ

The Gamma, Chi Square, and Normal approximations all seek to utilize the existing data to generate an inference for μ under parametric assumptions about the distribution of $\overline{X}$. However, none of these techniques directly considers the data's relative sparsity under high dispersion. When the probability of a zero value is high, many samples of data will include more than this expected proportion due to chance error. This overrepresentation of zeros exacerbates the difficulty of estimation in what is already a sparse data setting. For instance, in the case of $\mu = 10$ and $\theta = 0.025$, we expect 85.3% of the sample to be zeros. If the underlying experiment was repeated a large number of times, then roughly half of the samples would have more than 85.3% zeros. Furthermore, if $n = 30$ samples are drawn, a simulation experiment suggests that the resulting sample mean will be less than μ roughly 66% of the time.

We propose *growth estimators* for μ as a method of accounting for the potential overrepresentation of zeros in the data set. We will construct growth estimators using two separate procedures: adjusting the mean through a multiplicative growth factor and direct removal of some zero values from the data. The details of these procedures are provided in Sections 4.2 and 4.1.

These growth procedures are motivated by shrinkage estimators. For instance, in constructing a confidence interval for the Binomial proportion p, Agresti and Coull [24] suggest augmenting the existing data with two additional successes and two additional failures. A Normal approximation confidence interval for p based on these augmented data can be shown to perform measurably better than the Normal approximation alone. The method of Agresti and Coull [24] effectively shrinks the estimate of p toward the value 0.5 by adding a small amount of data. By contrast, we seek to grow the estimate of the Negative Binomial mean μ by removing a small amount of zero-valued data.

4.1. Growth by Adjustment (GBA). Suppose we believe that the data set contains approximately $k \in \mathbb{R}^+$ too many zero values. The removal of these zeros is tantamount to reweighting the sample mean by the *growth factor*

$$G = \frac{n}{n-k}. \quad (3)$$

Therefore, we estimate μ with the value

$$\widetilde{X} = \frac{1}{n-k}\sum_{i=1}^{n} X_i = \frac{n}{n-k}\overline{X} = G\overline{X}. \quad (4)$$

This growth estimator $\widetilde{X}$ has the expected value

$$E\left[\widetilde{X}\right] = \frac{n}{n-k}\mu = G\mu, \quad (5)$$

and the standard error is

$$\mathrm{SD}_{\mathrm{GBA}}\left(\widetilde{X}\right) = \frac{\sqrt{n}}{n-k}\sigma = \left[\frac{n}{n-k}\right]\frac{\sigma}{\sqrt{n}} = G \cdot \mathrm{SD}\left(\overline{X}\right). \quad (6)$$

4.2. Growth by Removal (GBR). The GBA method artificially inflates the sample mean by the growth factor G. However, this growth may also be achieved through direct removal of k zeros. We will refer to this procedure as *Growth by Removal* (GBR). As in the GBA method, this procedure depends upon the value of k. In general, k should be selected according to a predetermined rule, and it may not exceed the overall number of zeros in the data set. In practice, k should be less than this maximum because it is intended to remove only extraneous values.

In the original sample, the data's sum had the expected value $n\mu$. Since removing zero values does not change this sum, the mean of the $n - k$ remaining values has the expected value $G\mu$, as in (5). However, the associated standard error is computed differently from in the GBA method. The standard deviation of the remaining data is centered around $G\overline{X}$ rather than $\overline{X}$. The standard error then divides this quantity by $\sqrt{n-k}$. Assuming the data $X_1, \ldots, X_n$ are sorted in decreasing order, this is given by

$$\mathrm{SD}_{\mathrm{GBR}}\left(\widetilde{X}\right) = \sqrt{\frac{\sum_{i=1}^{n-k}\left(X_i - (n/(n-k))\overline{X}\right)^2}{(n-k-1)(n-k)}}. \quad (7)$$

4.3. Convergence to the Normal Approximation. Whether we employ the GBR or GBA methods, the mean may be adjusted by the growth factor G. For any fixed value of k, the growth estimator $G\overline{X}$ will asymptotically converge in sample size to $\overline{X}$. Since the central limit theorem applies, we propose a Normal approximation based upon this adjustment. We will rely upon the unbiased estimate s^2 of the variance σ^2 from the full data set (with no zeros removed). With z defined as the $(1 - \alpha/2)$th quantile of the standard Normal distribution, the GBA confidence interval will have the form

$$\mathrm{GBA}\colon G\overline{X} \pm zG\frac{s}{\sqrt{n}} = \frac{n}{n-k}\overline{X} \pm z\frac{\sqrt{n}}{n-k}s. \quad (8)$$

Meanwhile, the GBR method's interval applies its alternative computation of the standard error. That is,

$$\mathrm{GBR}\colon G\overline{X} \pm z \cdot \mathrm{SD}_{\mathrm{GBR}}\left(\widetilde{X}\right)$$
$$= \frac{n}{n-k}\overline{X} \pm z\sqrt{\frac{\sum_{i=1}^{n-k}\left(X_i - (n/(n-k))\overline{X}\right)^2}{(n-k-1)(n-k)}}. \quad (9)$$

4.4. Selection of k. Interestingly, the growth estimator adds both bias and variance to the Normal approximation of μ. The degree of additional error may be controlled through the selection of k, the number of zeros to remove from the data set. We emphasize that this selection should be made with extreme caution. In Section 5, we will explore how the Growth method's coverage probability is impacted by the choice of k. Based upon the results of these simulation experiments, we recommend the following default choices of k:

$$k = \begin{cases} \min\left(15, \dfrac{n}{10}\right), & \text{if } \widehat{\theta} \le 0.5 \\ \min\left(5, \dfrac{n}{10}\right), & \text{if } \widehat{\theta} > 0.5. \end{cases} \quad (10)$$

The intuition behind these choices is as follows: at small sample sizes, no more than one zero may be removed per ten data points. When the dispersion is high ($\theta \le 0.5$), we allow for more aggressive removal of zeros—up to a maximum of 15—to account for a higher degree of overrepresentation. At more moderate dispersions, we limit this removal to no more than 5 zeros. The value of θ may be estimated using maximum likelihood estimation or the method of moments when the former is not available.

Equation (10) may also be refined to incorporate a continuous function of θ. A simple linear interpolation between the respective maxima could be considered. However, deriving a more analytic relationship between k, n, and θ to optimize the coverage of the resulting confidence interval encounters a number of difficulties. The analytic distribution of growth method confidence intervals is a function of the negative binomial likelihood, which already has an extremely complex form. Additionally, the difficulty of estimating θ in small sample sizes may lead to misspecification of k. In combination, it would be very challenging to extrapolate the functional form of k to extremely small values of θ. In light of these concerns, we recommend using (10) as a default choice, and modifications may be considered in specific cases with additional simulation studies.

5. Simulation Studies

5.1. Experimental Design. We designed a simulation experiment to compare the Growth methods to the Bernstein, Gamma, Chi Square, and Normal confidence intervals. We did not include the bootstrap BCA method in the experiment due to its computational requirements and similarity to the Normal approximation in the simulations of Shilane et al. [2]. We selected an extensive set of parameters and sample sizes, with values displayed in Table 1. The most extreme case of $\mu = 10$ and $\theta = 0.025$ would roughly correspond to flipping a coin with a one-fourth of one percent chance of landing heads. Meanwhile, the most moderate case of $\mu = 2$ and $\theta = 1$ is equivalent to flipping a coin with a $1/3$ chance of heads.

Each combination of μ, θ, and n led to a unique and independent simulation experiment. Each experiment generated 10000 independent size n sets of i.i.d. $\mathrm{NB}(\mu, \theta)$ random variables in the **R** statistical programming language. On each

TABLE 1: Parameter values for the simulation experiments of Section 5.

Parameter	Values
μ	{2, 5, 10}
θ	{0.025, 0.05, 0.075, 0.1, 0.2, 0.3, ..., 1}
n	{5, 10, 15, 20, ..., 250, 300, 400, ..., 1000}
α	0.05
Trials	10000

size n data set, 95% confidence was generated according to each proposed method. We then approximated the coverage probability of each method by the empirical proportion of confidence intervals containing the true value of μ.

5.2. Coverage Probability Results. Figures 1, 2, and 3 provide examples of the simulation results at high, medium, and low dispersions. Each figure displays estimated coverage probabilities for the Normal, Gamma, Chi Square, Bernstein, GBA, and GBR methods as a function of sample size at particular combinations of μ and θ. Figure 1 depicts the case of $\mu = 10$ and $\theta = 0.025$, where the dispersion is the most extreme. Here even the Bernstein method requires a significantly large sample size to reach the desired 95% coverage probability. The Chi Square approximation performs as expected by reaching 95% coverage almost exactly when $\mu = 2n\theta$ and overcovering μ for larger sample sizes. Because θ is very small, we expect the Gamma approximation to outperform the normal. However, because of the extreme dispersion, both methods require an extremely large sample size to appropriately cover the mean. Even at $n = 250$, the Gamma method only covers μ 87.91% of the time, and the Normal only reaches a rate of 85.92%. Meanwhile, the Growth methods provide small but steady improvements over the Gamma approximation.

Figure 2 displays coverage probabilities for $\mu = 5$ and $\theta = 0.2$, where the dispersion is moderate and more typical of a Negative Binomial study. Both the Bernstein and Chi Square methods quickly overcover the mean, surpassing 95% by $n = 30$ and $n = 15$, respectively. The Gamma and Normal approximations are roughly in equipoise at this moderate level of dispersion. The Growth method improves upon their coverage by about 3% at sample sizes up to 100 and maintains at least a 1% improvement even out to $n = 250$. Despite the fairly moderate dispersion, the Normal and Gamma approximations appear to require 250 or more data points to approach convergence, whereas the Growth method reaches this point at about $n = 100$.

Figure 3 displays coverage probabilities in the case of $\mu = 2$ and $\theta = 1$. Because the dispersion is quite small, the Negative Binomial model is reasonably close to the Poisson distribution. We therefore expect the Normal approximation to perform quite well. Even in this case, the GBA method provides small improvements over the Normal at small and moderate sample sizes. The GBR method also provides early improvements. However, its coverage briefly dips below that of the Normal approximation at about $n = 50$. With relatively few zeros removed from the data, the Growth methods and Normal approximation quickly fall into agreement. By contrast, the Gamma approximation does not perform well n this scenario because its underlying limit theorem requires high dispersion relative to the sample size. Similarly, the Chi Square and Bernstein methods almost immediately overcover the mean.

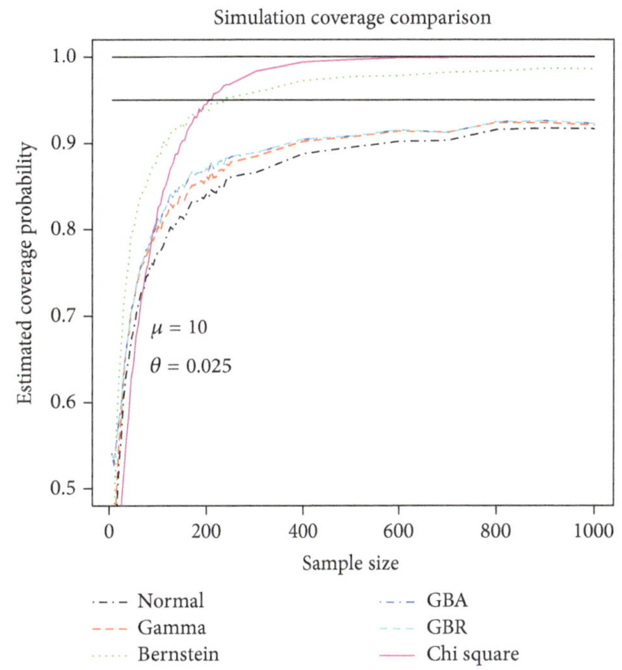

FIGURE 1: Simulation coverage probabilities for the proposed methods. With $\mu = 10$ and $\theta = 0.025$, this represents the most extreme dispersion among all simulation examples.

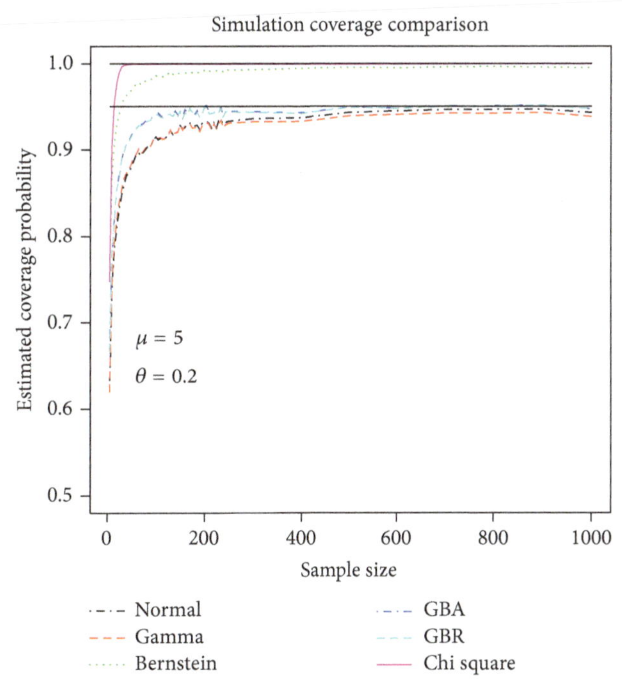

FIGURE 2: Simulation coverage probabilities for the proposed methods. With $\mu = 5$ and $\theta = 0.2$, this represents an intermediate dispersion among the range of simulation examples.

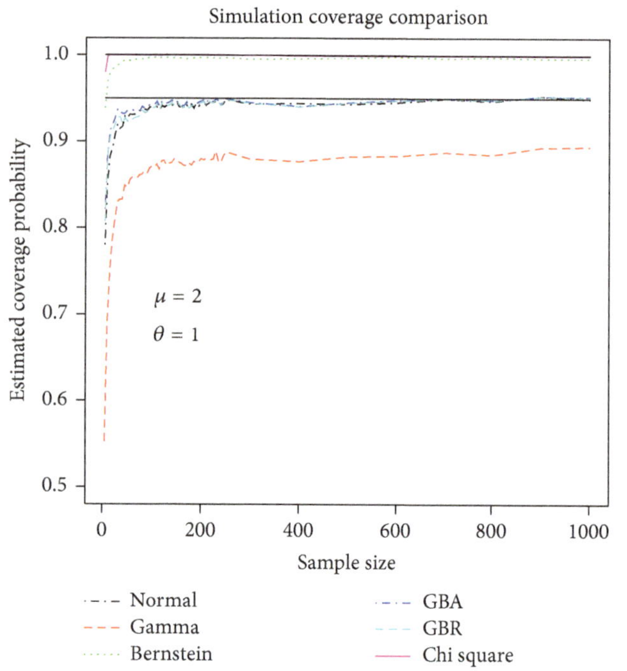

FIGURE 3: Simulation coverage probabilities for the proposed methods. With $\mu = 2$ and $\theta = 1$, this represents the most moderate dispersion among all simulation examples.

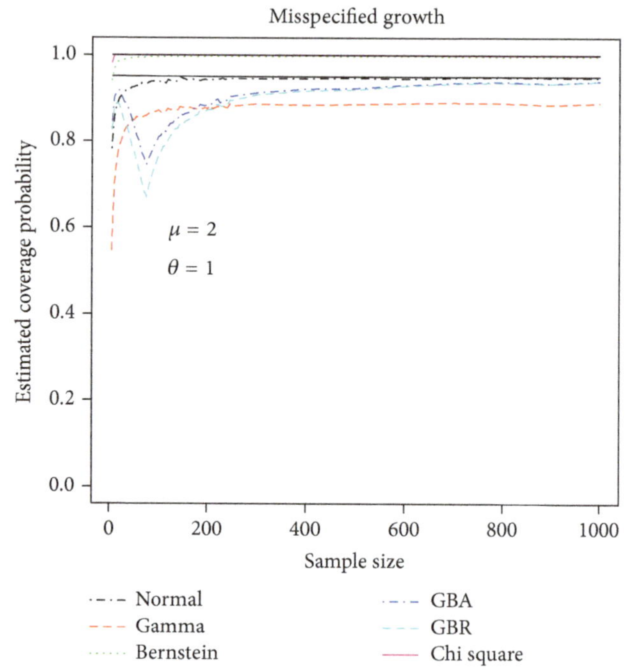

FIGURE 4: An example of misspecified growth. Despite low dispersion, zeros were aggressively removed at a rate of one for every 5 data points up to a maximum of 15. The GBA and GBR methods' coverage probabilities dip significantly until this maximum value is reached and then rebound toward the Normal approximation.

5.3. The Effect of Misspecified Growth. The Growth Method simulation results of Section 5.2 were obtained under the default selections of k given by (10). These recommended settings were obtained through a process of trial and error as applied to the simulation studies. The intuition of these recommendations is that zero values may be removed more aggressively under higher dispersions. We believe that the extensive range of parameters tested in the simulation study provide is reasonable evidence that these recommendations will generalize well. Other approaches to selecting k may also consider the observed sample mean or alter the gradations by sample size.

We urge the practitioner to exercise caution in selecting how many zeros to remove. An overzealous selection of k may lead to poor performance in the growth method's coverage probability. As an example, we repeated the simulation study in the case of $\mu = 2$ and $\theta = 1$ with $k = \min(15, n/5)$ instead of the recommended value of $k = \min(5, n/10)$. This aggressive approach removes zeros at double the rate and allows for a maximum that triples the recommendations for the low dispersion case of $\theta = 1$. Figure 4 displays the consequences of misspecified growth. Rather than the small improvement over or close agreement to the Normal approximation as seen in Figure 3, the growth method decreases significantly in coverage. This dip in performance continues until the maximum removal is reached at $n = 75$. For larger sample sizes, the Growth Method begins to rebound toward the Normal approximation. However, this case suggests that the practitioner should be careful not to select a value of k that is too large, especially when the dispersion is mild.

By contrast, higher dispersion settings allow for far more aggressive growth. We also repeated the simulation study in the case of $\mu = 10$ and $\theta = 0.025$, the most extreme dispersion considered. Figure 5 depicts the Growth Method's coverage when $k = \min(50, n/2)$, which effectively removes half the data points until sample size 100. In this example, the Growth method crosses the 90% coverage threshold by $n = 80$, a mark not reached by the Gamma or Normal approximations by $n = 250$. Indeed, the Growth Method accelerates in coverage even faster than the Bernstein Method.

Overall, the GBA method appears to perform slightly better than the GBR method across all simulation experiments. This difference may be attributed to the selection of k. The GBR method requires an integer value so that exactly k zeros may be removed. By contrast, the GBA method allows for adjustments using continuous values of k. The recommended selection procedure of (10) allows for fractional proportions of the overall sample size. This additional fraction allows for increased growth, which in turn leads to improved coverage. The GBA and GBR methods also differ in the computation of their standard errors. It appears that these standard errors are largely similar in value. We will substantiate this claim further in Section 5.4.

5.4. Confidence Interval Length Considerations. We have adopted coverage probability as our preferred metric of a confidence interval's quality. However, the length of these intervals is an important secondary consideration. Shorter intervals suggest greater precision in the estimator when the coverage probability is approximately equal. Figures 6 and 7 provide a comparison of the proposed methods' lengths. In each simulation experiment, we recorded the median

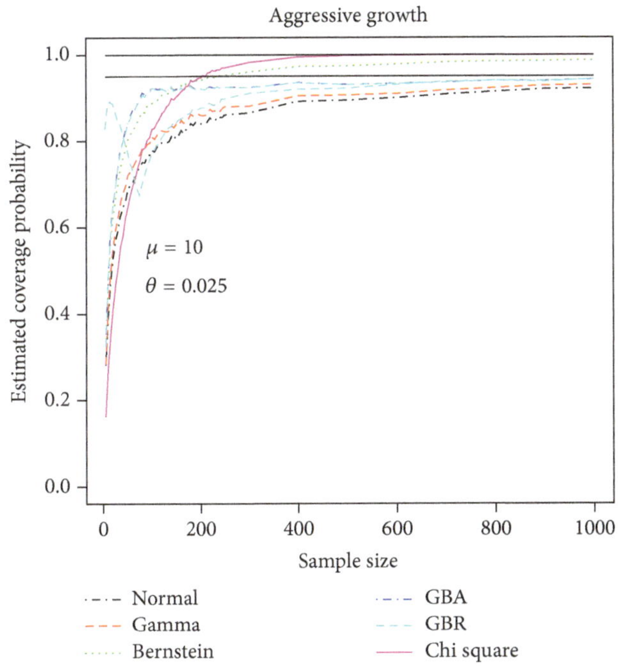

FIGURE 5: An example of aggressive growth. Under high dispersion, zeros were aggressively removed at a rate of one for every 2 data points up to a maximum of 50. The GBA and GBR methods' coverage probabilities accelerate quickly in this setting.

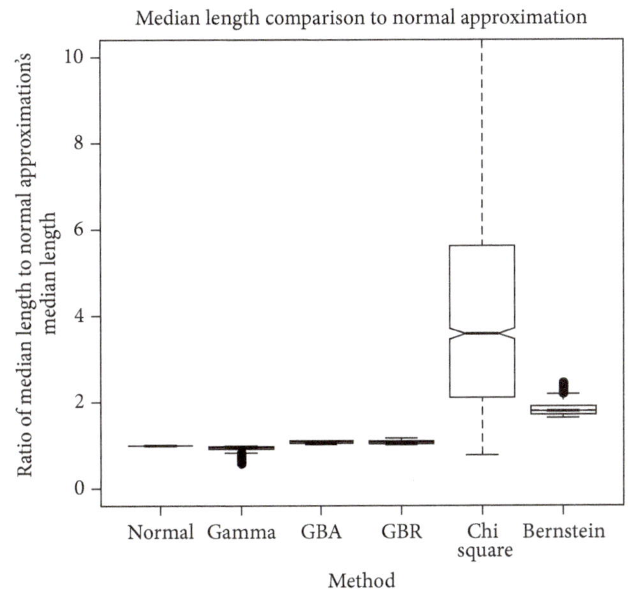

FIGURE 6: A comparison of median confidence interval length standardized by the Normal approximation's median values. Each simulation experiment computed the median interval length of each method. The values depicted are the ratios of these medians to that of the Normal approximation.

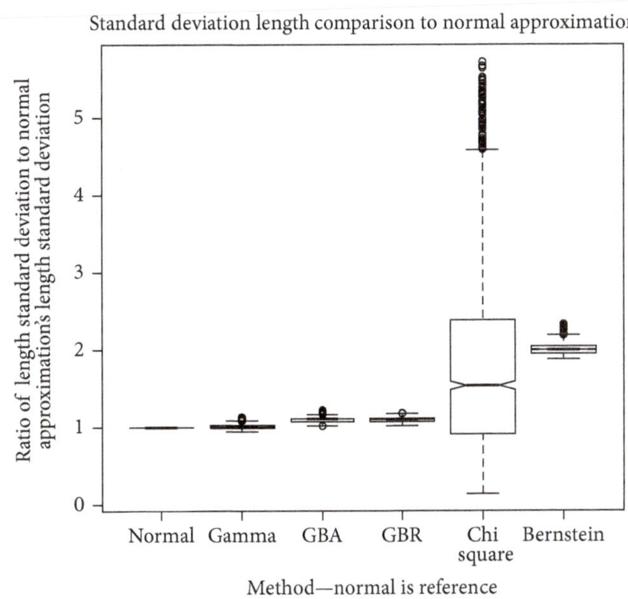

FIGURE 7: A comparison of confidence interval length stability. Each simulation experiment computed the standard deviation of interval length of each method. The values depicted are the ratios of these standard deviations to that of the Normal approximation.

and standard deviation of length for the 10000 confidence intervals generated by each method. We then computed the ratio of each method's median length to that of the Normal approximation in each experiment. Figure 6 displays the distribution of these ratios, and Figure 7 depicts the ratio of the standard deviation of length.

In general, it appears that the Bernstein method produces confidence intervals that are typically a factor of 1.8 longer than the corresponding Normal approximation. Because the GBA and GBR methods usually remove about one zero per ten data points, their median lengths were typically a factor of 10/9 larger than the Normal approximation's interval. Likewise, the same growth factor applies to the standard deviations of length in Figure 7. The Bernstein Method has a length variability that is roughly double that of the Normal approximation. This increased variability of length leads to the overcoverage in the method, as extremely long confidence intervals are far more likely to contain the mean. By contrast, the Gamma method typically produces a confidence interval that is 95% the length of the Normal approximation. This length depends on the degree of dispersion, with higher lengths (and improved coverage) occurring at smaller values of θ.

6. Discussion

The proposed growth methods provide improved confidence intervals for the mean of Negative Binomial random variables with unknown dispersion. Removing a small number of zeros from the data set bolsters the coverage probability at small and moderate sample sizes. Asymptotically, the GBA and GBR methods converge to the Normal approximation. Overall, applying a growth estimator produces intervals that are longer and more variable than the Normal approximation. The degree of increase may be controlled through a selection of the growth factor G, or, equivalently, the removal factor k. This selection depends on the sample size and degree of dispersion in the data. We emphasize that the number of zeros

k to remove should be selected cautiously to prevent coverage drop-offs such as the example depicted in Figure 4.

The GBA and GBR methods often provide similar coverage results, so other factors may be considered in selecting between them. First, the GBA growth factor $G = n/(n-k)$ (3) allows for noninteger values, while GBR must remove an integer number of zeros. The GBA method's possibility of continuous adjustments may present an opportunity for finer tuning of the method to the problem at hand. By contrast, a manual removal of zeros is more intuitively appealing. The remainder of the inference follows the same methods as the Normal approximation, which will be more familiar to many practitioners. When there are fewer than k overall zeros in the sample, it is also more reasonable to only remove the existing zeros rather than to adjust by a growth factor dependent on k. With all of this in mind, the GBR method may be a more practical choice even if the GBA method can provide slight improvements in the results.

The previous work of Shilane et al. [2] provided a piecewise solution to performing inference on the Negative Binomial mean. The Gamma, Chi Square, and Normal approximations performed well in largely complimentary settings, and Bernstein's inequality was used at smaller sample sizes and high dispersions. However, we have demonstrated that the GBA and GBR methods can perform well in a wide variety of settings. Using the relatively simple guidelines for selecting k in (10), these procedures outperformed the parametric approximations at both high and low dispersions. Because it allows for continuous values of k, the GBA method generally provided small improvements over the GBR results. A tail probability bound such as Bernstein's Inequality may still be considered at very small sample sizes and extremely high dispersions, but the growth method accelerates quickly as a function of sample size.

Future work on this problem may provide more solid theory for how the value of k should be selected. Growth estimators add both bias and variance to the Normal approximation, so a traditional bias-variance trade-off calculation does not apply. Indeed, a criterion such as the mean squared error would be optimized with the selection of $k = 0$, which is equivalent to the Normal approximation. The cautionary tale depicted in Figure 4 suggests that coverage is optimized at some intermediate value of k. However, the analytic coverage probability calculation is intractable because it depends upon the permutation distribution of the Negative Binomial sample.

The Growth Method may be extended to other sparse estimation problems. Shilane and Bean [25] previously examined the quality of the Normal approximation in two-sample Negative Binomial inference, and growth estimators may be easily adapted to this setting. We are also presently examining an application of the growth method for estimating the mean of Gamma random variables.

References

[1] R. R. Wilcox, *Robust Estimation and Hypothesis Testing*, Academic Press, Burlington, Mass, USA, 2005.

[2] D. Shilane, S. N. Evans, and A. E. Hubbard, "Confidence intervals for negative binomial random variables of high dispersion," *International Journal of Biostatistics*, vol. 6, no. 1, article 10, 2010.

[3] J. M. Hilbe, *Negative Binomial Regression*, Cambridge University Press, Cambridge, UK, 2007.

[4] S. J. Clark and J. N. Perry, "Estimation of the negative binomial parameter κ by maximum quasi-likelihood," *Biometrics*, vol. 45, no. 1, pp. 309–316, 1989.

[5] W. W. Piegorsch, "Maximum likelihood estimation for the negative binomial dispersion parameter," *Biometrics*, vol. 46, no. 3, pp. 863–867, 1990.

[6] J. Aragón, D. Eberly, and S. Eberly, "Existence and uniqueness of the maximum likelihood estimator for the two-parameter negative binomial distribution," *Statistics & Probability Letters*, vol. 15, no. 5, pp. 375–379, 1992.

[7] C. Ferreri, "On the ML-estimator of the positive and negative two-parameter binomial distribution," *Statistics & Probability Letters*, vol. 33, no. 2, pp. 129–134, 1997.

[8] E. P. Pieters, C. E. Gates, J. H. Matis, and W. L. Sterling, "Small sample comparison of different estimators of negative binomial parameters," *Biometrics*, vol. 33, no. 4, pp. 718–723, 1977.

[9] B. Efron and R. J. Tibshirani, *An Introduction to the Bootstrap*, vol. 57 of *Monographs on Statistics and Applied Probability*, Chapman and Hall, Boca Raton, Fla, USA, 1993.

[10] G. Casella and R. L. Berger, *Statistical Inference*, Wadsworth & BrooksSoftware, Pacific Grove, Calif, USA, 1990.

[11] C. J. Clopper and E. S. Pearson, "The use of confidence or fiducial limits illustrated in the case of the binomial," *Biometrika*, vol. 26, pp. 404–413, 1934.

[12] E. L. Crow and R. S. Gardner, "Confidence intervals for the expectation of a Poisson variable," *Biometrika*, vol. 46, pp. 441–453, 1959.

[13] T. E. Sterne, "Some remarks on confidence or fiducial limits," *Biometrika*, vol. 41, pp. 275–278, 1954.

[14] S. N. Bernstein, "Teoriya Veroiatnostei," Publisher Unknown, 1934, (Russian).

[15] G. Bennett, "Probability inequalities for the sum of independent random variables," *Journal of the American Statistical Association*, vol. 57, pp. 33–45, 1962.

[16] G. Bennett, "On the probability of large deviations from the expectation for sums of bounded, independent random variables," *Biometrika*, vol. 50, pp. 528–535, 1963.

[17] W. Hoeffding, "Probability inequalities for sums of bounded random variables," *Journal of the American Statistical Association*, vol. 58, pp. 13–30, 1963.

[18] A. C. Berry, "The accuracy of the Gaussian approximation to the sum of independent variates," *Transactions of the American Mathematical Society*, vol. 49, pp. 122–136, 1941.

[19] C.-G. Esseen, "On the Liapounoff limit of error in the theory of probability," *Arkiv för Matematik*, vol. 28, no. 9, pp. 1–19, 1942.

[20] C. G. Esseen, "A moment inequality with an application to the central limit theorem," *Scandinavian Actuarial Journal*, vol. 39, pp. 160–170, 1956.

[21] P. van Beek, "An application of Fourier methods to the problem of sharpening the Berry-Esseen inequality," *Zeitschrift für Wahrscheinlichkeitstheorie und verwandte Gebiete*, vol. 23, pp. 187–196, 1972.

[22] M. A. Rosenblum and M. J. van der Laan, "Confidence intervals for the population mean tailored to small sample sizes, with applications to survey sampling," *International Journal of Biostatistics*, vol. 5, article 10, 2009.

[23] M. J. van der Laan, D. B. Rubin et al., "Estimating function based crossvalidation and learning," Tech. Rep. 180, Division of Biostatistics, University of California-Berkeley, 2005.

[24] A. Agresti and B. A. Coull, "Approximate is better than "exact" for interval estimation of binomial proportions," *The American Statistician*, vol. 52, no. 2, pp. 119–126, 1998.

[25] D. Shilane and D. Bean, "Two-sample inference in highly dispersed negative binomial models," *The American Statistician*, 2011.

11

Exact Group Sequential Methods for Estimating a Binomial Proportion

Zhengjia Chen[1] **and Xinjia Chen**[2]

[1] *Department of Biostatistics and Bioinformatics, Emory University, Atlanta, GA 30322, USA*
[2] *Department of Electrical Engineering, Southern University and A&M College, Baton Rouge, LA 70813, USA*

Correspondence should be addressed to Zhengjia Chen; zchen38@emory.edu

Academic Editor: Xuelin Huang

We first review existing sequential methods for estimating a binomial proportion. Afterward, we propose a new family of group sequential sampling schemes for estimating a binomial proportion with prescribed margin of error and confidence level. In particular, we establish the uniform controllability of coverage probability and the asymptotic optimality for such a family of sampling schemes. Our theoretical results establish the possibility that the parameters of this family of sampling schemes can be determined so that the prescribed level of confidence is guaranteed with little waste of samples. Analytic bounds for the cumulative distribution functions and expectations of sample numbers are derived. Moreover, we discuss the inherent connection of various sampling schemes. Numerical issues are addressed for improving the accuracy and efficiency of computation. Computational experiments are conducted for comparing sampling schemes. Illustrative examples are given for applications in clinical trials.

1. Introduction

Estimating a binomial proportion is a problem of ubiquitous significance in many areas of engineering and sciences. For economical reasons and other concerns, it is important to use as fewer as possible samples to guarantee the required reliability of estimation. To achieve this goal, sequential sampling schemes can be very useful. In a sequential sampling scheme, the total number of observations is not fixed in advance. The sampling process is continued stage by stage until a prespecified stopping rule is satisfied. The stopping rule is evaluated with accumulated observations. In many applications, for administrative feasibility, the sampling experiment is performed in a group fashion. Similar to group sequential tests [1, Section 8], [2], an estimation method based on taking samples by groups and evaluating them sequentially is referred to as a group sequential estimation method. It should be noted that group sequential estimation methods are general enough to include fixed-sample-size and fully sequential procedures as special cases. Particularly, a fixed-sample-size method can be viewed as a group sequential procedure of only one stage. If the increment between the sample sizes of consecutive stages is equal to 1, then the group sequential method is actually a fully sequential method.

It is a common contention that statistical inference, as a unique science to quantify the uncertainties of inferential statements, should avoid errors in the quantification of uncertainties, while minimizing the sampling cost. That is, a statistical inferential method is expected to be exact and efficient. The conventional notion of exactness is that no approximation is involved, except the round-off error due to finite word length of computers. Existing sequential methods for estimating a binomial proportion are dominantly of asymptotic nature (see, e.g., [3–7] and the references therein). Undoubtedly, asymptotic techniques provide approximate solutions and important insights for the relevant problems. However, any asymptotic method inevitably introduces unknown error in the resultant approximate solution due to the necessary use of a finite number of samples. In the direction of nonasymptotic sequential estimation, the primary goal is to ensure that the true coverage probability is above the prespecified confidence level for any value of the associated parameter, while the required sample size is as low as possible. In this direction, Mendo and Hernando [8] developed an inverse binomial sampling scheme for estimating a binomial proportion with relative precision. Tanaka [9] developed a rigorous method for constructing fixed-width sequential confidence intervals for a binomial proportion. Although no approximation is involved, Tanaka's method is very conservative due to the bounding techniques employed in the derivation of sequential confidence intervals. Franzén [10] studied the construction of fixed-width sequential confidence intervals for a binomial proportion. However, no effective method for defining stopping rules is proposed in [10]. In his later paper [11], Franzén proposed to construct fixed-width confidence intervals based on sequential probability ratio tests (SPRTs) invented by Wald [12]. His method can generate fixed-sample-size confidence intervals based on SPRTs. Unfortunately, he made a fundamental flaw by mistaking that if the width of the fixed-sample-size confidence interval decreases to be smaller than the prespecified length as the number of samples is increasing, then the fixed-sample-size confidence interval at the termination of sampling process is the desired fixed-width sequential confidence interval guaranteeing the prescribed confidence level. More recently, Frey published a paper [13] in The American Statistician (TAS) on the classical problem of sequentially estimating a binomial proportion with prescribed margin of error and confidence level. Before Frey submitted his original manuscript to TAS in July 2009, a general framework of multistage parameter estimation had been established by Chen [14–18], which provides exact methods for estimating parameters of common distributions with various error criterion. This framework is also proposed in [19]. The approach of Frey [13] is similar to that of Chen [14–18] for the specific problem of estimating a binomial proportion with prescribed margin of error and confidence level.

In this paper, our primary interests are in the exact sequential methods for the estimation of a binomial proportion with prescribed margin of error and confidence level. We first introduce the exact approach established in [14–18]. In particular, we introduce the inclusion principle proposed in [18] and its applications to the construction of concrete stopping rules. We investigate the connection among various stopping rules. Afterward, we propose a new family of stopping rules which are extremely simple and accommodate some existing stopping rules as special cases. We provide rigorous justification for the feasibility and asymptotic optimality of such stopping rules. We prove that the prescribed confidence level can be guaranteed uniformly for all values of a binomial proportion by choosing appropriate parametric values for the stopping rule. We show that as the margin of error tends to be zero, the sample size tends to the attainable minimum as if the binomial proportion were exactly known. We derive analytic bounds for distributions and expectations of sample numbers. In addition, we address some critical computational issues and propose methods to improve the accuracy and efficiency of numerical calculation. We conduct extensive numerical experiment to study the performance of various stopping rules. We determine parametric values for the proposed stopping rules to achieve unprecedentedly efficiency while guaranteeing prescribed confidence levels. We attempt to make our proposed method as user-friendly as possible so that it can be immediately applicable even for layer persons.

The remainder of the paper is organized as follows. In Section 2, we introduce the exact approach proposed in [14–18]. In Section 3, we discuss the general principle of constructing stopping rules. In Section 4, we propose a new family of sampling schemes and investigate their feasibility, optimality, and analytic bounds of the distribution and expectation of sample numbers. In Section 5, we compare various computational methods. In particular, we illustrate why the natural method of evaluating coverage probability based on gridding parameter space is neither rigorous nor efficient. In Section 6, we present numerical results for various sampling schemes. In Section 7, we illustrate the applications of our group sequential method in clinical trials. Section 8 is the conclusion. The proofs of theorems are given in appendices. Throughout this paper, we shall use the following notations. The empty set is denoted by $\emptyset$. The set of positive integers is denoted by $\mathbb{N}$. The ceiling function is denoted by $\lceil\cdot\rceil$. The notation $\Pr\{E \mid \theta\}$ denotes the probability of the event E associated with parameter θ. The expectation of a random variable is denoted by $\mathbb{E}[\cdot]$. The standard normal distribution is denoted by $\Phi(\cdot)$. For $\alpha \in (0,1)$, the notation $\mathcal{Z}_\alpha$ denotes the critical value such that $\Phi(\mathcal{Z}_\alpha) = 1-\alpha$. For $n \in \mathbb{N}$, in the case that $X_1,\dots,X_n$ are i.i.d. samples of X, we denote the sample mean $(\sum_{i=1}^n X_i)/n$ by $\overline{X}_n$, which is also called the relative frequency when X is a Bernoulli random variable. The other notations will be made clear as we proceed.

2. How Can It Be Exact?

In many areas of scientific investigation, the outcome of an experiment is of dichotomy nature and can be modeled as a Bernoulli random variable X, defined in probability space $(\Omega, \Pr, \mathscr{F})$, such that

$$\Pr\{X=1\} = 1 - \Pr\{X=0\} = p \in (0,1), \quad (1)$$

where p is referred to as a binomial proportion. In general, there is no analytic method for evaluating the binomial proportion p. A frequently used approach is to estimate p based on i.i.d. samples $X_1, X_2, \dots$ of X. To reduce the sampling cost, it is appropriate to estimate p by a multistage sampling procedure. More formally, let $\epsilon \in (0,1)$ and $1-\delta$, with $\delta \in (0,1)$, be the prespecified margin of error and confidence level, respectively. The objective is to construct a sequential estimator $\widehat{\mathbf{p}}$ for p based on a multistage sampling scheme such that

$$\Pr\{|\widehat{\mathbf{p}} - p| < \epsilon \mid p\} \geq 1-\delta, \quad (2)$$

for any $p \in (0,1)$. Throughout this paper, the probability $\Pr\{|\widehat{\mathbf{p}} - p| < \epsilon \mid p\}$ is referred to as the *coverage probability*. Accordingly, the probability $\Pr\{|\widehat{\mathbf{p}} - p| \geq \epsilon \mid p\}$ is referred to as the *complementary coverage probability*. Clearly, a complete construction of a multistage estimation scheme needs to determine the number of stages, the sample sizes for all stages, the stopping rule, and the estimator for p. Throughout this paper, we let s denote the number of stages and let n_ℓ denote the number of samples at the ℓth stages. That is, the sampling process consists of s stages with sample sizes $n_1 < n_2 < \cdots < n_s$. For $\ell = 1, 2, \ldots, s$, define $K_\ell = \sum_{i=1}^{n_\ell} X_i$ and $\widehat{\mathbf{p}}_\ell = K_\ell / n_\ell$. The stopping rule is to be defined in terms of $\widehat{\mathbf{p}}_\ell$, $\ell = 1, \ldots, s$. Of course, the index of stage at the termination of the sampling process, denoted by $\mathbf{l}$, is a random number. Accordingly, the number of samples at the termination of the experiment, denoted by $\mathbf{n}$, is a random number which equals $n_\mathbf{l}$. Since for each ℓ, $\widehat{\mathbf{p}}_\ell$ is a maximum-likelihood and minimum-variance unbiased estimator of p, the sequential estimator for p is taken as

$$\widehat{\mathbf{p}} = \widehat{\mathbf{p}}_\mathbf{l} = \frac{\sum_{i=1}^{n_\mathbf{l}} X_i}{n_\mathbf{l}} = \frac{\sum_{i=1}^{\mathbf{n}} X_i}{\mathbf{n}}. \quad (3)$$

In the above discussion, we have outlined the general characteristics of a multistage sampling scheme for estimating a binomial proportion. It remains to determine the number of stages, the sample sizes for all stages, and the stopping rule so that the resultant estimator $\widehat{\mathbf{p}}$ satisfies (2) for any $p \in (0,1)$.

Actually, the problem of sequential estimation of a binomial proportion has been treated by Chen [14–18] in a general framework of multistage parameter estimation. The techniques of [14–18] are sufficient to offer exact solutions for a wide range of sequential estimation problems, including the estimation of a binomial proportion as a special case. The central idea of the approach in [14–18] is the control of coverage probability by a single parameter ζ, referred to as the *coverage tuning parameter*, and the adaptive rigorous checking of coverage guarantee by virtue of bounds of coverage probabilities. It is recognized in [14–18] that, due to the discontinuity of the coverage probability on parameter space, the conventional method of evaluating the coverage probability for a finite number of parameter values is neither rigorous not computationally efficient for checking the coverage probability guarantee.

As mentioned in the introduction, Frey published an article [13] in TAS on the sequential estimation of a binomial proportion with prescribed margin of error and confidence level. For clarity of presentation, the comparison of the works of Chen and Frey is given in Section 5.4. In the remainder of this section, we shall only introduce the idea and techniques of [14–18], which had been precedentially developed by Chen before Frey submitted his original manuscript to TAS in July 2009. We will introduce the approach of [14–18] with a focus on the special problem of estimating a binomial proportion with prescribed margin of error and confidence level.

2.1. Four Components Suffice. The exact methods of [14–18] for multistage parameter estimation have four main components as follows.

(i) Stopping rules parameterized by the coverage tuning parameter $\zeta > 0$ such that the associated coverage probabilities can be made arbitrarily close to 1 by choosing $\zeta > 0$ to be a sufficiently small number.

(ii) Recursively computable lower and upper bounds for the complementary coverage probability for a given ζ and an interval of parameter values.

(iii) Adapted branch and bound algorithm.

(iv) Bisection coverage tuning.

Without looking at the technical details, one can see that these four components are sufficient for constructing a sequential estimator so that the prescribed confidence level is guaranteed. The reason is as follows. As lower and upper bounds for the complementary coverage probability are available, the global optimization technique, branch and bound (B&B) algorithm [20], can be used to compute exactly the maximum of complementary coverage probability on the whole parameter space. Thus, it is possible to check rigorously whether the coverage probability associated with a given ζ is no less than the prespecified confidence level. Since the coverage probability can be controlled by ζ, it is possible to determine ζ as large as possible to guarantee the desired confidence level by a bisection search. This process is referred to as bisection coverage tuning in [14–18]. Since a critical subroutine needed for bisection coverage tuning is to check whether the coverage probability is no less than the prespecified confidence level, it is not necessary to compute exactly the maximum of the complementary coverage probability. Therefore, Chen revised the standard B&B algorithm to reduce the computational complexity and called the improved algorithm as the adapted B&B Algorithm. The idea is to adaptively partition the parameter space as many subintervals. If for all subintervals, the upper bounds of the complementary coverage probability are no greater than δ, then declare that the coverage probability is guaranteed. If there exists a subinterval for which the lower bound of the complementary coverage probability is greater than δ, then declare that the coverage probability is not guaranteed. Continue partitioning the parameter space if no decision can be made. The four components are illustrated in the sequel under the headings of stopping rules, interval bounding, adapted branch and bound, and bisection coverage tuning.

2.2. Stopping Rules. The first component for the exact sequential estimation of a binomial proportion is the stopping rule for constructing a sequential estimator such that the coverage probability can be controlled by the coverage tuning parameter ζ. For convenience of describing some concrete stopping rules, define

$$\mathscr{M}(z,\theta)=\begin{cases} z\ln\dfrac{\theta}{z} \\ \quad +(1-z)\ln\dfrac{1-\theta}{1-z} & \text{for } z\in(0,1), \\ & \theta\in(0,1), \\ \ln(1-\theta) & \text{for } z=0,\ \theta\in(0,1), \\ \ln\theta & \text{for } z=1,\ \theta\in(0,1), \\ -\infty & \text{for } z\in[0,1], \\ & \theta\notin(0,1), \end{cases}$$

$$S(k,l,n,p)=\begin{cases} \sum_{i=k}^{l}\binom{n}{i}p^i(1-p)^{n-i} & \text{for } p\in(0,1), \\ 0 & \text{for } p\notin(0,1), \end{cases} \tag{4}$$

where k and l are integers such that $0\le k\le l\le n$. Assume that $0<\zeta\delta<1$. For the purpose of controlling the coverage probability $\Pr\{|\widehat{\mathbf{p}}-p|<\epsilon\mid p\}$ by the coverage tuning parameter, Chen has proposed four stopping rules as follows.

Stopping Rule A. Continue sampling until $\mathscr{M}((1/2)-|(1/2)-\widehat{\mathbf{p}}_\ell|,(1/2)-|(1/2)-\widehat{\mathbf{p}}_\ell|+\epsilon)\le(\ln(\zeta\delta))/n_\ell$ for some $\ell\in\{1,\dots,s\}$.

Stopping Rule B. Continue sampling until $(|\widehat{\mathbf{p}}_\ell-(1/2)|-(2/3)\epsilon)^2\ge(1/4)+(\epsilon^2 n_\ell/2\ln(\zeta\delta))$ for some $\ell\in\{1,\dots,s\}$.

Stopping Rule C. Continue sampling until $S(K_\ell,n_\ell,n_\ell,\widehat{\mathbf{p}}_\ell-\epsilon)\le\zeta\delta$ and $S(0,K_\ell,n_\ell,\widehat{\mathbf{p}}_\ell+\epsilon)\le\zeta\delta$ for some $\ell\in\{1,\dots,s\}$.

Stopping Rule D. Continue sampling until $n_\ell\ge\widehat{\mathbf{p}}_\ell(1-\widehat{\mathbf{p}}_\ell)(2/\epsilon^2)\ln(1/\zeta\delta)$ for some $\ell\in\{1,\dots,s\}$.

Stopping Rule A was first proposed in [14, Theorem 7] and restated in [15, Theorem 16]. Stopping Rule B was first proposed in [16, Theorem 1] and represented as the third stopping rule in [21, Section 4.1.1]. Stopping Rule C originated from [17, Theorem 1] and was restated as the first stopping rule in [21, Section 4.1.1]. Stopping Rule D was described in the remarks following Theorem 7 of [22]. All these stopping rules can be derived from the general principles proposed in [18, Section 3] and [19, Section 2.4].

Given that a stopping rule can be expressed in terms of $\widehat{\mathbf{p}}_\ell$ and n_ℓ for $\ell=1,\dots,s$, it is possible to find a bivariate function $\mathscr{D}(\cdot,\cdot)$ on $\{(z,n):z\in[0,1],\ n\in\mathbb{N}\}$, taking values from $\{0,1\}$, such that the stopping rule can be stated as the following: continue sampling until $\mathscr{D}(\widehat{\mathbf{p}}_\ell,n_\ell)=1$ for some $\ell\in\{1,\dots,s\}$. It can be checked that such representation applies to Stopping Rules A, B, C, and D. For example, Stopping Rule B can be expressed in this way by virtue of function $\mathscr{D}(\cdot,\cdot)$ such that

$$\mathscr{D}(z,n)=\begin{cases} 1 & \text{if } \left(\left|z-\dfrac{1}{2}\right|-\dfrac{2}{3}\epsilon\right)^2\ge\dfrac{1}{4}+\dfrac{\epsilon^2 n}{2\ln(\zeta\delta)}, \\ 0 & \text{otherwise.}\end{cases} \tag{5}$$

The motivation of introducing function $\mathscr{D}(\cdot,\cdot)$ is to parameterize the stopping rule in terms of design parameters. Function $\mathscr{D}(\cdot,\cdot)$ determines the form of the stopping rule and, consequently, the sample sizes for all stages can be chosen as functions of design parameters. Specifically, let

$$N_{\min}=\min\left\{n\in\mathbb{N}:\mathscr{D}\left(\frac{k}{n},n\right)=1 \text{ for some nonnegative integer } k \text{ not exceeding } n\right\}, \tag{6}$$

$$N_{\max}=\min\left\{n\in\mathbb{N}:\mathscr{D}\left(\frac{k}{n},n\right)=1 \text{ for all nonnegative integer } k \text{ not exceeding } n\right\}. \tag{7}$$

To avoid unnecessary checking of the stopping criterion and thus reduce administrative cost, there should be a possibility that the sampling process is terminated at the first stage. Hence, the minimum sample size n_1 should be chosen to ensure that $\{\mathbf{n}=n_1\}\ne\emptyset$. This implies that the sample size n_1 for the first stage can be taken as $N_{\min}$. On the other hand, since the sampling process must be terminated at or before the sth stage, the maximum sample size n_s should be chosen to guarantee that $\{\mathbf{n}>n_s\}=\emptyset$. This implies that the sample size n_s for the last stage can be taken as $N_{\max}$. If the number of stages s is given, then the sample sizes for stages in between 1 and s can be chosen as $s-2$ integers between $N_{\min}$ and $N_{\max}$. Particularly, if the group sizes are expected to be approximately equal, then the sample sizes can be taken as

$$n_\ell=\left\lceil N_{\min}+\frac{\ell-1}{s-1}(N_{\max}-N_{\min})\right\rceil,\quad \ell=1,\dots,s. \tag{8}$$

Since the stopping rule is associated with the coverage tuning parameter ζ, it follows that the number of stages s and the sample sizes $n_1,n_2,\dots,n_s$ can be expressed as functions of ζ. In this sense, it can be said that the stopping rule is parameterized by the coverage tuning parameter ζ. The above method of parameterizing stopping rules has been used in [14–17] and proposed in [21, Section 2.1, page 9].

2.3. Interval Bounding. The second component for the exact sequential estimation of a binomial proportion is the method of bounding the complementary coverage probability $\Pr\{|\widehat{\mathbf{p}}-p|\ge\epsilon\mid p\}$ for p in an interval $[a,b]$ contained by interval $(0,1)$. Applying Theorem 8 of [15] to the special case of a Bernoulli distribution immediately yields

$$\begin{aligned}&\Pr\{\widehat{\mathbf{p}}\le a-\epsilon\mid b\}+\Pr\{\widehat{\mathbf{p}}\ge b+\epsilon\mid a\}\\ &\le\Pr\{|\widehat{\mathbf{p}}-p|\ge\epsilon\mid p\}\\ &\le\Pr\{\widehat{\mathbf{p}}\le b-\epsilon\mid a\}+\Pr\{\widehat{\mathbf{p}}\ge a+\epsilon\mid b\},\end{aligned} \tag{9}$$

∇ Let $k \leftarrow 0$, $l_0 \leftarrow \Psi_{\mathrm{lb}}(\mathscr{I}_{\mathrm{init}})$ and $u_0 \leftarrow \Psi_{\mathrm{ub}}(\mathscr{I}_{\mathrm{init}})$.
∇ Let $\mathscr{S}_0 \leftarrow \{\mathscr{I}_{\mathrm{init}}\}$ if $u_0 > \delta$. Otherwise, let $\mathscr{S}_0$ be empty.
∇ While $\mathscr{S}_k$ is nonempty, $l_k < \delta$ and u_k is greater than $\max\{l_k + \eta, \delta\}$, do the following:
◇ Split each interval in $\mathscr{S}_k$ as two new intervals of equal length.
Let S_k denote the set of all new intervals obtained from this splitting procedure.
◇ Eliminate any interval $\mathscr{I}$ from S_k such that $\Psi_{\mathrm{ub}}(\mathscr{I}) \leq \delta$.
◇ Let $\mathscr{S}_{k+1}$ be the set S_k processed by the above elimination procedure.
◇ Let $l_{k+1} \leftarrow \max_{\mathscr{I} \in \mathscr{S}_{k+1}} \Psi_{\mathrm{lb}}(\mathscr{I})$ and $u_{k+1} \leftarrow \max_{\mathscr{I} \in \mathscr{S}_{k+1}} \Psi_{\mathrm{ub}}(\mathscr{I})$. Let $k \leftarrow k + 1$.
∇ If $\mathscr{S}_k$ is empty and $l_k < \delta$, then declare $\max \Psi(\mathscr{I}_{\mathrm{init}}) \leq \delta$.
Otherwise, declare $\max \Psi(\mathscr{I}_{\mathrm{init}}) > \delta$.

Algorithm 1

for all $p \in [a,b] \subseteq (0,1)$. The bounds of (9) can be shown as follows. Note that $\Pr\{\widehat{\mathbf{p}} \leq a - \epsilon \mid p\} + \Pr\{\widehat{\mathbf{p}} \geq b + \epsilon \mid p\} \leq \Pr\{|\widehat{\mathbf{p}} - p| \geq \epsilon \mid p\} = \Pr\{\widehat{\mathbf{p}} \leq p - \epsilon \mid p\} + \Pr\{\widehat{\mathbf{p}} \geq p + \epsilon \mid p\} \leq \Pr\{\widehat{\mathbf{p}} \leq b - \epsilon \mid p\} + \Pr\{\widehat{\mathbf{p}} \geq a + \epsilon \mid p\}$ for $p \in [a,b] \subseteq (0,1)$. As a consequence of the monotonicity of $\Pr\{\widehat{\mathbf{p}} \geq \vartheta \mid p\}$ and $\Pr\{\widehat{\mathbf{p}} \leq \vartheta \mid p\}$ with respect to p, where ϑ is a real number independent of p, the lower and upper bounds of $\Pr\{|\widehat{\mathbf{p}} - p| \geq \epsilon \mid p\}$ for $p \in [a,b] \subseteq (0,1)$ can be given as $\Pr\{\widehat{\mathbf{p}} \leq a - \epsilon \mid b\} + \Pr\{\widehat{\mathbf{p}} \geq b + \epsilon \mid a\}$ and $\Pr\{\widehat{\mathbf{p}} \leq b - \epsilon \mid a\} + \Pr\{\widehat{\mathbf{p}} \geq a + \epsilon \mid b\}$, respectively.

In page 15, equation (1) of [15], Chen proposed to apply the recursive method of Schultz et al. [23, Section 2] to compute the lower and upper bounds of $\Pr\{|\widehat{\mathbf{p}} - p| \geq \epsilon \mid p\}$ given by (9). It should be pointed out that such lower and upper bounds of $\Pr\{|\widehat{\mathbf{p}} - p| \geq \epsilon \mid p\}$ can also be computed by the recursive path-counting method of Franzén [10, page 49].

2.4. Adapted Branch and Bound. The third component for the exact sequential estimation of a binomial proportion is the adapted B&B algorithm, which was proposed in [15, Section 2.8], for quick determination of whether the coverage probability is no less than $1 - \delta$ for any value of the associated parameter. Such a task of checking the coverage probability is also referred to as checking the coverage probability guarantee. Given that lower and upper bounds of the complementary coverage probability on an interval of parameter values can be obtained by the interval bounding techniques, this task can be accomplished by applying the B&B algorithm [20] to compute exactly the maximum of the complementary coverage probability on the parameter space. However, in our applications, it suffices to determine whether the maximum of the complementary coverage probability $\Pr\{|\widehat{\mathbf{p}} - p| \geq \epsilon \mid p\}$ with respect to $p \in (0,1)$ is greater than the confidence parameter δ. For fast checking whether the maximal complementary coverage probability exceeds δ, Chen proposed to reduce the computational complexity by revising the standard B&B algorithm as the Adapted B&B Algorithm in [15, Section 2.8]. To describe this algorithm, let $\mathscr{I}_{\mathrm{init}}$ denote the parameter space $(0,1)$. For an interval $\mathscr{I} \subseteq \mathscr{I}_{\mathrm{init}}$, let $\max \Psi(\mathscr{I})$ denote the maximum of the complementary coverage probability $\Pr\{|\widehat{\mathbf{p}} - p| \geq \epsilon \mid p\}$ with respect to $p \in \mathscr{I}$. Let $\Psi_{\mathrm{lb}}(\mathscr{I})$ and $\Psi_{\mathrm{ub}}(\mathscr{I})$ be, respectively, the lower and upper bounds of $\Psi(\mathscr{I})$, which can be obtained by the interval bounding techniques introduced in Section 2.3. Let $\eta > 0$ be a prespecified tolerance, which is much smaller than δ. The adapted B&B algorithm of [15] is represented with a slight modification as in Algorithm 1.

It should be noted that for a sampling scheme of symmetrical stopping boundary, the initial interval $\mathscr{I}_{\mathrm{init}}$ may be taken as $(0, 1/2)$ for the sake of efficiency. In Section 5.1, we will illustrate why the adapted B&B algorithm is superior than the direct evaluation based on gridding parameter space. As will be seen in Section 5.2, the objective of the adapted B&B algorithm can also be accomplished by the Adaptive Maximum Checking Algorithm due to Chen [21, Section 3.3] and rediscovered by Frey [13, Appendix]. An explanation is given in Section 5.3 for the advantage of working with the complementary coverage probability.

2.5. Bisection Coverage Tuning. The fourth component for the exact sequential estimation of a binomial proportion is Bisection Coverage Tuning. Based on the adaptive rigorous checking of coverage probability, Chen proposed in [14, Section 2.7] and [15, Section 2.6] to apply a bisection search method to determine maximal ζ such that the coverage probability is no less than $1 - \delta$ for any value of the associated parameter. Moreover, Chen has developed asymptotic results in [15, page 21, Theorem 18] for determining the initial interval of ζ needed for the bisection search. Specifically, if the complementary coverage probability $\Pr\{|\widehat{\mathbf{p}} - p| \geq \epsilon \mid p\}$ associated with $\zeta = \zeta_0$ tends to δ as $\epsilon \to 0$, then the initial interval of ζ can be taken as $[\zeta_0 2^i, \zeta_0 2^{i+1}]$, where i is the largest integer such that the complementary coverage probability associated with $\zeta = \zeta_0 2^i$ is no greater than δ for all $p \in (0,1)$. By virtue of a bisection search, it is possible to obtain $\zeta^* \in [\zeta_0 2^i, \zeta_0 2^{i+1}]$ such that the complementary coverage probability associated with $\zeta = \zeta^*$ is guaranteed to be no greater than δ for all $p \in (0,1)$.

3. Principle of Constructing Stopping Rules

In this section, we shall illustrate the inherent connection between various stopping rules. It will be demonstrated that a lot of stopping rules can be derived by virtue of the inclusion principle proposed by Chen [18, Section 3].

3.1. Inclusion Principle. The problem of estimating a binomial proportion can be considered as a special case of parameter estimation for a random variable X parameterized by $\theta \in \Theta$, where the objective is to construct a sequential estimator $\widehat{\theta}$ for θ such that $\Pr\{|\widehat{\theta} - \theta| < \epsilon \mid \theta\} \geq 1 - \delta$ for any $\theta \in \Theta$. Assume that the sampling process consists of s stages with sample sizes $n_1 < n_2 < \cdots < n_s$. For $\ell = 1, \ldots, s$, define an estimator $\widehat{\theta}_\ell$ for θ in terms of samples $X_1, \ldots, X_{n_\ell}$ of X. Let $[L_\ell, U_\ell]$, $\ell = 1, 2, \ldots, s$ be a sequence of confidence intervals such that for any ℓ, $[L_\ell, U_\ell]$ is defined in terms of $X_1, \ldots, X_{n_\ell}$ and that the coverage probability $\Pr\{L_\ell \leq \theta \leq U_\ell \mid \theta\}$ can be made arbitrarily close to 1 by choosing $\zeta > 0$ to be a sufficiently small number. In Theorem 2 of [18], Chen proposed the following general stopping rule:

$$\text{Continue sampling until } U_\ell - \epsilon \leq \widehat{\theta}_\ell \leq L_\ell + \epsilon \quad \text{for some } \ell \in \{1, \ldots, s\}. \tag{10}$$

At the termination of the sampling process, a sequential estimator for θ is taken as $\widehat{\theta} = \widehat{\theta}_{\mathbf{l}}$, where $\mathbf{l}$ is the index of stage at the termination of sampling process.

Clearly, the general stopping rule (10) can be restated as follows.

Continue sampling until the confidence interval $[L_\ell, U_\ell]$ is included by interval $[\widehat{\theta}_\ell - \epsilon, \widehat{\theta}_\ell + \epsilon]$ for some $\ell \in \{1, \ldots, s\}$.

The sequence of confidence intervals are parameterized by ζ for purpose of controlling the coverage probability $\Pr\{|\widehat{\theta} - \theta| < \epsilon \mid \theta\}$. Due to the inclusion relationship $[L_\ell, U_\ell] \subseteq [\widehat{\theta}_\ell - \epsilon, \widehat{\theta}_\ell + \epsilon]$, such a general methodology of using a sequence of confidence intervals to construct a stopping rule for controlling the coverage probability is referred to as the *inclusion principle*. It is asserted by Theorem 2 of [18] that

$$\Pr\left\{\left|\widehat{\theta} - \theta\right| < \epsilon \mid \theta\right\} \geq 1 - s\zeta\delta, \quad \forall \theta \in \Theta, \tag{11}$$

provided that $\Pr\{L_\ell < \theta < U_\ell \mid \theta\} \geq 1 - \zeta\delta$ for $\ell = 1, \ldots, s$ and $\theta \in \Theta$. This demonstrates that if the number of stages s is bounded respective to ζ, then the coverage probability $\Pr\{|\widehat{\theta} - \theta| < \epsilon \mid \theta\}$ associated with the stopping rule derived from the inclusion principle can be controlled by ζ. Actually, before explicitly proposing the inclusion principle in [18], Chen had extensively applied the inclusion principle in [14–17] to construct stopping rules for estimating parameters of various distributions such as binomial, Poisson, geometric, hypergeometric, and normal distributions. A more general version of the inclusion principle is proposed in [19, Section 2.4]. For simplicity of the stopping rule, Chen had made effort to eliminate the computation of confidence limits.

In the context of estimating a binomial proportion p, the inclusion principle immediately leads to the following general stopping rule:

$$\text{Continue sampling until } \widehat{\mathbf{p}}_\ell - \epsilon \leq L_\ell \leq U_\ell \leq \widehat{\mathbf{p}}_\ell + \epsilon \quad \text{for some } \ell \in \{1, \ldots, s\}. \tag{12}$$

Consequently, the sequential estimator for p is taken as $\widehat{\mathbf{p}}$ according to (3). It should be pointed out that the stopping rule (12) had been rediscovered by Frey in Section 2, the 1st paragraph of [13]. The four stopping rules considered in his paper follow immediately from applying various confidence intervals to the general stopping rule (12).

In the sequel, we will illustrate how to apply (12) to the derivation of Stopping Rules A, B, C, and D introduced in Section 2.2 and other specific stopping rules.

3.2. Stopping Rule from Wald Intervals. By virtue of Wald's method of interval estimation for a binomial proportion p, a sequence of confidence intervals $[L_\ell, U_\ell]$, $\ell = 1, \ldots, s$ for p can be constructed such that

$$L_\ell = \widehat{\mathbf{p}}_\ell - \mathcal{Z}_{\zeta\delta}\sqrt{\frac{\widehat{\mathbf{p}}_\ell(1 - \widehat{\mathbf{p}}_\ell)}{n_\ell}}, \quad U_\ell = \widehat{\mathbf{p}}_\ell + \mathcal{Z}_{\zeta\delta}\sqrt{\frac{\widehat{\mathbf{p}}_\ell(1 - \widehat{\mathbf{p}}_\ell)}{n_\ell}}, \quad \ell = 1, \ldots, s, \tag{13}$$

and that $\Pr\{L_\ell \leq p \leq U_\ell \mid p\} \approx 1 - 2\zeta\delta$ for $\ell = 1, \ldots, s$ and $p \in (0, 1)$. Note that, for $\ell = 1, \ldots, s$, the event $\{\widehat{\mathbf{p}}_\ell - \epsilon \leq L_\ell \leq U_\ell \leq \widehat{\mathbf{p}}_\ell + \epsilon\}$ is the same as the event $\{(\widehat{\mathbf{p}}_\ell - 1/2)^2 \geq (1/4) - n_\ell(\epsilon/\mathcal{Z}_{\zeta\delta})^2\}$. So, applying this sequence of confidence intervals to (12) results in the stopping rule "continue sampling until $(\widehat{\mathbf{p}}_\ell - 1/2)^2 \geq (1/4) - n_\ell(\epsilon/\mathcal{Z}_{\zeta\delta})^2$ for some $\ell \in \{1, \ldots, s\}$". Since for any $\zeta \in (0, 1/\delta)$, there exists a unique number $\zeta' \in (0, 1/\delta)$ such that $\mathcal{Z}_{\zeta\delta} = \sqrt{2\ln(1/\zeta'\delta)}$, this stopping rule is equivalent to "Continue sampling until $(\widehat{\mathbf{p}}_\ell - 1/2)^2 \geq (1/4) + (\epsilon^2 n_\ell / 2\ln(\zeta\delta))$ for some $\ell \in \{1, \ldots, s\}$." This stopping rule is actually the same as Stopping Rule D, since $\{(\widehat{\mathbf{p}}_\ell - 1/2)^2 \geq (1/4) + (\epsilon^2 n_\ell / 2\ln(\zeta\delta))\} = \{n_\ell \geq \widehat{\mathbf{p}}_\ell(1 - \widehat{\mathbf{p}}_\ell)(2/\epsilon^2)\ln(1/\zeta\delta)\}$ for $\ell \in \{1, \ldots, s\}$.

3.3. Stopping Rule from Revised Wald Intervals. Define $\widetilde{\mathbf{p}}_\ell = (n_\ell \widehat{\mathbf{p}}_\ell + a)/(n_\ell + 2a)$ for $\ell = 1, \ldots, s$, where a is a positive number. Inspired by Wald's method of interval estimation for p, a sequence of confidence intervals $[L_\ell, U_\ell]$, $\ell = 1, \ldots, s$ can be constructed such that

$$L_\ell = \widetilde{\mathbf{p}}_\ell - \mathcal{Z}_{\zeta\delta}\sqrt{\frac{\widetilde{\mathbf{p}}_\ell(1 - \widetilde{\mathbf{p}}_\ell)}{n_\ell}}, \quad U_\ell = \widetilde{\mathbf{p}}_\ell + \mathcal{Z}_{\zeta\delta}\sqrt{\frac{\widetilde{\mathbf{p}}_\ell(1 - \widetilde{\mathbf{p}}_\ell)}{n_\ell}} \tag{14}$$

and that $\Pr\{L_\ell \leq p \leq U_\ell \mid p\} \approx 1 - 2\zeta\delta$ for $\ell = 1, \ldots, s$ and $p \in (0, 1)$. This sequence of confidence intervals was applied by Frey [13] to the general stopping rule (12). As a matter of fact, such idea of revising Wald interval $[\,\overline{X}_n - \mathcal{Z}_{\zeta\delta}\sqrt{(\overline{X}_n(1 - \overline{X}_n))/n},\ \overline{X}_n + \mathcal{Z}_{\zeta\delta}\sqrt{(\overline{X}_n(1 - \overline{X}_n))/n}\,]$ by replacing the relative frequency $\overline{X}_n = (\sum_{i=1}^{n} X_i)/n$ involved in the confidence limits with $\widetilde{p}_a = (n\overline{X}_n + a)/(n + 2a)$ had been proposed by Chen [24, Section 4].

As can be seen from Section 2, page 243, of Frey [13], applying (12) with the sequence of revised Wald intervals yields the stopping rule "Continue sampling until

$(\widetilde{\mathbf{p}}_\ell - 1/2)^2 \geq (1/4) + (\epsilon^2 n_\ell / 2\ln(\zeta\delta))$ for some $\ell \in \{1, \ldots, s\}$." Clearly, replacing $\widehat{\mathbf{p}}_\ell$ in Stopping Rule D with $\widetilde{\mathbf{p}}_\ell = (a + n_\ell \widehat{\mathbf{p}}_\ell)/(n_\ell + 2a)$ also leads to this stopping rule.

3.4. Stopping Rule from Wilson's Confidence Intervals. Making use of the interval estimation method of Wilson [25], one can obtain a sequence of confidence intervals $[L_\ell, U_\ell]$, $\ell = 1, \ldots, s$ for p such that

$$L_\ell = \max\left\{0, \frac{\widehat{\mathbf{p}}_\ell + (\mathscr{Z}_{\zeta\delta}^2 / 2n_\ell)}{1 + (\mathscr{Z}_{\zeta\delta}^2 / n_\ell)} - \frac{\mathscr{Z}_{\zeta\delta}\sqrt{(\widehat{\mathbf{p}}_\ell(1-\widehat{\mathbf{p}}_\ell)/n_\ell) + (\mathscr{Z}_{\zeta\delta}/2n_\ell)^2}}{1 + (\mathscr{Z}_{\zeta\delta}^2 / n_\ell)}\right\},$$

$$U_\ell = \min\left\{1, \frac{\widehat{\mathbf{p}}_\ell + (\mathscr{Z}_{\zeta\delta}^2 / 2n_\ell)}{1 + (\mathscr{Z}_{\zeta\delta}^2 / n_\ell)} + \frac{\mathscr{Z}_{\zeta\delta}\sqrt{(\widehat{\mathbf{p}}_\ell(1-\widehat{\mathbf{p}}_\ell)/n_\ell) + (\mathscr{Z}_{\zeta\delta}/2n_\ell)^2}}{1 + (\mathscr{Z}_{\zeta\delta}^2 / n_\ell)}\right\} \quad (15)$$

and that $\Pr\{L_\ell \leq p \leq U_\ell \mid p\} \approx 1 - 2\zeta\delta$ for $\ell = 1, \ldots, s$ and $p \in (0, 1)$. It should be pointed out that the sequence of Wilson's confidence intervals has been applied by Frey [13, Section 2, page 243] to the general stopping rule (12) for estimating a binomial proportion.

Since a stopping rule directly involves the sequence of Wilson's confidence intervals is cumbersome, it is desirable to eliminate the computation of Wilson's confidence intervals in the stopping rule. For this purpose, we need to use the following result.

Theorem 1. *Assume that $0 < \zeta\delta < 1$ and $0 < \epsilon < 1/2$. Then, Wilson's confidence intervals satisfy $\{\widehat{\mathbf{p}}_\ell - \epsilon \leq L_\ell \leq U_\ell \leq \widehat{\mathbf{p}}_\ell + \epsilon\} = \{(|\widehat{\mathbf{p}}_\ell - 1/2| - \epsilon)^2 \geq (1/4) - n_\ell(\epsilon/\mathscr{Z}_{\zeta\delta})^2\}$ for $\ell = 1, \ldots, s$.*

See Appendix A for a proof. As a consequence of Theorem 1 and the fact that for any $\zeta \in (0, 1/\delta)$, there exists a unique number $\zeta' \in (0, 1/\delta)$ such that $\mathscr{Z}_{\zeta\delta} = \sqrt{2\ln(1/\zeta'\delta)}$, applying the sequence of Wilson's confidence intervals to (12) leads to the following stopping rule.

Continue sampling until

$$\left(\left|\widehat{\mathbf{p}}_\ell - \frac{1}{2}\right| - \epsilon\right)^2 \geq \frac{1}{4} + \frac{\epsilon^2 n_\ell}{2\ln(\zeta\delta)}, \quad (16)$$

for some $\ell \in \{1, \ldots, s\}$.

3.5. Stopping Rule from Clopper-Pearson Confidence Intervals. Applying the interval estimation method of Clopper-Pearson [26], a sequence of confidence intervals $[L_\ell, U_\ell]$, $\ell = 1, \ldots, s$ for p can be obtained such that $\Pr\{L_\ell \leq p \leq U_\ell \mid p\} \geq 1 - 2\zeta\delta$ for $\ell = 1, \ldots, s$ and $p \in (0, 1)$, where the upper confidence limit U_ℓ satisfies the equation $S(0, K_\ell, n_\ell, U_\ell) = \zeta\delta$ if $K_\ell < n_\ell$; and the lower confidence limit L_ℓ satisfies the equation $S(K_\ell, n_\ell, n_\ell, L_\ell) = \zeta\delta$ if $K_\ell > 0$. The well-known equation (10.8) in [27, page 173] implies that $S(0, k, n, p)$, with $0 \leq k < n$, is decreasing with respect to $p \in (0, 1)$ and that $S(k, n, n, p)$, with $0 < k \leq n$, is increasing with respect to $p \in (0, 1)$. It follows that

$$\begin{aligned}
\{\widehat{\mathbf{p}}_\ell - \epsilon \leq L_\ell\} &= \{0 < \widehat{\mathbf{p}}_\ell - \epsilon \leq L_\ell\} \cup \{\widehat{\mathbf{p}}_\ell \leq \epsilon\} \\
&= \{\widehat{\mathbf{p}}_\ell > \epsilon,\ S(K_\ell, n_\ell, n_\ell, \widehat{\mathbf{p}}_\ell - \epsilon) \leq \zeta\delta\} \\
&\quad \cup \{\widehat{\mathbf{p}}_\ell \leq \epsilon\} \\
&= \{\widehat{\mathbf{p}}_\ell > \epsilon,\ S(K_\ell, n_\ell, n_\ell, \widehat{\mathbf{p}}_\ell - \epsilon) \leq \zeta\delta\} \\
&\quad \cup \{\widehat{\mathbf{p}}_\ell \leq \epsilon,\ S(K_\ell, n_\ell, n_\ell, \widehat{\mathbf{p}}_\ell - \epsilon) \leq \zeta\delta\} \\
&= \{S(K_\ell, n_\ell, n_\ell, \widehat{\mathbf{p}}_\ell - \epsilon) \leq \zeta\delta\}, \\
\{\widehat{\mathbf{p}}_\ell + \epsilon \geq U_\ell\} &= \{1 > \widehat{\mathbf{p}}_\ell + \epsilon \geq U_\ell\} \cup \{\widehat{\mathbf{p}}_\ell \geq 1 - \epsilon\} \\
&= \{\widehat{\mathbf{p}}_\ell < 1 - \epsilon,\ S(0, K_\ell, n_\ell, \widehat{\mathbf{p}}_\ell + \epsilon) \leq \zeta\delta\} \\
&\quad \cup \{\widehat{\mathbf{p}}_\ell \geq 1 - \epsilon\} \\
&= \{\widehat{\mathbf{p}}_\ell < 1 - \epsilon,\ S(0, K_\ell, n_\ell, \widehat{\mathbf{p}}_\ell + \epsilon) \leq \zeta\delta\} \\
&\quad \cup \{\widehat{\mathbf{p}}_\ell \geq 1 - \epsilon,\ S(0, K_\ell, n_\ell, \widehat{\mathbf{p}}_\ell + \epsilon) \leq \zeta\delta\} \\
&= \{S(0, K_\ell, n_\ell, \widehat{\mathbf{p}}_\ell + \epsilon) \leq \zeta\delta\},
\end{aligned} \quad (17)$$

for $\ell = 1, \ldots, s$. Consequently,

$$\begin{aligned}
&\{\widehat{\mathbf{p}}_\ell - \epsilon \leq L_\ell \leq U_\ell \leq \widehat{\mathbf{p}}_\ell + \epsilon\} \\
&\quad = \{S(K_\ell, n_\ell, n_\ell, \widehat{\mathbf{p}}_\ell - \epsilon) \leq \zeta\delta,\ S(0, K_\ell, n_\ell, \widehat{\mathbf{p}}_\ell + \epsilon) \leq \zeta\delta\},
\end{aligned} \quad (18)$$

for $\ell = 1, \ldots, s$. This demonstrates that applying the sequence of Clopper-Pearson confidence intervals to the general stopping rule (12) gives Stopping Rule C.

It should be pointed out that Stopping Rule C was rediscovered by Frey as the third stopping rule in Section 2, page 243 of his paper [13].

3.6. Stopping Rule from Fishman's Confidence Intervals. By the interval estimation method of Fishman [28], a sequence of confidence intervals $[L_\ell, U_\ell]$, $\ell = 1, \ldots, s$ for p can be obtained such that

$$L_\ell = \begin{cases} 0 & \text{if } \widehat{\mathbf{p}}_\ell = 0, \\ \left\{\theta_\ell \in (0, \widehat{\mathbf{p}}_\ell) : \mathscr{M}(\widehat{\mathbf{p}}_\ell, \theta_\ell) = \dfrac{\ln(\zeta\delta)}{n_\ell}\right\} & \text{if } \widehat{\mathbf{p}}_\ell > 0, \end{cases}$$

$$U_\ell = \begin{cases} 1 & \text{if } \widehat{\mathbf{p}}_\ell = 1, \\ \left\{\theta_\ell \in (\widehat{\mathbf{p}}_\ell, 1) : \mathscr{M}(\widehat{\mathbf{p}}_\ell, \theta_\ell) = \dfrac{\ln(\zeta\delta)}{n_\ell}\right\} & \text{if } \widehat{\mathbf{p}}_\ell < 1. \end{cases} \quad (19)$$

Under the assumption that $0 < \zeta\delta < 1$ and $0 < \epsilon < 1/2$, by similar techniques as the proof of Theorem 7 of [22], it can be shown that $\{\widehat{\mathbf{p}}_\ell - \epsilon \le L_\ell \le U_\ell \le \widehat{\mathbf{p}}_\ell + \epsilon\} = \{\mathscr{M}((1/2) - |(1/2) - \widehat{\mathbf{p}}_\ell|, (1/2) - |(1/2) - \widehat{\mathbf{p}}_\ell| + \epsilon) \le (\ln(\zeta\delta))/n_\ell\}$ for $\ell = 1, \ldots, s$. Therefore, applying the sequence of confidence intervals of Fishman to the general stopping rule (12) gives Stopping Rule A.

It should be noted that Fishman's confidence intervals are actually derived from the Chernoff bounds of the tailed probabilities of the sample mean of Bernoulli random variable. Hence, Stopping Rule A is also referred to as the stopping rule from Chernoff bounds in this paper.

3.7. Stopping Rule from Confidence Intervals of Chen et al. Using the interval estimation method of Chen et al. [29], a sequence of confidence intervals $[L_\ell, U_\ell]$, $\ell = 1, \ldots, s$ for p can be obtained such that

$$L_\ell = \max\left\{0, \widehat{\mathbf{p}}_\ell + \frac{3}{4} \times \frac{1 - 2\widehat{\mathbf{p}}_\ell - \sqrt{1 + (9n_\ell/2\ln(1/\zeta\delta))\widehat{\mathbf{p}}_\ell(1-\widehat{\mathbf{p}}_\ell)}}{1 + (9n_\ell/8\ln(1/\zeta\delta))}\right\},$$

$$U_\ell = \min\left\{1, \widehat{\mathbf{p}}_\ell + \frac{3}{4} \times \frac{1 - 2\widehat{\mathbf{p}}_\ell + \sqrt{1 + (9n_\ell/2\ln(1/\zeta\delta))\widehat{\mathbf{p}}_\ell(1-\widehat{\mathbf{p}}_\ell)}}{1 + (9n_\ell/8\ln(1/\zeta\delta))}\right\} \quad (20)$$

and that $\Pr\{L_\ell \le p \le U_\ell \mid p\} \ge 1 - 2\zeta\delta$ for $\ell = 1, \ldots, s$ and $p \in (0, 1)$. Under the assumption that $0 < \zeta\delta < 1$ and $0 < \epsilon < 1/2$, by similar techniques as the proof of Theorem 1 of [30], it can be shown that $\{\widehat{\mathbf{p}}_\ell - \epsilon \le L_\ell \le U_\ell \le \widehat{\mathbf{p}}_\ell + \epsilon\} = \{(|\widehat{\mathbf{p}}_\ell - 1/2| - (2/3)\epsilon)^2 \ge (1/4) + (\epsilon^2 n_\ell / 2\ln(\zeta\delta))\}$ for $\ell = 1, \ldots, s$. This implies that applying the sequence of confidence intervals of Chen et al. to the general stopping rule (12) leads to Stopping Rule B.

Actually, the confidence intervals of Chen et al. [29] are derived from Massart's inequality [31] on the tailed probabilities of the sample mean of Bernoulli random variable. For this reason, Stopping Rule B is also referred to as the stopping rule from Massart's inequality in [21, Section 4.1.1].

4. Double-Parabolic Sequential Estimation

From Sections 2.2, 3.2, and 3.7, it can be seen that, by introducing a new parameter $\rho \in [0, 1]$ and letting ρ take values 2/3 and 0, respectively, Stopping Rules B and D can be accommodated as special cases of the following general stopping rule.

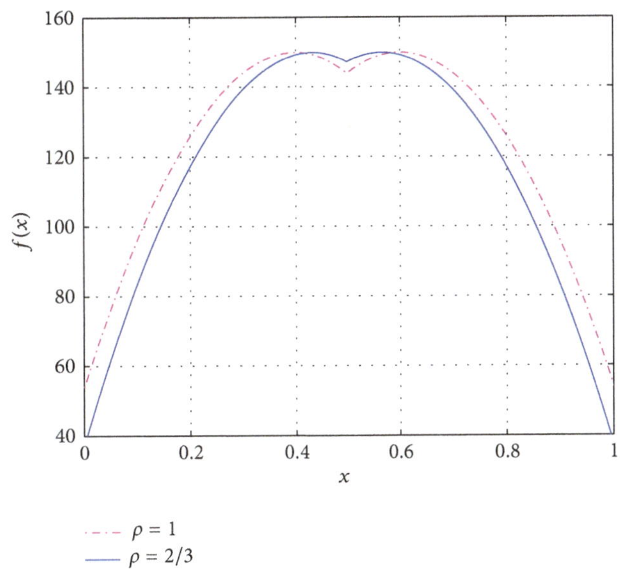

FIGURE 1: Double-parabolic sampling.

Continue the sampling process until

$$\left(\left|\widehat{\mathbf{p}}_\ell - \frac{1}{2}\right| - \rho\epsilon\right)^2 \ge \frac{1}{4} + \frac{\epsilon^2 n_\ell}{2\ln(\zeta\delta)} \quad (21)$$

for some $\ell \in \{1, 2, \ldots, s\}$, where $\zeta \in (0, 1/\delta)$.

Moreover, as can be seen from (16), the stopping rule derived from applying Wilson's confidence intervals to (12) can also be viewed as a special case of such general stopping rule with $\rho = 1$.

From the stopping condition (21), it can be seen that the stopping boundary is associated with the double-parabolic function $f(x) = (2/\epsilon^2)\ln(\zeta\delta)[(1/4) - (|x - 1/2| - \rho\epsilon)^2]$ such that x and $f(x)$ correspond to the sample mean and sample size, respectively. For $\epsilon = 0.1$, $\delta = 0.05$, and $\zeta = 1$, stopping boundaries with various ρ are shown by Figure 1.

For fixed ϵ and δ, the parameters ρ and ζ affect the shape of the stoping boundary in a way as follows. As ρ increases, the span of stopping boundary is increasing in the axis of sample mean. By decreasing ζ, the stopping boundary can be dragged toward the direction of increasing sample size. Hence, the parameter ρ is referred to as the *dilation coefficient*. The parameter ζ is referred to as the *coverage tuning parameter*. Since the stopping boundary consists of two parabolas, this approach of estimating a binomial proportion is referred to as the *double-parabolic sequential estimation* method.

4.1. Parametrization of the Sampling Scheme. In this section, we shall parameterize the double-parabolic sequential sampling scheme by the method described in Section 2.2. From the stopping condition (21), the stopping rule can be restated as follows. Continue sampling until $\mathscr{D}(\widehat{\mathbf{p}}_\ell, n_\ell) = 1$

for some $\ell \in \{1, \dots, s\}$, where the function $\mathscr{D}(\cdot, \cdot)$ is defined by

$$\mathscr{D}(z, n) = \begin{cases} 1 & \text{if } \left(\left|z - \frac{1}{2}\right| - \rho\epsilon\right)^2 \geq \frac{1}{4} + \frac{\epsilon^2 n}{2\ln(\zeta\delta)}, \\ 0 & \text{otherwise.} \end{cases} \quad (22)$$

Clearly, the function $\mathscr{D}(\cdot, \cdot)$ associated with the double-parabolic sequential sampling scheme depends on the design parameters ρ, ζ, ϵ and δ. Applying the function $\mathscr{D}(\cdot, \cdot)$ defined by (22) to (6) yields

$$N_{\min} = \min\left\{ n \in \mathbb{N} : \left(\left|\frac{k}{n} - \frac{1}{2}\right| - \rho\epsilon\right)^2 \geq \frac{1}{4} + \frac{\epsilon^2 n}{2\ln(\zeta\delta)} \text{ for some nonnegative integer } k \text{ not exceeding } n \right\}. \quad (23)$$

Since ϵ is usually small in practical applications, we restrict ϵ to satisfy $0 < \rho\epsilon \leq 1/4$. As a consequence of $0 \leq \rho\epsilon \leq 1/4$ and the fact that $|z - 1/2| \leq 1/2$ for any $z \in [0, 1]$, it must be true that $(|z - 1/2| - \rho\epsilon)^2 \leq ((1/2) - \rho\epsilon)^2$ for any $z \in [0, 1]$. It follows from (23) that $((1/2) - \rho\epsilon)^2 \geq (1/4) + (\epsilon^2 N_{\min}/2\ln(\zeta\delta))$, which implies that the minimum sample size can be taken as

$$N_{\min} = \left\lceil 2\rho\left(\frac{1}{\epsilon} - \rho\right) \ln \frac{1}{\zeta\delta} \right\rceil. \quad (24)$$

On the other hand, applying the function $\mathscr{D}(\cdot, \cdot)$ defined by (22) to (7) gives

$$N_{\max} = \min\left\{ n \in \mathbb{N} : \left(\left|\frac{k}{n} - \frac{1}{2}\right| - \rho\epsilon\right)^2 \geq \frac{1}{4} + \frac{\epsilon^2 n}{2\ln(\zeta\delta)} \text{ for all nonnegative integer } k \text{ not exceeding } n \right\}. \quad (25)$$

Since $(|z - 1/2| - \rho\epsilon)^2 \geq 0$ for any $z \in [0, 1]$, it follows from (25) that $(1/4) + (\epsilon^2 N_{\max}/2\ln(\zeta\delta)) \leq 0$, which implies that maximum sample size can be taken as

$$N_{\max} = \left\lceil \frac{1}{2\epsilon^2} \ln \frac{1}{\zeta\delta} \right\rceil. \quad (26)$$

Therefore, the sample sizes $n_1, \dots, n_s$ can be chosen as functions of ρ, ζ, ϵ, and δ which satisfy the following constraint:

$$N_{\min} \leq n_1 < \cdots < n_{s-1} < N_{\max} \leq n_s. \quad (27)$$

In particular, if the number of stages s is given and the group sizes are expected to be approximately equal, then the sample sizes, $n_1, \dots, n_s$, for all stages can be obtained by substituting $N_{\min}$ defined by (24) and $N_{\max}$ defined by (26) into (8). For example, if the values of design parameters are $\epsilon = 0.05$, $\delta = 0.05$, $\rho = 3/4$, $\zeta = 2.6759$ and $s = 7$, then the sample sizes of this sampling scheme are calculated as

$$n_1 = 59, \quad n_2 = 116, \quad n_3 = 173, \quad n_4 = 231, \quad n_5 = 288, \quad n_6 = 345, \quad n_7 = 403. \quad (28)$$

The stopping rule is completely determined by substituting the values of design parameters into (21).

4.2. Uniform Controllability of Coverage Probability. Clearly, for prespecified ϵ, δ, and ρ, the coverage probability $\Pr\{|\widehat{\mathbf{p}} - p| < \epsilon \mid p\}$ depends on the parameter ζ, the number of stages s, and the sample sizes $n_1, \dots, n_s$. As illustrated in Section 4.1, the number of stages s and the sample sizes $n_1, \dots, n_s$ can be defined as functions of $\zeta \in (0, 1/\delta)$. That is, the stopping rule can be parameterized by ζ. Accordingly, for any $p \in (0, 1)$, the coverage probability $\Pr\{|\widehat{\mathbf{p}} - p| < \epsilon \mid p\}$ becomes a function of ζ. The following theorem shows that it suffices to choose $\zeta \in (0, 1/\delta)$ small enough to guarantee the prespecified confidence level.

Theorem 2. *Let $\epsilon, \delta \in (0, 1)$ and $\rho \in (0, 1]$ be fixed. Assume that the number of stages s and the sample sizes $n_1, \dots, n_s$ are functions of $\zeta \in (0, 1/\delta)$ such that the constraint (27) is satisfied. Then, $\Pr\{|\widehat{\mathbf{p}} - p| < \epsilon \mid p\}$ is no less than $1 - \delta$ for any $p \in (0, 1)$ provided that*

$$0 < \zeta \leq \frac{1}{\delta} \exp\left(\frac{\ln(\delta/2) + \ln\left[1 - \exp\left(-2\epsilon^2\right)\right]}{4\epsilon\rho(1 - \rho\epsilon)} \right). \quad (29)$$

See Appendix B for a proof. For Theorem 2 to be valid, the choice of sample sizes is very flexible. Particularly, the sample sizes can be arithmetic or geometric progressions or any others, as long as the constraint (27) is satisfied. It can be seen that for the coverage probability to be uniformly controllable, the dilation coefficient ρ must be greater than 0. Theorem 2 asserts that there exists $\zeta > 0$ such that the coverage probability is no less than $1 - \delta$, regardless of the associated binomial proportion p. For the purpose of reducing sampling cost, we want to have a value of ζ as large as possible such that the prespecified confidence level is guaranteed for any $p \in (0, 1)$. This can be accomplished by the technical components introduced in Sections 2.1, 2.3, 2.4, and 2.5. Clearly, for every value of ρ, we can obtain a corresponding value of ζ (as large as possible) to ensure the desired confidence level. However, the performance of resultant stopping rules are different. Therefore, we can try a number of values of ρ and pick the best resultant stopping rule for practical use.

4.3. Asymptotic Optimality of Sampling Schemes. Now we shall provide an important reason why we propose the sampling scheme of that structure by showing its asymptotic optimality. Since the performance of a group sampling scheme will be close to its fully sequential counterpart, we investigate the optimality of the fully sequential sampling scheme. In this scenario, the sample sizes $n_1, n_2, \ldots, n_s$ are consecutive integers such that

$$\left\lceil 2\rho\left(\frac{1}{\epsilon}-\rho\right)\ln\frac{1}{\zeta\delta}\right\rceil = n_1 < n_2 < \cdots < n_{s-1} < n_s = \left\lceil \frac{1}{2\epsilon^2}\ln\frac{1}{\zeta\delta}\right\rceil. \quad (30)$$

The fully sequential sampling scheme can be viewed as a special case of a group sampling scheme of $s = n_s - n_1 + 1$ stages and group size 1. Clearly, if δ, ζ and ρ are fixed, the sampling scheme is dependent only on ϵ. Hence, for any $p \in (0,1)$, if we allow ϵ to vary in $(0,1)$, then the coverage probability $\Pr\{|\widehat{\mathbf{p}} - p| < \epsilon \mid p\}$ and the average sample number $\mathbb{E}[\mathbf{n}]$ are functions of ϵ. We are interested in knowing the asymptotic behavior of these functions as $\epsilon \to 0$, since ϵ is usually small in practical situations. The following theorem provides us the desired insights.

Theorem 3. *Assume that $\delta \in (0,1)$, $\zeta \in (0, 1/\delta)$ and $\rho \in (0,1]$ are fixed. Define $N(p,\epsilon,\delta,\zeta) = (2p(1-p)\ln(1/\zeta\delta))/\epsilon^2$ for $p \in (0,1)$ and $\epsilon \in (0,1)$. Then,*

$$\Pr\left\{\lim_{\epsilon\to 0}\frac{\mathbf{n}}{N(p,\epsilon,\delta,\zeta)} = 1 \mid p\right\} = 1, \quad (31)$$

$$\lim_{\epsilon\to 0}\Pr\{|\widehat{\mathbf{p}} - p| < \epsilon \mid p\} = 2\Phi\left(\sqrt{2\ln\frac{1}{\zeta\delta}}\right) - 1, \quad (32)$$

$$\lim_{\epsilon\to 0}\frac{\mathbb{E}[\mathbf{n}]}{N(p,\epsilon,\delta,\zeta)} = 1, \quad (33)$$

for any $p \in (0,1)$.

See Appendix C for a proof. From (32), it can be seen that $\lim_{\epsilon\to 0}\Pr\{|\widehat{\mathbf{p}} - p| < \epsilon \mid p\} = 1-\delta$ for any $p \in (0,1)$ if $\zeta = (1/\delta)\exp(-(1/2)\mathscr{Z}_{\delta/2}^2)$. Such value can be taken as an initial value for the coverage tuning parameter ζ. In addition to providing guidance on the coverage tuning techniques, Theorem 3 also establishes the optimality of the sampling scheme. To see this, let $\mathscr{N}(p,\epsilon,\delta)$ denote the minimum sample size n required for a fixed-sample-size procedure to guarantee that $\Pr\{|\overline{X}_n - p| < \epsilon \mid p\} \geq 1-\delta$ for any $p \in (0,1)$, where $\overline{X}_n = (\sum_{i=1}^n X_i)/n$. It is well known that from the central limit theorem,

$$\lim_{\epsilon\to 0}\frac{\mathscr{N}(p,\epsilon,\delta)}{p(1-p)\left(\mathscr{Z}_{\delta/2}/\epsilon\right)^2} = 1. \quad (34)$$

Applying (33), (34), and letting $\zeta = (1/\delta)\exp(-(1/2)\mathscr{Z}_{\delta/2}^2)$, we have $\lim_{\epsilon\to 0}(\mathscr{N}(p,\epsilon,\delta)/N(p,\epsilon,\delta,\zeta)) = 1$ for $p \in (0,1)$ and $\delta \in (0,1)$, which implies the asymptotic optimality of the double-parabolic sampling scheme. By virtue of (33), an approximate formula for computing the average sample number is given as follows:

$$\mathbb{E}[\mathbf{n}] \approx N(p,\epsilon,\delta,\zeta) = \frac{2p(1-p)\ln(1/\zeta\delta)}{\epsilon^2}, \quad (35)$$

for $p \in (0,1)$ and $\epsilon \in (0,1)$. From (34), one obtains $\mathscr{N}(p,\epsilon,\delta) \approx p(1-p)(\mathscr{Z}_{\delta/2}/\epsilon)^2$, which is a well-known result in statistics. In situations that no information of p is available, one usually uses

$$N_{\text{normal}} \stackrel{\text{def}}{=} \left\lceil \frac{1}{4}\left(\frac{\mathscr{Z}_{\delta/2}}{\epsilon}\right)^2\right\rceil \quad (36)$$

as the sample size for estimating the binomial proportion p with prescribed margin of error ϵ and confidence level $1-\delta$. Since the sample size formula (36) can lead to under-coverage, researchers in many areas are willing to use a more conservative but rigorous sample size formula

$$N_{\text{ch}} \stackrel{\text{def}}{=} \left\lceil \frac{\ln(2/\delta)}{2\epsilon^2}\right\rceil, \quad (37)$$

which is derived from the Chernoff-Hoeffding bound [32, 33]. Comparing (35) and (37), one can see that under the premise of guaranteeing the prescribed confidence level $1-\delta$, the double-parabolic sampling scheme can lead to a substantial reduction of sample number when the unknown binomial proportion p is close to 0 or 1.

4.4. Bounds on Distribution and Expectation of Sample Number. We shall derive analytic bounds for the cumulative distribution function and expectation of the sample number $\mathbf{n}$ associated with the double-parabolic sampling scheme. In this direction, we have obtained the following results.

Theorem 4. *Let $p \in (0,1/2]$. Define $a_\ell = (1/2) - \rho\epsilon - \sqrt{(1/4) + (\epsilon^2 n_\ell / 2\ln(\zeta\delta))}$ for $\ell = 1,\ldots,s$. Let τ denote the index of stage such that $a_{\tau-1} \leq p < a_\tau$. Then, $\Pr\{\mathbf{n} > n_\ell \mid p\} \leq \exp(n_\ell \mathscr{M}(a_\ell, p))$ for $\tau \leq \ell < s$. Moreover, $\mathbb{E}[\mathbf{n}] \leq n_\tau + \sum_{\ell=\tau}^{s-1}(n_{\ell+1} - n_\ell)\exp(n_\ell\mathscr{M}(a_\ell, p))$.*

See Appendix D for a proof. By the symmetry of the double-parabolic sampling scheme, similar analytic bounds for the distribution and expectation of the sample number can be derived for the case that $p \in [1/2, 1)$.

5. Comparison of Computational Methods

In this section, we shall compare various computational methods. First, we will illustrate why a frequently used method of evaluating the coverage probability based on gridding the parameter space is not rigorous and is less efficient as compared to the adapted B&B algorithm. Second, we will introduce the Adaptive Maximum Checking Algorithm of [21] which has better computational efficiency as compared to the adapted B&B algorithm. Third, we will explain that it is more advantageous in terms of numerical accuracy to work with the complementary coverage probability as compared to direct evaluation of the coverage probability. Finally, we will compare the computational methods of Chen [14–18] and Frey [13] for the design of sequential procedures for estimating a binomial proportion.

∇ Choose initial step size $d > \eta$.
∇ Let $F \leftarrow 0$, $T \leftarrow 0$ and $b \leftarrow \overline{\theta}$.
∇ While $F = T = 0$, do the following:
◇ Let st $\leftarrow 0$ and $\ell \leftarrow 2$;
◇ While st $= 0$, do the following:
⋆ Let $\ell \leftarrow \ell - 1$ and $d \leftarrow d2^{\ell}$.
⋆ If $b - d > \underline{\theta}$, then let $a \leftarrow b - d$ and $T \leftarrow 0$. Otherwise, let $a \leftarrow \underline{\theta}$ and $T \leftarrow 1$.
⋆ If $\overline{C}(a, b) < \delta$, then let st $\leftarrow 1$ and $b \leftarrow a$.
⋆ If $d < \eta$, then let st $\leftarrow 1$ and $F \leftarrow 1$.
∇ Return F.

ALGORITHM 2

5.1. Verifying Coverage Guarantee without Gridding Parameter Space. For purpose of constructing a sampling scheme so that the prescribed confidence level $1 - \delta$ is guaranteed, an essential task is to determine whether the coverage probability $\Pr\{|\widehat{\mathbf{p}} - p| < \epsilon \mid p\}$ associated with a given stopping rule is no less than $1 - \delta$. In other words, it is necessary to compare the infimum of coverage probability with $1 - \delta$. To accomplish such a task of checking coverage guarantee, a natural method is to evaluate the infimum of coverage probability as follows:

(i) choose m grid points $p_1, \ldots, p_m$ from parameter space $(0, 1)$;

(ii) compute $c_j = \Pr\{|\widehat{\mathbf{p}} - p| < \epsilon \mid p_j\}$ for $j = 1, \ldots, m$;

(iii) Take $\min\{c_1, \ldots, c_m\}$ as $\inf_{p\in(0,1)} \Pr\{|\widehat{\mathbf{p}} - p| < \epsilon \mid p\}$.

This method can be easily mistaken as an exact approach and has been frequently used for evaluating coverage probabilities in many problem areas.

It is not hard to show that if the sample size $\mathbf{n}$ of a sequential procedure has a support $\mathscr{S}$, then the coverage probability $\Pr\{|\widehat{\mathbf{p}} - p| < \epsilon \mid p\}$ is discontinuous at $p \in \mathscr{P} \cap (0, 1)$, where $\mathscr{P} = \{(k/n) \pm \epsilon : k$ is a nonnegative integer no greater than $n \in \mathscr{S}\}$. The set $\mathscr{P}$ typically has a large number of parameter values. Due to the discontinuity of the coverage probability as a function of p, the coverage probabilities can differ significantly for two parameter values which are extremely close. This implies that an intolerable error can be introduced by taking the minimum of coverage probabilities of a finite number of parameter values as the infimum of coverage probability on the whole parameter space. So, if one simply uses the minimum of the coverage probabilities of a finite number of parameter values as the infimum of coverage probability to check the coverage guarantee, the sequential estimator $\widehat{\mathbf{p}}$ of the resultant stopping rule will fail to guarantee the prescribed confidence level.

In addition to the lack of rigorousness, another drawback of checking coverage guarantee based on the method of gridding parameter space is its low efficiency. A critical issue is on the choice of the number, m, of grid points. If the number m is too small, the induced error can be substantial. On the other hand, choosing a large number for m results in high computational complexity.

In contrast to the method based on gridding parameter space, the adapted B&B algorithm is a rigorous approach for checking coverage guarantee as a consequence of the mechanism for comparing the bounds of coverage probability with the prescribed confidence level. The algorithm is also efficient due to the mechanism of pruning branches.

5.2. Adaptive Maximum Checking Algorithm. As illustrated in Section 2, the techniques developed in [14–18] are sufficient to provide exact solutions for a wide range of sequential estimation problems. However, one of the four components, the adapted B&B algorithm, requires computing both the lower and upper bounds of the complementary coverage probability. To further reduce the computational complexity, it is desirable to have a checking algorithm which needs only one of the lower and upper bounds. For this purpose, Chen had developed the Adaptive Maximum Checking Algorithm (AMCA) in [21, Section 3.3] and [19, Section 2.7]. In the following introduction of the AMCA, we shall follow the description of [21]. The AMCA can be applied to a wide class of computational problems dependent on the following critical subroutine.

Determine whether a function $C(\theta)$ is smaller than a prescribed number δ for every value of θ contained in interval $[\underline{\theta}, \overline{\theta}]$.

Particularly, for checking the coverage guarantee in the context of estimating a binomial proportion, the parameter θ is the binomial proportion p and the function $C(\theta)$ is actually the complementary coverage probability. In many situations, it is impossible or very difficult to evaluate $C(\theta)$ for every value of θ in interval $[\underline{\theta}, \overline{\theta}]$, since the interval may contain infinitely many or an extremely large number of values. Similar to the adapted B&B algorithm, the purpose of AMCA is to reduce the computational complexity associated with the problem of determining whether the maximum of $C(\theta)$ over $[\underline{\theta}, \overline{\theta}]$ is less than δ. The only assumption required for AMCA is that, for any interval $[a, b] \subseteq [\underline{\theta}, \overline{\theta}]$, it is possible to compute an upper bound $\overline{C}(a, b)$ such that $C(\theta) \leq \overline{C}(a, b)$ for any $\theta \in [a, b]$ and that the upper bound converges to $C(\theta)$ as the interval width $b - a$ tends to 0. The backward AMCA proceeds as in Algorithm 2.

The output of the backward AMCA is a binary variable F such that "$F = 0$" means "$C(\theta) < \delta$" and "$F = 1$" means "$C(\theta) \geq \delta$." An intermediate variable T is introduced in the description of AMCA such that "$T = 1$" means that the left endpoint of the interval is reached. The backward AMCA starts from the right endpoint of the interval (i.e., $b = \overline{\theta}$) and attempts to find an interval $[a, b]$ such that $\overline{C}(a, b) < \delta$. If such an interval is available, then, attempt to go backward to find the next consecutive interval with twice width. If doubling the interval width fails to guarantee $\overline{C}(a, b) < \delta$, then try to repeatedly cut the interval width in half to ensure that $\overline{C}(a, b) < \delta$. If the interval width becomes smaller than a prescribed tolerance η, then AMCA declares that "$F = 1$." For our relevant statistical problems, if $C(\theta) \geq \delta$ for some $\theta \in [\underline{\theta}, \overline{\theta}]$, it is sure that "$F = 1$" will be declared. On the other hand, it is possible that "$F = 1$" is declared even though $C(\theta) < \delta$ for any $\theta \in [\underline{\theta}, \overline{\theta}]$. However, such situation can be made extremely rare and immaterial if we choose η to be a very small number. Moreover, this will only introduce negligible conservativeness in the evaluation of $C(\theta)$ if η is chosen to be sufficiently small (e.g., $\eta = 10^{-15}$). Clearly, the backward AMCA can be easily modified as forward AMCA. Moreover, the AMCA can also be easily modified as Adaptive Minimum Checking Algorithm (forward and backward). For checking the maximum of complementary coverage probability $\Pr\{|\widehat{\mathbf{p}} - p| \geq \epsilon \mid p\}$, one can use the AMCA with $C(p) = \Pr\{|\widehat{\mathbf{p}} - p| \geq \epsilon \mid p\}$ over interval $[0, 1/2]$. We would like to point out that, in contrast to the adapted B&B algorithm, it seems difficult to generalize the AMCA to problems involving multidimensional parameter spaces.

5.3. Working with Complementary Coverage Probability. We would like to point out that, instead of evaluating the coverage probability as in [13], it is better to evaluate the complementary coverage probability for purpose of reducing numerical error. The advantage of working on the complementary coverage probability can be explained as follows. Note that, in many cases, the coverage probability is very close to 1 and the complementary coverage probability is very close to 0. Since the absolute precision for computing a number close to 1 is much lower than the absolute precision for computing a number close to 0, the method of directly evaluating the coverage probability will lead to intolerable numerical error for problems involving small δ. As an example, consider a situation that the complementary coverage probability is in the order of 10^{-5}. Direct computation of the coverage probability can easily lead to an absolute error of the order of 10^{-5}. However, the absolute error of computing the complementary coverage probability can be readily controlled at the order of 10^{-9}.

5.4. Comparison of Approaches of Chen and Frey. As mentioned in the introduction, Frey published a paper [13] in The American Statistician (TAS) on the sequential estimation of a binomial proportion with prescribed margin of error and confidence level. The approaches of Chen and Frey are based on the same strategy as follows. First, construct a family of stopping rules parameterized by γ (and possibly other design parameters) so that the associated coverage probability $\Pr\{|\widehat{\mathbf{p}} - p| < \epsilon \mid p\}$ can be controlled by parameter γ in the sense that the coverage probability can be made arbitrarily close to 1 by increasing γ. Second, apply a bisection search method to determine the parameter γ so that the coverage probability is no less than the prescribed confidence level $1 - \delta$ for any $p \in (0, 1)$.

For the purpose of controlling the coverage probability, Frey [13] applied the inclusion principle previously proposed in [18, Section 3] and used in [14–17]. As illustrated in Section 3, the central idea of inclusion principle is to use a sequence of confidence intervals to construct stopping rules so that the sampling process is continued until a confidence interval is included by an interval defined in terms of the estimator and margin of error. Due to the inclusion relationship, the associated coverage probability can be controlled by the confidence coefficients of the sequence of confidence intervals. The critical value γ used by Frey plays the same role for controlling coverage probabilities as that of the coverage tuning parameter ζ used by Chen. Frey [13] stated stopping rules in terms of confidence limits. This way of expressing stopping rules is straightforward and insightful, since one can readily see the principle behind the construction. For convenience of practical use, Chen proposed to eliminate the necessity of computing confidence limits.

Similar to the AMCA proposed in [21, Section 3.3], the algorithm of Frey [13, Appendix] for checking coverage guarantee adaptively scans the parameter space based on interval bounding. The adaptive method used by Frey for updating step size is essentially the same as that of the AMCA. Ignoring the number 0.01 in Frey's expression "$\varepsilon_i = \min\{0.01, 2(p_{i-1} - p_{i-2})\}$," which has very little impact on the computational efficiency, Frey's step size ε_i can be identified as the adaptive step size d in the AMCA. The operation associated with "$\varepsilon_i = \min\{0.01, 2(p_{i-1} - p_{i-2})\}$" has a similar function as that of the command "Let st $\leftarrow$ 0 and $\ell \leftarrow 2$" in the outer loop of the AMCA. The operation associated with Frey's expression "$p_{i-1} + \varepsilon_i/2^j,\ j \geq 0$" is equivalent to that of the command "Let $\ell \leftarrow \ell - 1$ and $d \leftarrow d2^{\ell}$" in the inner loop of the AMCA. Frey proposed to declare a failure of coverage guarantee if "the distance from p_{i-1} to the candidate value for p_i falls below 10^{-14}." The number "10^{-14}" actually plays the same role as "η" in the AMCA, where "$\eta = 10^{-15}$" is recommended by [21].

6. Numerical Results

In this section, we shall illustrate the proposed double-parabolic sampling scheme through examples. As demonstrated in Sections 2.2 and 4, the double-parabolic sampling scheme can be parameterized by the dilation coefficient ρ and the coverage tuning parameter ζ. Hence, the performance of the resultant stopping rule can be optimized with respect to $\rho \in (0, 1]$ and ζ by choosing various values of ρ from interval $(0, 1]$ and determining the corresponding values of

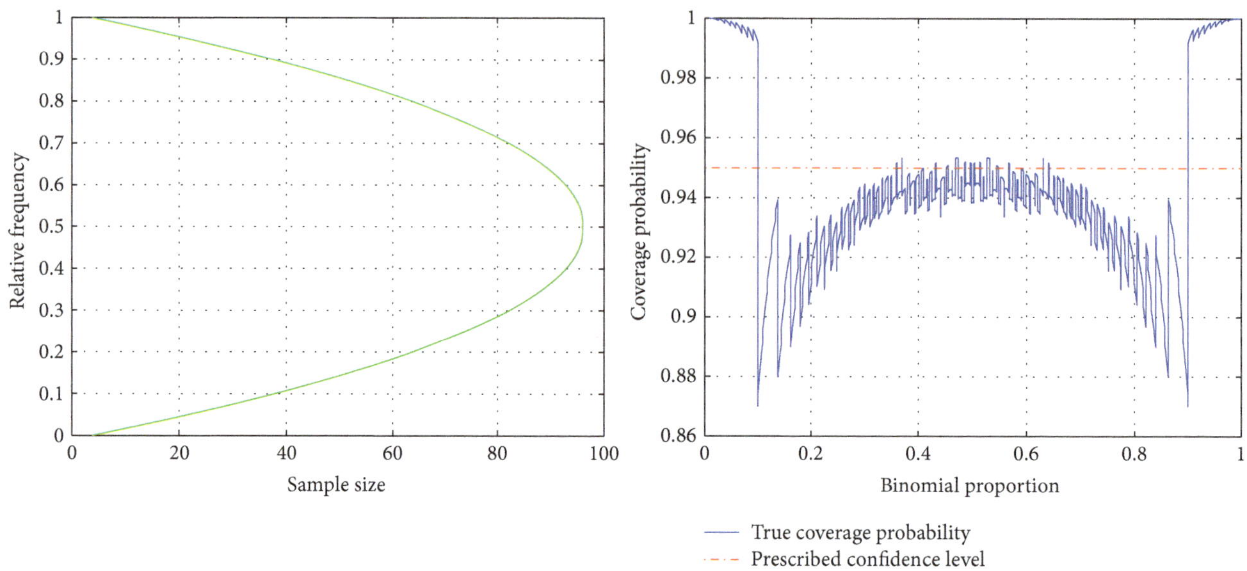

Figure 2: Double-parabolic sampling with $\epsilon = 0.1$, $\delta = 0.05$, $\rho = 1/10$, and $\zeta = 2.93$.

Table 1: Coverage tuning parameter.

ϵ	δ	ζ	ϵ	δ	ζ	ϵ	δ	ζ
0.1	0.1	2.0427	0.1	0.05	2.4174	0.1	0.01	3.0608
0.05	0.1	2.0503	0.05	0.05	2.5862	0.05	0.01	3.3125
0.02	0.1	2.1725	0.02	0.05	2.5592	0.02	0.01	3.4461
0.01	0.1	2.1725	0.01	0.05	2.5592	0.01	0.01	3.4461

ζ by the computational techniques introduced in Section 2 to guarantee the desired confidence interval.

6.1. Asymptotic Analysis May Be Inadequate. For fully sequential cases, we have evaluated the double-parabolic sampling scheme with $\epsilon = 0.1$, $\delta = 0.05$, $\rho = 0.1$, and $\zeta = (1/\delta)\exp(-(1/2)\mathcal{Z}^2_{\delta/2}) \approx 2.93$. The stopping boundary is displayed in the left side of Figure 2. The function of coverage probability with respect to the binomial proportion is shown in the right side of Figure 2, which indicates that the coverage probabilities are generally substantially lower than the prescribed confidence level $1 - \delta = 0.05$. By considering $\epsilon = 0.1$ as a small number and applying the asymptotic theory, the coverage probability associated with the sampling scheme is expected to be close to 0.95. This numerical example demonstrates that although the asymptotic method is insightful and involves virtually no computation, it may not be adequate.

In general, the main drawback of an asymptotic method is that there is no guarantee of coverage probability. Although an asymptotical method asserts that if the margin of error ϵ tends to 0, the coverage probability will tend to the prespecified confidence level $1 - \delta$, it is difficult to determine how small the margin of error ϵ is sufficient for the asymptotic method to be applicable. Note that $\epsilon \to 0$ implies the average sample size tends to ∞. However, in reality, the sample sizes must be finite. Consequently, an asymptotic method inevitably introduces unknown statistical error. Since an asymptotic method does not necessarily guarantee the prescribed confidence level, it is not fair to compare its associated sample size with that of an exact method, which guarantees the prespecified confidence level.

This example also indicates that, due to the discrete nature of the problem, the coverage probability is a discontinuous and erratic function of p, which implies that Monte Carlo simulation is not suitable for evaluating the coverage performance.

6.2. Parametric Values of Fully Sequential Schemes. For fully sequential cases, to allow direct application of our double-parabolic sequential method, we have obtained values of coverage tuning parameter ζ, which guarantee the prescribed confidence levels, for double-parabolic sampling schemes with $\rho = 3/4$ and various combinations of (ϵ, δ) as shown in Table 1. We used the computational techniques introduced in Section 2 to obtain this table.

To illustrate the use of Table 1, suppose that one wants a fully sequential sampling procedure to ensure that $\Pr\{|\widehat{\mathbf{p}} - p| < 0.1 \mid p\} > 0.95$ for any $p \in (0, 1)$. This means that one can choose $\epsilon = 0.1$, $\delta = 0.05$ and the range of sample size is given by (30). From Table 1, it can be seen that the value of ζ corresponding to $\epsilon = 0.1$ and $\delta = 0.05$ is 2.4174. Consequently, the stopping rule is completely determined by substituting the values of design parameters $\epsilon = 0.1$,

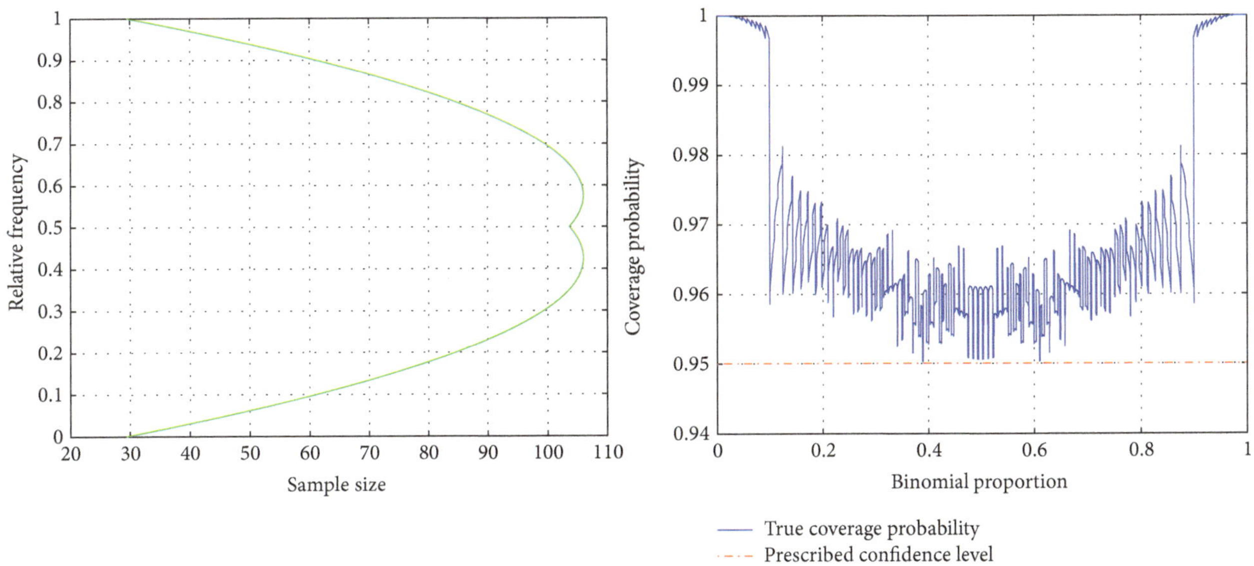

FIGURE 3: Double-parabolic sampling with $\epsilon = 0.1$, $\delta = 0.05$, $\rho = 3/4$, and $\zeta = 2.4174$.

TABLE 2: Coverage tuning parameter.

	$s = 3$	$s = 4$	$s = 5$	$s = 6$	$s = 7$	$s = 8$	$s = 9$	$s = 10$
$\epsilon = 0.1$	2.6583	2.6583	2.5096	2.5946	2.4459	2.6512	2.5096	2.4459
$\epsilon = 0.05$	2.6759	2.6759	2.6759	2.6759	2.6759	2.6759	2.6759	2.6759
$\epsilon = 0.02$	2.6725	2.6725	2.6725	2.6725	2.6725	2.6725	2.6725	2.6725
$\epsilon = 0.01$	2.6796	2.6796	2.6796	2.6796	2.6796	2.5875	2.6796	2.6796

$\delta = 0.05$, $\rho = 3/4$, and $\zeta = 2.4174$ into its definition. The stopping boundary of this sampling scheme is displayed in the left side of Figure 3. The function of coverage probability with respect to the binomial proportion is shown in the right side of Figure 3.

6.3. Parametric Values of Group Sequential Schemes. In many situations, especially in clinical trials, it is desirable to use group sequential sampling schemes. In Tables 2 and 3, assuming that sample sizes satisfy (8) for the purpose of having approximately equal group sizes, we have obtained parameters for concrete schemes by the computational techniques introduced in Section 2.

For dilation coefficient $\rho = 3/4$ and confidence parameter $\delta = 0.05$, we have obtained values of coverage tuning parameter ζ, which guarantee the prescribed confidence level 0.95, for double-parabolic sampling schemes, with the number of stages s ranging from 3 to 10, as shown in Table 2.

For dilation coefficient $\rho = 3/4$ and confidence parameter $\delta = 0.01$, we have obtained values of coverage tuning parameter ζ, which guarantee the prescribed confidence level 0.99, for double-parabolic sampling schemes, with the number of stages s ranging from 3 to 10, as shown in Table 3.

To illustrate the use of these tables, suppose that one wants a ten-stage sampling procedure of approximately equal group sizes to ensure that $\Pr\{|\widehat{\mathbf{p}} - p| < 0.01 \mid p\} > 0.99$ for any $p \in (0, 1)$. This means that one can choose $\epsilon = \delta = 0.01$, $s = 10$ and sample sizes satisfying (8). To obtain appropriate parameter values for the sampling procedure, one can look at Table 3 to find the coverage tuning parameter ζ corresponding to $\epsilon = 0.01$ and $s = 10$. From Table 3, it can be seen that ζ can be taken as 3.5753. Consequently, the stopping rule is completely determined by substituting the values of design parameters $\epsilon = 0.01$, $\delta = 0.01$, $\rho = 3/4$, $\zeta = 3.5753$, and $s = 10$ into its definition and (8). The stopping boundary of this sampling scheme and the function of coverage probability with respect to the binomial proportion are displayed, respectively, in the left and right sides of Figure 4.

6.4. Comparison of Sampling Schemes. We have conducted numerical experiments to investigate the impact of dilation coefficient ρ on the performance of our double-parabolic sampling schemes. Our computational experiences indicate that the dilation coefficient $\rho = 3/4$ is frequently a good choice in terms of average sample number and coverage probability. For example, consider the case that the margin of error is given as $\epsilon = 0.1$ and the prescribed confidence level is $1 - \delta$ with $\delta = 0.05$. For the double-parabolic sampling scheme with the dilation coefficient ρ chosen as $2/3, 3/4$, and 1, we have determined that, to ensure the prescribed confidence level $1 - \delta = 0.95$, it suffices to set the coverage tuning parameter ζ as 2.1, 2.4 and 2.4, respectively. The average sample numbers of these sampling schemes and the coverage probabilities as functions of the binomial proportion are shown, respectively, in the left and right sides

TABLE 3: Coverage tuning parameter.

	$s = 3$	$s = 4$	$s = 5$	$s = 6$	$s = 7$	$s = 8$	$s = 9$	$s = 10$
$\epsilon = 0.1$	3.3322	3.3322	3.3322	3.3322	3.3322	3.2709	3.0782	3.3322
$\epsilon = 0.05$	3.5074	3.5074	3.5074	3.5074	3.5074	3.5074	3.5074	3.5074
$\epsilon = 0.02$	3.5430	3.5430	3.5430	3.5430	3.5430	3.5430	3.5430	3.5430
$\epsilon = 0.01$	3.5753	3.5753	3.5753	3.5753	3.5753	3.5753	3.5753	3.5753

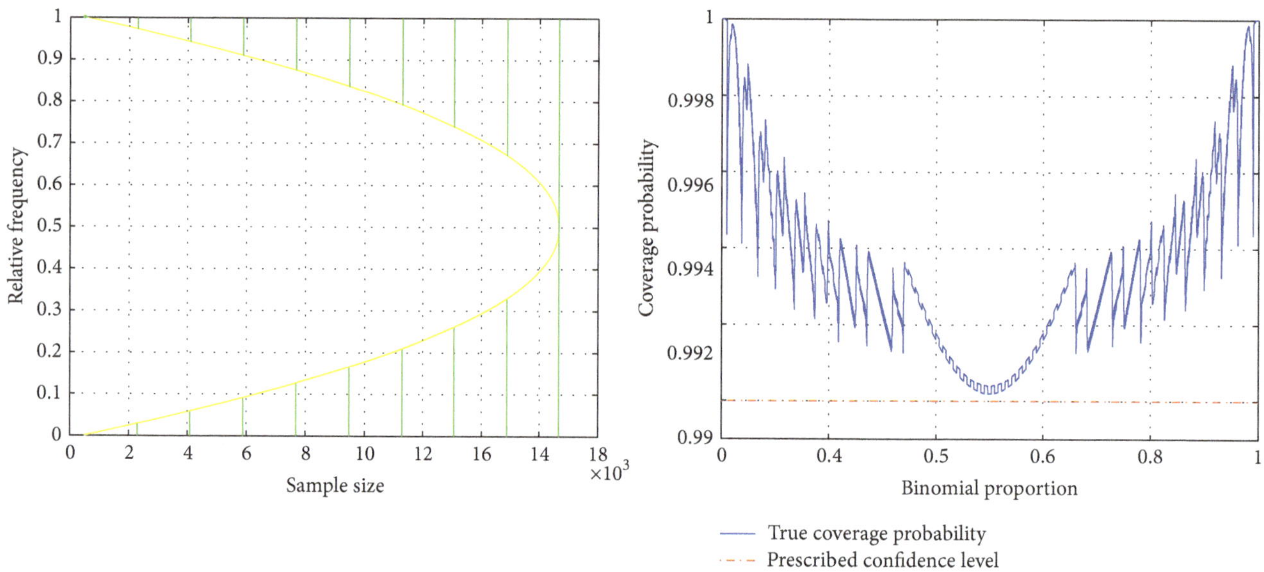

FIGURE 4: Double-parabolic sampling with $\epsilon = \delta = 0.01$, $s = 10$, $\rho = 3/4$, and $\zeta = 3.5753$.

of Figure 5. From Figure 5, it can be seen that a double-parabolic sampling scheme with dilation coefficient $\rho = 3/4$ has better performance in terms of average sample number and coverage probability as compared to that of the double-parabolic sampling scheme with smaller or larger values of dilation coefficient.

We have investigated the impact of confidence intervals on the performance of fully sequential sampling schemes constructed from the inclusion principle. We have observed that the stopping rule derived from Clopper-Pearson intervals generally outperforms the stopping rules derived from other types of confidence intervals. However, via appropriate choice of the dilation coefficient, the double-parabolic sampling scheme can perform uniformly better than the stopping rule derived from Clopper-Pearson intervals. To illustrate, consider the case that $\epsilon = 0.1$ and $\delta = 0.05$. For stopping rules derived from Clopper-Pearson intervals, Fishman's intervals, Wilson's intervals, and revised Wald intervals with $a = 4$, we have determined that to guarantee the prescribed confidence level $1 - \delta = 0.95$, it suffices to set the coverage tuning parameter ζ as $0.5, 1, 2.4$, and 0.37, respectively. For the stopping rule derived from Wald intervals, we have determined $\zeta = 0.77$ to ensure the confidence level, under the condition that the minimum sample size is taken as $\lceil (1/\epsilon)\ln(1/\zeta\delta)\rceil$. Recall that for the double-parabolic sampling scheme with $\rho = 3/4$, we have obtained $\zeta = 2.4$ for purpose of guaranteeing the confidence level. The average sample numbers of these sampling schemes are shown in Figure 6. From these plots, it can be seen that as compared to the stopping rule derived from Clopper-Pearson intervals, the stopping rule derived from the revised Wald intervals performs better in the region of p close to 0 or 1, but performs worse in the region of p in the middle of $(0, 1)$. The performance of stopping rules from Fishman's intervals (i.e., from Chernoff bound) and Wald intervals are obviously inferior as compared to that of the stopping rule derived from Clopper-Pearson intervals. It can be observed that the double-parabolic sampling scheme uniformly outperforms the stopping rule derived from Clopper-Pearson intervals.

6.5. Estimation with High Confidence Level. In some situations, we need to estimate a binomial proportion with a high confidence level. For example, one might want to construct a sampling scheme such that, for $\epsilon = 0.05$ and $\delta = 10^{-10}$, the resultant sequential estimator $\widehat{\mathbf{p}}$ satisfies $\Pr\{|\widehat{\mathbf{p}} - p| < \epsilon \mid p\} > 1 - \delta$ for any $p \in (0, 1)$. By working with the complementary coverage probability, we determined that it suffices to let the dilation coefficient $\rho = 3/4$ and the coverage tuning parameter $\zeta = 7.65$. The stopping boundary and the function of coverage probability with respect to the binomial proportion are displayed, respectively, in the left and right sides of Figure 7. As addressed in Section 5.3, it should be noted that it is impossible to obtain such a sampling scheme without working with the complementary coverage probability.

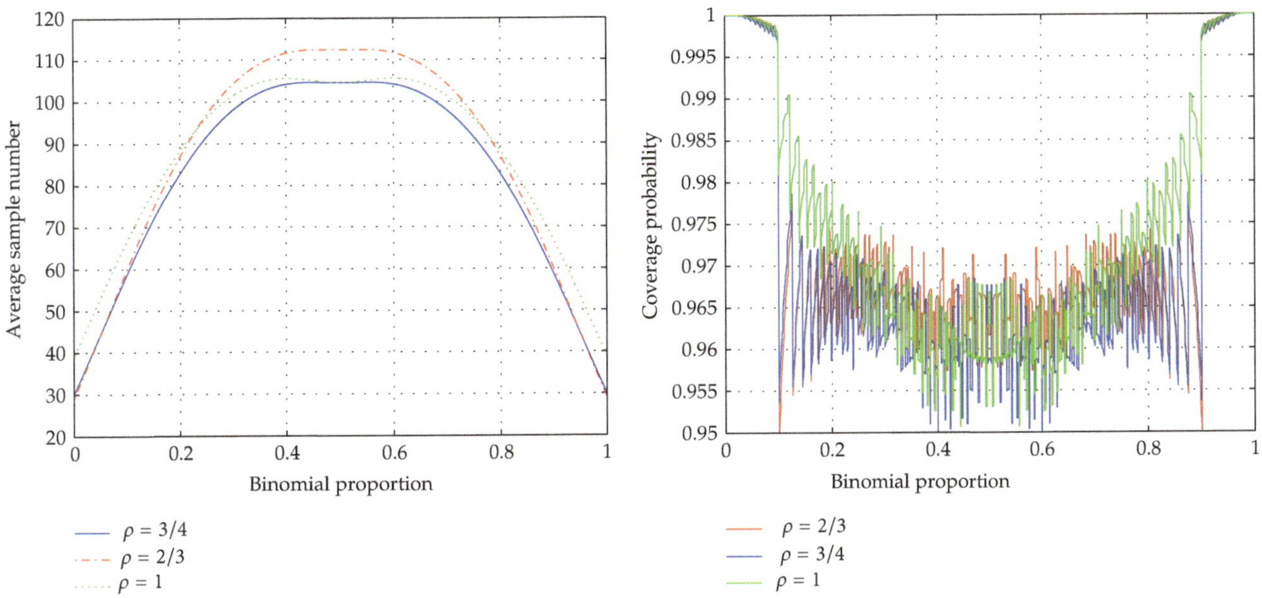

FIGURE 5: Double-parabolic sampling with various dilation coefficients.

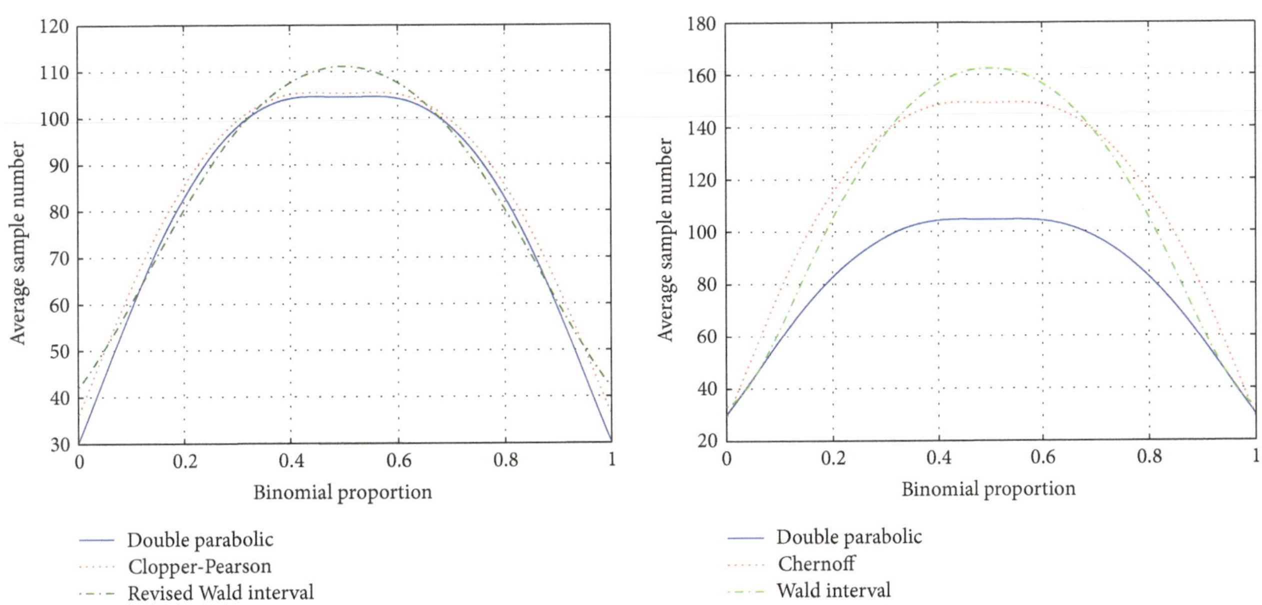

FIGURE 6: Comparison of average sample numbers.

7. Illustrative Examples for Clinical Trials

In this section, we shall illustrate the applications of our double-parabolic group sequential estimation method in clinical trials.

An example of our double-parabolic sampling scheme can be illustrated as follows. Assume that $\epsilon = \delta = 0.05$ is given and that the sampling procedure is expected to have 7 stages with sample sizes satisfying (8). Choosing $\rho = 3/4$, we have determined that it suffices to take $\zeta = 2.6759$ to guarantee that the coverage probability is no less than $1 - \delta = 0.95$ for all $p \in (0, 1)$. Accordingly, the sample sizes of this sampling scheme are calculated as 59, 116, 173, 231, 288, 345, and 403. This sampling scheme, with a sample path, is shown in the left side of Figure 8. In this case, the stopping rule can be equivalently described by virtue of Figure 8 as the following: continue sampling until $(\widehat{\mathbf{p}}_\ell, n_\ell)$ hit a green line at some stage. The coverage probability is shown in the right side of Figure 8.

To apply this estimation method in a clinical trial for estimating the proportion p of a binomial response with margin of error 0.05 and confidence level 95%, we can have seven groups of patients with group sizes 59, 57, 57, 58, 57, 57, and 58. In the first stage, we conduct experiment with

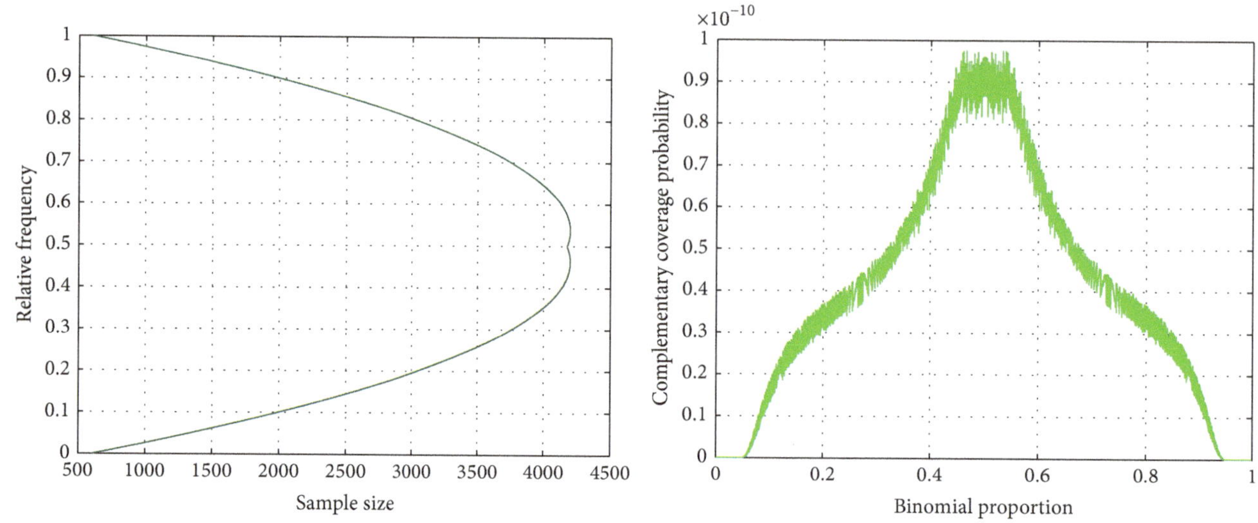

FIGURE 7: Double-parabolic sampling with $\epsilon = 0.05$, $\delta = 10^{-10}$, $\rho = 3/4$, and $\zeta = 7.65$.

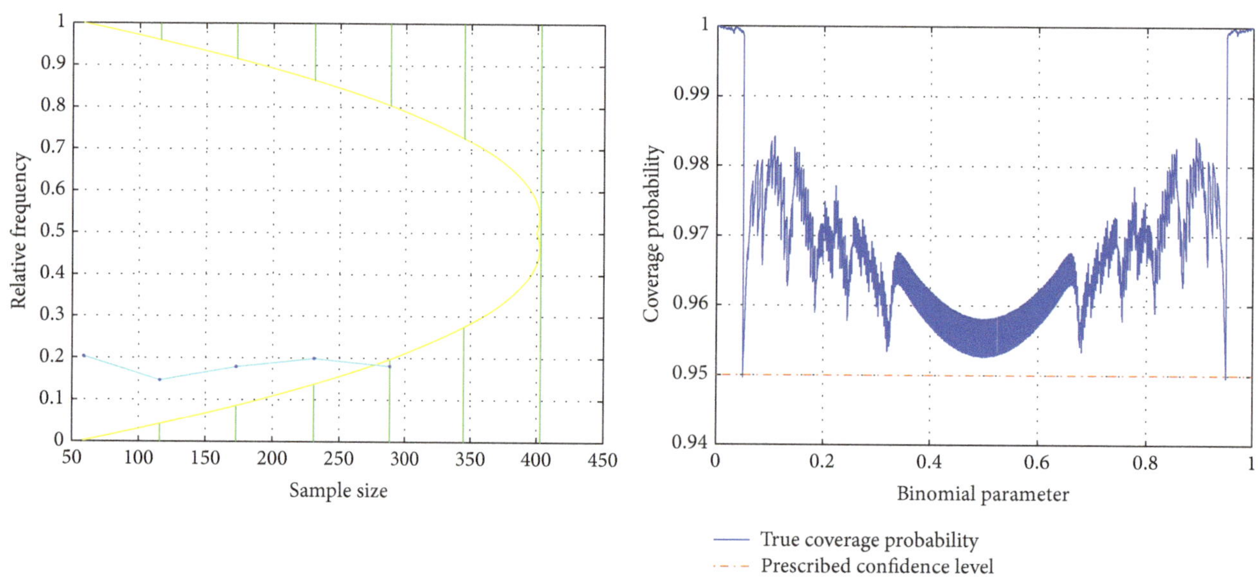

FIGURE 8: Double-parabolic sampling with $\epsilon = \delta = 0.05$, $s = 7$, $\rho = 3/4$, and $\zeta = 2.6759$.

the 59 patients of the first group. We observe the relative frequency of response and record it as $\widehat{\mathbf{p}}_1$. Suppose that there are 12 patients having positive responses, then the relative frequency at the first stage is $\widehat{\mathbf{p}}_1 = 12/59 = 0.2034$. With the values of $(\widehat{\mathbf{p}}_1, n_1) = (0.2034, 59)$, we check if the stopping rule is satisfied. This is equivalent to see if the point $(\widehat{\mathbf{p}}_1, n_1)$ hits a green line at the first stage. For such value of $(\widehat{\mathbf{p}}_1, n_1)$, it can be seen that the stopping condition is not fulfilled. So, we need to conduct the second stage of experiment with the 57 patients of the second group. We observe the response of these 57 patients. Suppose we observe that 5 patients among this group have positive responses. Then, we add 5 with 12, the number of positive responses before the second stage, to obtain 17 positive responses among $n_2 = 59 + 57 = 116$ patients. So, at the second stage, we get the relative frequency $\widehat{\mathbf{p}}_2 = 17/116 = 0.1466$. Since the stopping rule is not satisfied with the values of $(\widehat{\mathbf{p}}_2, n_2) = (0.1466, 116)$, we need to conduct the third stage of experiment with the 57 patients of the third group. Suppose we observe that 14 patients among this group have positive responses. Then, we add 14 with 17, the number of positive responses before the third stage, to get 31 positive responses among $n_3 = 59 + 57 + 57 = 173$ patients. So, at the third stage, we get the relative frequency $\widehat{\mathbf{p}}_3 = 31/173 = 0.1792$. Since the stopping rule is not satisfied with the values of $(\widehat{\mathbf{p}}_3, n_3) = (0.1792, 173)$, we need to conduct the fourth stage of experiment with the 58 patients of the fourth group. Suppose we observe that 15 patients among this group have positive responses. Then, we add 15 with 31, the number of positive responses before the fourth stage, to get 46 positive responses among $n_4 = 59 + 57 + 57 + 58 = 231$ patients. So, at the fourth stage, we get the relative frequency

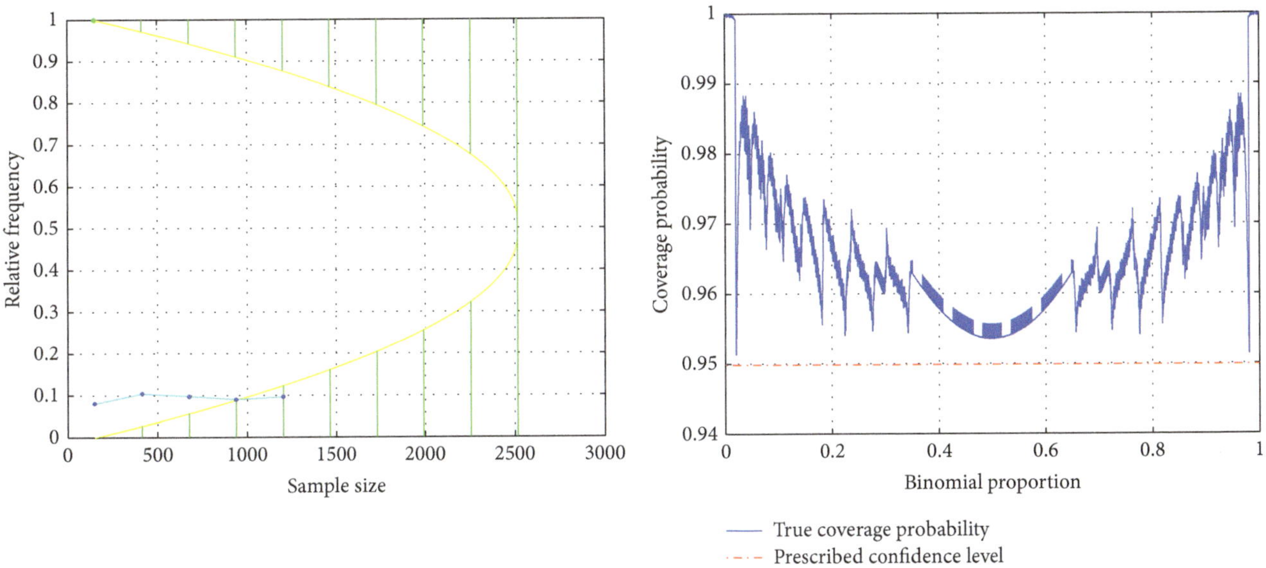

FIGURE 9: Double-parabolic sampling with $\epsilon = 0.02$, $\delta = 0.05$, $s = 10$, $\rho = 3/4$, and $\zeta = 2.6725$.

$\widehat{\mathbf{p}}_4 = 46/231 = 0.1991$. Since the stopping rule is not satisfied with the values of $(\widehat{\mathbf{p}}_4, n_4) = (0.1991, 231)$, we need to conduct the fifth stage of experiment with the 57 patients of the fifth group. Suppose we observe that 6 patients among this group have positive responses. Then, we add 6 with 46, the number of positive responses before the fifth stage, to get 52 positive responses among $n_5 = 59 + 57 + 57 + 58 + 57 = 288$ patients. So, at the fifth stage, we get the relative frequency $\widehat{\mathbf{p}}_5 = 52/288 = 0.1806$. It can be seen that the stopping rule is satisfied with the values of $(\widehat{\mathbf{p}}_5, n_5) = (0.1806, 288)$. Therefore, we can terminate the sampling experiment and take $\widehat{\mathbf{p}} = 52/288 = 0.1806$ as an estimate of the proportion of the whole population having positive responses. With a 95% confidence level, one can believe that the difference between the true value of p and its estimate $\widehat{\mathbf{p}} = 0.1806$ is less than 0.05.

In this experiment, we only use 288 samples to obtain the estimate for p. Except the round-off error, there is no other source of error for reporting statistical accuracy, since no asymptotic approximation is involved. As compared to fixed-sample-size procedure, we achieved a substantial save of samples. To see this, one can check that using the rigorous formula (37) gives a sample size 738, which is overly conservative. From the classical approximate formula (35), the sample size is determined as 385, which has been known to be insufficient to guarantee the prescribed confidence level 95%. The exact method of [34] shows that at least 391 samples are needed. As compared to the best-fixed-sample size obtained by the method of [34], the reduction of sample sizes resulted from our double-parabolic sampling scheme is $391 - 288 = 103$. It can be seen that the fixed-sample-size procedure wastes $103/288 = 35.76\%$ samples as compared to our group sequential method, which is also an exact method. This percentage may not be serious if it were a save of a number of simulation runs. However, as the number count is for patients, the reduction of samples is important for ethical and economical reasons. Using our group sequential method, the worst-case sample size is equal to 403, which is only 12 more than the minimum sample size of fixed-sample procedure. However, a lot of samples can be saved in the average case.

As ϵ or δ become smaller, the reduction of samples is more significant. For example, let $\epsilon = 0.02$ and $\delta = 0.05$, we have a double-parabolic sample scheme with 10 stages. The sampling scheme, with a sample path, is shown in the left side of Figure 9. The coverage probability is shown in the right side of Figure 9.

8. Conclusion

In this paper, we have reviewed recent development of group sequential estimation methods for a binomial proportion. We have illustrated the inclusion principle and its applications to various stopping rules. We have introduced computational techniques in the literature, which suffice for determining parameters of stopping rules to guarantee desired confidence levels. Moreover, we have proposed a new family of sampling schemes with stopping boundary of double-parabolic shape, which are parameterized by the coverage tuning parameter and the dilation coefficient. These parameters can be determined by the exact computational techniques to reduce the sampling cost, while ensuring prescribed confidence levels. The new family of sampling schemes are extremely simple in structure and asymptotically optimal as the margin of error tends to 0. We have established analytic bounds for the distribution and expectation of the sample number at the termination of the sampling process. We have obtained parameter values via the exact computational techniques for the proposed sampling schemes such that the confidence levels are guaranteed and that the sampling schemes are generally more efficient as compared to existing ones.

Appendices

A. Proof of Theorem 1

Consider function $g(x,z) = (x-z)^2/x(1-x)$ for $x \in (0,1)$ and $z \in [0,1]$. It can be checked that $\partial g(x,z)/\partial x = (x-z)[z(1-x)+x(1-z)][x(1-x)]^{-2}$, which shows that for any fixed $z \in [0,1]$, $-g(x,z)$ is a unimodal function of $x \in (0,1)$, with a maximum attained at $x = z$. By such a property of $g(x,z)$ and the definition of Wilson's confidence intervals, we have

$$\begin{aligned}
\{\widehat{\mathbf{p}}_\ell - \epsilon \le L_\ell\} &= \{0 < \widehat{\mathbf{p}}_\ell - \epsilon \le L_\ell\} \cup \{\widehat{\mathbf{p}}_\ell \le \epsilon\} \\
&= \left\{0 < \widehat{\mathbf{p}}_\ell - \epsilon \le L_\ell \le \widehat{\mathbf{p}}_\ell,\ g(L_\ell, \widehat{\mathbf{p}}_\ell) = \frac{\mathcal{Z}_{\zeta\delta}^2}{n_\ell}\right\} \\
&\quad \cup \{\widehat{\mathbf{p}}_\ell \le \epsilon\} \\
&= \left\{\widehat{\mathbf{p}}_\ell > \epsilon,\ \frac{\epsilon^2}{(\widehat{\mathbf{p}}_\ell - \epsilon)[1-(\widehat{\mathbf{p}}_\ell - \epsilon)]} \ge \frac{\mathcal{Z}_{\zeta\delta}^2}{n_\ell}\right\} \\
&\quad \cup \{\widehat{\mathbf{p}}_\ell \le \epsilon\}, \\
\{\widehat{\mathbf{p}}_\ell + \epsilon \ge U_\ell\} &= \{1 > \widehat{\mathbf{p}}_\ell + \epsilon \ge U_\ell\} \cup \{\widehat{\mathbf{p}}_\ell + \epsilon \ge 1\} \\
&= \left\{1 > \widehat{\mathbf{p}}_\ell + \epsilon \ge U_\ell \ge \widehat{\mathbf{p}}_\ell,\ g(U_\ell, \widehat{\mathbf{p}}_\ell) = \frac{\mathcal{Z}_{\zeta\delta}^2}{n_\ell}\right\} \\
\cup \{\widehat{\mathbf{p}}_\ell + \epsilon \ge 1\} &= \left\{\widehat{\mathbf{p}}_\ell < 1-\epsilon,\ \frac{\epsilon^2}{(\widehat{\mathbf{p}}_\ell + \epsilon)[1-(\widehat{\mathbf{p}}_\ell + \epsilon)]} \ge \frac{\mathcal{Z}_{\zeta\delta}^2}{n_\ell}\right\} \\
&\quad \cup \{\widehat{\mathbf{p}}_\ell \ge 1 - \epsilon\},
\end{aligned} \tag{A.1}$$

for $\ell = 1, \ldots, s$, where we have used the fact that $\{\widehat{\mathbf{p}}_\ell > \epsilon\} \subseteq \{L_\ell > 0\}$, $\{\widehat{\mathbf{p}}_\ell < 1-\epsilon\} \subseteq \{U_\ell < 1\}$ and $0 \le L_\ell \le \widehat{\mathbf{p}}_\ell \le U_\ell \le 1$. Recall that $0 < \epsilon < 1/2$. It follows that

$$\begin{aligned}
&\{\widehat{\mathbf{p}}_\ell - \epsilon \le L_\ell \le U_\ell \le \widehat{\mathbf{p}}_\ell + \epsilon\} \\
&= \left\{\epsilon < \widehat{\mathbf{p}}_\ell < 1-\epsilon,\ \frac{\epsilon^2}{(\widehat{\mathbf{p}}_\ell - \epsilon)[1-(\widehat{\mathbf{p}}_\ell - \epsilon)]} \ge \frac{\mathcal{Z}_{\zeta\delta}^2}{n_\ell},\right. \\
&\qquad \left.\frac{\epsilon^2}{(\widehat{\mathbf{p}}_\ell + \epsilon)[1-(\widehat{\mathbf{p}}_\ell + \epsilon)]} \ge \frac{\mathcal{Z}_{\zeta\delta}^2}{n_\ell}\right\} \\
&\cup \left\{\widehat{\mathbf{p}}_\ell \le \epsilon,\ \frac{\epsilon^2}{(\widehat{\mathbf{p}}_\ell + \epsilon)[1-(\widehat{\mathbf{p}}_\ell + \epsilon)]} \ge \frac{\mathcal{Z}_{\zeta\delta}^2}{n_\ell}\right\} \\
&\cup \left\{\widehat{\mathbf{p}}_\ell \ge 1-\epsilon,\ \frac{\epsilon^2}{(\widehat{\mathbf{p}}_\ell - \epsilon)[1-(\widehat{\mathbf{p}}_\ell - \epsilon)]} \ge \frac{\mathcal{Z}_{\zeta\delta}^2}{n_\ell}\right\} \\
&= \left\{\epsilon < \widehat{\mathbf{p}}_\ell < 1-\epsilon,\ \left(\left|\widehat{\mathbf{p}}_\ell - \frac{1}{2}\right| - \epsilon\right)^2 \ge \frac{1}{4} - n_\ell\left(\frac{\epsilon}{\mathcal{Z}_{\zeta\delta}}\right)^2\right\} \\
&\cup \left\{\widehat{\mathbf{p}}_\ell \le \epsilon,\ \left(\left|\widehat{\mathbf{p}}_\ell - \frac{1}{2}\right| - \epsilon\right)^2 \ge \frac{1}{4} - n_\ell\left(\frac{\epsilon}{\mathcal{Z}_{\zeta\delta}}\right)^2\right\} \\
&\cup \left\{\widehat{\mathbf{p}}_\ell \ge 1-\epsilon,\ \left(\left|\widehat{\mathbf{p}}_\ell - \frac{1}{2}\right| - \epsilon\right)^2 \ge \frac{1}{4} - n_\ell\left(\frac{\epsilon}{\mathcal{Z}_{\zeta\delta}}\right)^2\right\} \\
&= \left\{\left(\left|\widehat{\mathbf{p}}_\ell - \frac{1}{2}\right| - \epsilon\right)^2 \ge \frac{1}{4} - n_\ell\left(\frac{\epsilon}{\mathcal{Z}_{\zeta\delta}}\right)^2\right\},
\end{aligned} \tag{A.2}$$

for $\ell = 1, \ldots, s$. This completes the proof of the theorem.

B. Proof of Theorem 2

By the assumption that $n_s \ge (1/2\epsilon^2)\ln(1/\zeta\delta)$, we have $(1/4) + (\epsilon^2 n_s/2\ln(\zeta\delta)) \le 0$ and, consequently, $\Pr\{(|\widehat{\mathbf{p}}_s - 1/2| - \rho\epsilon)^2 \ge (1/4) + (\epsilon^2 n_s/2\ln(\zeta\delta))\} = 1$. It follows from the definition of the sampling scheme that the sampling process must stop at or before the sth stage. In other words, $\Pr\{\mathbf{l} \le s\} = 1$. This allows one to write

$$\begin{aligned}
\Pr\{|\widehat{\mathbf{p}} - p| \ge \epsilon \mid p\} &= \sum_{\ell=1}^{s} \Pr\{|\widehat{\mathbf{p}} - p| \ge \epsilon,\ \mathbf{l} = \ell \mid p\} \\
&= \sum_{\ell=1}^{s} \Pr\{|\widehat{\mathbf{p}}_\ell - p| \ge \epsilon,\ \mathbf{l} = \ell \mid p\} \\
&\le \sum_{\ell=1}^{s} \Pr\{|\widehat{\mathbf{p}}_\ell - p| \ge \epsilon \mid p\},
\end{aligned} \tag{B.1}$$

for $p \in (0,1)$. By virtue of the well-known Chernoff-Hoeffding bound [32, 33], we have

$$\Pr\{|\widehat{\mathbf{p}}_\ell - p| \ge \epsilon \mid p\} \le 2\exp(-2n_\ell\epsilon^2), \tag{B.2}$$

for $\ell = 1, \ldots, s$. Making use of (B.1), (B.2), and the fact that $n_1 \ge 2\rho((1/\epsilon) - \rho)\ln(1/\zeta\delta)$ as can be seen from (30), we have

$$\begin{aligned}
&\Pr\{|\widehat{\mathbf{p}} - p| \ge \epsilon \mid p\} \\
&\le 2\sum_{\ell=1}^{s} \exp(-2n_\ell\epsilon^2) \\
&\le 2\sum_{m=n_1}^{\infty} \exp(-2m\epsilon^2) = \frac{2\exp(-2n_1\epsilon^2)}{1-\exp(-2\epsilon^2)} \\
&\le \frac{2\exp(-2\epsilon^2 \times 2\rho((1/\epsilon) - \rho)\ln(1/\zeta\delta))}{1-\exp(-2\epsilon^2)} \\
&= \frac{2\exp(4\epsilon\rho(1-\rho\epsilon)\ln(\zeta\delta))}{1-\exp(-2\epsilon^2)},
\end{aligned} \tag{B.3}$$

for any $p \in (0,1)$. Therefore, to guarantee that $\Pr\{|\widehat{\mathbf{p}} - p| < \epsilon \mid p\} \ge 1 - \delta$ for any $p \in (0,1)$, it is sufficient to choose ζ

such that $2\exp(4\epsilon p(1-p\epsilon)\ln(\zeta\delta)) \le \delta[1-\exp(-2\epsilon^2)]$. This inequality can be written as $4\epsilon p(1-p\epsilon)\ln(\zeta\delta) \le \ln(\delta/2) + \ln[1-\exp(-2\epsilon^2)]$ or, equivalently, $\zeta \le (1/\delta)\exp((\ln(\delta/2) + \ln[1-\exp(-2\epsilon^2)])/4\epsilon p(1-p\epsilon))$. The proof of the theorem is thus completed.

C. Proof of Theorem 3

First, we need to show that $\Pr\{\lim_{\epsilon\to 0}(\mathbf{n}/N(p,\epsilon,\delta,\zeta)) = 1 \mid p\} = 1$ for any $p \in (0,1)$. Clearly, the sample number $\mathbf{n}$ is a random number dependent on ϵ. Note that for any $\omega \in \Omega$, the sequences $\{\overline{X}_{\mathbf{n}(\omega)}(\omega)\}_{\epsilon\in(0,1)}$ and $\{\overline{X}_{\mathbf{n}(\omega)-1}(\omega)\}_{\epsilon\in(0,1)}$ are subsets of $\{\overline{X}_m(\omega)\}_{m=1}^{\infty}$. By the strong law of large numbers, for almost every $\omega \in \Omega$, the sequence $\{\overline{X}_m(\omega)\}_{m=1}^{\infty}$ converges to p. Since every subsequence of a convergent sequence must converge, it follows that the sequences $\{\overline{X}_{\mathbf{n}(\omega)}(\omega)\}_{\epsilon\in(0,1)}$ and $\{\overline{X}_{\mathbf{n}(\omega)-1}(\omega)\}_{\epsilon\in(0,1)}$ converge to p as $\epsilon \to 0$ provided that $\mathbf{n}(\omega) \to \infty$ as $\epsilon \to 0$. Since it is certain that $\mathbf{n} \ge 2p((1/\epsilon)-p)\ln(1/\zeta\delta) \to \infty$ as $\epsilon \to 0$, we have that $\{\lim_{\epsilon\to 0}((\mathbf{n}-1)/\mathbf{n}) = 1\}$ is a sure event. It follows that $B = \{\lim_{\epsilon\to 0}\overline{X}_{\mathbf{n}-1} = p,\ \lim_{\epsilon\to 0}\overline{X}_{\mathbf{n}} = p,\ \lim_{\epsilon\to 0}((\mathbf{n}-1)/\mathbf{n}) = 1\}$ is an almost sure event. By the definition of the sampling scheme, we have

$$A = \left\{\left(\left|\overline{X}_{\mathbf{n}-1} - \frac{1}{2}\right| - p\epsilon\right)^2 < \frac{1}{4} + \frac{\epsilon^2(\mathbf{n}-1)}{2\ln(\zeta\delta)},\right.$$
$$\left.\left(\left|\overline{X}_{\mathbf{n}} - \frac{1}{2}\right| - p\epsilon\right)^2 \ge \frac{1}{4} + \frac{\epsilon^2\mathbf{n}}{2\ln(\zeta\delta)}\right\} \tag{C.1}$$

as a sure event. Hence, $A \cap B$ is an almost sure event. Define $C = \{\lim_{\epsilon\to 0}(\mathbf{n}/N(p,\epsilon,\delta,\zeta)) = 1\}$. We need to show that C is an almost sure event. For this purpose, we let $\omega \in A \cap B$ and expect to show that $\omega \in C$. As a consequence of $\omega \in A \cap B$,

$$\frac{\mathbf{n}(\omega)}{N(p,\epsilon,\delta,\zeta)} < \frac{\mathbf{n}(\omega)}{\mathbf{n}(\omega)-1} \times \frac{\left[(1/4) - \left(\left|\overline{X}_{\mathbf{n}(\omega)-1}(\omega) - (1/2)\right| - p\epsilon\right)^2\right]}{p(1-p)},$$
$$\lim_{\epsilon\to 0}\overline{X}_{\mathbf{n}(\omega)-1}(\omega) = p, \qquad \lim_{\epsilon\to 0}\frac{\mathbf{n}(\omega)-1}{\mathbf{n}(\omega)} = 1. \tag{C.2}$$

By the continuity of the function $|x-1/2| - p\epsilon$ with respect to x and ϵ, we have

$$\limsup_{\epsilon\to 0}\frac{\mathbf{n}(\omega)}{N(p,\epsilon,\delta,\zeta)} \le \lim_{\epsilon\to 0}\frac{\mathbf{n}(\omega)}{\mathbf{n}(\omega)-1} \times \frac{\left[(1/4)-\left(\left|\lim_{\epsilon\to 0}\overline{X}_{\mathbf{n}(\omega)-1}(\omega)-(1/2)\right|-\lim_{\epsilon\to 0}p\epsilon\right)^2\right]}{p(1-p)} = 1. \tag{C.3}$$

On the other hand, as a consequence of $\omega \in A \cap B$,

$$\frac{\mathbf{n}(\omega)}{N(p,\epsilon,\delta,\zeta)} \ge \frac{\left[(1/4) - \left(\left|\overline{X}_{\mathbf{n}(\omega)}(\omega) - (1/2)\right| - p\epsilon\right)^2\right]}{p(1-p)},$$
$$\lim_{\epsilon\to 0}\overline{X}_{\mathbf{n}(\omega)}(\omega) = p. \tag{C.4}$$

Making use of the continuity of the function $|x-1/2| - p\epsilon$ with respect to x and ϵ, we have

$$\liminf_{\epsilon\to 0}\frac{\mathbf{n}(\omega)}{N(p,\epsilon,\delta,\zeta)} \ge \frac{\left[(1/4) - \left(\left|\lim_{\epsilon\to 0}\overline{X}_{\mathbf{n}(\omega)}(\omega) - (1/2)\right| - \lim_{\epsilon\to 0}p\epsilon\right)^2\right]}{p(1-p)} = 1. \tag{C.5}$$

Combining (C.3) and (C.5) yields $\lim_{\epsilon\to 0}(\mathbf{n}(\omega)/N(p,\epsilon,\delta,\zeta)) = 1$ and thus $A\cap B \subseteq C$. This implies that C is an almost sure event and thus $\Pr\{\lim_{\epsilon\to 0}(\mathbf{n}/N(p,\epsilon,\delta,\zeta)) = 1 \mid p\} = 1$ for $p \in (0,1)$.

Next, we need to show that $\lim_{\epsilon\to 0}\Pr\{|\widehat{\mathbf{p}} - p| < \epsilon \mid p\} = 2\Phi(\sqrt{2\ln(1/\zeta\delta)}) - 1$ for any $p \in (0,1)$. For simplicity of notations, let $\sigma = \sqrt{p(1-p)}$ and $a = \sqrt{2\ln(1/\zeta\delta)}$. Note that $\Pr\{|\widehat{\mathbf{p}}-p| < \epsilon \mid p\} = \Pr\{|\overline{X}_{\mathbf{n}}-p| < \epsilon \mid p\} = \Pr\{\sqrt{\mathbf{n}}|\overline{X}_{\mathbf{n}}-p|/\sigma < \epsilon\sqrt{\mathbf{n}}/\sigma\}$. Clearly, for any $\eta \in (0,a)$,

$$\begin{aligned}
&\Pr\left\{\frac{\sqrt{\mathbf{n}}\left|\overline{X}_{\mathbf{n}} - p\right|}{\sigma} < \frac{\epsilon\sqrt{\mathbf{n}}}{\sigma}\right\} \\
&\le \Pr\left\{\frac{\sqrt{\mathbf{n}}\left|\overline{X}_{\mathbf{n}} - p\right|}{\sigma} < \frac{\epsilon\sqrt{\mathbf{n}}}{\sigma},\ \frac{\epsilon\sqrt{\mathbf{n}}}{\sigma} \in [a-\eta, a+\eta]\right\} + \Pr\left\{\frac{\epsilon\sqrt{\mathbf{n}}}{\sigma} \notin [a-\eta, a+\eta]\right\} \\
&\le \Pr\left\{\frac{\sqrt{\mathbf{n}}\left|\overline{X}_{\mathbf{n}} - p\right|}{\sigma} < a+\eta,\ \frac{\epsilon\sqrt{\mathbf{n}}}{\sigma} \in [a-\eta, a+\eta]\right\} + \Pr\left\{\frac{\epsilon\sqrt{\mathbf{n}}}{\sigma} \notin [a-\eta, a+\eta]\right\} \\
&\le \Pr\left\{\frac{\sqrt{\mathbf{n}}\left|\overline{X}_{\mathbf{n}} - p\right|}{\sigma} < a+\eta\right\} + \Pr\left\{\frac{\epsilon\sqrt{\mathbf{n}}}{\sigma} \notin [a-\eta, a+\eta]\right\},
\end{aligned} \tag{C.6}$$

$$\begin{aligned}
&\Pr\left\{\frac{\sqrt{\mathbf{n}}\left|\overline{X}_{\mathbf{n}}-p\right|}{\sigma}<\frac{\epsilon\sqrt{\mathbf{n}}}{\sigma}\right\}\\
&\geq \Pr\left\{\frac{\sqrt{\mathbf{n}}\left|\overline{X}_{\mathbf{n}}-p\right|}{\sigma}<\frac{\epsilon\sqrt{\mathbf{n}}}{\sigma},\ \frac{\epsilon\sqrt{\mathbf{n}}}{\sigma}\in[a-\eta,a+\eta]\right\}\\
&\geq \Pr\left\{\frac{\sqrt{\mathbf{n}}\left|\overline{X}_{\mathbf{n}}-p\right|}{\sigma}<a-\eta,\ \frac{\epsilon\sqrt{\mathbf{n}}}{\sigma}\in[a-\eta,a+\eta]\right\}\\
&\geq \Pr\left\{\frac{\sqrt{\mathbf{n}}\left|\overline{X}_{\mathbf{n}}-p\right|}{\sigma}<a-\eta\right\}\\
&\quad -\Pr\left\{\frac{\epsilon\sqrt{\mathbf{n}}}{\sigma}\notin[a-\eta,a+\eta]\right\}.
\end{aligned} \tag{C.7}$$

Recall that we have established that $\mathbf{n}/N(p,\epsilon,\delta,\zeta)\rightarrow 1$ almost surely as $\epsilon\rightarrow 0$. This implies that $(\epsilon\sqrt{\mathbf{n}}/\sigma)\rightarrow a$ and $\mathbf{n}/N(p,\epsilon,\delta,\zeta)\rightarrow 1$ in probability as ϵ tends to zero. It follows from Anscombe's random central limit theorem [35] that as ϵ tends to zero, $\sqrt{\mathbf{n}}(\overline{X}_{\mathbf{n}}-p)/\sigma$ converges in distribution to a Gaussian random variable with zero mean and unit variance. Hence, from (C.6),

$$\begin{aligned}
&\limsup_{\epsilon\rightarrow 0}\Pr\left\{\frac{\sqrt{\mathbf{n}}\left|\overline{X}_{\mathbf{n}}-p\right|}{\sigma}<\frac{\epsilon\sqrt{\mathbf{n}}}{\sigma}\right\}\\
&\leq \lim_{\epsilon\rightarrow 0}\Pr\left\{\frac{\sqrt{\mathbf{n}}\left|\overline{X}_{\mathbf{n}}-p\right|}{\sigma}<a+\eta\right\}\\
&\quad+\lim_{\epsilon\rightarrow 0}\Pr\left\{\frac{\epsilon\sqrt{\mathbf{n}}}{\sigma}\notin[a-\eta,a+\eta]\right\}\\
&=2\Phi(a+\eta)-1
\end{aligned} \tag{C.8}$$

and from (C.7),

$$\begin{aligned}
&\liminf_{\epsilon\rightarrow 0}\Pr\left\{\frac{\sqrt{\mathbf{n}}\left|\overline{X}_{\mathbf{n}}-p\right|}{\sigma}<\frac{\epsilon\sqrt{\mathbf{n}}}{\sigma}\right\}\\
&\geq \lim_{\epsilon\rightarrow 0}\Pr\left\{\frac{\sqrt{\mathbf{n}}\left|\overline{X}_{\mathbf{n}}-p\right|}{\sigma}<a-\eta\right\}\\
&\quad-\lim_{\epsilon\rightarrow 0}\Pr\left\{\frac{\epsilon\sqrt{\mathbf{n}}}{\sigma}\notin[a-\eta,a+\eta]\right\}\\
&=2\Phi(a-\eta)-1.
\end{aligned} \tag{C.9}$$

Since this argument holds for arbitrarily small $\eta\in(0,a)$, it must be true that

$$\begin{aligned}
&\liminf_{\epsilon\rightarrow 0}\Pr\left\{\frac{\sqrt{\mathbf{n}}\left|\overline{X}_{\mathbf{n}}-p\right|}{\sigma}<\frac{\epsilon\sqrt{\mathbf{n}}}{\sigma}\right\}\\
&=\limsup_{\epsilon\rightarrow 0}\Pr\left\{\frac{\sqrt{\mathbf{n}}\left|\overline{X}_{\mathbf{n}}-p\right|}{\sigma}<\frac{\epsilon\sqrt{\mathbf{n}}}{\sigma}\right\}\\
&=2\Phi(a)-1.
\end{aligned} \tag{C.10}$$

So, $\lim_{\epsilon\rightarrow 0}\Pr\{|\widehat{\mathbf{p}}-p|<\epsilon\mid p\}=\lim_{\epsilon\rightarrow 0}\Pr\{\sqrt{\mathbf{n}}|\overline{X}_{\mathbf{n}}-p|/\sigma<\epsilon\sqrt{\mathbf{n}}/\sigma\}=2\Phi(a)-1=2\Phi(\sqrt{2\ln(1/(\zeta\delta))})-1$ for any $p\in(0,1)$.

Now, we focus our attention to show that $\lim_{\epsilon\rightarrow 0}(\mathbb{E}[\mathbf{n}]/N(p,\epsilon,\delta,\zeta))=1$ for any $p\in(0,1)$. For this purpose, it suffices to show that

$$\begin{aligned}
1-\eta&\leq\liminf_{\epsilon\rightarrow 0}\frac{\mathbb{E}[\mathbf{n}]}{N(p,\epsilon,\delta,\zeta)}\\
&\leq\limsup_{\epsilon\rightarrow 0}\frac{\mathbb{E}[\mathbf{n}]}{N(p,\epsilon,\delta,\zeta)}\\
&\leq 1+\eta,\quad\forall p\in(0,1),
\end{aligned} \tag{C.11}$$

for any $\eta\in(0,1)$. For simplicity of notations, we abbreviate $N(p,\epsilon,\delta,\zeta)$ as N in the sequel. Since we have established $\Pr\{\lim_{\epsilon\rightarrow 0}(\mathbf{n}/N(p,\epsilon,\delta,\zeta))=1\}=1$, we can conclude that

$$\lim_{\epsilon\rightarrow 0}\Pr\{(1-\eta)N\leq\mathbf{n}\leq(1+\eta)N\}=1. \tag{C.12}$$

Noting that

$$\begin{aligned}
\mathbb{E}[\mathbf{n}]&=\sum_{m=0}^{\infty}m\Pr\{\mathbf{n}=m\}\\
&\geq\sum_{(1-\eta)N\leq m\leq(1+\eta)N}m\Pr\{\mathbf{n}=m\}\\
&\geq(1-\eta)N\sum_{(1-\eta)N\leq m\leq(1+\eta)N}\Pr\{\mathbf{n}=m\},
\end{aligned} \tag{C.13}$$

we have

$$\mathbb{E}[\mathbf{n}]\geq(1-\eta)N\Pr\{(1-\eta)N\leq\mathbf{n}\leq(1+\eta)N\}. \tag{C.14}$$

Combining (C.12) and (C.14) yields

$$\begin{aligned}
&\liminf_{\epsilon\rightarrow 0}\frac{\mathbb{E}[\mathbf{n}]}{N(p,\epsilon,\delta,\zeta)}\\
&\quad\geq(1-\eta)\lim_{\epsilon\rightarrow 0}\Pr\{(1-\eta)N\leq\mathbf{n}\leq(1+\eta)N\}\\
&\quad=1-\eta.
\end{aligned} \tag{C.15}$$

On the other hand, using $\mathbb{E}[\mathbf{n}]=\sum_{m=0}^{\infty}\Pr\{\mathbf{n}>m\}$, we can write

$$\begin{aligned}
\mathbb{E}[\mathbf{n}]&=\sum_{0\leq m<(1+\eta)N}\Pr\{\mathbf{n}>m\}+\sum_{m\geq(1+\eta)N}\Pr\{\mathbf{n}>m\}\\
&\leq\lceil(1+\eta)N\rceil+\sum_{m\geq(1+\eta)N}\Pr\{\mathbf{n}>m\}.
\end{aligned} \tag{C.16}$$

Since $\limsup_{\epsilon\rightarrow 0}(\lceil(1+\eta)N\rceil/N(p,\epsilon,\delta,\zeta))=1+\eta$, for the purpose of establishing $\limsup_{\epsilon\rightarrow 0}(\mathbb{E}[\mathbf{n}]/N(p,\epsilon,\delta,\zeta))\leq 1+\eta$, it remains to show that

$$\limsup_{\epsilon\rightarrow 0}\frac{\sum_{m\geq(1+\eta)N}\Pr\{\mathbf{n}>m\}}{N(p,\epsilon,\delta,\zeta)}=0. \tag{C.17}$$

Consider functions $f(x) = (1/4) - (|x - 1/2| - \rho\epsilon)^2$ and $g(x) = x(1-x)$ for $x \in [0,1]$. Note that

$$\begin{aligned} |f(x) - g(x)| &= \left|(x-1/2)^2 - (|x-1/2| - \rho\epsilon)^2\right| \\ &= \rho\epsilon\left||2x-1| - \rho\epsilon\right| \le \rho\epsilon(1+\rho\epsilon), \end{aligned} \tag{C.18}$$

for all $x \in [0,1]$. For $p \in (0,1)$, there exists a positive number $\gamma < \min\{p, 1-p\}$ such that $|g(x) - g(p)| < (\eta/2)p(1-p)$ for any $x \in (p-\gamma, p+\gamma)$, since $g(x)$ is a continuous function of x. From now on, let $\epsilon > 0$ be sufficiently small such that $\rho\epsilon(1+\rho\epsilon) < (\eta/2)p(1-p)$. Then,

$$\begin{aligned} f(x) &\le g(x) + \rho\epsilon(1+\rho\epsilon) \\ &< g(p) + \frac{\eta}{2}p(1-p) + \rho\epsilon(1+\rho\epsilon) \\ &< (1+\eta)\,p(1-p), \end{aligned} \tag{C.19}$$

for all $x \in (p-\gamma, p+\gamma)$. This implies that

$$\begin{aligned} &\left\{\overline{X}_m \in (p-\gamma, p+\gamma)\right\} \\ &\subseteq \left\{(1+\eta)\,p(1-p) \ge \frac{1}{4} - \left(\left|\overline{X}_m - \frac{1}{2}\right| - \rho\epsilon\right)^2\right\} \end{aligned} \tag{C.20}$$

for all $m > 0$. Taking complementary events on both sides of (C.20) leads to

$$\begin{aligned} &\left\{(1+\eta)\,p(1-p) < \frac{1}{4} - \left(\left|\overline{X}_m - \frac{1}{2}\right| - \rho\epsilon\right)^2\right\} \\ &\subseteq \left\{\overline{X}_m \notin (p-\gamma, p+\gamma)\right\}, \end{aligned} \tag{C.21}$$

for all $m > 0$. Since $(1+\eta)p(1-p) = ((1+\eta)N\epsilon^2/2\ln(1/\zeta\delta)) \le (m\epsilon^2/2\ln(1/\zeta\delta))$ for all $m \ge (1+\eta)N$, it follows that

$$\begin{aligned} &\left\{\frac{m\epsilon^2}{2\ln(1/\zeta\delta)} < \frac{1}{4} - \left(\left|\overline{X}_m - \frac{1}{2}\right| - \rho\epsilon\right)^2\right\} \\ &\subseteq \left\{\overline{X}_m \notin (p-\gamma, p+\gamma)\right\}, \end{aligned} \tag{C.22}$$

for all $m \ge (1+\eta)N$. Therefore, we have shown that if ϵ is sufficiently small, then there exists a number $\gamma > 0$ such that

$$\begin{aligned} \{\mathbf{n} > m\} &\subseteq \left\{\left(\left|\overline{X}_m - \frac{1}{2}\right| - \rho\epsilon\right)^2 < \frac{1}{4} + \frac{m\epsilon^2}{2\ln(\zeta\delta)}\right\} \\ &\subseteq \left\{\overline{X}_m \notin (p-\gamma, p+\gamma)\right\}, \end{aligned} \tag{C.23}$$

for all $m \ge (1+\eta)N$. Using this inclusion relationship and the Chernoff-Hoeffding bound [32, 33], we have

$$\Pr\{\mathbf{n} > m\} \le \Pr\left\{\overline{X}_m \notin (p-\gamma, p+\gamma)\right\} \le 2\exp\left(-2m\gamma^2\right), \tag{C.24}$$

for all $m \ge (1+\eta)N$ provided that $\epsilon > 0$ is sufficiently small. Letting $k = \lceil(1+\eta)N\rceil$ and using (C.24), we have

$$\begin{aligned} \sum_{m \ge (1+\eta)N} \Pr\{\mathbf{n} > m\} &= \sum_{m \ge k} \Pr\{\mathbf{n} > m\} \\ &\le \sum_{m \ge k} 2\exp\left(-2m\gamma^2\right) \\ &= \frac{2\exp\left(-2k\gamma^2\right)}{1 - \exp\left(-2\gamma^2\right)}, \end{aligned} \tag{C.25}$$

provided that ϵ is sufficiently small. Consequently,

$$\begin{aligned} &\limsup_{\epsilon \to 0} \frac{\sum_{m \ge (1+\eta)N} \Pr\{\mathbf{n} > m\}}{N(p,\epsilon,\delta,\zeta)} \\ &\le \limsup_{\epsilon \to 0} \frac{2}{N} \frac{\exp\left(-2k\gamma^2\right)}{1 - \exp(-2\gamma^2)} = 0, \end{aligned} \tag{C.26}$$

since $k \to \infty$ and $N \to \infty$ as $\epsilon \to 0$. So, we have established (C.11). Since the argument holds for arbitrarily small $\eta > 0$, it must be true that $\lim_{\epsilon \to 0}(\mathbb{E}[\mathbf{n}]/N(p,\epsilon,\delta,\zeta)) = 1$ for any $p \in (0,1)$. This completes the proof of the theorem.

D. Proof of Theorem 4

Recall that $\mathbf{l}$ denotes the index of stage at the termination of the sampling process. Observing that

$$\begin{aligned} n_s - n_1 \Pr\{\mathbf{l} = 1\} &= n_s \Pr\{\mathbf{l} \le s\} - n_1 \Pr\{\mathbf{l} \le 1\} \\ &= \sum_{\ell=2}^{s} \left(n_\ell \Pr\{\mathbf{l} \le \ell\} - n_{\ell-1} \Pr\{\mathbf{l} < \ell\}\right) \\ &= \sum_{\ell=2}^{s} n_\ell \left(\Pr\{\mathbf{l} \le \ell\} - \Pr\{\mathbf{l} < \ell\}\right) \\ &\quad + \sum_{\ell=2}^{s} \left(n_\ell - n_{\ell-1}\right) \Pr\{\mathbf{l} < \ell\} \\ &= \sum_{\ell=2}^{s} n_\ell \Pr\{\mathbf{l} = \ell\} \\ &\quad + \sum_{\ell=1}^{s-1} \left(n_{\ell+1} - n_\ell\right) \Pr\{\mathbf{l} \le \ell\}, \end{aligned} \tag{D.1}$$

we have $n_s - \sum_{\ell=1}^{s} n_\ell \Pr\{\mathbf{l} = \ell\} = \sum_{\ell=1}^{s-1} (n_{\ell+1} - n_\ell)\Pr\{\mathbf{l} \leq \ell\}$. Making use of this result and the fact $n_s = n_1 + \sum_{\ell=1}^{s-1}(n_{\ell+1} - n_\ell)$, we have

$$\begin{aligned}\mathbb{E}[\mathbf{n}] &= \sum_{\ell=1}^{s} n_\ell \Pr\{\mathbf{l} = \ell\} = n_s - \left(n_s - \sum_{\ell=1}^{s} n_\ell \Pr\{\mathbf{l} = \ell\} \right) \\ &= n_1 + \sum_{\ell=1}^{s-1} (n_{\ell+1} - n_\ell) - \sum_{\ell=1}^{s-1} (n_{\ell+1} - n_\ell) \Pr\{\mathbf{l} \leq \ell\} \\ &= n_1 + \sum_{\ell=1}^{\tau-1} (n_{\ell+1} - n_\ell) \Pr\{\mathbf{l} > \ell\} \\ &\quad + \sum_{\ell=\tau}^{s-1} (n_{\ell+1} - n_\ell) \Pr\{\mathbf{l} > \ell\}.\end{aligned} \tag{D.2}$$

By the definition of the stopping rule, we have

$$\begin{aligned}\{\mathbf{l} > \ell\} &\subseteq \left\{ \left(\left| \widehat{\mathbf{p}}_\ell - \frac{1}{2} \right| - \rho\epsilon \right)^2 < \frac{1}{4} + \frac{\epsilon^2 n_\ell}{2\ln(\zeta\delta)} \right\} \\ &= \left\{ \rho\epsilon - \sqrt{\frac{1}{4} + \frac{\epsilon^2 n_\ell}{2\ln(\zeta\delta)}} < \left| \widehat{\mathbf{p}}_\ell - \frac{1}{2} \right| \right. \\ &\qquad \left. < \rho\epsilon + \sqrt{\frac{1}{4} + \frac{\epsilon^2 n_\ell}{2\ln(\zeta\delta)}} \right\} \\ &= \left\{ \rho\epsilon - \sqrt{\frac{1}{4} + \frac{\epsilon^2 n_\ell}{2\ln(\zeta\delta)}} < \frac{1}{2} - \widehat{\mathbf{p}}_\ell \right. \\ &\qquad \left. < \rho\epsilon + \sqrt{\frac{1}{4} + \frac{\epsilon^2 n_\ell}{2\ln(\zeta\delta)}},\ \widehat{\mathbf{p}}_\ell \leq \frac{1}{2} \right\} \\ &\quad \cup \left\{ \rho\epsilon - \sqrt{\frac{1}{4} + \frac{\epsilon^2 n_\ell}{2\ln(\zeta\delta)}} < \widehat{\mathbf{p}}_\ell - \frac{1}{2} \right. \\ &\qquad \left. < \rho\epsilon + \sqrt{\frac{1}{4} + \frac{\epsilon^2 n_\ell}{2\ln(\zeta\delta)}},\ \widehat{\mathbf{p}}_\ell > \frac{1}{2} \right\} \\ &\subseteq \{a_\ell < \widehat{\mathbf{p}}_\ell < b_\ell\} \cup \{1 - b_\ell < \widehat{\mathbf{p}}_\ell < 1 - a_\ell\},\end{aligned} \tag{D.3}$$

for $1 \leq \ell < s$, where $b_\ell = (1/2) - \rho\epsilon + \sqrt{(1/4) + (\epsilon^2 n_\ell / 2\ln(\zeta\delta))}$ for $\ell = 1, \ldots, s-1$. By the assumption that ϵ and ρ are nonnegative, we have $1 - b_\ell - a_\ell = 2\rho\epsilon \geq 0$ for $\ell = 1, \ldots, s-1$. It follows from (D.3) that $\{\mathbf{l} > \ell\} \subseteq \{\widehat{\mathbf{p}}_\ell > a_\ell\}$ for $\ell = 1, \ldots, s-1$. By the definition of τ, we have $p < a_\ell$ for $\tau \leq \ell < s$. Making use of this fact, the inclusion relationship $\{\mathbf{l} > \ell\} \subseteq \{\widehat{\mathbf{p}}_\ell > a_\ell\}$, $\ell = 1, \ldots, s-1$, and Chernoff-Hoeffding bound [32, 33], we have

$$\begin{aligned}\Pr\{\mathbf{n} > n_\ell \mid p\} &= \Pr\{\mathbf{l} > \ell \mid p\} \\ &\leq \Pr\{\widehat{\mathbf{p}}_\ell > a_\ell \mid p\} \leq \exp(n_\ell \mathscr{M}(a_\ell, p))\end{aligned} \tag{D.4}$$

for $\tau \leq \ell < s$. It follows from (D.2) and (D.4) that

$$\begin{aligned}\mathbb{E}[\mathbf{n}] &\leq n_1 + \sum_{\ell=1}^{\tau-1} (n_{\ell+1} - n_\ell) + \sum_{\ell=\tau}^{s-1} (n_{\ell+1} - n_\ell) \Pr\{\mathbf{l} > \ell\} \\ &= n_\tau + \sum_{\ell=\tau}^{s-1} (n_{\ell+1} - n_\ell) \Pr\{\mathbf{l} > \ell\} \\ &\leq n_\tau + \sum_{\ell=\tau}^{s-1} (n_{\ell+1} - n_\ell) \exp(n_\ell \mathscr{M}(a_\ell, p)).\end{aligned} \tag{D.5}$$

This completes the proof of the theorem.

Acknowledgment

This paper is supported in part by NIH/NCI Grants no. 1 P01 CA116676, P30 CA138292-01 and 5 P50 CA128613.

References

[1] S.-C. Chow, J. Shao, and H. Wang, *Sample Size Calculations in Clinical Research*, vol. 20 of *Chapman & Hall/CRC Biostatistics Series*, Chapman & Hall/CRC, Boca Raton, Fla, USA, 2nd edition, 2008.

[2] C. Jennison and B. W. Turnbull, *Group Sequential Methods with Applications to Clinical Trials*, Chapman & Hall/CRC, Boca Raton, Fla, USA, 2000.

[3] Y. S. Chow and H. Robbins, "On the asymptotic theory of fixed-width sequential confidence intervals for the mean," *Annals of Mathematical Statistics*, vol. 36, pp. 457–462, 1965.

[4] B. K. Ghosh and P. K. Sen, *Handbook of Sequential Analysis*, vol. 118 of *Statistics: Textbooks and Monographs*, Marcel Dekker, New York, NY, USA, 1991.

[5] M. Ghosh, N. Mukhopadhyay, and P. K. Sen, *Sequential Estimation*, Wiley Series in Probability and Statistics: Probability and Statistics, John Wiley & Sons, New York, NY, USA, 1997.

[6] T. L. Lai, "Sequential analysis: some classical problems and new challenges," *Statistica Sinica*, vol. 11, no. 2, pp. 303–408, 2001.

[7] D. Siegmund, *Sequential Analysis*, Springer Series in Statistics, Springer, New York, NY, USA, 1985.

[8] L. Mendo and J. M. Hernando, "Improved sequential stopping rule for Monte Carlo simulation," *IEEE Transactions on Communications*, vol. 56, no. 11, pp. 1761–1764, 2008.

[9] M. Tanaka, "On a confidence interval of given length for the parameter of the binomial and the Poisson distributions," *Annals of the Institute of Statistical Mathematics*, vol. 13, pp. 201–215, 1961.

[10] S. Franzén, "Fixed length sequential confidence intervals for the probability of response," *Sequential Analysis*, vol. 20, no. 1-2, pp. 45–54, 2001.

[11] S. Franzén, "SPRT fixed length confidence intervals," *Communications in Statistics. Theory and Methods*, vol. 33, no. 2, pp. 305–319, 2004.

[12] A. Wald, *Sequential Analysis*, John Wiley & Sons, New York, NY, USA, 1947.

[13] J. Frey, "Fixed-width sequential confidence intervals for a proportion," *The American Statistician*, vol. 64, no. 3, pp. 242–249, 2010.

[14] X. Chen, "A new framework of multistage estimation," http://arxiv.org/abs/0809.1241v1, September 2008.

[15] X. Chen, "A new framework of multistage estimation," http://arxiv.org/abs/0809.1241v12, April 2009.

[16] X. Chen, "Multistage estimation of bounded-variable means," http://arxiv.org/abs/0809.4679v1, September 2008.

[17] X. Chen, "Estimating the parameters of binomial and Poisson distributions via multistage sampling," http://arxiv.org/abs/0810.0430v1, October 2008.

[18] X. Chen, "Confidence interval for the mean of a bounded random variable and its applications in point estimation," http://arxiv.org/abs/0802.3458v2, April 2009.

[19] X. Chen, "A new framework of multistage parametric inference," in *Sensors, and Command, Control, Communications, and Intelligence (C3I) Technologies for Homeland Security and Homeland Defense IX*, usa, April 2010.

[20] A. H. Land and A. G. Doig, "An automatic method of solving discrete programming problems," *Econometrica*, vol. 28, pp. 497–520, 1960.

[21] X. Chen, "A new framework of multistage estimation," http://arxiv.org/abs/0809.1241v16, November 2009.

[22] X. Chen, "A new framework of multistage estimation," http://arxiv.org/abs/0809.1241v4, December 2008.

[23] J. R. Schultz, F. R. Nichol, G. L. Elfring, and S. D. Weed, "Multiple stage procedures for drug screening," *Biometrics*, vol. 29, no. 2, pp. 293–300, 1973.

[24] H. Chen, "The accuracy of approximate intervals for a binomial parameter," *Journal of the American Statistical Association*, vol. 85, no. 410, pp. 514–518, 1990.

[25] E. B. Wilson, "Probable inference, the law of succession, and statistical inference," *Journal of the American Statistical Association*, vol. 22, pp. 209–212, 1927.

[26] C. J. Clopper and E. S. Pearson, "The use of confidence or fiducial limits illustrated in the case of the binomial," *Biometrika*, vol. 26, pp. 404–413, 1934.

[27] W. Feller, *An Introduction to Probability Theory and Its Applications. Vol. I*, John Wiley & Sons, New York, NY, USA, 3rd edition, 1968.

[28] G. S. Fishman, "Confidence intervals for the mean in the bounded case," *Statistics & Probability Letters*, vol. 12, no. 3, pp. 223–227, 1991.

[29] X. Chen, K. Zhou, and J. L. Aravena, "Explicit formula for constructing binomial confidence interval with guaranteed coverage probability," *Communications in Statistics. Theory and Methods*, vol. 37, no. 8–10, pp. 1173–1180, 2008.

[30] X. Chen, "Multistage estimation of bounded-variable means," http://arxiv.org/abs/0809.4679v2, October 2008.

[31] P. Massart, "The tight constant in the Dvoretzky-Kiefer-Wolfowitz inequality," *The Annals of Probability*, vol. 18, no. 3, pp. 1269–1283, 1990.

[32] H. Chernoff, "A measure of asymptotic efficiency for tests of a hypothesis based on the sum of observations," *Annals of Mathematical Statistics*, vol. 23, pp. 493–507, 1952.

[33] W. Hoeffding, "Probability inequalities for sums of bounded random variables," *Journal of the American Statistical Association*, vol. 58, pp. 13–30, 1963.

[34] X. Chen, "Exact computation of minimum sample size for estimation of binomial parameters," *Journal of Statistical Planning and Inference*, vol. 141, no. 8, pp. 2622–2632, 2011.

[35] F. J. Anscombe, "Sequential estimation," *Journal of the Royal Statistical Society. Series B*, vol. 15, pp. 1–21, 1953.

12

On-Line Selection of c-Alternating Subsequences from a Random Sample

Robert W. Chen,[1] Larry A. Shepp,[2] and Justín Ju-Chen Yang[3]

[1] *Department of Mathematics, University of Miami, Coral Gables, FL 33124, USA*
[2] *Department of Statistics, Wharton School, University of Pennsylvania, Philadelphia, PA 19104, USA*
[3] *Department of Statistics, Harvard University, 1 Oxford Street, Cambridge, MA 02138-2901, USA*

Correspondence should be addressed to Robert W. Chen; chen@math.miami.edu

Academic Editor: Shein-chung Chow

A sequence $X_1, X_2, \ldots, X_n$ is a c-alternating sequence if any odd term is less than or equal to the next even term $-c$ and the any even term is greater than or equal to the next odd term $+c$, where c is a nonnegative constant. In this paper, we present an optimal on-line procedure to select a c-alternating subsequence from a symmetric distributed random sample. We also give the optimal selection rate when the sample size goes to infinity.

1. Introduction

Given a finite (or infinite) sequence $x = \{x_1, x_2, \ldots, x_n, \ldots\}$ of real numbers, we say that a subsequence $x_{i_1}, x_{i_2}, \ldots, x_{i_k}, \ldots$ with $1 \le i_1 < i_2 < \cdots < i_k < \cdots$ is c-alternating if we have $x_{i_1} + c < x_{i_2} > x_{i_3} + c < x_{i_4} \cdots$, where c is a nonnegative real number. When $c = 0$, $x_{i_1}, x_{i_2}, \ldots, x_{i_k}, \ldots$ is alternating. We are mainly concerned with the length $\alpha(x)$ of the longest c-alternating subsequence of x. Here, we study the problem of making on-line selection of a c-alternating subsequence. That is now we regard the sequence $x_1, x_2, \ldots$ as being available to us sequentially, and, at time i when x_i is available, we must choose to include x_i as a term of our subsequence or reject x_i as a member of our subsequence.

We will consider the sequence to be given by independent, identically distributed, symmetric random variables over the interval $[0, 1]$. In [1], Arlotto et al. studied the case that the sequence to be given by independent, identically distributed, uniform random variables over the interval $[0, 1]$ and $c = 0$. So this paper can be considered as an extension of their paper.

Now we need to be more explicit about the set II of feasible strategies for on-line selection. At time i, when X_i is presented to us, we must decide to select X_i based on its value, the value of earlier members of the sequence, and the actions we have taken in the past. All of this information can be captured by saying that i_k, the index of the kth selection, must be a stopping time with respect to the increasing sequence of σ-fields, $F_i = \sigma\{X_1, X_2, \ldots, X_i\}$, $i = 1, 2, \ldots$. Given any feasible policy π in II the random variable of most interest here is $A_n^o(\pi)$, the number of selections made by the policy π up to and including time n. In other words, $A_n^o(\pi)$ is equal to the largest k for which there are stopping times $1 \le i_1 < i_2 < \cdots < i_k \le n$ such that $\{X_{i_1}, X_{i_2}, \ldots, X_{i_k}\}$ is a c-alternating subsequence of the sequence $\{X_1, X_2, \ldots, X_n\}$. In this paper, we are interested in the optimal selection and the asymptotic rate of the optimal selection. That is, we have a selection policy π_n^* in II such that

$$\lim_{n \to \infty} E\left[A_n^o(\pi_n^*)\right] = \lim_{n \to \infty} \sup_{\pi \in II} E\left[A_n^o(\pi)\right]. \quad (1)$$

2. Main Results

For each $0 \le c < 1$ and each $0 \le \lambda \le (1 - c)/2$, we define a threshold function f^* as follows: $f^*(y) = \max\{c + \lambda, c + y\}$ for all $0 \le y \le 1 - c$. We now recursively define random variables $\{Y_i : i = 1, 2, \ldots\}$ by setting $Y_0 = y$ and taking $Y_i = Y_{i-1}$ if $X_i < f^*(Y_{i-1})$, $Y_i = 1 - X_i$ if $X_i \ge f^*(Y_{i-1})$. Introduce a value function $V(\lambda, y, \rho) = E\{\sum_{i=1}^{\infty} \rho^{i-1} I[X_i \ge f^*(Y_{i-1})] \mid Y_0 = y\}$, where $0 < \rho < 1$ is a constant and $I[X_i \geq f^*(Y_{i-1})]$ is the indicator function of the event $[X_i \geq f^*(Y_{i-1})]$.

Let G be the distribution function and g the probability density function of X_i. If X_i is not a uniform random variable over the interval $[0,1]$, then we will assume that g' exists and is nonzero. Since X_i is symmetric over the interval $[0,1]$, $G(1-x) = 1 - G(x)$ and $g(x) = g(1-x)$ for all $0 \leq x \leq 1$.

It is easy to see that

$$\begin{aligned} V(\lambda, y, \rho) &= \rho G[\max(\lambda, y) + c] V(\lambda, y, \rho) \\ &\quad + \int_{\max(\lambda,y)+c}^{1} [1 + \rho V(\lambda, 1-x, \rho)] g(x) \, dx \\ &= \rho G[\max(\lambda, y) + c] V(\lambda, y, \rho) \\ &\quad + \int_{0}^{1-\max(\lambda,y)-c} [1 + \rho V(\lambda, x, \rho)] g(x) \, dx \end{aligned} \quad (2)$$

since $g(x) = g(1-x)$ for all $0 \leq x \leq 1$. For simplicity, we will let $V(y)$ denote $V(\lambda, y, \rho)$ for fixed λ and ρ. It is easy to check that for all $0 \leq y \leq \lambda$, $V(y) = \rho G(\lambda + c)V(y) + \int_0^{1-\lambda-c} [1 + \rho V(x)]g(x)dx$, for all $\lambda < y \leq 1-c$, $V(y) = \rho G(y+c)V(y) + \int_0^{1-y-c} [1 + \rho V(x)]g(x)dx$, and for all $1-c < y \leq 1$, $V(y) = 0$. Since for all $0 \leq y \leq \lambda$, $V(y) = V(\lambda)$, $V(y) = \rho G(y+c)V(y) + [1 + \rho V(\lambda)]G(1-y-c)$ for all $1-\lambda-c \leq y \leq 1-c$. Since $G(1-y-c) = 1 - G(y+c)$, $V(y) = \rho G(y+c)V(y) + [1 + \rho V(\lambda)][1 - G(y+c)]$ for all $1-\lambda-c \leq y \leq 1-c$. From now on, we will let $V'(\lambda)$ denote the right derivative of the function V at λ and $V'(1-c)$ denote the left derivative of the function V at $1-c$.

For $\lambda < y < 1-c$, when we differentiate $V(y)$ we have the following differential equation:

$$\begin{aligned} V'(y)\{1 - \rho G(y+c)\} &= \rho g(y+c) V(y) \\ &\quad - [1 + \rho V(1-y-c)] g(1-y-c). \end{aligned} \quad (3)$$

Since $g(1-y-c) = g(y+c)$ and $g(1-y) = g(y)$. Now we have the following differential equations:

$$\begin{aligned} V'(y)\{1 - \rho G(y+c)\} &= \rho g(y+c) V(y) - [1 + \rho V(1-y-c)] g(y+c) \\ &= g(y+c)\{\rho [V(y) - V(1-y-c)] - 1\}, \end{aligned} \quad (4)$$

$$\begin{aligned} V'(1-y-c)\{1 - \rho G(1-y)\} &= \rho g(y) V(1-y-c) - [1 + \rho V(y)] g(y) \\ &= g(y)\{\rho [V(1-y-c) - V(y)] - 1\}. \end{aligned} \quad (5)$$

Add (3) and (4) together, we have

$$V'(1-y-c) \frac{1 - \rho G(1-y)}{g(y)} = -V'(y) \frac{1 - \rho G(y+c)}{g(y+c)} - 2. \quad (6)$$

In summary, we have the following equations:

$$\begin{aligned} V(1-\lambda-c)[1-\rho+\rho G(\lambda)] &= [1+\rho V(\lambda)] G(\lambda), \\ V'(\lambda)[1 - \rho G(\lambda+c)] &= g(\lambda+c)\{\rho [V(\lambda) - V(1-\lambda-c)] - 1\}, \\ V'(1-\lambda-c)[1-\rho+\rho G(\lambda)] &= g(\lambda)\{\rho [V(1-\lambda-c) - V(\lambda)] - 1\}. \end{aligned} \quad (7)$$

For all $\lambda \leq y \leq 1-c$, let us define

$$h(y) = \frac{[1 - \rho G(y+c)]^2 (1 - \rho + \rho G(y))}{g(y+c)}. \quad (8)$$

Then we have the following equation:

$$\begin{aligned} V'(1-\lambda-c) &= V'(\lambda) \frac{h(\lambda)}{h(1-\lambda-c)} \\ &\quad - \frac{2\rho}{h(1-\lambda-c)} \int_{\lambda}^{1-\lambda-c} (1-\rho+\rho G(x-c)) g(x) \, dx. \end{aligned} \quad (9)$$

Proof of (9). Differentiate (4). Again, we have

$$\begin{aligned} V''(y)[1 - \rho G(y+c)] &= \rho g(y+c) [2V'(y) + V'(1-y-c)] \\ &\quad + g'(y+c)\{\rho [V(y) - V(1-y-c)] - 1\}. \end{aligned} \quad (10)$$

Replace $\rho[V(y) - V(1-y-c)] - 1$ by $V'(y)((1-\rho G(y+c))/g(y+c))$ and replace $V'(1-y-c)$ by $\{-2 - V'(y)((1-\rho G(y+c))/g(y+c))\}\{g(y)/(1-\rho+\rho G(y))\}$. And after the simplification, we have the following equation:

$$\begin{aligned} V''(y)[1 - \rho G(y+c)] &= \rho g(y+c) \frac{-2g(y)}{1-\rho+\rho G(y)} + V'(y) \\ &\quad \times \left\{ g'(y+c) \frac{1-\rho G(y+c)}{g(y+c)} \right. \\ &\quad \left. + \rho g(y+c) \left[2 - \frac{g(y)[1-\rho G(y+c)]}{g(y+c)[1-\rho+\rho G(y)]} \right] \right\} \\ &\quad + \rho g(y+c). \end{aligned} \quad (11)$$

Multiplying both sides of (11) by $[1-\rho+\rho G(y)][1-\rho G(y+c)]/g(y+c)$, we obtain the following equation:

$$V''(y)\frac{[1-\rho G(y+c)]^2[1-\rho+\rho G(y)]}{g(y+c)}$$
$$=-2\rho g(y)[1-\rho G(y+c)]$$
$$+V'(y)g'(y+c)\left[\frac{1-\rho G(y+c)}{g(y+c)}\right]^2[1-\rho+\rho G(y)]$$
$$+V'(y)\rho g(y+c)\left\{2\frac{[1-\rho+\rho G(y)][1-\rho G(y+c)]}{g'(y+c)}\right.$$
$$\left.-g(y)\left[\frac{1-\rho G(y+c)}{g(y+c)}\right]^2\right\}. \tag{12}$$

Notice that

$$-h'(y)=g'(y+c)\left[\frac{1-\rho G(y+c)}{g(y+c)}\right]^2[1-\rho+\rho G(y)]$$
$$+\rho g(y+c)\left\{2\frac{[1-\rho+\rho G(y)][1-\rho G(y+c)]}{g'(y+c)}\right.$$
$$\left.-g(y)\left[\frac{1-\rho G(y+c)}{g(y+c)}\right]^2\right\}. \tag{13}$$

Equation (12) can be rewritten as

$$V''(y)h(y)+V'(y)h'(y)=-2\rho[1-\rho G(y+c)]g(y). \tag{14}$$

By integrating both sides of (14), we have

$$V'(y)=V'(\lambda)\frac{h(\lambda)}{h(y)}-\frac{2\rho}{h(y)}\int_\lambda^y[1-\rho G(z+c)]g(z)\,dz. \tag{15}$$

□

Therefore, we have the following theorem.

Theorem 1.

(i) $V(1-\lambda-c)[1-\rho+\rho G(\lambda)]=[1+\rho V(\lambda)]G(\lambda)$,

(ii) $V'(\lambda)[1-\rho G(\lambda+c)]$

$$=g(\lambda+c)\{\rho[V(\lambda)-V(1-\lambda-c)]-1\},$$

(iii) $V'(1-\lambda-c)[1-\rho+\rho G(\lambda)]$

$$=g(\lambda)\{\rho[V(1-\lambda-c)-V(\lambda)]-1\},$$

(iv) $V'(1-\lambda-c)$

$$=V'(\lambda)\frac{h(\lambda)}{h(1-\lambda-c)}-\frac{2\rho}{h(1-\lambda-c)}$$
$$\times\int_{\lambda+c}^{1-\lambda}[1-\rho+\rho G(x-c)]g(x)\,dx. \tag{16}$$

Now we have four unknown variables $V(\lambda)$, $V(1-\lambda-c)$, and $V'(\lambda)$, $V'(1-\lambda-c)$ and also have four linear equations involving these four unknown variables. We solve these four linear equations and obtain the following solutions.

Theorem 2.

(i) $$V(\lambda)=\left(\rho G(\lambda)[1-\rho G(\lambda+c)]\right.$$
$$\left.+\rho\int_{\lambda+c}^{1-\lambda}[1-\rho+\rho G(x-c)]g(x)\,dx\right)$$
$$\times\left(\rho(1-\rho)[1-\rho G(\lambda+c)]\right)^{-1},$$

(ii) $$V'(\lambda)=\left(g(\lambda+c)\right.$$
$$\times\left\{\rho\int_{\lambda+c}^{1-\lambda}[1-\rho+\rho G(x-c)]g(x)\,dx\right.$$
$$-[1-\rho G(\lambda+c)]$$
$$\left.\left.\times[1-\rho+\rho G(\lambda)]\right\}\right)$$
$$\times\left([1-\rho G(\lambda+c)]^2[1-\rho+\rho G(\lambda)]\right)^{-1},$$

(iii) $V(1-\lambda-c)$

$$=G(\lambda)\left\{\left([1-\rho+\rho G(\lambda)][1-\rho G(\lambda+c)]\right.\right.$$
$$\left.-\rho\int_{\lambda+c}^{1-\lambda}[1-\rho+\rho G(x-c)]g(x)\,dx\right)$$
$$\times\left((1-\rho)[1-\rho G(\lambda+c)]\right.$$
$$\left.\left.\times[1-\rho+\rho G(\lambda)]\right)^{-1}\right\}. \tag{17}$$

By Theorem 2, $V'(\lambda)=0$ if and only if

$$\rho\int_{\lambda+c}^{1-\lambda}[1-\rho+\rho G(x-c)]g(x)\,dx$$
$$=[1-\rho G(\lambda+c)][1-\rho+\rho G(\lambda)] \tag{18}$$

since $g(\lambda+c)$ and $[1-\rho G(\lambda+c)]^2[1-\rho+\rho G(\lambda)]$ are positive. For each $0<\rho<1$, let $\lambda(\rho)$ denote a solution of $V'(\lambda)=0$.

The next theorem indicates that when $\rho < 1$ but close enough to 1, $\lambda(\rho)$ is unique and $0 \le \lambda(\rho) < (1-c)/2$.

Theorem 3. *When $\rho < 1$ but close enough to 1, $\lambda(\rho)$ is unique and $0 \le \lambda(\rho) < (1-c)/2$.*

Proof. For all $0 \le \lambda < (1-c)/2$, let

$$K(\lambda) = \rho \int_{\lambda+c}^{1-\lambda} [1-\rho+\rho G(x-c)]\, g(x)\, dx - [1-\rho G(\lambda+c)][1-\rho+\rho G(\lambda)]. \tag{19}$$

Then

$$K'(\lambda) = -\rho\{g(1-\lambda)[1-\rho+\rho G(1-\lambda-c)] + g(\lambda)[1-\rho G(\lambda+c)]\} < 0,$$

$$K\left(\frac{1-c}{2}\right) = -\left[1-\rho+\rho G\left(\frac{1-c}{2}\right)\right]\left[1-\rho G\left(\frac{1+c}{2}\right)\right] < 0,$$

$$K(0) = \rho^2 \int_c^1 G(x-c)\, g(x)\, dx - (1-\rho)^2 > 0 \tag{20}$$

if ρ is close enough to 1. Therefore, $0 \le \lambda(\rho) < (1-c)/2$ and $\lambda(\rho)$ is unique if ρ is close enough to 1. This completes the proof of Theorem 3. □

A routine calculation is as follows:

$$V''(\lambda(\rho)) = \frac{-2\rho g(\lambda(\rho))\, g(\lambda(\rho)+c)}{[1-\rho+\rho G(\lambda(\rho))][1-\rho G(\lambda(\rho)+c)]} < 0. \tag{21}$$

So we have found the maximum of the function V. After the simplification,

$$V(\lambda(\rho), \lambda(\rho), \rho) = \frac{1-\rho+2\rho G(\lambda(\rho))}{\rho(1-\rho)} \tag{22}$$

if ρ is close enough to 1. Therefore, $(1-\rho)V(\lambda(\rho),\lambda(\rho),\rho) \to 2G(\lambda(1))$ as $\rho \uparrow 1$, where

$$\int_{\lambda(1)+c}^{1-\lambda(1)} G(x-c)\, g(x)\, dx = G(\lambda(1))[1-G(\lambda(1)+c)]. \tag{23}$$

Example 4. When $c = 0$, then $2G(\lambda(1)) = 2 - \sqrt{2}$.

Example 5. When $G(x) = x$ for all $0 \le x \le 1$, then $2G(\lambda(1)) = (1-c)(2-\sqrt{2})$.

In [1], the following two strategies are mentioned:

(I) the maximally timid strategy which can be described as follows: at the start, accept the first observation which is less than $1/2$, then accept the next one which is greater than $1/2$, then accept the next one which is less than $1/2$. Continue this way until we observe n observations, then we stop;

(II) the purely greedy strategy which can be described as follows: at the start, accept the first observation, then accept the next one which is greater than the first one, accept the next one which is less than the second selected one, then accept the next one which is greater than the third selected one. Continue this way until we observe n observations, then we stop.

Now we define these two strategies for the c-alternating subsequence as follows:

(I′) the maximally c-timid strategy which can be described as follows: at the start, accept the first observation which is less than $(1-c)/2$, accept the next one which is greater than $(1+c)/2$, accept the next one which is less than $(1-c)/2$, then accept the next one which is greater than $(1+c)/2$. Continue this way until we observe n observations, then we stop;

(II′) the purely c-greedy strategy which can be described as follows: at the start, accept the first observation which is less than $1-c$, accept the next one which is greater than the first selected one $+c$, accept the next one which is less than the second selected one $-c$, then accept the next one which is greater than the third selected one $+c$. Continue this way until we observe n observations, then we stop.

When $c = 0$, the maximally c-timid strategy is the maximally timid strategy and the purely c-greedy strategy is the purely greedy strategy. In fact, the maximally c-timid strategy is the case when $\lambda = (1-c)/2$ and the purely c-greedy strategy is the case when $\lambda = 0$.

The asymptotic selection rate for the maximally c-timid strategy is $G((1-c)/2)$ and the asymptotic selection rate for the purely c-greedy strategy is $\int_0^{1-c} G(x)g(x+c)dx/(1-G(c))$.

Example 6. When $c = 0$, then the asymptotic selection rate for both the maximally timid strategy and the purely greedy strategy is $1/2$. These results are the same as those in [1].

Example 7. When $G(x) = x$ for all $0 \le x \le 1$, then the asymptotic selection rate for both the maximally c-timid strategy and the purely c-greedy strategy is $(1/2)(1-c)$. It is easy to see these results are consistent with the result of Example 5.

If the random variables $X_1, X_2, \ldots, X_n$ are independent, identically distributed symmetric random variables over the interval $[a,b]$, where $a < b$ and a, b are finite, then we can change X_i into Z_i by $Z_i = (X_i - a)/(b-a)$. Then $Z_1, Z_2, \ldots, Z_n$ are independent, identically distributed symmetric random variables over the interval $[0,1]$. Let $c' = c/(b-a)$. Then selecting a c-alternating subsequence from the random sample $X_1, X_2, \ldots, X_n$ is exactly the same to select a c'-alternating subsequence from the random sample $Z_1, Z_2, \ldots, Z_n$. So the asymptotic selection rate is still

TABLE 1: The asymptotic optimal selection rates for various distributions and c.

c	$G_1(x)$	$G_2(x)$	$G_3(x)$	$G_4(x)$	$G_5(x)$
0.0	$2-\sqrt{2}$	$2-\sqrt{2}$	$2-\sqrt{2}$	$2-\sqrt{2}$	$2-\sqrt{2}$
0.1	0.540492	0.545962	0.527208	0.499942	0.481120
0.2	0.509190	0.506516	0.468629	0.414948	0.380891
0.3	0.489881	0.466834	0.410051	0.332852	0.291619
0.4	0.466034	0.426295	0.351472	0.255622	0.214252
0.5	0.424335	0.384147	0.292893	0.185176	0.148786
0.6	0.366680	0.339473	0.234315	0.123399	0.095223
0.7	0.294696	0.290641	0.175736	0.072153	0.053563
0.8	0.209273	0.234738	0.117157	0.033284	0.023806
0.9	0.110936	0.164273	0.058579	0.008624	0.005952
1.0	0	0	0	0	0

the same. Or we can find $\lambda(1)$ directly by solving the following equation:

$$\int_{a+\lambda(1)+c}^{b-a-\lambda(1)} G(x-c)\, g(x)\, dx = G(a+\lambda(1))\,[1-G(a+\lambda(1)+c)]. \tag{24}$$

For $0 \le c \le (b-a)$, $a \le y \le (b-c)$, and $a \le \lambda < a+(b-a-c)/2$ the threshold function f^* is defined by $f^*(y) = \max(\lambda, y)+c$ for all $a \le y \le (b-c)$. This time, we recursively define random variables $\{Y_i : i = 1, 2, \ldots\}$ by setting $Y_0 = y$ and taking $Y_i = Y_{i-1}$ if $X_i < f^*(Y_{i-1})$, $Y_i = b - X_i$ if $X_i \ge f^*(Y_{i-1})$. Here $a \le y \le (b-c)$,

$$\begin{aligned}
G_1(x) &= \begin{cases} 2x(1-x) & \text{if } 0 \le x \le \dfrac{1}{2}, \\ 1-2x+2x^2 & \text{if } \dfrac{1}{2} < x \le 1, \end{cases} \\
G_2(x) &= \frac{2}{\pi}\sin^{-1}(\sqrt{x}), \\
G_3(x) &= x, \\
G_4(x) &= 3x^2 - 2x^3, \\
G_5(x) &= \begin{cases} 2x^2 & \text{if } 0 \le x \le \dfrac{1}{2}, \\ -2x^2+4x-1 & \text{if } \dfrac{1}{2} < x \le 1. \end{cases}
\end{aligned} \tag{25}$$

From Table 1, it seems that when the distribution has higher chances on the tails, then the asymptotic optimal selection rate is higher. On the other hand, when the distribution has higher chance near the center, then the asymptotic optimal selection rate is lower. However, we do not have a proof for this statement.

We are now considering the case when $X_1, X_2, \ldots, X_n$ have an arbitrary distribution. We have made some progress, but it is still in the premature state. We hope to be able to find the asymptotic optimal selection rate in a forth coming paper.

References

[1] A. Arlotto, R. W. Chen, L. A. Shepp, and J. M. Steele, "Online selection of alternating subsequences from a random sample," *Journal of Applied Probability*, vol. 48, no. 4, pp. 1114–1132, 2011.

Testing for Main Random Effects in Two-Way Random and Mixed Effects Models: Modifying the F Statistic

Trent Gaugler[1] and Michael G. Akritas[2]

[1] *Department of Statistics, Carnegie Mellon University, 5000 Forbes Avenue, 132B Baker Hall, Pittsburgh, PA 15213, USA*
[2] *Department of Statistics, Penn State University, 325 Thomas Building, University Park, PA 16802, USA*

Correspondence should be addressed to Trent Gaugler; tgaugler@andrew.cmu.edu

Academic Editor: Jose Sarabia

A procedure for testing the significance of the main random effect is proposed under a model which does not require the traditional assumptions of symmetry, homoscedasticity, and normality for the error term and random effects. To accommodate this level of model generality, and also unbalanced designs, suitable adjustments to the F-test are made. The extensive simulations performed under the random effects model, and the unrestricted and restricted versions of the mixed effects model, indicate that the classical F procedure is extremely liberal under heteroscedasticity and unbalancedness. The proposed test procedure performs well in all settings and is comparable to the classical F-test when the classical assumptions are met. An analysis of a dataset from the Mussel Watch Project is presented.

1. Introduction

Consider a two-factor mixed or random effects design, and let Y_{ijk} denote the kth observation at level i of the row factor and level j of the column factor. In the case of a mixed effects design we assume that the row factor is fixed.

The classical two-way model, compare [1–5], uses the decomposition

$$Y_{ijk} = \mu + \alpha_i + \beta_j + \gamma_{ij} + \epsilon_{ijk}. \qquad (1)$$

For the random effects model, the α_i's, β_j's, γ_{ij}'s, and ϵ_{ijk}'s are mutually independent, the α_i's are iid $N(0,\sigma_\alpha^2)$, the β_j's are iid $N(0,\sigma_\beta^2)$, the γ_{ij}'s are iid $N(0,\sigma_\gamma^2)$, and the ϵ_{ijk}'s are iid $N(0,\sigma_\epsilon^2)$.

For the mixed effects model, there are two common definitions of the effects. The *unrestricted* version of the model assumes

$$\sum_i \alpha_i = 0, \quad \beta_j \text{ are iid } N\left(0,\sigma_\beta^2\right), \quad \gamma_{ij} \text{ are iid } N\left(0,\sigma_\gamma^2\right), \qquad (2)$$

and the β_j, γ_{ij} are all independent. The *restricted* version of the model keeps the above assumptions but requires that

$$\sum_i \gamma_{ij} = 0, \qquad (3)$$

implying that $\gamma_{ij}, \gamma_{i'j}$ are correlated. Both versions assume the ϵ_{ijk} to be iid $N(0,\sigma_\epsilon^2)$ and independent from the other random effects. There is a dichotomy of opinion in the statistical literature as to which model should be used. Cornfield and Tukey [6], Scheffé [1], Winer [7], and Khuri et al. [5], among others, advocate the restricted version. Searle [2], Hocking [8], and Searle et al. [4] advocate the unrestricted version. While the statistic for testing for no main random effect differs in the two versions, both use what Scheffé [1, page 264] calls the *symmetry assumption*. Basically, this is the assumption of independence of the random main and interaction effects. This assumption was criticized in [9, 10] as unrealistic in most practical situations. More importantly, simulations demonstrated that the classical F-test for testing the significance of the main fixed effect is very sensitive to departures from the symmetry assumption even in balanced designs with homoscedastic errors. Additional simulations showed the classical F-test for testing the significance of the interaction effect to also be very sensitive to unbalancedness and heteroscedasticity.

This paper deals with testing the significance of the main random effect in two-factor mixed and random effects designs. Simulations suggest that the classical F-test for this hypothesis is also very sensitive to departures from the symmetry assumption, as well as to unbalancedness and heteroscedasticity. New test procedures are presented under fully nonparametric models for the two-factor mixed and random effects designs. Unbalanced mixed and random effects designs are incorporated in the modeling by considering the sample sizes n_{ij} to be random. The models and test procedures pertain to both continuous and discrete data and allow heteroscedasticity in the error and interaction terms, as well as nonnormality. The interaction term is not assumed independent from the main effect (so the symmetry assumption is not made), though, under the random effects model, the two are shown to be uncorrelated.

The asymptotic theory of the test statistics is derived under the Neyman-Scott framework, in the sense that the number of levels of both factors is large but the group sizes can remain fixed. In such cases, test statistics are constructed by taking the difference (as opposed to the ratio) of the appropriate mean squares; compare [9–12] and references therein. In accordance to this literature, the proposed test statistics for testing for no main random effects are scaled versions of MSB-MSE* and MSB-MSAB, for the mixed and random effects designs, respectively, where the precise definition of these mean squares is given in Section 3.

The novelty of the proposed test statistics is twofold. First, the mean squares are based on unweighted cell averages to accommodate unbalanced designs. Second, in order to accommodate heteroscedasticity, MSE is replaced by a different linear combination of the cell sample variances, denoted by MSE*, which, under the null hypothesis, has the same expected value as MSB (ensuring zero mean value of the test statistic under the null hypothesis). A modified version of the MSE for accommodating heteroscedasticity was first used in Akritas and Papadatos [13] in the context of a high-dimensional one-way fixed effects design.

The rest of the paper is organised as follows. Section 2 reviews the fully nonparametric random and mixed effects models. In Section 3 we derive the test statistics and present their asymptotic distributions. Proofs of the propositions in Section 3 appear in the appendix. Section 4 discusses the estimation of the limiting null distributions and also presents results of simulations comparing our testing procedures to the usual F-tests. Finally, in Section 5 we present the analysis of a dataset from NOAA's NS&T Mussel Watch Program.

2. Fully Nonparametric Models

Here we review the fully nonparametric random and mixed effects models developed in [9, 10].

2.1. The Random Effects Model. Consider a two-way random effects design. The random levels of the row factor are obtained by random sampling from the population $\mathscr{A}_S$, while the random levels of the column factor are obtained by random sampling from the population $\mathscr{A}_T$. Let S denote a row level randomly selected from $\mathscr{A}_S$, and T denote a column level randomly selected from $\mathscr{A}_T$. Assume that the selections are independent. For each factor-level combination (S, T), observations Y_{STk}, $k = 1, \ldots, n_{ST}$, are generated from a distribution function, F_{ST}. F_{ST} can be any arbitrary distribution function, which depends only on the factor levels S, T.

Decompose Y_{STk} into a random mean and error term as follows:

$$\begin{aligned} Y_{STk} &= E(Y_{STk} \mid S, T) + [Y_{STk} - E(Y_{STk} \mid S, T)] \\ &= \mu_{ST} + \epsilon_{STk}, \end{aligned} \tag{4}$$

where μ_{ST} and ϵ_{STk} are defined implicitly in (4). The definitions of μ_{ST} and ϵ_{STk} imply

$$\begin{aligned} E(\epsilon_{STk}) &= 0, \qquad E(\epsilon_{STk} \mid S) = 0, \\ E(\epsilon_{STk} \mid T) &= 0, \qquad E(\epsilon_{STk} \mid S, T) = 0. \end{aligned} \tag{5}$$

Conditioning on S, T and using the last relation in (5) it further follows that

$$\begin{gathered} E(\mu_{ST}\epsilon_{STk}) = 0, \qquad E(\mu_{ST}\epsilon_{STk} \mid S) = 0, \\ E(\mu_{ST}\epsilon_{STk} \mid T) = 0. \end{gathered} \tag{6}$$

Next, decompose the random means as

$$\mu_{ST} = \mu + \alpha_S + \beta_T + \gamma_{ST}, \tag{7}$$

where the overall mean and the random effects α_S, β_T, and γ_{ST} are defined as

$$\begin{aligned} \mu &= E(Y_{STk}) = E(\mu_{ST}), \\ \alpha_S &= E(Y_{STk} \mid S) - \mu = E(\mu_{ST} \mid S) - \mu, \\ \beta_T &= E(Y_{STk} \mid T) - \mu = E(\mu_{ST} \mid T) - \mu, \\ \gamma_{ST} &= \mu_{ST} - \mu - \alpha_S - \beta_T. \end{aligned} \tag{8}$$

Combining (4) and (7) we obtain

$$Y_{STk} = \mu + \alpha_S + \beta_T + \gamma_{ST} + \epsilon_{STk}, \tag{9}$$

where the random effects have zero means, are both marginally and conditionally on S and T uncorrelated from the error terms (this follows from (6)), and satisfy

$$\begin{gathered} E(\alpha_S \mid T) = E(\beta_T \mid S) = E(\gamma_{ST} \mid S) = E(\gamma_{ST} \mid T) = 0, \\ E(\alpha_S\beta_T) = E(\alpha_S\gamma_{ST}) = E(\beta_T\gamma_{ST}) = 0, \end{gathered} \tag{10}$$

as is easily seen using the independence of S, T. Note that even though the main and interaction effects are uncorrelated, in general they will not be independent. This is because the interaction term is typically related to the two main effects by a multiplicative relation rendering the interaction effect nonnormal even in the presence of normal main effects. Therefore the symmetry assumption is typically violated in random effects designs.

Relation (9) with the associated properties (5), (6), and (10) is derived using only the assumption that S and T are independent. Thus, it is a nonparametric version of the classical random effects model, in the sense that it allows for arbitrary (rather than normal) distributions of the random effects and error term, the variance (as well as the entire distribution) of the error term can depend on the random factor levels, and the effects are uncorrelated, and uncorrelated to the error term, as opposed to the assumed independence of the classical model.

2.2. The Mixed Effects Model. Consider a design where the rows correspond to the fixed effects and the columns correspond to the random effects. The levels of the random effects are obtained by simple random sampling from a population $\mathscr{A}$. Let T denote a randomly selected element from $\mathscr{A}$, i a fixed effect level, and Y_{iT} an observation obtained from the cell (i, T). We place no restrictions on the form of the distribution function, F_{iT}, of Y_{iT} other than the obvious restriction that it has to depend on i and T. In particular, F_{iT} can be a discrete distribution function. Since T is obtained from a random selection, F_{iT}, its mean μ_{iT} and variance σ_{iT}^2 are all random. The nonparametric model consists of a decomposition of Y_{iT} in terms of μ_{iT} plus an error term, and further decomposition of μ_{iT} to define the main effects and interaction terms. In particular, write

$$\begin{aligned} Y_{iT} &= E(Y_{iT} \mid T) + (Y_{iT} - E(Y_{iT} \mid T)) \\ &= \mu_{iT} + (Y_{iT} - \mu_{iT}) = \mu_{iT} + \epsilon_{iT}, \end{aligned} \tag{11}$$

where the error term ϵ_{iT} is defined implicitly in the above relation. An easy consequence of the definition of ϵ_{iT} is

$$E(\epsilon_{iT}) = E(\epsilon_{iT} \mid T) = 0, \qquad E(\epsilon_{iT}\mu_{iT}) = E(\epsilon_{iT}\mu_{iT} \mid T) = 0. \tag{12}$$

Next decompose μ_{iT} into main and interaction effects as

$$\mu_{iT} = \mu + \alpha_i + \beta_T + \gamma_{iT}, \tag{13}$$

where $\mu = (1/a)\sum_i E(\mu_{iT})$ is the overall mean and

$$\begin{gathered} \alpha_i = E(\mu_{iT}) - \mu, \qquad \beta_T = \frac{1}{a}\sum_i \mu_{iT} - \mu, \\ \gamma_{iT} = \mu_{iT} - \mu - \alpha_i - \beta_T, \end{gathered} \tag{14}$$

are the main and interaction effects. An immediate implication of (14) is

$$\sum_i \alpha_i = 0, \qquad E(\beta_T) = 0, \qquad \sum_i \gamma_{iT} = 0, \qquad E(\gamma_{iT}) = 0. \tag{15}$$

Using (12) it further follows that

$$\begin{aligned} E(\epsilon_{iT}\beta_T) &= E(\epsilon_{iT}\beta_T \mid T) = 0, \\ E(\epsilon_{iT}\gamma_{iT}) &= E(\epsilon_{iT}\gamma_{iT} \mid T) = 0. \end{aligned} \tag{16}$$

Combining (11) and (13) we have

$$Y_{iT} = \mu + \alpha_i + \beta_T + \gamma_{iT} + \epsilon_{iT}. \tag{17}$$

While model (17) has the appearance of the usual mixed effects model, it does not require normality (or even continuity) of the observations, the variance of the random interaction effects can depend on i, and the random effects β_T and γ_{iT} are not uncorrelated without further assumptions (which we do not make) about constant variance of the random means and constant covariance for any two distinct random means.

Suppose now that there are a fixed levels, so $i = 1, \ldots, a$, and let T_j, $j = 1, \ldots, b$, denote the levels of the random effect. Recall that the T_j are obtained by simple random sampling from $\mathscr{A}$. Also let Y_{iT_jk}, $k = 1, \ldots, n_{ij}$, denote the replications (iid observations) in factor-level combination (i, T_j). For notational simplicity we write Y_{ijk} for Y_{iT_jk} and j for T_j. Thus, conditionally on j (or T_j), the Y_{ijk} are iid F_{ij}. From (17) we have

$$Y_{ijk} = \mu + \alpha_i + \beta_j + \gamma_{ij} + \epsilon_{ijk}, \tag{18}$$

where the effects and error terms satisfy (14), (15), and (16). In addition, γ_{ij_1} and γ_{ij_2} are independent for $j_1 \neq j_2$, ϵ_{ijk}, $k = 1, \ldots, n_{ij}$, are iid conditionally on j, $\epsilon_{i_1j_1k_1}$, and $\epsilon_{i_2j_2k_2}$ are independent for $j_1 \neq j_2$, and $\epsilon_{i_1jk_1}$ and $\epsilon_{i_2jk_2}$ are uncorrelated for $i_1 \neq i_2$. These imply

$$\begin{aligned} E(\epsilon_{i_1jk_1}\epsilon_{i_2jk_2}) &= 0, \quad \text{for } k_1 \neq k_2, \\ E(\epsilon_{i_1j_1k_1}\epsilon_{i_2j_2k_2}) &= 0, \quad \text{for } j_1 \neq j_2, \\ E(\epsilon_{i_1jk_1}\epsilon_{i_2jk_2}) &= 0, \quad \text{for } i_1 \neq i_2, \\ E(\gamma_{ij_1}\gamma_{ij_2}) &= 0, \quad \text{for } j_1 \neq j_2. \end{aligned} \tag{19}$$

We note that relation (14) implies that γ_{ij} is a function of the random level T_j. Since the random levels are iid, we have that the variances of γ_{ij}, $j = 1, \ldots, b$, are equal. Arguing similarly, we have that the variances of the ϵ_{ijk}, $j = 1, \ldots, b$, are equal. However the model allows heteroscedasticity across different levels of the fixed effect, that is, $\sigma^2(\gamma_{i_1j_1}) \neq \sigma^2(\gamma_{i_2j_2})$ and $\sigma^2(\epsilon_{i_1j_1k_1}) \neq \sigma^2(\epsilon_{i_2j_2k_2})$, for $i_1 \neq i_2$. Moreover, the conditional variances of the error term given different levels of the random effect can be different. Summarizing, we have

$$\sigma^2(\gamma_{ij_1}) = \sigma^2(\gamma_{ij_2}), \qquad \sigma^2(\epsilon_{ij_1k_1}) = \sigma^2(\epsilon_{ij_2k_2}), \tag{20}$$

but the model allows

$$\begin{aligned} \sigma^2(\gamma_{i_1j_1}) &\neq \sigma^2(\gamma_{i_2j_2}), \quad \text{for } i_1 \neq i_2, \\ \sigma^2(\epsilon_{i_1j_1k_1}) &\neq \sigma^2(\epsilon_{i_2j_2k_2}), \quad \text{for } i_1 \neq i_2, \\ \operatorname{Var}(\epsilon_{ij_1k_1} \mid T_{j_1}) &\neq \operatorname{Var}(\epsilon_{ij_2k_2} \mid T_{j_2}). \end{aligned} \tag{21}$$

3. The Test Statistics and Their Asymptotic Distributions

Before proceeding, we first define our notation:

$$\overline{Y}_{ij\cdot} = \frac{1}{n_{ij}}\sum_{k=1}^{n_{ij}} Y_{ijk}, \quad \widetilde{Y}_{i\cdot\cdot} = \frac{1}{b}\sum_{j=1}^{b}\overline{Y}_{ij\cdot}, \quad \widetilde{Y}_{\cdot j\cdot} = \frac{1}{a}\sum_{i=1}^{a}\overline{Y}_{ij\cdot},$$

$$\widetilde{Y}_{\cdots} = \frac{1}{a}\sum_{i=1}^{a}\widetilde{Y}_{i\cdot\cdot} = \frac{1}{b}\sum_{j=1}^{b}\widetilde{Y}_{\cdot j\cdot}, \quad \overline{\beta}_{\cdot} = \frac{1}{b}\sum_{j=1}^{b}\beta_j,$$

$$\overline{\gamma}_{i\cdot} = \frac{1}{b}\sum_{j=1}^{b}\gamma_{ij}, \quad \overline{\epsilon}_{ij\cdot} = \frac{1}{n_{ij}}\sum_{k=1}^{n_{ij}}\epsilon_{ijk}, \quad \widetilde{\epsilon}_{i\cdot\cdot} = \frac{1}{b}\sum_{j=1}^{b}\overline{\epsilon}_{ij\cdot},$$

$$\widetilde{\epsilon}_{\cdot j\cdot} = \frac{1}{a}\sum_{i=1}^{a}\overline{\epsilon}_{ij\cdot}, \quad \widetilde{\epsilon}_{\cdots} = \frac{1}{b}\sum_{j=1}^{b}\widetilde{\epsilon}_{\cdot j\cdot} = \frac{1}{a}\sum_{i=1}^{a}\widetilde{\epsilon}_{i\cdot\cdot}. \tag{22}$$

The usual F statistic in the balanced case in the restricted version of the mixed model is given by

$$F = \frac{\text{MSB}}{\text{MSE}} = \frac{a^2 bn(n-1)\sum_j\left(\overline{Y}_{\cdot j\cdot} - \overline{Y}_{\cdots}\right)^2}{(b-1)\sum_i\sum_j\sum_k\left(Y_{ijk} - \overline{Y}_{ij\cdot}\right)^2}, \tag{23}$$

while the usual F statistic in the balanced case in the random effects model is given by

$$F = \frac{\text{MSB}}{\text{MSAB}} = \frac{a(a-1)\sum_j\left(\overline{Y}_{\cdot j\cdot} - \overline{Y}_{\cdots}\right)^2}{\sum_i\sum_j\left(\overline{Y}_{ij\cdot} - \overline{Y}_{i\cdot\cdot} - \overline{Y}_{\cdot j\cdot} + \overline{Y}_{\cdots}\right)^2}. \tag{24}$$

In the unbalanced case there is no theoretically justified procedure, but Searle [14] does have a number of recommendations which are implemented in software packages.

The test statistics we propose for testing the hypothesis of no main random effect, $H_0(B) : \sigma^2(\beta_j) = 0$, $j = 1,\ldots,b$, are modeled after the usual F statistics given in (23) and (24), except that we take the difference rather than the ratio, as explained in the Introduction. Moreover, in order to accommodate unbalanced designs, $\overline{Y}_{i\cdot\cdot}$, $\overline{Y}_{\cdot j\cdot}$, and $\overline{Y}_{\cdots}$ are replaced by $\widetilde{Y}_{i\cdot\cdot}$, $\widetilde{Y}_{\cdot j\cdot}$, and $\widetilde{Y}_{\cdots}$, respectively. As mentioned in the Introduction, our asymptotic theory requires that the difference has mean value zero. While this is the case for the random effects design, the expected value of MSB-MSE is not zero for the mixed effects design with heteroscedastic errors. Therefore we replace MSE with a different linear combination of the cell sample variances:

$$\text{MSE}^* = \frac{1}{a^3 b}\sum_{i=1}^{a}\sum_{j=1}^{b}\frac{n_{\cdot j}}{n_{ij}}\frac{1}{n_{ij}-1}\sum_{k=1}^{n_{ij}}\left(Y_{ijk} - \overline{Y}_{ij\cdot}\right)^2. \tag{25}$$

Thus, the proposed test statistics for the mixed effects and random effects models are, respectively,

$$S_{B,\text{mixed}} = \frac{1}{a(b-1)}\sum_{i=1}^{a}\sum_{j=1}^{b}\sum_{k=1}^{n_{ij}}\left(\widetilde{Y}_{\cdot j\cdot} - \widetilde{Y}_{\cdots}\right)^2 - \text{MSE}^*, \tag{26}$$

$$S_{B,\text{rand}} = \frac{1}{a^2 b}\sum_{i=1}^{a}\sum_{j=1}^{b}\left[(a-1)\left(\widetilde{Y}_{\cdot j\cdot} - \widetilde{Y}_{\cdots}\right)^2 - R_{ij}^2\right], \tag{27}$$

where $R_{ij} = \overline{Y}_{ij\cdot} - \widetilde{Y}_{\cdot j\cdot} - \widetilde{Y}_{i\cdot\cdot} + \widetilde{Y}_{\cdots}$.

Proposition 1. *Let $S_{B,mixed}$ be as defined in (26) and let $S_{B,rand}$ be as defined in (27), with the observations Y_{ijk} generated according to the model of Section 2. Then,*

(i) *if the null hypothesis $H_0(B) : \sigma^2(\beta_j) = 0$, $j = 1,\ldots,b$ holds,*

$$E\left(S_{B,mixed}\right) = E\left(S_{B,rand}\right) = 0, \tag{28}$$

(ii) *if the null hypothesis $H_0(B) : \sigma^2(\beta_j) = 0$, $j = 1,\ldots,b$ does not hold,*

$$E\left(S_{B,mixed}\right) > 0, \qquad E\left(S_{B,rand}\right) > 0. \tag{29}$$

The second item in Proposition 1 suggests that the null hypothesis should be rejected at level α whenever the test statistic takes a value larger than the $100(1-\alpha)$th percentile of the null distribution. The next proposition establishes asymptotic representations of $S_{B,\text{mixed}}$ and $S_{B,\text{rand}}$ that are useful for establishing their asymptotic distributions.

Proposition 2. *Under $H_0(B)$,*

(i)

$$\sqrt{b}S_{B,mixed} = \frac{1}{a\sqrt{b}}\sum_j n_{\cdot j}\left[\left(\widetilde{\epsilon}_{\cdot j\cdot} - \widetilde{\epsilon}_{\cdots}\right)^2 - E\left(\widetilde{\epsilon}_{\cdot j\cdot} - \widetilde{\epsilon}_{\cdots}\right)^2\right] + o_p(1), \tag{30}$$

(ii)

$$aS_{B,rand} = \frac{1}{ab}\sum_i\sum_{i_1\neq i}\sum_j\gamma_{ij}\gamma_{i_1 j} + o_p(1). \tag{31}$$

Theorem 3. *Let $s_b^2 = \text{Var}((1/\sqrt{b})\sum_j n_{ij}[(\widetilde{\epsilon}_{\cdot j\cdot} - \widetilde{\epsilon}_{\cdots})^2 - E(\widetilde{\epsilon}_{\cdot j\cdot} - \widetilde{\epsilon}_{\cdots})^2])$. Then under $H_0(B)$*

$$\sqrt{b}\frac{S_{B,mixed}}{s_b} \longrightarrow N(0,1), \tag{32}$$

in distribution as $a, b \to \infty$.

Proof. Letting $T_b = (1/\sqrt{b})\sum_j n_{ij}[(\widetilde{\epsilon}_{\cdot j\cdot} - \widetilde{\epsilon}_{\cdots})^2 - E(\widetilde{\epsilon}_{\cdot j\cdot} - \widetilde{\epsilon}_{\cdots})^2]$, then, as stated above,

$$\sqrt{b}\frac{S_{B,\text{mixed}}}{s_b} \approx \frac{T_b}{s_b}. \tag{33}$$

The proof that $T_b/s_b \to N(0,1)$ in distribution follows from an application of the Lindeberg-Feller CLT. We first rewrite our statement in the notation of the Lindeberg-Feller CLT and then show that the Lindeberg condition holds after a straightforward application of the Dominated Convergence Theorem. Now let $Y_j^{a,b} = (1/\sqrt{b})n_{ij}[(\widetilde{\epsilon}_{\cdot j\cdot} - \widetilde{\epsilon}_{\cdots})^2 - E(\widetilde{\epsilon}_{\cdot j\cdot} - \widetilde{\epsilon}_{\cdots})^2]$, so that $T_b = \sum_j Y_j^{a,b}$. Also note that we can write

$$s_b^2 = \operatorname{Var}(T_b) = \frac{1}{b}\sum_j n_{ij}^2 \operatorname{Var}\left(\left(\widetilde{\epsilon}_{\cdot j\cdot} - \widetilde{\epsilon}_{\cdots}\right)^2\right), \qquad (34)$$

which is finite because we have assumed a finite fourth moment for the ϵ terms. Moreover, this implies that

$$\operatorname{Var}\left(\sqrt{b}Y_j^{a,b}\right) = n_{ij}^2 \operatorname{Var}\left(\left(\widetilde{\epsilon}_{\cdot j\cdot} - \widetilde{\epsilon}_{\cdots}\right)^2\right), \qquad (35)$$

which is again finite. Now in order to use the Lindeberg-Feller CLT, we need to verify the Lindeberg condition:

$$\forall \epsilon > 0, \quad \frac{1}{s_b^2}\sum_{j=1}^{b} E\left[\left(Y_j^{a,b}\right)^2 I\left\{\left|Y_j^{a,b}\right| > \epsilon s_b\right\}\right] \longrightarrow 0. \qquad (36)$$

Note that

$$\begin{aligned}
&\frac{1}{s_b^2}\sum_{j=1}^{b} E\left[\left(Y_j^{a,b}\right)^2 I\left\{\left|Y_j^{a,b}\right| > \epsilon s_b\right\}\right] \\
&= \frac{1}{b s_b^2}\sum_{j=1}^{b} E\left[\left(\sqrt{b}Y_j^{a,b}\right)^2 I\left\{\left|\sqrt{b}Y_j^{a,b}\right| > \sqrt{b}\epsilon s_b\right\}\right] \qquad (37)\\
&= \frac{1}{s_b^2} E\left[\left(\sqrt{b}Y_1^{a,b}\right)^2 I\left\{\left|\sqrt{b}Y_1^{a,b}\right| > \sqrt{b}\epsilon s_b\right\}\right].
\end{aligned}$$

Because $E(\sqrt{b}Y_1^{a,b}) = 0$ and $\operatorname{Var}(\sqrt{b}Y_1^{a,b})$ is finite, the Dominated Convergence Theorem implies that $E[(\sqrt{b}Y_1^{a,b})^2 I\{|\sqrt{b}Y_1^{a,b}| > \sqrt{b}\epsilon s_b\}] \to 0$, and hence the Lindeberg condition is verified. This completes the proof of the theorem. □

Theorem 4. *Define the operator $H_b : g \to H_b g$, for $g \in L_2([0,1], \mathscr{B}_{[0,1]}, \lambda_{[0,1]})$, where $\mathscr{B}_{[0,1]}$ is the class of Borel subsets of $[0,1]$ and $\lambda_{[0,1]}$ is the Lebesgue measure on $\mathscr{B}_{[0,1]}$, by*

$$(H_b g)(s) = \int h_b(s,t)\, g(t)\, dt, \qquad (38)$$

where $h_b(s,t) = (1/b)\sum_{j=1}^{b} \gamma_{sj}\gamma_{tj}$. Let λ_ν^b, f_ν^b, $\nu = 1,2,\ldots$, be the eigenvalues and eigenfunctions of the integral equation

$$(H_b f)(s) = \lambda f(s), \qquad (39)$$

and assume that the eigenvalues satisfy

$$\sum_{\nu=1}^{\infty}\left(\sup_b \lambda_\nu^b\right)^2 < \infty. \qquad (40)$$

Then, if γ_{ij} and ϵ_{ij} have finite fourth moments and the covariances $\operatorname{Cov}(\gamma_{ij}, \gamma_{ij'})$, $\operatorname{Cov}(\gamma_{ij}^2, \widetilde{\epsilon}_{ij\cdot}^2)$, $\operatorname{Cov}(\gamma_{ij}^2, \gamma_{ij'}^2)$, and $\operatorname{Cov}(\epsilon_{ij}^2, \epsilon_{ij'}^2)$ are bounded uniformly in j, j', the asymptotic null distribution of $S_{B,rand}$, as $a \to \infty$, $b \to \infty$, is the same as the asymptotic distribution of $\sum_{\nu=1}^{\infty}\lambda_\nu^b(Z_\nu^2 - 1)$, as $b \to \infty$, where Z_ν, $\nu = 1,2,\ldots$, are independent standard normal random variables.

Proof. The proof proceeds by first conditioning on the levels of the column factor $\mathbf{T} = (T_1,\ldots,T_b)$ and then applying the arguments used in the proof of Theorem 3.3 of [9]. Indeed, the proof of the aforementioned theorem uses an expansion similar to (31) of Proposition 2 (with the indices i, j interchanged) which is modeled as a U-statistic with kernel depending on a. The same arguments establish the limiting distribution of the test statistic conditionally on $\mathbf{T} = (T_1,\ldots,T_b)$. As $b \to \infty$, the strong law of large numbers guarantees that the kernel $h_b(s,t)$ converges to the same limit, and thus the conditional limiting distribution does not depend on $\mathbf{T} = (T_1,\ldots,T_b)$. Thus, the test statistic converges to the same distribution unconditionally. □

4. Test Procedures and Simulation Results

In testing for the random main effect in the mixed effects model, we only need to estimate the variance term s_b^2 in Theorem 3. According to (30) of Proposition 2 we need to estimate the variance of

$$\frac{1}{\sqrt{b}}\sum_j n_{ij}\left(\widetilde{\epsilon}_{\cdot j\cdot} - \widetilde{\epsilon}_{\cdots}\right)^2. \qquad (41)$$

It is easy to see that

$$\frac{1}{\sqrt{b}}\sum_j n_{ij}\left(\widetilde{\epsilon}_{\cdot j\cdot} - \widetilde{\epsilon}_{\cdots}\right)^2 - \frac{1}{\sqrt{b}}\sum_j n_{ij}\widetilde{\epsilon}_{\cdot j\cdot}^2 \longrightarrow 0, \qquad (42)$$

in probability. Thus, we need to estimate the variance of the independent terms $\widetilde{\epsilon}_{\cdot j\cdot}^2$. By the Central Limit Theorem, as $a \to \infty$, $\widetilde{\epsilon}_{\cdot j\cdot}$ is approximately Normally distributed with mean zero and variance $(1/a^2)\sum_i(\sigma_{ij}^2/n_{ij})$. Therefore, the approximate variance of $\widetilde{\epsilon}_{\cdot j\cdot}^2$ is $2((1/a^2)\sum_i(\sigma_{ij}^2/n_{ij}))^2$. As a result, we estimate s_b^2 by

$$\widehat{s}_b^2 = \frac{2}{a^6 b}\sum_j\left(n_{\cdot j}\sum_i \frac{s_{ij}^2}{n_{ij}}\right)^2. \qquad (43)$$

Then, according to Proposition 1 and Theorem 3, $H_0(B)$ is rejected at level α when

$$b^{1/2}\frac{S_{B,\text{mixed}}}{\widehat{s}_b} > z_\alpha, \qquad (44)$$

where z_α is the $(1-\alpha)100$th percentile of the standard normal distribution.

Testing for main effects in the random effects model involves estimation of the asymptotic distribution of Theorem 4. Koltchinskii and Giné [15] show that the infinite spectrum of a Hilbert-Schmidt operator H can be approximated by the finite spectrum of the empirical matrix version of the operator. We will use this result to approximate the spectrum of the operator H_b in Theorem 4 by its empirical version, $\widetilde{H}_a$, defined below. To this end, the following representation of $S_{B,\text{rand}}$ as a U-statistic is useful:

$$aS_{B,\text{rand}} = \frac{1}{ab}\sum_{i_1}\sum_{i_2 \neq i_1}\sum_{j}\left(\overline{Y}_{i_1 j\cdot} - \overline{Y}_{i_1\cdot\cdot}\right)\left(\overline{Y}_{i_2 j\cdot} - \overline{Y}_{i_2\cdot\cdot}\right) = \frac{a-1}{b}U_a, \tag{45}$$

where the U-statistic U_a is defined by the kernel

$$h\left(S_{i_1}, S_{i_2}\right) = \sum_{j}\left(\overline{Y}_{i_1 j\cdot} - \overline{Y}_{i_1\cdot\cdot}\right)\left(\overline{Y}_{i_2 j\cdot} - \overline{Y}_{i_2\cdot\cdot}\right). \tag{46}$$

Under the null hypothesis $H_0(B)$, the empirical version, $\widetilde{H}_a$, of the operator H_b is the $a \times a$ matrix with entries $\widetilde{H}_{i,i'} = (1/a)h(S_i, S_{i'})$. The a eigenvalues $\lambda_1^a, \ldots, \lambda_a^a$ of $\widetilde{H}_a$ are then used to estimate the asymptotic distribution of $S_{B,\text{rand}}$ as $\sum_{i=1}^{a} \lambda_i^a (Z_i^2 - 1)$, where the Z_i^2 are randomly sampled from the χ_1^2 distribution. To approximate the estimated distribution, we randomly generate a large number B of sets $Z_1^2, \ldots, Z_a^2$, and use the formula above to obtain B realizations from the distribution. Using this approximate distribution, for a level α test we reject $H_0(B)$ when $S_{B,\text{rand}}$ is greater than the $(1-\alpha)$100th percentile.

We now present simulation results to compare our test procedures for the main random effect in the mixed and random effects models to the usual F-tests given by (23) and (24). Note that these statistics are only valid for balanced designs. In the unbalanced designs, the datasets used to obtain the achieved size and power are written out to files and input to SAS for the F-test to be carried out in SAS PROC MIXED. The achieved size and power are calculated from the P values given from SAS.

Using the decomposition in relation (18), we generate our data for the mixed effects model as follows.

(1) Set μ equal to any constant.

(2) Generate a vector of main random effects T_j, $j = 1, \ldots, b$ from the standard normal distribution.

(3) Independently generate a vector of fixed "functions" λ_i, $i = 1, \ldots, a$, from the standard exponential distribution and set $\gamma_i^* = \lambda_i - \overline{\lambda}$.

(4) Generate Y_{ijk} from a normal distribution with variance σ_i^2 and mean $\mu + \alpha_i + T_j + T_j * \gamma_i^*$, where the last two terms are the main random and interaction effects, respectively.

Similarly, using the decomposition in relation (9), we generate our data for the random effects model as follows.

(1) Set μ equal to any constant.

(2) Generate a vector of main row random effects S_i, $i = 1, \ldots, a$ from the standard normal distribution.

(3) Independently generate a vector of main column random effects T_j, $j = 1, \ldots, b$ from the standard normal distribution, and set $T_j^* = \tau T_j$, where τ is used to set the degree of deviation from the null hypothesis.

(4) Set the interaction effects $\gamma_{ij} = S_i T_j$.

(5) Generate Y_{ijk} from a normal distribution with variance σ_i^2 and mean $\mu + S_i + T_j^* + S_i T_j$.

In Table 1, each simulation used $a = b = 20$. For the unbalanced setups, rows $i = 1, \ldots, 15$ had 4 observations per cell, whereas rows $i = 16, \ldots, 20$ had 2 observations per cell. For the heteroscedastic setups, the first 15 rows all had variance 1, while the remaining rows had variance 5; that is, $\sigma_1^2 = \cdots = \sigma_{15}^2 = 1$, $\sigma_{16}^2 = \cdots = \sigma_{20}^2 = 5$.

5. Data Analysis

One real-world application for our methodology can be found through the National Oceanic and Atmospheric Administration's National Status and Trends Program. In 1986, this division undertook a very large scale project to monitor the levels of numerous chemical contaminants and organic chemical constituents in marine sediment and bivalve (mollusk) tissue samples. This project (http://ccma.nos.noaa.gov/about/coast/nsandt/musselwatch.aspx), dubbed the Mussel Watch Project, is still on-going and there are no apparent plans to discontinue it in the near future. There are currently over 300 coastal sites at which sediment and bivalve samples are collected and analysed for the project. Each site is categorised as being within a certain estuarine drainage area (EDA). For our data analysis, we chose to analyse the Mercury concentrations in tissue samples of the *Crassostrea virginica*, or Eastern Oyster. There was a high degree of missingness of Mercury concentrations in the dataset, so we used imputations to generate these measurements. Because of how we performed our imputations, only EDAs that did not have missing measurements for two consecutive years and EDAs that had measurements in 1986 were included in the analysis. Once we had a complete imputed dataset, we employed our methodology to test for effects due to EDA.

5.1. Imputations. Because of the $1/(n_{ij} - 1)$ coefficient that is used in the statistic $S_{B,\text{mixed}}$, each cell needs to have at least 2 observations. To facilitate the imputations, we identified 195 instances of 2 consecutive measurements falling into the following groupings: 1986-1987, 1988-1989, 1990-1991, 1992-1993, 1994-1995, 1996-1997, 1998-1999, 2000-2001, 2002-2003, and 2004-2005. If any cell had more than one observation, they were averaged. We used these 195 pairs to build the regression model $\widehat{CV}_{t,j} = 0.046424 + 0.65852CV_{t-1,j}$, where we predict a value of Mercury concentration at time t from the Mercury concentration at time $t-1$. For this model, the P value for the test of $H_0 : \beta_1 = 0$ versus $H_A : \beta_1 \neq 0$ was 0.000 and the $R^2 = 36.8\%$. When fitting the regression model in Minitab, we also saved the residuals from the model and the actual imputations that we used were $\widehat{CV}_{t,j} + R^*$, where R^* is a randomly selected residual from the fitting of the model. We separately treat imputations for cells with only

TABLE 1: Comparison of achieved type I error rates and power for $S_{B,\text{mixed}}$, $S_{B,\text{rand}}$, and F with $N = 1000$ tests for the random and mixed effects models.

	Random effects model								Mixed effects model							
	Balanced				Unbalanced				Balanced				Unbalanced			
	$\alpha = .05$		$\alpha = .10$		$\alpha = .05$		$\alpha = .10$		$\alpha = .05$		$\alpha = .10$		$\alpha = .05$		$\alpha = .10$	
	S_B	F	S_B	F	S_B	F	S_B	F	S_B	F	S_B	F	S_B	F	S_B	F
Homo.																
H_0	.054	.220	.111	.247	.049	.167	.092	.198	.065	.047	.112	.099	.063	.045	.109	.086
A_1	.053	.229	.123	.265	.069	NA	.140	NA	.227	.187	.318	.296	.184	.138	.260	.228
A_2	.125	.337	.221	.367	.113	NA	.201	NA	.761	.719	.820	.804	.542	.489	.634	.638
A_3	.236	.532	.369	.563	.234	NA	.360	NA	.975	.963	.982	.982	.888	.863	.920	.917
Hetero.																
H_0	.045	.150	.089	.182	.015	.230	.042	.303	.070	.056	.106	.100	.062	.935	.107	.952
A_1	.045	.160	.088	.199	.021	NA	.053	NA	.077	.060	.129	.121	.074	NA	.124	NA
A_2	.114	.273	.190	.316	.047	NA	.094	NA	.151	.129	.221	.215	.104	NA	.152	NA
A_3	.195	.430	.318	.467	.117	NA	.215	NA	.270	.235	.353	.341	.131	NA	.189	NA

one observation and imputations for empty cells. When cells are empty, we use the regression equation

$$\widehat{CV}_{t,j} = 0.046424 + 0.65852CV_{t-1,j} + R^*, \quad (47)$$

twice to impute two values of Hg concentration at time t. When a cell has only one observation, we use a different technique.

(1) If the EDA has any cells with more than one observation in any year, we calculate residuals for each year with more than one observation as $e_{ijk} = Y_{ijk} - \overline{Y}_{ij\cdot}$, $k = 1, \ldots, n_{ij}$. Then for the EDA and year of interest, we impute a second observation as $Y_{ij2} = Y_{ij1} + e^*_{ijk}$, where e^*_{ijk} denotes a randomly selected residual.

(2) If the EDA of interest has at most one observation in each cell, we identify the closest neighboring EDA (denoted by level j^*) and calculate residuals for each year with more than one observation in the neighboring EDA as $e_{ij^*k} = Y_{ij^*k} - \overline{Y}_{ij^*\cdot}$, $k = 1, \ldots, n_{ij^*}$. Then for the EDA and year of interest, we impute a second observation as $Y_{ij2} = Y_{ij1} + e^*_{ij^*k}$, where $e^*_{ij^*k}$ denotes a randomly selected residual.

After performing the imputations, there are a total of $N = 1808$ observations of Hg concentration in CV tissue samples. The 20 years included are 1986–2005, and there are 39 EDAs included in the dataset. The data are rather unbalanced, with most EDAs having only two observations per year, while others such as the Galveston Bay EDA have as many as six observations per year. Moreover, the observations do not seem to be homoscedastic. These points are illustrated in Figure 1, where we show scatterplots of Hg concentrations for six EDAs after imputation.

5.2. The Analysis. In our analysis of this dataset under the mixed effects model, we take the years as the fixed effect and the EDAs as the random effect. We can think of the EDAs as a random effect because we are only analyzing a very small subset of a much larger group of EDAs and the location effect is not of specific interest. Some may view the time effect as being of specific interest, while others may not. Whether one views the time effect as a fixed or random factor would precipitate analysis using a mixed or random effects model.

We want to model the Hg concentration with an interaction effect between year and location, so we use the model in relation (18). Because we are dealing with compositional data, we employ the common practice of analyzing the data after taking a log transformation; compare [16].

When we apply our proposed test procedure based on the test statistic S_C, we obtain a P value of 0.0588. When we apply the F-test computed from SAS PROC MIXED, we obtain a P value of 0.0016.

The P value for the test using S_C is simply calculated as $1 - \Phi(\cdot)$, where Φ is the distribution function of the standard normal random variable. The P value for the SAS F-test is calculated as $1 - F_{\nu_1,\nu_2}(\cdot)$, where ν_1 and ν_2 denote the numerator and denominator degrees of freedom. In the test for the random interaction effect, SAS uses $\nu_1 = 722$, $\nu_2 = 1028$. The simulation results shown in Table 1 suggest that the significance of the interaction effect indicated by the F-test cannot be trusted. Indeed, according to these simulation results, we expect the F-test to detect interaction in almost any unbalanced and heteroscedastic dataset. On the other hand, the simulation results suggest that the marginal significance indicated by the P value obtained from the proposed test procedure can be trusted. According to this we recommend that the interaction effect be included in the modeling of this dataset.

6. Summary

The classical F statistic for testing the significance of main random effects in two-factor mixed and random effects model is shown to be very sensitive to violations of the assumptions under which it is derived, in particular those of symmetry, homoscedasticity, and balancedness. Two new

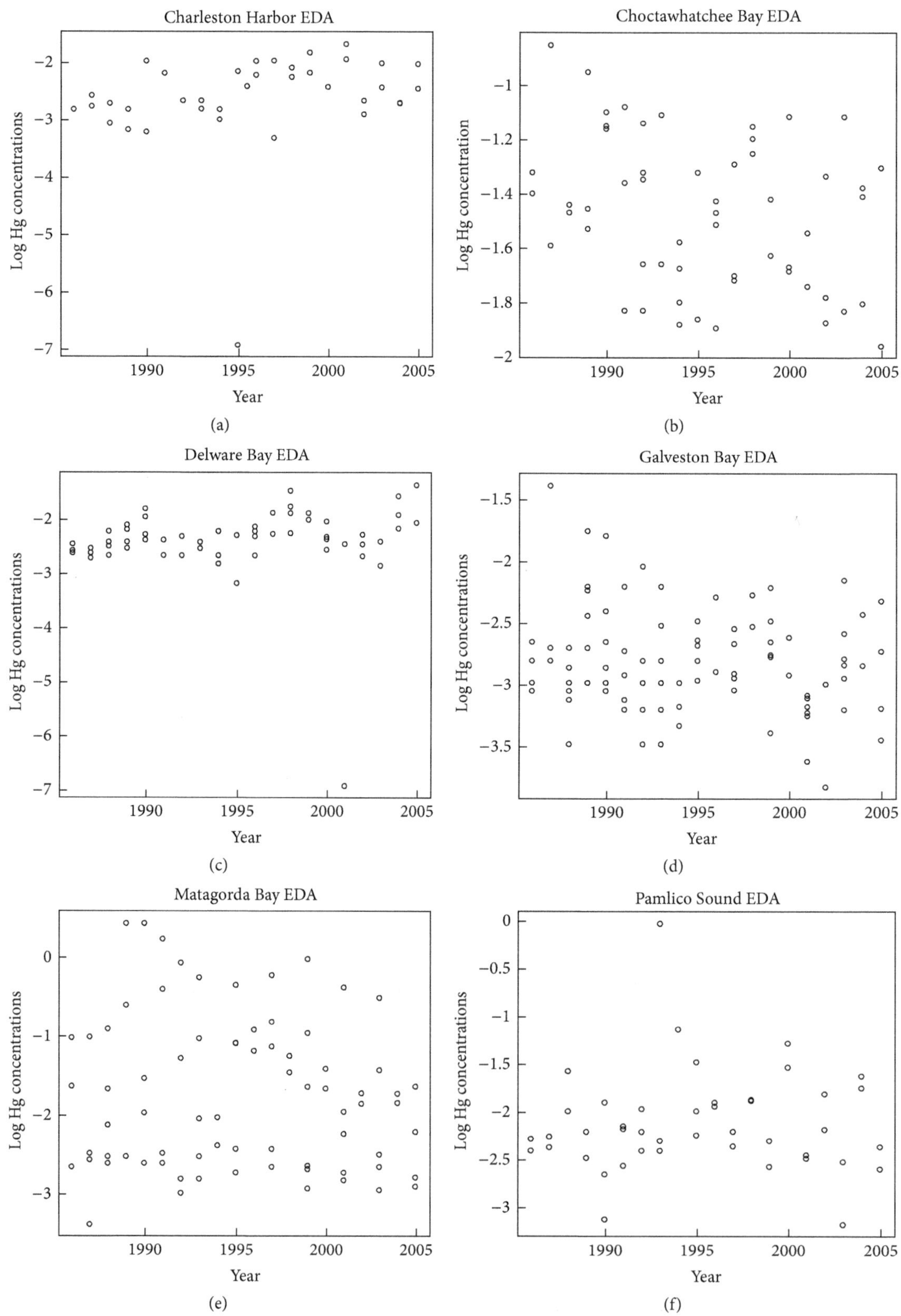

FIGURE 1: Scatterplot of Mercury concentrations in Eastern Oysters by year after imputation for six EDAs.

test procedures for testing the hypothesis of no main random effects are proposed under fully nonparametric modeling of the mixed and random effects designs. The test statistics are defined as differences (as opposed to ratios) of suitably defined mean squares, and their asymptotic theory is derived as the number of levels tends to infinity. Simulations suggest that the new procedures perform reasonably well in situations where the F statistic breaks down and have comparable performance to it when the assumptions it requires are met. The practical value of the proposed procedures is demonstrated with the analysis of a real dataset.

Appendices

A. Proof of Proposition 1

We first present the proof for the statistic under the mixed effects model, and then for the statistic under the random effects model.

Proof for $S_{B,\text{mixed}}$. Using relations (15) and (18) we write

$$S_{B,\text{mixed}} = \frac{1}{a(b-1)}\sum_{i=1}^{a}\sum_{j=1}^{b}\sum_{k=1}^{n_{ij}}\left[\left(\beta_j-\overline{\beta}_.\right)+\left(\overline{\epsilon}_{.j.}-\overline{\epsilon}_{...}\right)\right]^2-\text{MSE}^*. \tag{A.1}$$

The conditional expectation of the first part given the vector $\mathbf{T}$ of random column levels is

$$\begin{aligned}
&E\left[\frac{1}{a(b-1)}\sum_{i=1}^{a}\sum_{j=1}^{b}n_{ij}\left[\left(\beta_j-\overline{\beta}_.\right)+\left(\overline{\epsilon}_{.j.}-\overline{\epsilon}_{...}\right)\right]^2\mid\mathbf{T}\right]\\
&=\frac{1}{a(b-1)}\sum_i\sum_j n_{ij}\left[\left(\beta_j-\overline{\beta}_.\right)^2+\left(1-\frac{1}{b}\right)\text{Var}\left(\overline{\epsilon}_{.j.}\mid T_j\right)\right]\\
&=\frac{1}{a(b-1)}\\
&\quad\times\sum_i\sum_j n_{ij}\left[\left(\beta_j-\overline{\beta}_.\right)^2+\left(1-\frac{1}{b}\right)\frac{1}{a^2}\sum_{i_1}\frac{\text{Var}\left(\epsilon_{i_1jk}\mid T_j\right)}{n_{i_1j}}\right].
\end{aligned} \tag{A.2}$$

Therefore, under the null hypothesis that $\beta_j=0$, a.s., for all j, the expected value of $S_{B,\text{mixed}}$ is zero conditionally on $\mathbf{T}$, and thus also unconditionally. Moreover, it is clear that under the alternative the expected value of $S_{B,\text{mixed}}$ is positive.

Proof for $S_{B,\text{rand}}$. Using relation (9), we write

$$\begin{aligned}
S_{B,\text{rand}} &= \frac{1}{a^2b}\sum_{i=1}^{a}\sum_{j=1}^{b}(a-1)\\
&\quad\times\left(\left(\beta_j-\overline{\beta}_.\right)+\left(\overline{\gamma}_{.j}-\overline{\gamma}_{..}\right)+\left(\overline{\epsilon}_{.j.}-\overline{\epsilon}_{...}\right)\right)^2\\
&\quad-\frac{1}{a^2b}\sum_{i=1}^{a}\sum_{j=1}^{b}\left(\left(\gamma_{ij}-\overline{\gamma}_{.j}-\overline{\gamma}_{i.}+\overline{\gamma}_{..}\right)\right.\\
&\qquad\left.+\left(\overline{\epsilon}_{ij.}-\overline{\epsilon}_{.j.}-\overline{\epsilon}_{i..}+\overline{\epsilon}_{...}\right)\right)^2.
\end{aligned} \tag{A.3}$$

Thus, the proof of this part of the proposition will follow if we show

$$E\left[\frac{1}{a^2b}\sum_{i=1}^{a}\sum_{j=1}^{b}(a-1)\left(\overline{\gamma}_{.j}-\overline{\gamma}_{..}\right)^2\right] = E\left[\frac{1}{a^2b}\sum_{i=1}^{a}\sum_{j=1}^{b}\left(\gamma_{ij}-\overline{\gamma}_{.j}-\overline{\gamma}_{i.}+\overline{\gamma}_{..}\right)^2\right], \tag{A.4}$$

$$E\left[\frac{1}{a^2b}\sum_{i=1}^{a}\sum_{j=1}^{b}(a-1)\left(\overline{\epsilon}_{.j.}-\overline{\epsilon}_{...}\right)^2\right] = E\left[\frac{1}{a^2b}\sum_{i=1}^{a}\sum_{j=1}^{b}\left(\overline{\epsilon}_{ij.}-\overline{\epsilon}_{.j.}-\overline{\epsilon}_{i..}+\overline{\epsilon}_{...}\right)^2\right]. \tag{A.5}$$

Relation (A.4) follows from

$$\begin{aligned}
E\left[(a-1)\left(\overline{\gamma}_{.j}-\overline{\gamma}_{..}\right)^2\right] &= E\left\{E\left[(a-1)\left(\overline{\gamma}_{.j}-\overline{\gamma}_{..}\right)^2\mid\mathbf{T}\right]\right\}\\
&=\frac{(a-1)(b-1)}{ab}E\left[\text{Var}\left(\gamma_{ij}\mid T_j\right)\right],\\
E\left[\left(\gamma_{ij}-\overline{\gamma}_{.j}-\overline{\gamma}_{i.}+\overline{\gamma}_{..}\right)^2\right]&\\
&=E\left[\left(\gamma_{ij}-\overline{\gamma}_{i.}\right)^2-2\left(\gamma_{ij}-\overline{\gamma}_{i.}\right)\left(\overline{\gamma}_{.j}-\overline{\gamma}_{..}\right)+\left(\overline{\gamma}_{.j}-\overline{\gamma}_{..}\right)^2\right]\\
&=E\left[\left(1-\frac{2}{a}\right)\left(\gamma_{ij}-\overline{\gamma}_{i.}\right)^2+\frac{1}{a^2}\sum_{i_1}\left(\gamma_{i_1j}-\overline{\gamma}_{i_1.}\right)^2\right]\\
&=\frac{a-1}{a}E\left[\left(\gamma_{ij}-\overline{\gamma}_{i.}\right)^2\right]=\frac{(a-1)(b-1)}{ab}E\left[\text{Var}\left(\gamma_{ij}\mid T_j\right)\right],
\end{aligned} \tag{A.6}$$

where the second equality follows from the fact that $E(\gamma_{i_1j}-\overline{\gamma}_{i_1.})(\gamma_{i_2j}-\overline{\gamma}_{i_2.})=0$ for $i_1\neq i_2$. Relation (A.5) follows similarly.

B. Proof of Proposition 2

We first show the proof for the statistic under the mixed effects model, and then for the statistic under the random effects model.

Proof for $S_{B,\text{mixed}}$. Because we have shown above in the proof of Proposition 1 that, under $H_0(B)$, the statistic has mean value zero, it suffices to show that with data generated as in Section 2 and effects defined in (18), the following convergence holds:

$$\frac{1}{a^2\sqrt{b}}\sum_{i=1}^{a}\sum_{j=1}^{b}\sum_{k=1}^{n_{ij}}\left[\frac{1}{n_{ij}-1}\left(\epsilon_{ijk}-\bar{\epsilon}_{ij\cdot}\right)^2 - E\left(\frac{1}{n_{ij}-1}\left(\epsilon_{ijk}-\bar{\epsilon}_{ij\cdot}\right)^2\right)\right] \longrightarrow 0, \quad (B.1)$$

in probability as $a, b \rightarrow \infty$. Now, first note that the above expression is centered, so we only need to show that its variance tends to zero as a, b tend to ∞. This is easily done using the facts, given in relation (19), that the ϵ terms are uncorrelated for different levels of the fixed effect, independent for different levels of the random effect, and, conditionally on the level of the random effect, iid for $k = 1, \ldots, n_{ij}$. Note that, using these relations,

$$\begin{aligned}\operatorname{Var}&\left[\frac{1}{a^2\sqrt{b}}\sum_{i=1}^{a}\sum_{j=1}^{b}\sum_{k=1}^{n_{ij}}\left[\frac{\left(\epsilon_{ijk}-\bar{\epsilon}_{ij\cdot}\right)^2}{n_{ij}-1} - E\left(\frac{\left(\epsilon_{ijk}-\bar{\epsilon}_{ij\cdot}\right)^2}{n_{ij}-1}\right)\right] \mid \vec{S},\vec{T}\right]\\ &= \frac{1}{a^4 b}\sum_{i=1}^{a}\sum_{j=1}^{b} n_{ij}\operatorname{Var}\left[\frac{1}{n_{ij}-1}\left(\epsilon_{ijk}-\bar{\epsilon}_{ij\cdot}\right)^2 - E\left(\frac{1}{n_{ij}-1}\left(\epsilon_{ijk}-\bar{\epsilon}_{ij\cdot}\right)^2\right) \mid \vec{S},\vec{T}\right].\end{aligned} \quad (B.2)$$

Because we assume a finite fourth moment for the ϵ terms and a finite covariance for any two distinct squared ϵ terms, this conditional variance tends to zero. Therefore, by the Dominated Convergence Theorem, the unconditional variance also tends to zero, thus proving the proposition.

Proof for $S_{B,\mathrm{rand}}$. First, note that after some straightforward algebra we can write

$$\begin{aligned} aS_{B,\mathrm{rand}} &= \frac{1}{ab}\sum_{i}\sum_{i_1\neq i}\sum_{j}\left[\left(\gamma_{ij}-\bar{\gamma}_{i\cdot}\right)\left(\gamma_{i_1 j}-\bar{\gamma}_{i_1\cdot}\right) + \left(\bar{\epsilon}_{ij\cdot}-\bar{\epsilon}_{i\cdot\cdot}\right)\left(\bar{\epsilon}_{i_1 j\cdot}-\bar{\epsilon}_{i_1\cdot\cdot}\right)\right]\\ &+ \frac{1}{ab}\sum_{i}\sum_{i_1\neq i}\sum_{j}\left[\frac{2(a-1)}{a}\left(\gamma_{ij}-\bar{\gamma}_{i\cdot}\right)\left(\bar{\epsilon}_{i_1 j\cdot}-\bar{\epsilon}_{i_1\cdot\cdot}\right) + \frac{2}{ab}\left(\gamma_{i_1 j}-\bar{\gamma}_{i_1\cdot}\right)\left(\bar{\epsilon}_{ij\cdot}-\bar{\epsilon}_{i\cdot\cdot}\right)\right].\end{aligned} \quad (B.3)$$

To complete the proof, we claim that $(1/ab)\sum_i\sum_{i_1\neq i}\sum_j(\bar{\epsilon}_{ij\cdot}-\bar{\epsilon}_{i\cdot\cdot})(\bar{\epsilon}_{i_1 j\cdot}-\bar{\epsilon}_{i_1\cdot\cdot})$ converges to zero after showing that the leading term of this product tends to zero. We proceed similarly for the last two terms in $S_{B,\mathrm{rand}}$. Finally, we show that $(1/ab)\sum_i\sum_{i_1\neq i}\sum_j \gamma_{ij}\bar{\gamma}_{i_1\cdot}$ tends to zero to finish the proof. Note that each of these terms has expected value zero, so to demonstrate these facts, we only need to show that the variances tend to zero. To this end, note that

$$\begin{aligned}
(1)\ \operatorname{Var}&\left(\frac{1}{ab}\sum_{i}\sum_{i_1\neq i}\sum_{j}\bar{\epsilon}_{ij\cdot}\bar{\epsilon}_{i_1 j\cdot}\right)\\
&= \frac{1}{a^2b^2}\sum_{i_1}\sum_{i_2\neq i_1}\sum_{j_1}\sum_{i_3}\sum_{i_4\neq i_3}\sum_{j_2}E\left(\bar{\epsilon}_{i_1 j_1\cdot}\bar{\epsilon}_{i_2 j_1\cdot}\bar{\epsilon}_{i_3 j_2\cdot}\bar{\epsilon}_{i_4 j_2\cdot}\right)\\
&= \frac{1}{a^2b^2}\sum_{i_1}\sum_{i_2\neq i_1}\sum_{j}E\left(\bar{\epsilon}^2_{i_1 j\cdot}\bar{\epsilon}^2_{i_2 j\cdot}\right) \longrightarrow 0,\\
(2)\ \operatorname{Var}&\left(\frac{2(a-1)}{a^2b}\sum_{i}\sum_{i_1\neq i}\sum_{j}\gamma_{ij}\bar{\epsilon}_{i_1 j\cdot}\right)\\
&= \frac{4(a-1)^2}{a^4b^2}\sum_{i_1}\sum_{i_2\neq i_1}\sum_{j_1}\sum_{i_3}\sum_{i_4\neq i_3}\sum_{j_2}E\left(\gamma_{i_1 j_1}\bar{\epsilon}_{i_2 j_1\cdot}\gamma_{i_3 j_2}\bar{\epsilon}_{i_4 j_2\cdot}\right)\\
&= \frac{4(a-1)^2}{a^4b^2}\sum_{i_1}\sum_{i_2\neq i_1}\sum_{j}E\left(\gamma^2_{i_1 j}\bar{\epsilon}^2_{i_2 j\cdot}\right) \longrightarrow 0,\\
(3)\ \operatorname{Var}&\left(\frac{1}{ab}\sum_{i}\sum_{i_1\neq i}\sum_{j}\gamma_{ij}\bar{\gamma}_{i_1\cdot}\right)\\
&= \operatorname{Var}\left(\frac{1}{ab^2}\sum_{i}\sum_{i_1\neq i}\sum_{j}\gamma_{ij}\gamma_{i_1 j}\right) + \operatorname{Var}\left(\frac{1}{ab^2}\sum_{i}\sum_{i_1\neq i}\sum_{j}\sum_{j_1\neq j}\gamma_{ij}\gamma_{i_1 j_1}\right)\\
&= \frac{1}{a^2b^4}\sum_{i_1}\sum_{i_2\neq i_1}\sum_{j_1}\sum_{i_3}\sum_{i_4\neq i_3}\sum_{j_2}E\left(\gamma_{i_1 j_1}\gamma_{i_2 j_1}\gamma_{i_3 j_2}\gamma_{i_4 j_2}\right)\\
&+ \frac{1}{a^2b^4}\times\sum_{i_1}\sum_{i_2\neq i_1}\sum_{j_1}\sum_{j_2\neq j_1}\sum_{i_3}\sum_{i_4\neq i_3}\sum_{j_3}\sum_{j_4\neq j_3}E\left(\gamma_{i_1 j_1}\gamma_{i_2 j_2}\gamma_{i_3 j_3}\gamma_{i_4 j_4}\right)\\
&= \frac{1}{a^2b^4}\sum_{i_1}\sum_{i_2\neq i_1}\sum_{j}E\left(\gamma^2_{i_1 j}\gamma^2_{i_2 j}\right) + \frac{1}{a^2b^4}\sum_{i_1}\sum_{i_2\neq i_1}\sum_{j_1}\sum_{j_2\neq j_1}E\left(\gamma^2_{i_1 j_1}\gamma^2_{i_2 j_2}\right) \longrightarrow 0.
\end{aligned} \quad (B.4)$$

This completes the proof of the proposition.

Acknowledgments

This research was supported in part by NSF Grants SES-0318200 and DMS-0805598. This research was initiated during the second author's sabbatical visit at the Australian National University in the Spring of 2006. Helpful discussions with Peter Hall are gratefully acknowledged.

References

[1] H. Scheffé, *The Analysis of Variance*, John Wiley & Sons, New York, NY, USA, 1959.

[2] S. R. Searle, *Linear Models*, John Wiley & Sons, New York, NY, USA, 1971.

[3] S. F. Arnold, *The Theory of Linear Models and Multivariate Analysis*, John Wiley & Sons, New York, NY, USA, 1981.

[4] S. R. Searle, G. Casella, and C. E. McCulloch, *Variance Components*, John Wiley & Sons, New York, NY, USA, 1992.

[5] A. I. Khuri, T. Mathew, and B. K. Sinha, *Statistical Tests for Mixed Linear Models*, John Wiley & Sons, New York, NY, USA, 1998.

[6] J. Cornfield and J. W. Tukey, "Average values of mean squares in factorials," *Annals of Mathematical Statistics*, vol. 27, pp. 907–949, 1956.

[7] B. Winer, *Statistical Principles in Experimental Design*, McGraw-Hill, New York, NY, USA, 1971.

[8] R. R. Hocking, "A discussion of the two-way mixed model," *The American Statistician*, vol. 27, pp. 148–152, 1973.

[9] T. Gaugler and M. G. Akritas, "Mixed effects designs: the symmetry assumption and missing data," *Journal of the American Statistical Association*, vol. 107, no. 499, pp. 1230–1238, 2012.

[10] T. Gaugler and M. G. Akritas, "Testing for interaction in two-way random and mixed effects models: the fully nonparametric approach," *Biometrics*, vol. 67, no. 4, pp. 1314–1320, 2011.

[11] A. Bathke, "The ANOVA *F* test can still be used in some balanced designs with unequal variances and nonnormal data," *Journal of Statistical Planning and Inference*, vol. 126, no. 2, pp. 413–422, 2004.

[12] A. K. Gupta, S. W. Harrar, and Y. Fujikoshi, "Asymptotics for testing hypothesis in some multivariate variance components model under non-normality," *Journal of Multivariate Analysis*, vol. 97, no. 1, pp. 148–178, 2006.

[13] M. G. Akritas and N. Papadatos, "Heteroscedastic one-way ANOVA and lack-of-fit tests," *Journal of the American Statistical Association*, vol. 99, no. 466, pp. 368–382, 2004.

[14] S. R. Searle, *Linear Models for Unbalanced Data*, John Wiley & Sons, New York, NY, USA, 1987.

[15] V. Koltchinskii and E. Giné, "Random matrix approximation of spectra of integral operators," *Bernoulli*, vol. 6, no. 1, pp. 113–167, 2000.

[16] J. Aitchison, *The Statistical Analysis of Compositional Data*, Chapman & Hall, London, UK, 1986.

14

Nonlinear Survival Regression Using Artificial Neural Network

Akbar Biglarian,[1] Enayatollah Bakhshi,[1] Ahmad Reza Baghestani,[2] Mahmood Reza Gohari,[3] Mehdi Rahgozar,[1] and Masoud Karimloo[1]

[1] *Department of Biostatistics, University of Social Welfare and Rehabilitation Sciences (USWRS), Tehran 1985713834, Iran*
[2] *Department of Biostatistics, Faculty of Paramedical Sciences, Shahid Beheshti University of Medical Sciences, Tehran 1971653313, Iran*
[3] *Hospital Management Research Center, Tehran University of Medical Sciences (TUMS), Tehran 1996713883, Iran*

Correspondence should be addressed to Akbar Biglarian; abiglarian@gmail.com

Academic Editor: Shein-chung Chow

Survival analysis methods deal with a type of data, which is waiting time till occurrence of an event. One common method to analyze this sort of data is Cox regression. Sometimes, the underlying assumptions of the model are not true, such as nonproportionality for the Cox model. In model building, choosing an appropriate model depends on complexity and the characteristics of the data that effect the appropriateness of the model. One strategy, which is used nowadays frequently, is artificial neural network (ANN) model which needs a minimal assumption. This study aimed to compare predictions of the ANN and Cox models by simulated data sets, which the average censoring rate were considered 20% to 80% in both simple and complex model. All simulations and comparisons were performed by R 2.14.1.

1. Introduction

Many different parametric, nonparametric, and semiparametric regression methods are increasingly examined to explore the relationship between a response variable and a set of covariates. The choice of an appropriate method for modeling depends on the methodology of the survey and the nature of the outcome and explanatory variables.

A common research question in medical research is to determine whether a set of covariates are correlated with the survival or failure times. Two major characteristics of survival data are censoring and violation of normal assumption for ordinary least squares multiple regressions. These two characteristics of time variable are reasons that straightforward multiple regression techniques cannot be used. Different parametric and semiparametric models in survival regression were introduced which model survival or hazard function. Parametric models, for instance, exponential or weibull, predict survival function while accelerated failure time models are parametric regression methods with logarithm failure time as dependent variable [1, 2].

Choosing an appropriate model for the analysis of the survival data depends on some conditions which are called the underlying assumptions of the model. Sometimes, these assumptions may not be true, for example: (a) lack of independence between consequent waiting times to occurrence of an event or nonproportionality of hazards in semiparametric models, (b) lack of independency of censoring or the distribution of failure times in the case of parametric models [1–3].

Although, the Cox regression model is an efficient strategy in analyzing survival data, but when the assumptions of this model are fail, the free assumption methods could be suitable.

Artificial neural network (ANN) models, which are completely nonparametric, have been used increasingly in different areas of sciences. Although analyzing the data using ANN methodology is usually more complex than traditional approaches, ANN models are more flexible and efficient when our main aim is prediction or classification of an outcome using different explanatory variables [4–17].

Note that when several covariates and complex interactions are of concern the best method is ANN; otherwise, based on model assumptions simple regression models can be appropriately used.

In this study, simulated data sets with different rates of censoring were used to predict the outcome using ANN and traditional Cox regression models, and then the results of predictions were compared.

2. Methods

2.1. Cox Regression Model. Suppose that T denotes a continuous nonnegative random variable describing the failure time of an event (i.e., time-to-event) in a system. The probability density function of t; that is, the actual survival time is $f(t)$. The survival function, $S(t)$, is probability that the failure occurs later than time t. The related hazard function, $h(t)$, denotes the probability density of an event occurring around time t, given that it has not occurred prior to time t.

As we know, an inherent characteristic of survival data is censoring. Right censored data which is the commonest form of censoring occurs when survival times are greater than some defined time point [1, 2]. The generated data used in this study contains right-censored data.

Proportional hazards model, which also called Cox regression, is a popular method in analysis of survival data. This model is presented as

$$h(t \mid x) = h_0(t)\exp\left(\boldsymbol{\beta}'\mathbf{x}\right). \tag{1}$$

In this model, $\boldsymbol{\beta}$ is the regression coefficients vector, $\mathbf{x}$ is a vector of covariates, and $h_0(t)$ is the baseline hazard which is unspecified and a function of time. The likelihood and partial likelihood function for right censored survival data [18], given by (2) and (3):

$$\begin{aligned} L &= \prod_{i=1}^{N} f(t_i, x_i)^{\delta_i} S(t_i, x_i)^{1-\delta_i} \\ &= \prod_{i=1}^{N} \left(\frac{f(t_i, x_i)}{S(t_i, x_i)}\right)^{\delta_i} S(t_i, x_i) \\ &= \prod_{i=1}^{N} h(t_i, x_i)^{\delta_i} S(t_i, x_i), \end{aligned} \tag{2}$$

$$L = \prod_{i=1}^{D} \frac{\exp\left(\sum_{k=1}^{p} \beta_k X_{(i)k}\right)}{\sum_{j\in R(t_i)} \exp\left(\sum_{k=1}^{p} \beta_k X_{(i)k}\right)}, \tag{3}$$

δ_i, is censoring status where $\delta_i = 1$ if the observation is complete and $\delta_i = 0$ if it is censored. $R(t_i)$ is defined as the risk set at time t_i.

To fit the Cox regression and estimate $\boldsymbol{\beta}$, the partial likelihood in (3) is maximized using iteratively reweighted least squares to implement the Newton-Raphson method. However, in the high-dimensional case, this approach cannot be used to estimate $\boldsymbol{\beta}$; even $\boldsymbol{\beta}$ is not unique.

2.2. Neural Networks Model. An ANN consists of several layers. Layers are interconnected group of artificial neurons. In addition, each layer has a weight that indicates the amount of the effect of neurons on each other. Usually an ANN model has three layers that called input, hidden (middle), and output. The input layer contains the predictors. The hidden layer contains unobservable nodes and applies a nonlinear transformation to the linear combination of input layer. The value of each hidden node is a function of the predictors. The output layer contains the outcome which is some functions of the hidden units. In hidden and output layers, the exact form of the function depends on the network type and user definition (based on response variable).

There are different methods for learning in the NN. For example, in multiple layers perceptron (MLP), which is the most commonly used, the learning performs with minimization of the mean square error of the output and by back-propagation learning algorithm [16, 19, 20].

In this paper, we use the activation transfer function (g_h) as sigmoid function in hidden and in output layers (g_o). The ith response y_i for the predictor values H_{ih} is a nonlinear function as

$$\begin{aligned} y_i &= g_o\left(\mathbf{H}_i'\boldsymbol{\beta}\right) + \varepsilon_i, \\ H_{ih} &= g_h\left(\mathbf{X}_i'\boldsymbol{\alpha}_h\right). \end{aligned} \tag{4}$$

Note that $\mathbf{X}_i'$ is ith row of the input data matrix $\mathbf{X}$, H_{ih} is a nonlinear function of linear combination of input data, $\boldsymbol{\beta}$ is the vector weights of the hidden to the output units, and $\boldsymbol{\alpha}$ is the matrix weights of the input to the hidden units. Equation (4) together yield the MLP model:

$$y_i = g_o\left(\mathbf{H}_i'\boldsymbol{\beta}\right) + \varepsilon_i = g_o\left[\beta_0 + \sum_{h=1}^{H-1} \beta_h g_h\left(\mathbf{X}_i'\boldsymbol{\alpha}_h\right)\right] + \varepsilon_i. \tag{5}$$

By the sigmoid activation function, (5) can be written as below which is a nonlinear regression:

$$\begin{aligned} y_i &= \left[1 + \exp\left(-\mathbf{H}_i'\boldsymbol{\beta}\right)\right]^{-1} + \varepsilon_i \\ &= \left[1 + \exp\left(-\beta_0 - \sum_{h=1}^{H-1} \beta_h\left[1 + \exp\left(-\mathbf{X}_i'\boldsymbol{\alpha}_j\right)\right]^{-1}\right)\right]^{-1} + \varepsilon_i \\ &= g\left(\mathbf{X}_i, \boldsymbol{\beta}, \boldsymbol{\alpha}_1, \boldsymbol{\alpha}_2, \dots, \boldsymbol{\alpha}_H\right) + \varepsilon_i, \end{aligned} \tag{6}$$

where: $\boldsymbol{\beta}, \boldsymbol{\alpha}_1, \boldsymbol{\alpha}_2, \dots, \boldsymbol{\alpha}_H$ are unknown parameter vectors, $\mathbf{X}_i$ is a vector of known constants, and ε_i are residuals. The parameters (weights) can be estimated by optimizing some criterion function such as maximizing the log-likelihood function or minimizing the sum of squared errors.

In an MLP framework, a serious problem is overfitting. To control of the overfitting, usually a penalty term is added to the optimization criterion. To this, penalized least squares criterion for parameter estimation is given by [1]:

$$\begin{aligned} E^* &= \sum_{i=1}^{n} (y_i - \hat{y}_i)^2 + p_\lambda\left(\boldsymbol{\beta}, \boldsymbol{\alpha}_1, \boldsymbol{\alpha}_2, \dots, \boldsymbol{\alpha}_H\right) \\ &= \sum_{i=1}^{n} \left(y_i - g\left(\mathbf{X}_i, \boldsymbol{\beta}, \boldsymbol{\alpha}_1, \boldsymbol{\alpha}_2, \dots, \boldsymbol{\alpha}_H\right)\right)^2 \\ &\quad + p_\lambda\left(\boldsymbol{\beta}, \boldsymbol{\alpha}_1, \boldsymbol{\alpha}_2, \dots, \boldsymbol{\alpha}_H\right), \end{aligned} \tag{7}$$

where the penalty term is $p_\lambda(\boldsymbol{\beta}, \boldsymbol{\alpha}_1, \boldsymbol{\alpha}_2, \ldots, \boldsymbol{\alpha}_H) = \lambda(\sum \beta_i^2 + \sum \alpha_{ij}^2)$.

In likelihood schema which is often used in shrinkage method, an adaption of (7) is [8, 21]

$$L = -\log \text{likelihood} + \lambda\left(\sum \beta_i^2 + \sum \alpha_{ij}^2\right). \quad (8)$$

The penalty weight λ regulates between over- and underfitting. A best value of λ is between 0.001 and 0.1 and is chosen by cross-validation [1, 8]. In this paper, we use (8) to get the parameter estimated. It is mentioned that, for an outcome (the response variable) with two classes $\delta = (0/1)$, p_i is probability of event for the ith patient, and the error function provides the cross-entropy error function as

$$E = -\sum \{\delta_i \log(\widehat{p}_i) + (1 - \delta_i)\log(1 - \widehat{p}_i)\}. \quad (9)$$

An ANN can be modeled as a generalized linear modeling with nonlinear predictors [8–11]. Bignazoli et al. [8] introduced a method called partial logistic ANN, and Lisboa [22] developed it with fit smooth estimates of the discrete time hazard in structure. It is similar to MLP [23] with additional covariate, namely, time as an input and given by

$$\frac{h_p(x_p, t_k)}{1 - h_p(x_p, t_k)} = \exp\left(\sum_{h=1}^{N_h} w_h g\left(\sum_{i=1}^{N_i} w_{ih}x_{pi} + w_k t_k + b_h\right) + b\right), \quad (10)$$

where N_i and N_h denote the number of the input and the hidden nodes, respectively. b_h and b denote bias term in the hidden and output layers, respectively. After the estimation of the network weights, $\mathbf{w}$, a single output node estimates conditional failure probability values from the connections with the middle units, and the survivorship is calculated from the estimated discrete time hazard by multiplying the conditionals for survival over time interval. Then −log(likelihood) statistics could be obtained as [8]

$$E = -\sum_{p=1}^{\text{no of patient}} \sum_{k=1}^{t_1} \left[\delta_{pk} \log\left(h_p(x_p, t_k)\right) + \left(1 - \delta_{pk}\right)\log\left(1 - h_p(x_p, t_k)\right)\right], \quad (11)$$

where δ is the censoring indicator function.

2.3. Model Fitting. The ultimate goal of the learning process is to minimize the error by net. In training step, to fit the model by a fixed number of hidden nodes, we use penalized likelihood as

$$E^* = E + \lambda\left(\sum \beta_i^2 + \sum \alpha_{ij}^2\right). \quad (12)$$

By using this, we improve the convergence of the optimization and also control overfitting problem [1, 8, 9, 16, 21].

To identify the number of the hidden nodes and then model selection, Bayesian Information Criterion (BIC) and Network Information Criterion (NIC) [8, 23, 24], that is, generalization of Akaike Information Criterion (AIC), are calculated:

$$\text{BIC} = -2 \times \log \text{likelihood} + \log(N) \times P,$$
$$\text{NIC} = 2E^* + 2P, \quad (13)$$

where P is the number of the parameters estimated, and N is the number of observations in training set. The best model is with the smallest value of these criterions. In addition, to assess prediction accuracy in validation (testing) group, we calculated classification accuracy and mean square error (MSE).

The best model is selected with the smallest value of MSE. The models considered 2, 3, 4, 5, 10, 15, and 20 hidden nodes. The weight decay was considered 0.012 which is chosen based on some empirical study [25].

At finally, in order to comparison of the Cox and ANN predictions, classification accuracy and concordance indexes were calculated. All simulations and comparisons were performed by R 2.14.1.

3. Simulation

In order to compare the accuracy of the predictions by ANN and Cox regression, four different simulation schemes, based on Monte Carlo simulation, were used. In each schema, hazard at any time t was considered as exponential form [26], namely, λ (Table 1). For each schema, 1,000 independent random observations were generated and then with the based on the relationship between exponential parameter and independent variables, survival times were generated. Afterward, the survival times were transformed as right censored. In this context, if generated time t_i is greater than the quantile of exponential function with parameter of λ, it is considered as censorship. This process was repeated 100 times. To access the accuracy of predictions, each sample is randomly divided to two parts. The first part, the training group, was consisting of 700 observations, and the 300 remainder observations were allocated to second group, that is, the testing group. Furthermore, in all simulation, the average rates of censorship were considered equal to 20%, 30%, 40%, 50%, 60%, 70%, and 80%. In addition, the models were considered with the main effects and without/with any interaction terms as simple and complex model, respectively.

In simulation 1 and 2, two covariates were used which was generated randomly from binomial and standard normal distributions. The models of these simulations were consisting of any and one interaction terms. In simulation 3, three covariates were used which was generated randomly from binomial and standard normal distributions, respectively. The model of this simulation has had two interaction terms. In simulation 4, four covariates were used which was generated randomly from binomial and standard normal distributions. The models of these simulations are complex and consist of two-, three- and four-interaction terms (Table 1).

TABLE 1: Describing of models* and its censoring status.

Model	Censoring status
Model (I): $\lambda = \exp(X_1 + 0.2X_2), X_1 \sim \text{Ber}(0.5), X_2 \sim N(0,1)$	Average censoring for training and testing sets equal to 20%, 30%, 40%, 50%, 60%, 70%, and 80%
Model (II): $\lambda = \exp(X_1 + 0.2X_2 + 0.2X_1X_2), X_1 \sim \text{Ber}(0.5), X_2 \sim N(0,1)$	Average censoring for training and testing sets equal to 20%, 30%, 40%, 50%, 60%, 70%, and 80%
Model (III): $\lambda = \exp(X_1 + X_2 + 0.5X_3 + 0.5X_1X_1 + 0.25X_1X_3 + 0.25X_1X_3)$, $X_1 \sim \text{Ber}(0.25), X_2 \sim \text{Ber}(0.5), X_3 \sim N(0,1)$	Average censoring for training and testing sets equal to 20%, 30%, 40%, 50%, 60%, 70%, and 80%
Model (IV): $\lambda = \exp[2(X_1 + X_2) + 1(X_3) + 0.5(X_4) + 1.0(X_1X_2 + X_1X_3 + X_1X_4)$ $+ 0.5(X_2X_3 + X_2X_4 + X_3X_4) + 0.25(X_1X_2X_3 + X_1X_2X_4 + X_1X_3X_4 + X_2X_3X_4)$ $+ 0.5(X_1X_2X_3X_4)], X_1 \sim \text{Ber}(0.25), X_2 \sim \text{Ber}(0.5), X_3 \sim N(0,1), X_4 \sim N(0,1)$,	Average censoring for training and testing sets equal to 20%, 30%, 40%, 50%, 60%, 70%, and 80%

*The hazard at any time t was considered as exponential form, and then the survival time was generated.

Table 2: Results of ANN models for simulated data.

# of H Model*	MLP (H = 2) BIC**	MLP (H = 2) MSE***	MLP (H = 3) BIC	MLP (H = 3) MSE	MLP (H = 4) BIC	MLP (H = 4) MSE	MLP (H = 5) BIC	MLP (H = 5) MSE	MLP (H = 10) BIC	MLP (H = 10) MSE	MLP (H = 15) BIC	MLP (H = 15) MSE	MLP (H = 20) BIC	MLP (H = 20) MSE
I:														
20%	0.213	1.055	0.203	1.148	0.205	1.252	0.214	1.302	0.223	1.405	0.254	1.387	0.278	2.235
30%	0.213	1.145	0.212	1.149	0.210	1.241	0.211	1.249	0.219	1.411	0.239	1.422	0.269	2.222
40%	0.212	1.134	0.217	1.163	0.209	1.321	0.209	1.392	0.228	1.433	0.249	1.473	0.283	2.314
50%	0.227	1.243	0.209	1.187	0.234	1.449	0.203	1.218	0.231	1.429	0.251	1.521	0.279	2.543
60%	0.256	1.289	0.217	1.251	0.239	1.443	0.211	1.325	0.241	1.511	0.262	1.611	0.281	2.468
70%	0.281	1.259	0.226	1.263	0.259	1.398	0.232	1.281	0.224	1.534	0.273	1.597	0.280	2.413
80%	0.278	1.337	0.232	1.258	0.266	1.527	0.223	1.216	0.249	1.617	0.261	2.018	0.266	2.731
II:														
20%	0.229	1.255	0.223	1.248	0.235	1.452	0.234	1.402	0.323	2.005	0.264	1.987	0.269	2.315
30%	0.254	1.431	0.282	1.428	0.235	1.665	0.243	1.723	0.259	1.835	0.263	2.650	0.276	2.123
40%	0.231	1.233	0.218	1.219	0.221	1.332	0.220	1.381	0.249	1.531	0.257	1.583	0.291	2.411
50%	0.237	1.342	0.222	1.317	0.241	1.539	0.213	1.318	0.251	1.689	0.266	1.591	0.280	2.561
60%	0.256	1.381	0.245	1.359	0.248	1.471	0.232	1.425	0.255	1.533	0.282	1.635	0.279	2.471
70%	0.281	1.259	0.236	1.263	0.263	1.411	0.223	1.279	0.263	1.576	0.291	1.608	0.291	2.409
80%	0.278	1.337	0.255	1.358	0.287	1.597	0.269	1.221	0.247	1.523	0.287	2.107	0.273	2.729
III:														
20%	0.283	1.261	0.235	1.274	0.250	1.408	0.241	1.299	0.240	1.539	0.277	1.607	0.291	2.433
30%	0.269	1.344	0.241	1.262	0.260	1.532	0.249	1.284	0.252	1.623	0.258	2.208	0.269	2.751
40%	0.367	1.204	0.259	1.888	0. 202	1.883	0.202	2.329	0.247	2.428	0.222	2.222	0.264	1.895
50%	0.241	1.358	0.232	1.321	0.252	1.541	0.220	1.327	0.263	1.699	0.277	1.603	0.288	2.593
60%	0.261	1.421	0.255	1.362	0.259	1.485	0.244	1.432	0.274	1.543	0.295	1.688	0.280	2.528
70%	0.289	1.371	0.267	1.274	0.271	1.422	0.253	1.269	0.292	1.552	0.286	1.662	0.303	2.687
80%	0.286	1.345	0.273	1.366	0.281	1.622	0.265	1.482	0.295	1.538	0.302	2.134	0.333	2.869
IV:														
20%	0.243	1.388	0.269	1.422	0.239	1.365	0.241	1.350	0.270	1.708	0.288	1.694	0.293	2.585
30%	0.267	1.435	0.259	1.471	0.230	1.488	0.237	1.499	0.280	1.638	0.285	1.679	0.311	2.569
40%	0.290	1.387	0.273	1.365	0.282	1.433	0.268	1.328	0.288	1.595	0.296	1.675	0.298	2.701
50%	0.292	1.667	0.281	1.621	0.283	1.560	0.271	1.634	0.279	1.807	0.288	1.632	0.302	2.325
60%	0.298	1.832	0.204	1.834	0.199	2.386	0.161	2.912	0.158	2.126	0.159	2.325	0.173	2.667
70%	0.238	2.181	0.239	2.174	0.247	2.392	0.237	2.403	0. 253	2.359	0.233	2.249	0.256	2.411
80%	0.300	2.087	0.301	1.902	0.321	2.099	0.292	2.077	0.295	2.364	0.289	2.630	0.322	2.716

*Four models with different rates of censoring.
**Average of BIC for learning set (700 cases with 100 replications).
***Average of MSE for testing set (300 cases with 100 replications).

TABLE 3: Results of concordance indexes of simulation study in testing subset (300 cases with 100 replications).

Model*	ANN	Cox Reg.
I:		
20%	0.821 ± 0.028	0.820 ± 0.031
30%	0.819 ± 0.027	0.820 ± 0.030
40%	0.824 ± 0.028	0.823 ± 0.030
50%	0.823 ± 0.027	0.823 ± 0.029
60%	0.823 ± 0.027	0.822 ± 0.030
70%	0.825 ± 0.027	0.822 ± 0.032
80%	0.826 ± 0.030	0.823 ± 0.031
II:		
20%	0.818 ± 0.029	0.817 ± 0.031
30%	0.817 ± 0.034	0.815 ± 0.046
40%	0.818 ± 0.032	0.818 ± 0.031
50%	0.816 ± 0.035	0.814 ± 0.032
60%	0.816 ± 0.031	0.813 ± 0.032
70%	0.814 ± 0.033	0.811 ± 0.031
80%	0.812 ± 0.034	0.808 ± 0.036
III:		
20%	0.808 ± 0.030	0.798 ± 0.035
30%	0.799 ± 0.033	0.790 ± 0.033
40%	0.812 ± 0.043	0.795 ± 0.038
50%	0.805 ± 0.028	0.795 ± 0.030
60%	0.804 ± 0.028	0.790 ± 0.033
70%	0.805 ± 0.028	0.791 ± 0.034
80%	0.803 ± 0.028	0.790 ± 0.032
IV:		
20%	0.764 ± 0.033	0.759 ± 0.036
30%	0.773 ± 0.029	0.762 ± 0.033
40%	0.777 ± 0.030	0.759 ± 0.035
50%	0.779 ± 0.035	0.764 ± 0.034
60%	0.812 ± 0.039	0.790 ± 0.036
70%	0.764 ± 0.036	0.744 ± 0.035
80%	0.766 ± 0.035	0.741 ± 0.037

*Four models with different rates of censoring.

The model selection is based on BIC for learning set and SSE criterion for the testing subset data as a verification. The results in Table 2 show that the simple model performs with less hidden node but complex model performs better with more hidden nodes. The MSE values confirm these results (Table 2).

In the next step, to compare of ANN and Cox regression predictions, concordance indexes were calculated from classification accuracy table in testing subset. Concordance index was reported as a generalization of the area under receiver operating characteristic curve for censored data [27, 28]. This index means that the proportion of the cases that are classified correctly in noncensored (event) and censored groups and 0 to 1 values indicated as the ability of the models accuracy. The concordance index of ANN and Cox regression models was reported in Table 3. The results of simulation study in simpler model showed that there was not any different between the predictions of Cox regression and NN models. But NN predictions were better than Cox regression predictions in complex model with high rates of censoring.

4. Conclusion

In this paper, we presented two approaches for modeling of survival data with different degrees of censoring: Cox regression and neural network models. A Monte-Carlo simulation study was performed to compare predictive accuracy of Cox and neural network models in simulation data sets.

In the simulation study, four different models were considered. The rate of censorship in each of these models was considered from 20% up 80%. These models were considered with the main effects and also with the interaction terms. Then the ability of these models in prediction was evaluated. As was seen, in simple models and with less censored cases, there was little difference in ANN and Cox regression models predictions. It seems that for simpler models, the levels of censorship have no effect on predictions, but the predictions in more complex models depend on the levels of censorship. The results showed that the NN model for more complex models was provided better predictions. But for simpler models predictions there was not any different in results. This result was consistent with the finding from Xiang's study [26]. Therefore, NN model is proposed in two cases of: (1) occurrence of high censorship (i.e., censoring rate of 60% and higher) and/or (2) in the complex models (i.e., with many covariates and any interaction terms). This is a very good result and can be used in practical issues which often are faced on with many numbers of variables and also many cases of censorship. For that reason, in these two cases the ANN strategy can be used as an alternate of traditional Cox model. Finally, it is mentioned that there are some flexible alternative methods, such as piecewise exponential and grouped time models which can be used for survival data and then its ability compared with ANN model.

Acknowledgment

The authors wish to express their special thanks to referees for their valuable comments.

References

[1] E. T. Lee and J. W. Wang, *Statistical Methods for Survival Data Analysis*, Wiley Series in Probability and Statistics, Wiley-Interscience, Hoboken, NJ, USA, 3rd edition, 2003.

[2] V. Lagani and I. Tsamardinos, "Structure-based variable selection for survival data," *Bioinformatics*, vol. 26, no. 15, pp. 1887–1894, 2010.

[3] M. H. Kutner, C. J. Nachtsheim, and J. Neter, *Applied Linear Regression Models*, McGraw-Hill/Irwin, New York, NY, USA, 4th edition, 2004.

[4] W. G. Baxt and J. Skora, "Prospective validation of artificial neural networks trained to identify acute myocardial infarction," *Lancet*, vol. 347, pp. 12–15, 1996.

[5] B. A. Mobley, E. Schecheer, and W. E. Moore, "Prediction of coronary artery stenosis by artificial networks," *Artificial Intelligence in Medicine*, vol. 18, pp. 187–203, 2000.

[6] D. West and V. West, "Model selection for medical diagnostic decision support system: breast cancer detection case," *Artificial Intelligence in Medicine*, vol. 20, pp. 183–204, 2000.

[7] F. Ambrogi, N. Lama, P. Boracchi, and E. Biganzoli, "Selection of artificial neural network models for survival analysis with genetic algorithms," *Computational Statistics & Data Analysis*, vol. 52, no. 1, pp. 30–42, 2007.

[8] E. Biganzoli, P. Boracchi, L. Mariani, and E. Marubini, "Feed forward neural networks for the analysis of censored survival data a partial logistic regression approach," *Statistics in Medicine*, vol. 17, pp. 1169–1186, 1998.

[9] E. Biganzoli, P. Boracchi, and E. Marubini, "A general framework for neural network models on censored survival data," *Neural Networks*, vol. 15, pp. 209–218, 2002.

[10] E. M. Bignazoli and P. Borrachi, "The Partial Logistic Artificial Neural Network (PLANN): a tool for the flexible modelling of censored survival data," in *Proceedings of the European Conference on Emergent Aspects in Clinical Data Analysis (EACDA '05)*, 2005.

[11] E. M. Bignazoli, P. Borrachi, F. Amborgini, and E. Marubini, "Artificial neural network for the joint modeling of discrete cause-specific hazards," *Artificial Intelligence in Medicine*, vol. 37, pp. 119–130, 2006.

[12] R. Bittern, A. Cuschieri, S. D. Dolgobrodov et al., "An artificial neural network for analysing the survival of patients with colorectal cancer," in *Proceedings of the European Symposium on Artificial Neural Networks (ESANN '05)*, Bruges, Belgium, April 2005.

[13] K. U. Chen and C. J. Christian, "Using back-propagation neural network to forecast the production values of the machinery industry in Taiwan," *Journal of American Academy of Business, Cambridge*, vol. 9, no. 1, pp. 183–190, 2006.

[14] C. L. Chia, W. Nick Street, and H. W. William, "Application of artificial neural network-based survival analysis on two breast cancer datasets," in *Proceedings of the American Medical Informatics Association Annual Symposium (AMIA '07)*, pp. 130–134, Chicago, Ill, USA, November 2007.

[15] A. Eleuteri, R. Tagliaferri, and L. Milano, "A novel neural network-based survival analysis model," *Neural Networks*, vol. 16, pp. 855–864, 2003.

[16] B. D. Ripley and R. M. Ripley, "Neural networks as statistical methods in survival analysis," in *Clinical Applications of Artificial Neural Networks*, pp. 237–255, Cambridge University Press, Cambridge, UK, 2001.

[17] B. Warner and M. Manavendra, "Understanding neural networks as statistical tools," *Amstat*, vol. 50, no. 4, pp. 284–293, 1996.

[18] J. P. Klein and M. L. Moeschberger, *Survival Analysis: Techniques for Censored and Truncated Data*, Springer, New York, NY, USA, 2th edition, 2003.

[19] J. W. Kay and D. M. Titterington, *Statistics and Neural Networks*, Oxford University Press, Oxford, UK, 1999.

[20] E. P. Goss and G. S. Vozikis, "Improving health care organizational management through neural network learning," *Health Care Management Science*, vol. 5, pp. 221–227, 2002.

[21] R. M. Ripley, A. L. Harris, and L. Tarassenko, "Non-linear survival analysis using neural networks," *Statistics in Medicine*, vol. 23, pp. 825–842, 2004.

[22] P. J. G. Lisboa, H. Wong, P. Harris, and R. Swindell, "A Bayesian neural network approach for modeling censored data with an application to prognosis after surgery for breast cancer," *Artificial Intelligence in Medicine*, vol. 28, no. 1, pp. 1–25, 2003.

[23] C. M. Bishop, *Neural Networks for Pattern Recognition*, The Clarendon Press Oxford University Press, New York, NY, USA, 1995.

[24] B. D. Ripley, *Pattern Recognition and Neural Networks*, Cambridge University Press, Cambridge, UK, 1996.

[25] N. Fallah, G. U. Hong, K. Mohammad et al., "Nonlinear Poisson regression using neural networks: a simulation study," *Neural Computing and Applications*, vol. 18, no. 8, pp. 939–943, 2009.

[26] A. Xiang, P. Lapuerta, A. Ryutov et al., "Comparison of the performance of neural network methods and Cox regression for censored survival data," *Computational Statistics & Data Analysis*, vol. 34, no. 2, pp. 243–257, 2000.

[27] F. E. Harrell, R. M. Califf, D. B. Pryor et al., "Evaluating the yield of medical tests," *The Journal of the American Medical Association*, vol. 247, pp. 2543–2546, 1982.

[28] F. E. Harrell, K. L. Lee, R. M. Califf et al., "Regression modeling strategies for improved prognostic prediction," *Statistics in Medicine*, vol. 3, pp. 143–152, 1984.

15

Estimation of Extreme Values by the Average Conditional Exceedance Rate Method

A. Naess,[1] O. Gaidai,[2] and O. Karpa[3]

[1] *Department of Mathematical Sciences and CeSOS, Norwegian University of Science and Technology, 7491 Trondheim, Norway*
[2] *Norwegian Marine Technology Research Institute, 7491 Trondheim, Norway*
[3] *Centre for Ships and Ocean Structures (CeSOS), Norwegian University of Science and Technology, 7491 Trondheim, Norway*

Correspondence should be addressed to A. Naess; arvidn@math.ntnu.no

Academic Editor: A. Thavaneswaran

This paper details a method for extreme value prediction on the basis of a sampled time series. The method is specifically designed to account for statistical dependence between the sampled data points in a precise manner. In fact, if properly used, the new method will provide statistical estimates of the exact extreme value distribution provided by the data in most cases of practical interest. It avoids the problem of having to decluster the data to ensure independence, which is a requisite component in the application of, for example, the standard peaks-over-threshold method. The proposed method also targets the use of subasymptotic data to improve prediction accuracy. The method will be demonstrated by application to both synthetic and real data. From a practical point of view, it seems to perform better than the POT and block extremes methods, and, with an appropriate modification, it is directly applicable to nonstationary time series.

1. Introduction

Extreme value statistics, even in applications, are generally based on asymptotic results. This is done either by assuming that the epochal extremes, for example, yearly extreme wind speeds at a given location, are distributed according to the generalized (asymptotic) extreme value distribution with unknown parameters to be estimated on the basis of the observed data [1, 2]. Or it is assumed that the exceedances above high thresholds follow a generalized (asymptotic) Pareto distribution with parameters that are estimated from the data [1–4]. The major problem with both of these approaches is that the asymptotic extreme value theory itself cannot be used in practice to decide to what extent it is applicable for the observed data. And since the statistical tests to decide this issue are rarely precise enough to completely settle this problem, the assumption that a specific asymptotic extreme value distribution is the appropriate distribution for the observed data is based more or less on faith or convenience.

On the other hand, one can reasonably assume that in most cases long time series obtained from practical measurements do contain values that are large enough to provide useful information about extreme events that are truly asymptotic. This cannot be strictly proved in general, of course, but the accumulated experience indicates that asymptotic extreme value distributions do provide reasonable, if not always very accurate, predictions when based on measured data. This is amply documented in the vast literature on the subject, and good references to this literature are [2, 5, 6]. In an effort to improve on the current situation, we have tried to develop an approach to the extreme value prediction problem that is less restrictive and more flexible than the ones based on asymptotic theory. The approach is based on two separate components which are designed to improve on two important aspects of extreme value prediction based on observed data. The first component has the capability to accurately capture and display the effect of statistical dependence in the data, which opens for the opportunity of using all the available data in the analysis. The second component is then constructed so as to make it possible to incorporate to a certain extent also the subasymptotic part of the data into the estimation of extreme values, which is of some importance for accurate estimation. We have used the proposed method on a wide variety of estimation problems, and our experience is that it represents a very powerful addition to the toolbox of methods for extreme value estimation. Needless to say, what is presented in this paper is by no means considered a closed chapter. It is a novel method, and it is to be expected that several aspects of the proposed approach will see significant improvements.

2. Cascade of Conditioning Approximations

In this section, a sequence of nonparametric distribution functions will be constructed that converges to the exact extreme value distribution for the time series considered. This constitutes the core of the proposed approach.

Consider a stochastic process $Z(t)$, which has been observed over a time interval, $(0, T)$ say. Assume that values $X_1, \ldots, X_N$, which have been derived from the observed process, are allocated to the discrete times $t_1, \ldots, t_N$ in $(0, T)$. This could be simply the observed values of $Z(t)$ at each t_j, $j = 1, \ldots, N$, or it could be average values or peak values over smaller time intervals centered at the t_j's. Our goal in this paper is to accurately determine the distribution function of the extreme value $M_N = \max\{X_j; j = 1, \ldots, N\}$. Specifically, we want to estimate $P(\eta) = \text{Prob}(M_N \le \eta)$ accurately for large values of η. An underlying premise for the development in this paper is that a rational approach to the study of the extreme values of the sampled time series is to consider exceedances of the individual random variables X_j above given thresholds, as in classical extreme value theory. The alternative approach of considering the exceedances by upcrossing of given thresholds by a continuous stochastic process has been developed in [7, 8] along lines similar to that adopted here. The approach taken in the present paper seems to be the appropriate way to deal with the recorded data time series of, for example, the hourly or daily largest wind speeds observed at a given location.

From the definition of $P(\eta)$ it follows that

$$\begin{aligned} P(\eta) &= \text{Prob}(M_N \le \eta) = \text{Prob}\{X_N \le \eta, \ldots, X_1 \le \eta\} \\ &= \text{Prob}\{X_N \le \eta \mid X_{N-1} \le \eta, \ldots, X_1 \le \eta\} \\ &\quad \cdot \text{Prob}\{X_{N-1} \le \eta, \ldots, X_1 \le \eta\} \\ &= \prod_{j=2}^{N} \text{Prob}\{X_j \le \eta \mid X_{j-1} \le \eta, \ldots, X_1 \le \eta\} \\ &\quad \cdot \text{Prob}(X_1 \le \eta). \end{aligned} \tag{1}$$

In general, the variables X_j are statistically dependent. Hence, instead of assuming that all the X_j are statistically independent, which leads to the classical approximation

$$P(\eta) \approx P_1(\eta) := \prod_{j=1}^{N} \text{Prob}(X_j \le \eta), \tag{2}$$

where := means "by definition", the following one-step memory approximation will, to a certain extent, account for the dependence between the X_j's,

$$\begin{aligned} &\text{Prob}\{X_j \le \eta \mid X_{j-1} \le \eta, \ldots, X_1 \le \eta\} \\ &\quad \approx \text{Prob}\{X_j \le \eta \mid X_{j-1} \le \eta\}, \end{aligned} \tag{3}$$

for $2 \le j \le N$. With this approximation, it is obtained that

$$\begin{aligned} P(\eta) &\approx P_2(\eta) \\ &:= \prod_{j=2}^{N} \text{Prob}\{X_j \le \eta \mid X_{j-1} \le \eta\} \, \text{Prob}(X_1 \le \eta). \end{aligned} \tag{4}$$

By conditioning on one more data point, the one-step memory approximation is extended to

$$\begin{aligned} &\text{Prob}\{X_j \le \eta \mid X_{j-1} \le \eta, \ldots, X_1 \le \eta\} \\ &\quad \approx \text{Prob}\{X_j \le \eta \mid X_{j-1} \le \eta, X_{j-2} \le \eta\}, \end{aligned} \tag{5}$$

where $3 \le j \le N$, which leads to the approximation

$$\begin{aligned} P(\eta) \approx P_3(\eta) &:= \prod_{j=3}^{N} \text{Prob}\{X_j \le \eta \mid X_{j-1} \le \eta, X_{j-2} \le \eta\} \\ &\quad \cdot \text{Prob}\{X_2 \le \eta \mid X_1 \le \eta\} \, \text{Prob}(X_1 \le \eta). \end{aligned} \tag{6}$$

For a general k, $2 \le k \le N$, it is obtained that

$$\begin{aligned} P(\eta) &\approx P_k(\eta) \\ &:= \prod_{j=k}^{N} \text{Prob}\{X_j \le \eta \mid X_{j-1} \le \eta, \ldots, X_{j-k+1} \le \eta\} \\ &\quad \cdot \prod_{j=2}^{k-1} \text{Prob}\{X_j \le \eta \mid X_{j-1} \le \eta \ \ldots, X_1 \le \eta\} \\ &\quad \cdot \text{Prob}(X_1 \le \eta), \end{aligned} \tag{7}$$

where $P(\eta) = P_N(\eta)$.

It should be noted that the one-step memory approximation adopted above is not a Markov chain approximation [9–11], nor do the k-step memory approximations lead to kth-order Markov chains [12, 13]. An effort to relinquish the Markov chain assumption to obtain an approximate distribution of clusters of extremes is reported in [14].

It is of interest to have a closer look at the values for $P(\eta)$ obtained by using (7) as compared to (2). Now, (2) can be rewritten in the form

$$P(\eta) \approx P_1(\eta) = \prod_{j=1}^{N} \left(1 - \alpha_{1j}(\eta)\right), \tag{8}$$

where $\alpha_{1j}(\eta) = \text{Prob}\{X_j > \eta\}$, $j = 1, \ldots, N$. Then the approximation based on assuming independent data can be written as

$$P(\eta) \approx F_1(\eta) := \exp\left(-\sum_{j=1}^{N}\alpha_{1j}(\eta)\right), \quad \eta \longrightarrow \infty. \tag{9}$$

Alternatively, (7) can be expressed as,

$$P(\eta) \approx P_k(\eta) = \prod_{j=k}^{N}\left(1-\alpha_{kj}(\eta)\right)\prod_{j=1}^{k-1}\left(1-\alpha_{jj}(\eta)\right), \tag{10}$$

where $\alpha_{kj}(\eta) = \text{Prob}\{X_j > \eta \mid X_{j-1} \le \eta, \ldots, X_{j-k+1} \le \eta\}$, for $j \ge k \ge 2$, denotes the exceedance probability conditional on $k-1$ previous nonexceedances. From (10) it is now obtained that

$$P(\eta) \approx F_k(\eta) := \exp\left(-\sum_{j=k}^{N}\alpha_{kj}(\eta) - \sum_{j=1}^{k-1}\alpha_{jj}(\eta)\right), \quad \eta \longrightarrow \infty, \tag{11}$$

and $F_k(\eta) \to P(\eta)$ as $k \to N$ with $F_N(\eta) = P(\eta)$ for $\eta \to \infty$.

For the cascade of approximations $F_k(\eta)$ to have practical significance, it is implicitly assumed that there is a cut-off value k_c satisfying $k_c \ll N$ such that effectively $F_{k_c}(\eta) = F_N(\eta)$. It may be noted that for k-dependent stationary data sequences, that is, for data where X_i and X_j are independent whenever $|j-i| > k$, then $P(\eta) = P_{k+1}(\eta)$ exactly, and, under rather mild conditions on the joint distributions of the data, $\lim_{N\to\infty} P_1(\eta) = \lim_{N\to\infty} P(\eta)$ [15]. In fact, it can be shown that $\lim_{N\to\infty} P_1(\eta) = \lim_{N\to\infty} P(\eta)$ is true for weaker conditions than k-dependence [16]. However, for finite values of N the picture is much more complex, and purely asymptotic results should be used with some caution. Cartwright [17] used the notion of k-dependence to investigate the effect on extremes of correlation in sea wave data time series.

Returning to (11), extreme value prediction by the conditioning approach described above reduces to estimation of (combinations) of the $\alpha_{kj}(\eta)$ functions. In accordance with the previous assumption about a cut-off value k_c, for all k-values of interest, $k \ll N$, so that $\sum_{j=1}^{k-1}\alpha_{jj}(\eta)$ is effectively negligible compared to $\sum_{j=k}^{N}\alpha_{kj}(\eta)$. Hence, for simplicity, the following approximation is adopted, which is applicable to both stationary and nonstationary data,

$$F_k(\eta) = \exp\left(-\sum_{j=k}^{N}\alpha_{kj}(\eta)\right), \quad k \ge 1. \tag{12}$$

Going back to the definition of $\alpha_{1j}(\eta)$, it follows that $\sum_{j=1}^{N}\alpha_{1j}(\eta)$ is equal to the expected number of exceedances of the threshold η during the time interval $(0, T)$. Equation (9) therefore expresses the approximation that the stream of exceedance events constitute a (nonstationary) Poisson process. This opens for an understanding of (12) by interpreting the expressions $\sum_{j=k}^{N}\alpha_{kj}(\eta)$ as the expected effective number of independent exceedance events provided by conditioning on $k-1$ previous observations.

3. Empirical Estimation of the Average Conditional Exceedance Rates

The concept of average conditional exceedance rate (ACER) of order k is now introduced as follows:

$$\varepsilon_k(\eta) = \frac{1}{N-k+1}\sum_{j=k}^{N}\alpha_{kj}(\eta), \quad k = 1, 2, \ldots. \tag{13}$$

In general, this ACER function also depends on the number of data points N.

In practice, there are typically two scenarios for the underlying process $Z(t)$. Either we may consider to be a stationary process, or, in fact, even an ergodic process. The alternative is to view $Z(t)$ as a process that depends on certain parameters whose variation in time may be modelled as an ergodic process in its own right. For each set of values of the parameters, the premise is that $Z(t)$ can then be modelled as an ergodic process. This would be the scenario that can be used to model long-term statistics [18, 19].

For both these scenarios, the empirical estimation of the ACER function $\varepsilon_k(\eta)$ proceeds in a completely analogous way by counting the total number of favourable incidents, that is, exceedances combined with the requisite number of preceding nonexceedances, for the total data time series and then finally dividing by $N-k+1 \approx N$. This can be shown to apply for the long-term situation.

A few more details on the numerical estimation of $\varepsilon_k(\eta)$ for $k \ge 2$ may be appropriate. We start by introducing the following random functions:

$$\begin{aligned} A_{kj}(\eta) &= \mathbf{1}\left\{X_j > \eta, X_{j-1} \le \eta, \ldots, X_{j-k+1} \le \eta\right\}, \\ &\qquad j = k, \ldots, N, \ k = 2, 3, \ldots, \\ B_{kj}(\eta) &= \mathbf{1}\left\{X_{j-1} \le \eta, \ldots, X_{j-k+1} \le \eta\right\}, \\ &\qquad j = k, \ldots, N, \ k = 2, \ldots, \end{aligned} \tag{14}$$

where $\mathbf{1}\{\mathscr{A}\}$ denotes the indicator function of some event $\mathscr{A}$. Then,

$$\alpha_{kj}(\eta) = \frac{\mathbb{E}\left[A_{kj}(\eta)\right]}{\mathbb{E}\left[B_{kj}(\eta)\right]}, \quad j = k, \ldots, N, \ k = 2, \ldots, \tag{15}$$

where $\mathbb{E}[\cdot]$ denotes the expectation operator. Assuming an ergodic process, then obviously $\varepsilon_k(\eta) = \alpha_{kk}(\eta) = \cdots = \alpha_{kN}(\eta)$, and by replacing ensemble means with corresponding time averages, it may be assumed that for the time series at hand

$$\varepsilon_k(\eta) = \lim_{N\to\infty}\frac{\sum_{j=k}^{N}a_{kj}(\eta)}{\sum_{j=k}^{N}b_{kj}(\eta)}, \tag{16}$$

where $a_{kj}(\eta)$ and $b_{kj}(\eta)$ are the realized values of $A_{kj}(\eta)$ and $B_{kj}(\eta)$, respectively, for the observed time series.

Clearly, $\lim_{\eta\to\infty}\mathbb{E}[B_{kj}(\eta)] = 1$. Hence, $\lim_{\eta\to\infty}\tilde{\varepsilon}_k(\eta)/\varepsilon_k(\eta) = 1$, where

$$\tilde{\varepsilon}_k(\eta) = \frac{\sum_{j=k}^{N}\mathbb{E}\left[A_{kj}(\eta)\right]}{N-k+1}. \tag{17}$$

The advantage of using the modified ACER function $\tilde{\varepsilon}_k(\eta)$ for $k \geq 2$ is that it is easier to use for nonstationary or long-term statistics than $\varepsilon_k(\eta)$. Since our focus is on the values of the ACER functions at the extreme levels, we may use any function that provides correct predictions of the appropriate ACER function at these extreme levels.

To see why (17) may be applicable for nonstationary time series, it is recognized that

$$P(\eta) \approx \exp\left(-\sum_{j=k}^{N} \alpha_{kj}(\eta)\right) = \exp\left(-\sum_{j=k}^{N} \frac{\mathbb{E}\left[A_{kj}(\eta)\right]}{\mathbb{E}\left[B_{kj}(\eta)\right]}\right) \underset{\eta\to\infty}{\simeq} \exp\left(-\sum_{j=k}^{N} \mathbb{E}\left[A_{kj}(\eta)\right]\right). \quad (18)$$

If the time series can be segmented into K blocks, such that $\mathbb{E}[A_{kj}(\eta)]$ remains approximately constant within each block and such that $\sum_{j\in C_i} \mathbb{E}[A_{kj}(\eta)] \approx \sum_{j\in C_i} a_{kj}(\eta)$ for a sufficient range of η-values, where C_i denotes the set of indices for block no. i, $i = 1, \ldots, K$, then $\sum_{j=k}^{N} \mathbb{E}[A_{kj}(\eta)] \approx \sum_{j=k}^{N} a_{kj}(\eta)$. Hence,

$$P(\eta) \approx \exp\left(-(N-k+1)\hat{\varepsilon}_k(\eta)\right), \quad (19)$$

where

$$\hat{\varepsilon}_k(\eta) = \frac{1}{N-k+1}\sum_{j=k}^{N} a_{kj}(\eta). \quad (20)$$

It is of interest to note what events are actually counted for the estimation of the various $\varepsilon_k(\eta)$, $k \geq 2$. Let us start with $\varepsilon_2(\eta)$. It follows from the definition of $\varepsilon_2(\eta)$ that $\varepsilon_2(\eta)\ (N-1)$ can be interpreted as the expected number of exceedances above the level η, satisfying the condition that an exceedance is counted only if it is immediately preceded by a non-exceedance. A reinterpretation of this is that $\varepsilon_2(\eta)\ (N-1)$ equals the average number of clumps of exceedances above η, for the realizations considered, where a clump of exceedances is defined as a maximum number of consecutive exceedances above η. In general, $\varepsilon_k(\eta)\ (N-k+1)$ then equals the average number of clumps of exceedances above η separated by at least $k-1$ nonexceedances. If the time series analysed is obtained by extracting local peak values from a narrow band response process, it is interesting to note the similarity between the ACER approximations and the envelope approximations for extreme value prediction [7, 20]. For alternative statistical approaches to account for the effect of clustering on the extreme value distribution, the reader may consult [21–26]. In these works, the emphasis is on the notion of an extremal index, which characterizes the clumping or clustering tendency of the data and its effect on the extreme value distribution. In the ACER functions, these effects are automatically accounted for.

Now, let us look at the problem of estimating a confidence interval for $\varepsilon_k(\eta)$, assuming a stationary time series. If R realizations of the requisite length of the time series is available, or, if one long realization can be segmented into R subseries, then the sample standard deviation $\hat{s}_k(\eta)$ can be estimated by the standard formula

$$\hat{s}_k(\eta)^2 = \frac{1}{R-1}\sum_{r=1}^{R}\left(\hat{\varepsilon}_k^{(r)}(\eta) - \hat{\varepsilon}_k(\eta)\right)^2, \quad (21)$$

where $\hat{\varepsilon}_k^{(r)}(\eta)$ denotes the ACER function estimate from realization no. r, and $\hat{\varepsilon}_k(\eta) = \sum_{r=1}^{R} \hat{\varepsilon}_k^{(r)}(\eta)/R$.

Assuming that realizations are independent, for a suitable number R, for example, $R \geq 20$, (21) leads to a good approximation of the 95% confidence interval CI $= (C^-(\eta), C^+(\eta))$ for the value $\varepsilon_k(\eta)$, where

$$C^{\pm}(\eta) = \hat{\varepsilon}_k(\eta) \pm \frac{1.96\,\hat{s}_k(\eta)}{\sqrt{R}}. \quad (22)$$

Alternatively, and which also applies to the non-stationary case, it is consistent with the adopted approach to assume that the stream of conditional exceedances over a threshold η constitute a Poisson process, possibly non-homogeneous. Hence, the variance of the estimator $\widehat{E}_k(\eta)$ of $\tilde{\varepsilon}_k(\eta)$, where

$$\widehat{E}_k(\eta) = \frac{\sum_{j=k}^{N} A_{kj}(\eta)}{N-k+1} \quad (23)$$

is $\mathrm{Var}[\widehat{E}_k(\eta)] = \tilde{\varepsilon}_k(\eta)$. Therefore, for high levels η, the approximate limits of a 95% confidence interval of $\tilde{\varepsilon}_k(\eta)$, and also $\varepsilon_k(\eta)$, can be written as

$$C^{\pm}(\eta) = \hat{\varepsilon}_k(\eta)\left(1 \pm \frac{1.96}{\sqrt{(N-k+1)\,\hat{\varepsilon}_k(\eta)}}\right). \quad (24)$$

4. Estimation of Extremes for the Asymptotic Gumbel Case

The second component of the approach to extreme value estimation presented in this paper was originally derived for a time series with an asymptotic extreme value distribution of the Gumbel type, compared with [27]. We have therefore chosen to highlight this case first, also because the extension of the asymptotic distribution to a parametric class of extreme value distribution tails that are capable of capturing to some extent subasymptotic behaviour is more transparent, and perhaps more obvious, for the Gumbel case. The reason behind the efforts to extend the extreme value distributions to the subasymptotic range is the fact that the ACER functions allow us to use not only asymptotic data, which is clearly an advantage since proving that observed extremes are truly asymptotic is really a nontrivial task.

The implication of the asymptotic distribution being of the Gumbel type on the possible subasymptotic functional forms of $\varepsilon_k(\eta)$ cannot easily be decided in any detail. However, using the asymptotic form as a guide, it is assumed that the behaviour of the mean exceedance rate in the tail is dominated by a function of the form $\exp\{-a(\eta-b)^c\}$ $(\eta \geq \eta_1 \geq b)$, where a, b, and c are suitable constants, and η_1 is an appropriately chosen tail marker. Hence, it will be assumed that,

$$\varepsilon_k(\eta) = q_k(\eta)\exp\left\{-a_k(\eta - b_k)^{c_k}\right\}, \quad \eta \ge \eta_1, \tag{25}$$

where the function $q_k(\eta)$ is slowly varying, compared with the exponential function $\exp\{-a_k(\eta - b_k)^{c_k}\}$ and a_k, b_k, and c_k are suitable constants, that in general will be dependent on k. Note that the value $c_k = q_k(\eta) = 1$ corresponds to the asymptotic Gumbel distribution, which is then a special case of the assumed tail behaviour.

From (25) it follows that

$$-\log\left|\log\left(\frac{\varepsilon_k(\eta)}{q_k(\eta)}\right)\right| = -c_k\log(\eta - b_k) - \log(a_k). \tag{26}$$

Therefore, under the assumptions made, a plot of $-\log|\log(\varepsilon_k(\eta)/q_k(\eta))|$ versus $\log(\eta - b_k)$ will exhibit a perfectly linear tail behaviour.

It is realized that if the function $q_k(\eta)$ could be replaced by a constant value, say q_k, one would immediately be in a position to apply a linear extrapolation strategy for deep tail prediction problems. In general, $q_k(\eta)$ is not constant, but its variation in the tail region is often sufficiently slow to allow for its replacement by a constant, possibly by adjusting the tail marker η_1. The proposed statistical approach to the prediction of extreme values is therefore based on the assumption that we can write,

$$\varepsilon_k(\eta) = q_k\exp\left\{-a_k(\eta - b_k)^{c_k}\right\}, \quad \eta \ge \eta_1, \tag{27}$$

where a_k, b_k, c_k, and q_k are appropriately chosen constants. In a certain sense, this is a minimal class of parametric functions that can be used for this purpose which makes it possible to achieve three important goals. Firstly, the parametric class contains the asymptotic form given by $c_k = q_k = 1$ as a special case. Secondly, the class is flexible enough to capture, to a certain extent, subasymptotic behaviour of any extreme value distribution, that is, asymptotically Gumbel. Thirdly, the parametric functions agree with a wide range of known special cases, of which a very important example is the extreme value distribution for a regular stationary Gaussian process, which has $c_k = 2$.

The viability of this approach has been successfully demonstrated by the authors for mean up-crossing rate estimation for extreme value statistics of the response processes related to a wide range of different dynamical systems, compared with [7, 8].

As to the question of finding the parameters a, b, c, q (the subscript k, if it applies, is suppressed), the adopted approach is to determine these parameters by minimizing the following mean square error function, with respect to all four arguments,

$$F(a,b,c,q) = \sum_{j=1}^{J} w_j\left|\log\hat{\varepsilon}(\eta_j) - \log q + a(\eta_j - b)^c\right|^2, \tag{28}$$

where $\eta_1 < \cdots < \eta_J$ denotes the levels where the ACER function has been estimated, w_j denotes a weight factor that puts more emphasis on the more reliably estimated $\hat{\varepsilon}(\eta_j)$. The choice of weight factor is to some extent arbitrary. We have previously used $w_j = (\log C^+(\eta_j) - \log C^-(\eta_j))^{-\theta}$ with $\theta = 1$ and 2, combined with a Levenberg-Marquardt least squares optimization method [28]. This has usually worked well provided reasonable and initial values for the parameters were chosen. Note that the form of w_j puts some restriction on the use of the data. Usually, there is a level η_j beyond which w_j is no longer defined, that is, $C^-(\eta_j)$ becomes negative. Hence, the summation in (28) has to stop before that happens. Also, the data should be preconditioned by establishing the tail marker η_1 based on inspection of the empirical ACER functions.

In general, to improve robustness of results, it is recommended to apply a nonlinearly constrained optimization [29]. The set of constraints is written as

$$\begin{gathered}\log q - a(\eta_i - b)^c \le 0,\\ 0 < q < +\infty,\\ \min_j X_j < b \le \eta_1,\\ 0 < a < +\infty,\\ 0 < c < 5.\end{gathered} \tag{29}$$

Here, the first nonlinear inequality constraint is evident, since under our assumption we have $\hat{\varepsilon}_k(\eta_i) = q\exp\{-a(\eta_i - b)^c\}$, and $\hat{\varepsilon}_k(\eta_i) < 1$ by definition.

A Note of Caution. When the parameter c is equal to 1.0 or close to it, that is, the distribution is close to the Gumbel distribution, the optimization problem becomes ill-defined or close to ill-defined. It is seen that when $c = 1.0$, there is an infinity of (b, q) values that gives exactly the same value of $F(a, b, c, q)$. Hence, there is no well-defined optimum in parameter space. There are simply too many parameters. This problem is alleviated by fixing the q-value, and the obvious choice is $q = 1$.

Although the Levenberg-Marquardt method generally works well with four or, when appropriate, three parameters, we have also developed a more direct and transparent optimization method for the problem at hand. It is realized by scrutinizing (28) that if b and c are fixed, the optimization problem reduces to a standard weighted linear regression problem. That is, with both b and c fixed, the optimal values of a and $\log q$ are found using closed form weighted linear regression formulas in terms of w_j, $y_j = \log\varepsilon(\eta_j)$ and $x_j = (\eta_j - b)^c$. In that light, it can also be concluded that the best linear unbiased estimators (BLUE) are obtained for $w_j = \sigma_{yj}^{-2}$, where $\sigma_{yj}^2 = \mathrm{Var}[y_j]$ (empirical) [30, 31]. Unfortunately, this is not a very practical weight factor for the kind of problem we have here because the summation in (28) then typically would have to stop at undesirably small values of η_j.

It is obtained that the optimal values of a and q are given by the relations

$$a^*(b,c) = -\frac{\sum_{j=1}^{N} w_j \left(x_j - \overline{x}\right)\left(y_j - \overline{y}\right)}{\sum_{j=1}^{N} w_j \left(x_j - \overline{x}\right)^2}, \tag{30}$$

$$\log q^*(b,c) = \overline{y} + a^*(b,c)\,\overline{x},$$

where $\overline{x} = \sum_{j=1}^{N} w_j x_j / \sum_{j=1}^{N} w_j$, with a similar definition of $\overline{y}$.

To calculate the final optimal set of parameters, one may use the Levenberg-Marquardt method on the function $\widetilde{F}(b,c) = F(a^*(b,c), b, c, q^*(b,c))$ to find the optimal values b^* and c^*, and then use (30) to calculate the corresponding a^* and q^*.

For a simple construction of a confidence interval for the predicted, deep tail extreme value given by a particular ACER function as provided by the fitted parametric curve, the empirical confidence band is reanchored to the fitted curve by centering the individual confidence intervals $\mathrm{CI}_{0.95}$ for the point estimates of the ACER function on the fitted curve. Under the premise that the specified class of parametric curves fully describes the behaviour of the ACER functions in the tail, parametric curves are fitted as described above to the boundaries of the reanchored confidence band. These curves are used to determine a first estimate of a 95% confidence interval of the predicted extreme value. To obtain a more precise estimate of the confidence interval, a bootstrapping method would be recommended. A comparison of estimated confidence intervals by both these methods will be presented in the section on extreme value prediction for synthetic data. As a final point, it has been observed that the predicted value is not very sensitive to the choice of η_1, provided it is chosen with some care. This property is easily recognized by looking at the way the optimized fitting is done. If the tail marker is in the appropriate domain of the ACER function, the optimal fitted curve does not change appreciably by moving the tail marker.

5. Estimation of Extremes for the General Case

For independent data in the general case, the ACER function $\varepsilon_1(\eta)$ can be expressed asymptotically as

$$\varepsilon_1(\eta) \underset{\eta\to\infty}{\simeq} \left[1 + \xi\left(a(\eta - b)\right)\right]^{-1/\xi}, \tag{31}$$

where $a > 0$, b, ξ are constants. This follows from the explicit form of the so-called generalized extreme value (GEV) distribution Coles [1].

Again, the implication of this assumption on the possible subasymptotic functional forms of $\varepsilon_k(\eta)$ in the general case is not a trivial matter. The approach we have chosen is to assume that the class of parametric functions needed for the prediction of extreme values for the general case can be modelled on the relation between the Gumbel distribution and the general extreme value distribution. While the extension of the asymptotic Gumbel case to the proposed class of subasymptotic distributions was fairly transparent, this is not equally so for the general case. However, using a similar kind of approximation, the behaviour of the mean exceedance rate in the subasymptotic part of the tail is assumed to follow a function largely of the form $[1 + \xi(a(\eta - b)^c)]^{-1/\xi}$ ($\eta \ge \eta_1 \ge b$), where $a > 0$, b, $c > 0$, and $\xi > 0$ are suitable constants, and η_1 is an appropriately chosen tail level. Hence, it will be assumed that [32]

$$\varepsilon_k(\eta) = q_k(\eta)\left[1 + \xi_k\left(a_k(\eta - b_k)^{c_k}\right)\right]^{-1/\xi_k}, \quad \eta \ge \eta_1, \tag{32}$$

where the function $q_k(\eta)$ is weakly varying, compared with the function $[1 + \xi_k(a_k(\eta - b_k)^{c_k})]^{-1/\xi_k}$ and $a_k > 0$, b_k, $c_k > 0$ and $\xi_k > 0$ are suitable constants, that in general will be dependent on k. Note that the values $c_k = 1$ and $q_k(\eta) = 1$ corresponds to the asymptotic limit, which is then a special case of the general expression given in (25). Another method to account for subasymptotic effects has recently been proposed by Eastoe and Tawn [33], building on ideas developed by Tawn [34], Ledford and Tawn [35] and Heffernan and Tawn [36]. In this approach, the asymptotic form of the marginal distribution of exceedances is kept, but it is modified by a multiplicative factor accounting for the dependence structure of exceedances within a cluster.

An alternative form to (32) would be to assume that

$$\varepsilon_k(\eta) = \left[1 + \xi_k\left(a_k(\eta - b_k)^{c_k} + d_k(\eta)\right)\right]^{-1/\xi_k}, \quad \eta \ge \eta_1, \tag{33}$$

where the function $d_k(\eta)$ is weakly varying compared with the function $a_k(\eta - b_k)^{c_k}$. However, for estimation purposes, it turns out that the form given by (25) is preferable as it leads to simpler estimation procedures. This aspect will be discussed later in the paper.

For practical identification of the ACER functions given by (32), it is expedient to assume that the unknown function $q_k(\eta)$ varies sufficiently slowly to be replaced by a constant. In general, $q_k(\eta)$ is not constant, but its variation in the tail region is assumed to be sufficiently slow to allow for its replacement by a constant. Hence, as in the Gumbel case, it is in effect assumed that $q_k(\eta)$ can be replaced by a constant for $\eta \ge \eta_1$, for an appropriate choice of tail marker η_1. For simplicity of notation, in the following we will suppress the index k on the ACER functions, which will then be written as

$$\varepsilon(\eta) = q\left[1 + \tilde{a}(\eta - b)^c\right]^{-\gamma}, \quad \eta \ge \eta_1, \tag{34}$$

where $\gamma = 1/\xi$, $\tilde{a} = a\xi$.

For the analysis of data, first the tail marker η_1 is provisionally identified from visual inspection of the log plot $(\eta, \ln\hat{\varepsilon}_k(\eta))$. The value chosen for η_1 corresponds to the beginning of regular tail behaviour in a sense to be discussed below.

The optimization process to estimate the parameters is done relative to the log plot, as for the Gumbel case. The mean square error function to be minimized is in the general case written as

$$F(\tilde{a}, b, c, q, \gamma) = \sum_{j=1}^{N} w_j \Big|\log\hat{\varepsilon}(\eta_j) - \log q + \gamma\log\left[1 + \tilde{a}(\eta_j - b)^c\right]\Big|^2, \tag{35}$$

where w_j is a weight factor as previously defined.

An option for estimating the five parameters $\tilde{a}$, b, c, q, γ is again to use the Levenberg-Marquardt least squares optimization method, which can be simplified also in this case by observing that if $\tilde{a}$, b, and c are fixed in (28), the optimization problem reduces to a standard weighted linear regression problem. That is, with $\tilde{a}$, b, and c fixed, the optimal values of γ and $\log q$ are found using closed form weighted linear regression formulas in terms of w_j, $y_j = \log\hat{\varepsilon}(\eta_j)$ and $x_j = 1 + \tilde{a}(\eta_j - b)^c$.

It is obtained that the optimal values of γ and $\log q$ are given by relations similar to (30). To calculate the final optimal set of parameters, the Levenberg-Marquardt method may then be used on the function $\tilde{F}(\tilde{a}, b, c) = F(\tilde{a}, b, c, q^*(\tilde{a}, b, c), \gamma^*(\tilde{a}, b, c))$ to find the optimal values $\tilde{a}^*$, b^*, and c^*, and then the corresponding γ^* and q^* can be calculated. The optimal values of the parameters may, for example, also be found by a sequential quadratic programming (SQP) method [37].

6. The Gumbel Method

To offer a comparison of the predictions obtained by the method proposed in this paper with those obtained by other methods, we will use the predictions given by the two methods that seem to be most favored by practitioners, the Gumbel method and the peaks-over-threshold (POT) method, provided, of course, that the correct asymptotic extreme value distribution is of the Gumbel type.

The Gumbel method is based on recording epochal extreme values and fitting these values to a corresponding Gumbel distribution [38]. By assuming that the recorded extreme value data are Gumbel distributed, then representing the obtained data set of extreme values as a Gumbel probability plot should ideally result in a straight line. In practice, one cannot expect this to happen, but on the premise that the data follow a Gumbel distribution, a straight line can be fitted to the data. Due to its simplicity, a popular method for fitting this straight line is the method of moments, which is also reasonably stable for limited sets of data. That is, writing the Gumbel distribution of the extreme value M_N as

$$\text{Prob}\,(M_N \le \eta) = \exp\{-\exp(-a(\eta - b))\}, \tag{36}$$

it is known that the parameters a and b are related to the mean value m_M and standard deviation σ_M of $M(T)$ as follows: $b = m_M - 0.57722a^{-1}$ and $a = 1.28255/\sigma_M$ [39]. The estimates of m_M and σ_M obtained from the available sample therefore provides estimates of a and b, which leads to the fitted Gumbel distribution by the moment method.

Typically, a specified quantile value of the fitted Gumbel distribution is then extracted and used in a design consideration. To be specific, let us assume that the requested quantile value is the $100(1 - \alpha)\%$ fractile, where α is usually a small number, for example, $\alpha = 0.1$. To quantify the uncertainty associated with the obtained $100(1 - \alpha)\%$ fractile value based on a sample of size $\tilde{N}$, the 95% confidence interval of this value is often used. A good estimate of this confidence interval can be obtained by using a parametric bootstrapping method [40, 41]. Note that the assumption that the initial $\tilde{N}$ extreme values are actually generated with good approximation from a Gumbel distribution cannot easily be verified with any accuracy in general, which is a drawback of this method. Compared with the POT method, the Gumbel method would also seem to use much less of the information available in the data. This may explain why the POT method has become increasingly popular over the past years, but the Gumbel method is still widely used in practice.

7. The Peaks-over-Threshold Method

7.1. The Generalized Pareto Distribution. The POT method for independent data is based on what is called the generalized Pareto (GP) distribution (defined below) in the following manner: it has been shown in [42] that asymptotically the excess values above a high level will follow a GP distribution if and only if the parent distribution belongs to the domain of attraction of one of the extreme value distributions. The assumption of a Poisson process model for the exceedance times combined with GP distributed excesses can be shown to lead to the generalized extreme value (GEV) distribution for corresponding extremes, see below. The expression for the GP distribution is

$$G(y) = G(y; a, c) = \text{Prob}\,(Y \le y) = 1 - \left(1 + c\,\frac{y}{a}\right)_+^{-1/c}. \tag{37}$$

Here $a > 0$ is a scale parameter and c $(-\infty < c < \infty)$ determines the shape of the distribution. $(z)_+ = \max(0, z)$.

The asymptotic result referred to above implies that (37) can be used to represent the conditional cumulative distribution function of the excess $Y = X - u$ of the observed variates X over the threshold u, given that $X > u$ for u is sufficiently large [42]. The cases $c > 0$, $c = 0$, and $c < 0$ correspond to Fréchet (Type II), Gumbel (Type I), and reverse Weibull (Type III) domains of attraction, respectively, compared with section below.

For $c = 0$, which corresponds to the Gumbel extreme value distribution, the expression between the parentheses in (37) is understood in a limiting sense as the exponential distribution

$$G(y) = G(y; a, 0) = \exp\left(-\frac{y}{a}\right). \tag{38}$$

Since the recorded data in practice are rarely independent, a declustering technique is commonly used to filter the data to achieve approximate independence [1, 2].

7.2. Return Periods. The return period R of a given value x_R of X in terms of a specified length of time τ, for example, a year, is defined as the inverse of the probability that the specified value will be exceeded in any time interval of length τ. If λ denotes the mean exceedance rate of the threshold u per length of time τ (i.e., the average number of data points above the threshold u per τ), then the return period R of the value of X corresponding to the level $x_R = u + y$ is given by the relation

$$R = \frac{1}{\lambda \text{Prob}\,(X > x_R)} = \frac{1}{\lambda \text{Prob}\,(Y > y)}. \tag{39}$$

Hence, it follows that

$$\text{Prob}(Y \le y) = 1 - \frac{1}{(\lambda R)}. \tag{40}$$

Invoking (1) for $c \neq 0$ leads to the result

$$x_R = u - \frac{a\left[1 - (\lambda R)^c\right]}{c}. \tag{41}$$

Similarly, for $c = 0$, it is found that,

$$x_R = u + a \ln(\lambda R), \tag{42}$$

where u is the threshold used in the estimation of c and a.

8. Extreme Value Prediction for Synthetic Data

In this section, we illustrate the performance of the ACER method and also the 95% CI estimation. We consider 20 years of synthetic wind speed data, amounting to 2000 data points, which is not much for detailed statistics. However, this case may represent a real situation when nothing but a limited data sample is available. In this case, it is crucial to provide extreme value estimates utilizing all data available. As we will see, the tail extrapolation technique proposed in this paper performs better than asymptotic methods such as POT or Gumbel.

The extreme value statistics will first be analyzed by application to synthetic data for which the exact extreme values can be calculated [43]. In particular, it is assumed that the underlying (normalized) stochastic process $Z(t)$ is stationary and Gaussian with mean value zero and standard deviation equal to one. It is also assumed that the mean zero up-crossing rate $\nu^+(0)$ is such that the product $\nu^+(0)T = 10^3$, where $T = 1$ year, which seems to be typical for the wind speed process. Using the Poisson assumption, the distribution of the yearly extreme value of $Z(t)$ is then calculated by the formula

$$F^{1\,\text{yr}}(\eta) = \exp\left\{-\nu^+(\eta)\,T\right\} = \exp\left\{-10^3 \exp\left(-\frac{\eta^2}{2}\right)\right\}, \tag{43}$$

where $T = 1$ year and $\nu^+(\eta)$ is the mean up-crossing rate per year, η is the scaled wind speed. The 100-year return period value $\eta^{100\,\text{yr}}$ is then calculated from the relation $F^{1\,\text{yr}}(\eta^{100\,\text{yr}}) = 1 - 1/100$, which gives $\eta^{100\,\text{yr}} = 4.80$.

The Monte Carlo simulated data to be used for the synthetic example are generated based on the observation that the peak events extracted from measurements of the wind speed process, are usually separated by 3-4 days. This is done to obtain approximately independent data, as required by the POT method. In accordance with this, peak event data are generated from the extreme value distribution

$$F^{3\,\text{d}}(\eta) = \exp\left\{-q \exp\left(-\frac{\eta^2}{2}\right)\right\}, \tag{44}$$

where $q = \nu^+(0)T = 10$, which corresponds to $T = 3.65$ days, and $F^{1\,\text{yr}}(\eta) = (F^{3\,\text{d}}(\eta))^{100}$.

Since the data points (i.e., $T = 3.65$ days maxima) are independent, $\varepsilon_k(\eta)$ is independent of k. Therefore, we put $k = 1$. Since we have 100 data from one year, the data amounts to 2000 data points. For estimation of a 95% confidence interval for each estimated value of the ACER function $\varepsilon_1(\eta)$ for the chosen range of η-values, the required standard deviation in (22) was based on 20 estimates of the ACER function using the yearly data. This provided a 95% confidence band on the optimally fitted curve based on 2000 data. From these data, the predicted 100-year return level is obtained from $\widehat{\varepsilon}_1(\eta^{100\,\text{yr}}) = 10^{-4}$. A nonparametric bootstrapping method was also used to estimate a 95% confidence interval based on 1000 resamples of size 2000.

The POT prediction of the 100-year return level was based on using maximum likelihood estimates (MLE) of the parameters in (37) for a specific choice of threshold. The 95% confidence interval was obtained from the parametrically bootstrapped PDF of the POT estimate for the given threshold. A sample of 1000 data sets was used. One of the unfortunate features of the POT method is that the estimated 100 year value may vary significantly with the choice of threshold. So also for the synthetic data. We have followed the standard recommended procedures for identifying a suitable threshold [1].

Note that in spite of the fact that the true asymptotic distribution of exceedances is the exponential distribution in (38), the POT method used here is based on adopting (37). The reason is simply that this is the recommended procedure [1], which is somewhat unfortunate but understandable. The reason being that the GP distribution provides greater flexibility in terms of curve fitting. If the correct asymptotic distribution of exceedances had been used on this example, poor results for the estimated return period values would be obtained. The price to pay for using the GP distribution is that the estimated parameters may easily lead to an asymptotically inconsistent extreme value distribution.

The 100-year return level predicted by the Gumbel method was based on using the method of moments for parameter estimation on the sample of 20 yearly extremes. This choice of estimation method is due to the small sample of extreme values. The 95% confidence interval was obtained from the parametrically bootstrapped PDF of the Gumbel prediction. This was based on a sample of size 10,000 data sets of 20 yearly extremes. The results obtained by the method of moments were compared with the corresponding results obtained by using the maximum likelihood method. While there were individual differences, the overall picture was one of very good agreement.

In order to get an idea about the performance of the ACER, POT, and Gumbel methods, 100 independent 20-year MC simulations as discussed above were done. Table 1 compares predicted values and confidence intervals for a selection of 10 cases together with average values over the 100 simulated cases. It is seen that the average of the 100 predicted 100-year return levels is slightly better for the ACER method than for both the POT and the Gumbel methods. But more significantly, the range of predicted 100-year return levels by the ACER method is 4.34–5.36, while the same for the POT method is 4.19–5.87 and for the Gumbel method is 4.41–5.71.

TABLE 1: 100-year return level estimates and 95% CI (BCI = CI by bootstrap) for A = ACER, G = Gumbel, and P = POT. Exact value = 4.80.

Sim. No.	A $\hat{\eta}^{100}$	A CI	A BCI	G $\hat{\eta}^{100}$	G BCI	P $\hat{\eta}^{100}$	P BCI
1	5.07	(4.67, 5.21)	(4.69, 5.42)	4.41	(4.14, 4.73)	4.29	(4.13, 4.52)
10	4.65	(4.27, 4.94)	(4.37, 5.03)	4.92	(4.40, 5.58)	4.88	(4.42, 5.40)
20	4.86	(4.49, 5.06)	(4.44, 5.19)	5.04	(4.54, 5.63)	5.04	(4.48, 5.74)
30	4.75	(4.22, 5.01)	(4.33, 5.02)	4.75	(4.27, 5.32)	4.69	(4.24, 5.26)
40	4.54	(4.20, 4.74)	(4.27, 4.88)	4.80	(4.31, 5.39)	4.73	(4.19, 5.31)
50	4.80	(4.35, 5.05)	(4.42, 5.14)	4.91	(4.41, 5.50)	4.79	(4.31, 5.34)
60	4.84	(4.36, 5.20)	(4.48, 5.19)	4.85	(4.36, 5.43)	4.71	(4.32, 5.23)
70	5.02	(4.47, 5.31)	(4.62, 5.36)	4.96	(4.47, 5.53)	4.97	(4.47, 5.71)
80	4.59	(4.33, 4.81)	(4.38, 4.98)	4.76	(4.31, 5.31)	4.68	(4.15, 5.27)
90	4.84	(4.49, 5.11)	(4.60, 5.30)	4.77	(4.34, 5.32)	4.41	(4.23, 4.64)
100	4.62	(4.29, 5.05)	(4.45, 5.09)	4.79	(4.31, 5.41)	4.53	(4.05, 4.88)
Av. 100	4.82	(4.41, 5.09)	(4.48, 5.18)	4.84	(4.37, 5.40)	4.72	(4.27, 5.23)

Hence, in this case the ACER method performs consistently better than both these methods. It is also observed from the estimated 95% confidence intervals that the ACER method, as implemented in this paper, provides slightly higher accuracy than the other two methods. Lastly, it is pointed out that the confidence intervals of the 100-year return level estimated by the ACER method obtained by either the simplified extrapolated confidence band approach or by nonparametric bootstrapping are very similar, except for a slight mean shift. As a final comparison, the 100 bootstrapped confidence intervals obtained for the ACER and Gumbel methods missed the target value three times, while for the POT method this number was 18.

An example of the ACER plot and results obtained for one set of data is presented in Figure 1. The predicted 100-year value is 4.85 with a predicted 95% confidence interval (4.45, 5.09). Figure 2 presents POT predictions based on MLE for different thresholds in terms of the number n of data points above the threshold. The predicted value is 4.7 at $n = 204$, while the 95% confidence interval is (4.25, 5.28). The same data set as in Figure 1 was used. This was also used for the Gumbel plot shown in Figure 3. In this case the predicted value based on the method of moments (MM) is $\eta_{\mathrm{MM}}^{100\,\mathrm{yr}} = 4.75$ with a parametric bootstrapped 95% confidence interval of (4.34, 5.27). Prediction based on the Gumbel-Lieblein BLUE method (GL), compared with for example, Cook [44], is $\eta_{\mathrm{GL}}^{100\,\mathrm{yr}} = 4.73$ with a parametric bootstrapped 95% confidence interval equal to (4.35, 5.14).

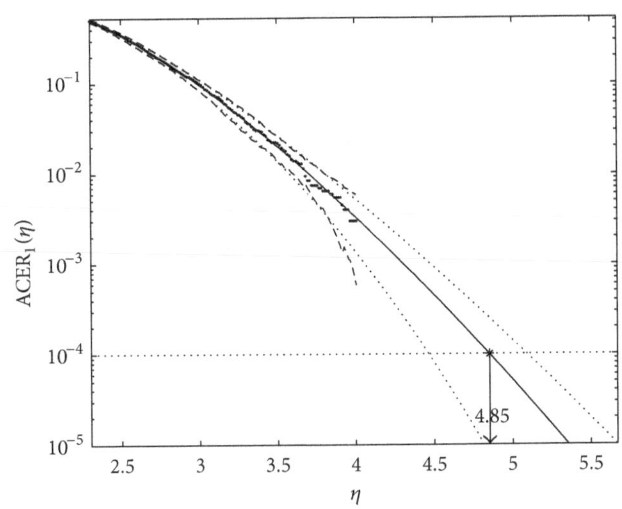

FIGURE 1: Synthetic data ACER $\hat{\varepsilon}_1$, Monte Carlo simulation ($*$); optimized curve fit (—); empirical 95% confidence band (- -); optimized confidence band (· · ·). Tail marker $\eta_1 = 2.3$.

9. Measured Wind Speed Data

In this section, we analyze real wind speed data, measured at two weather stations off the coast of Norway: at Nordøyan and at Hekkingen, see Figure 4. Extreme wind speed prediction is an important issue for design of structures exposed to the weather variations. Significant efforts have been devoted to the problem of predicting extreme wind speeds on the basis of measured data by various authors over several decades, see, for example, [45–48] for extensive references to previous work.

Hourly maximum gust wind was recorded during the 13 years 1999–2012 at Nordøyan and the 14 years 1998–2012 at Hekkingen. The objective is to estimate a 100-year wind speed. Variation in the wind speed caused by seasonal variations in the wind climate during the year makes the wind speed a non-stationary process on the scale of months. Moreover, due to global climate change, yearly statistics may vary on the scale of years. The latter is, however, a slow process, and for the purpose of long-term prediction we assume here that within a time span of 100 years a quasi-stationary model of the wind speeds applies. This may not be entirely true, of course.

9.1. Nordøyan. Figure 5 highlights the cascade of ACER estimates $\hat{\varepsilon}_1, \ldots, \hat{\varepsilon}_{96}$, for the case of 13 years of hourly data recorded at the Nordøyan weather station. Here, $\hat{\varepsilon}_{96}$ is considered to represent the final converged results. By "converged," we mean that $\hat{\varepsilon}_{96} \approx \hat{\varepsilon}_k$ for $k > 96$ in the tail, so that there is no

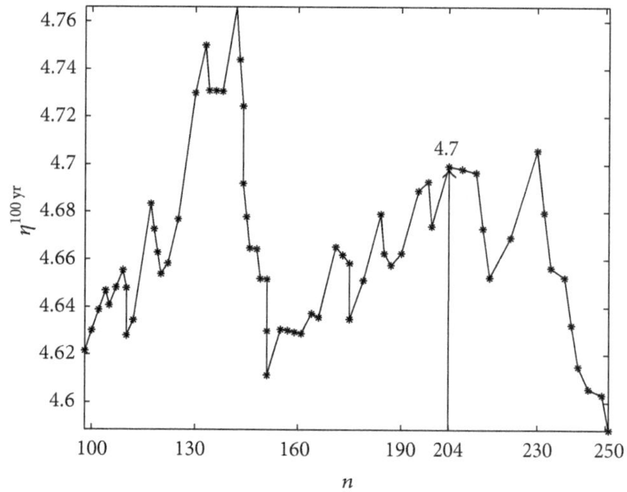

FIGURE 2: The point estimate $\tilde{\eta}^{100\,yr}$ of the 100-year return period value based on 20 years synthetic data as a function of the number n of data points above threshold. The return level estimate = 4.7 at $n = 204$.

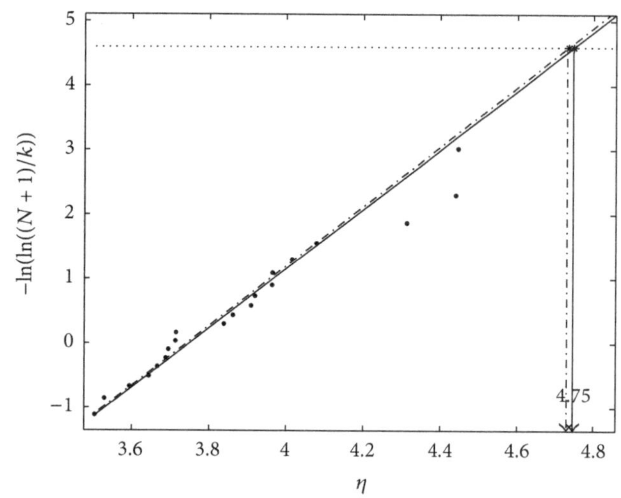

FIGURE 3: The point estimate $\tilde{\eta}^{100\,yr}$ of the 100-year return period value based on 20 years synthetic data. Lines are fitted by the method of moments—solid line (—) and the Gumbel-Lieblein BLUE method—dash-dotted lite (-··-). The return level estimate by the method of moments is 4.75, by the Gumbel-Lieblein BLUE method is 4.73.

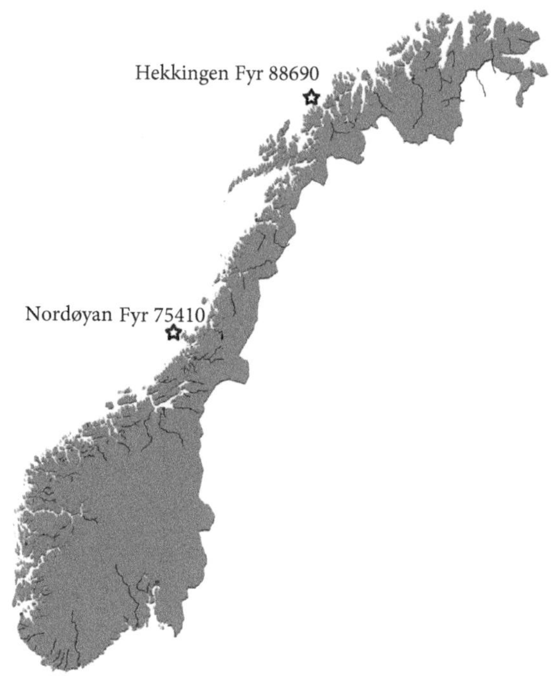

FIGURE 4: Wind speed measurement stations.

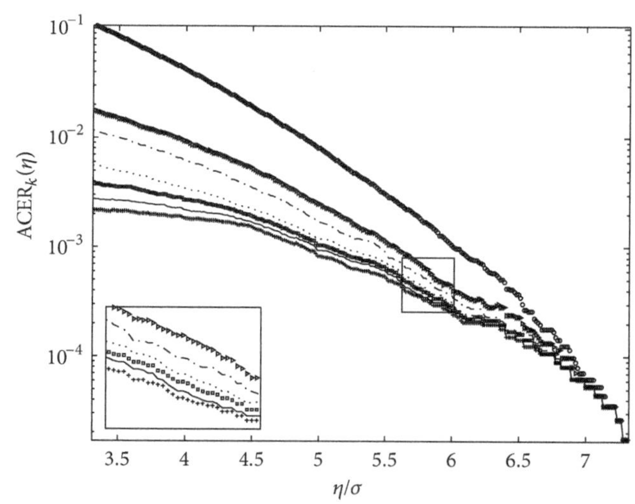

FIGURE 5: Nordøyan wind speed statistics, 13 years hourly data. Comparison between ACER estimates for different degrees of conditioning. $\sigma = 6.01$ m/s.

need to consider conditioning of an even higher order than 96. Figure 5 reveals a rather strong statistical dependence between consecutive data, which is clearly reflected in the effect of conditioning on previous data values. It is also interesting to observe that this effect is to some extent captured already by $\hat{\varepsilon}_2$, that is, by conditioning only on the value of the previous data point. Subsequent conditioning on more than one previous data point does not lead to substantial changes in ACER values, especially for tail values. On the other hand, to bring out fully the dependence structure of these data, it was necessary to carry the conditioning process to (at least) the 96th ACER function, as discussed above.

However, from a practical point of view, the most important information provided by the ACER plot of Figure 5 is that for the prediction of a 100-year value, one may use the first ACER function. The reason for this is that Figure 5 shows that all the ACER functions coalesce in the far tail. Hence, we may use any of the ACER functions for the prediction. Then, the obvious choice is to use the first ACER function, which allows us to use all the data in its estimation and thereby increase accuracy.

In Figure 6 is shown the results of parametric estimation of the return value and its 95% CI for 13 years of hourly

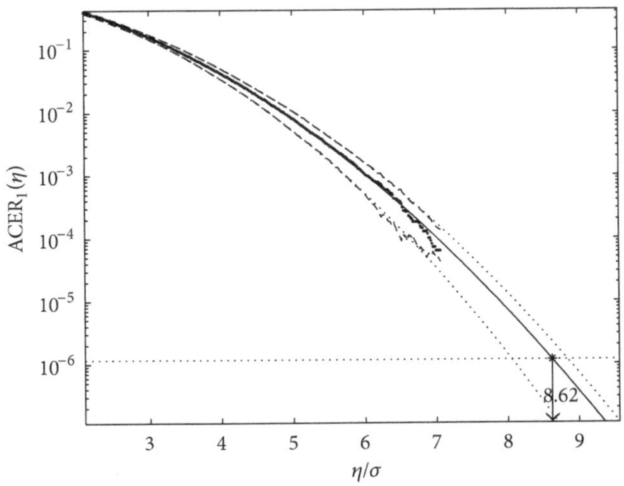

FIGURE 6: Nordøyan wind speed statistics, 13 years hourly data. $\hat{\varepsilon}_1(\eta)$ ($*$); optimized curve fit (—); empirical 95% confidence band (- -); optimized confidence band ($\cdots$). Tail marker $\eta_1 = 12.5$ m/s $= 2.08\sigma$ ($\sigma = 6.01$ m/s).

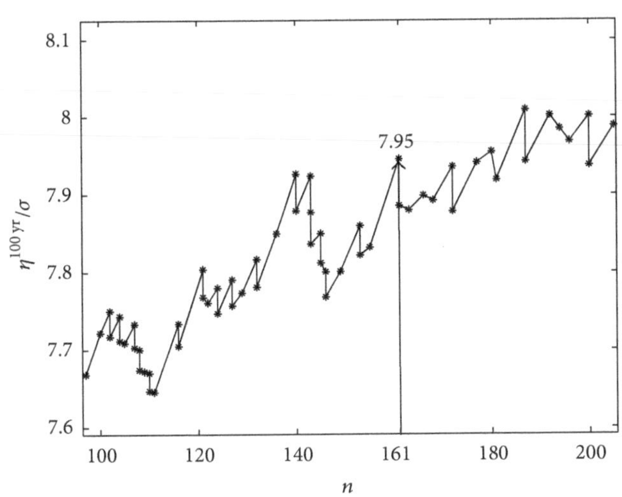

FIGURE 7: The point estimate $\eta^{100\,\mathrm{yr}}$ of the 100-year return level based on 13 years hourly data as a function of the number n of data points above threshold. $\sigma = 6.01$ m/s.

maxima. The predicted 100-year return speed is $\eta^{100\,\mathrm{yr}} = 51.85$ m/s with 95% confidence interval (48.4, 53.1). $R = 13$ years of data may not be enough to guarantee (22), since we required $R \geq 20$. Nevertheless, for simplicity, we use it here even with $R = 13$, accepting that it may not be very accurate.

Figure 7 presents POT predictions for different threshold numbers based on MLE. The POT prediction is $\eta^{100\,\mathrm{yr}} = 47.8$ m/s at threshold $n = 161$, while the bootstrapped 95% confidence interval is found to be (44.8, 52.7) m/s based on 10,000 generated samples. It is interesting to observe the unstable characteristics of the predictions over a range of threshold values, while they are quite stable on either side of this range giving predictions that are more in line with the results from the other two methods.

Figure 8 presents a Gumbel plot based on the 13 yearly extremes extracted from the 13 years of hourly data. The

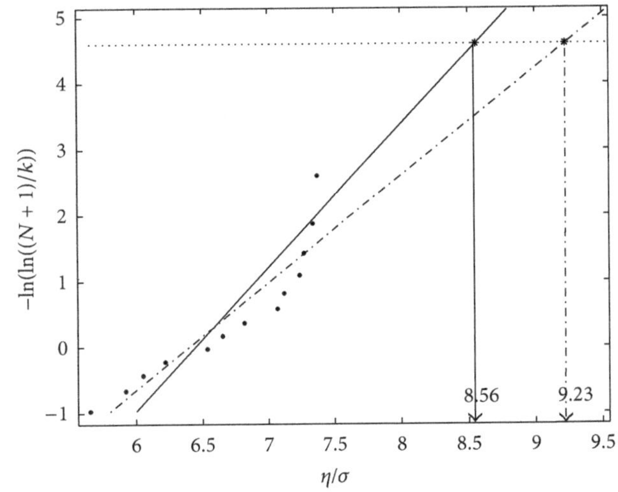

FIGURE 8: Nordøyan wind speed statistics, 13 years of hourly data. Gumbel plot of yearly extremes. Lines are fitted by the method of moments—solid line (—) and the Gumbel-Lieblein BLUE method—dash-dotted lite (- $\cdot$ -). $\sigma = 6.01$ m/s.

TABLE 2: Predicted 100-year return period levels for Nordøyan Fyr weather station by the ACER method for different degrees of conditioning, annual maxima, and POT methods, respectively.

Method	Spec	$\hat{\eta}^{100\,\mathrm{yr}}$, m/s	95% CI ($\hat{\eta}^{100\,\mathrm{yr}}$), m/s
ACER, various k	1	51.85	(48.4, 53.1)
	2	51.48	(46.1, 54.1)
	4	52.56	(46.7, 55.7)
	24	52.90	(47.0, 56.2)
	48	54.62	(47.7, 57.6)
	72	53.81	(46.9, 58.3)
	96	54.97	(47.5, 60.5)
Annual maxima	MM	51.5	(45.2, 59.3)
	GL	55.5	(48.0, 64.9)
POT	—	47.8	(44.8, 52.7)

Gumbel prediction based on the method of moments (MM) is $\eta_{\mathrm{MM}}^{100\,\mathrm{yr}} = 51.5$ m/s, with a parametric bootstrapped 95% confidence interval equal to (45.2, 59.3) m/s, while prediction based on the Gumbel-Lieblein BLUE method (GL) is $\eta_{\mathrm{GL}}^{100\,\mathrm{yr}} = 55.5$ m/s, with a parametric bootstrapped 95% confidence interval equal to (48.0, 64.9) m/s.

In Table 2 the 100-year return period values for the Nordøyan station are listed together with the predicted 95% confidence intervals for all methods.

9.2. Hekkingen. Figure 9 shows the cascade of estimated ACER functions $\hat{\varepsilon}_1, \ldots, \hat{\varepsilon}_{96}$ for the case of 14 years of hourly data. As for Nordøyan, $\hat{\varepsilon}_{96}$ is used to represent the final converged results. Figure 9 also reveals a rather strong statistical dependence between consecutive data at moderate wind speed levels. This effect is again to some extent captured already by $\hat{\varepsilon}_2$, so that subsequent conditioning on more than one previous data point does not lead to substantial changes in ACER values, especially for tail values.

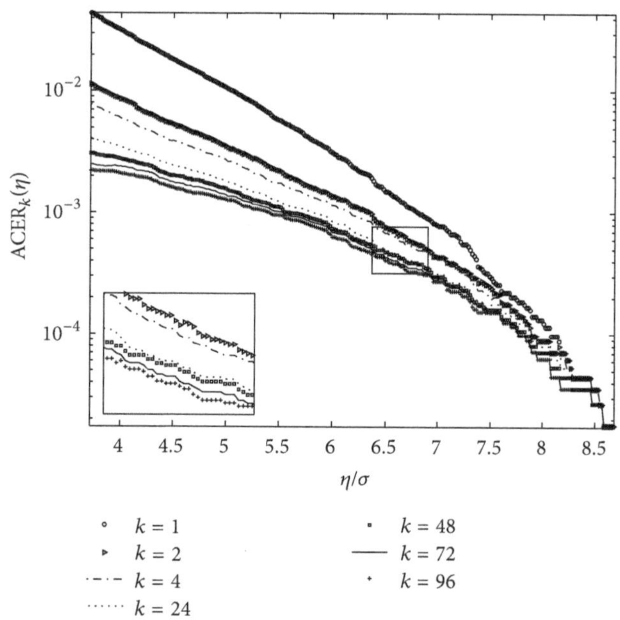

FIGURE 9: Hekkingen wind speed statistics, 14 years hourly data. Comparison between ACER estimates for different degrees of conditioning. $\sigma = 5.72$ m/s.

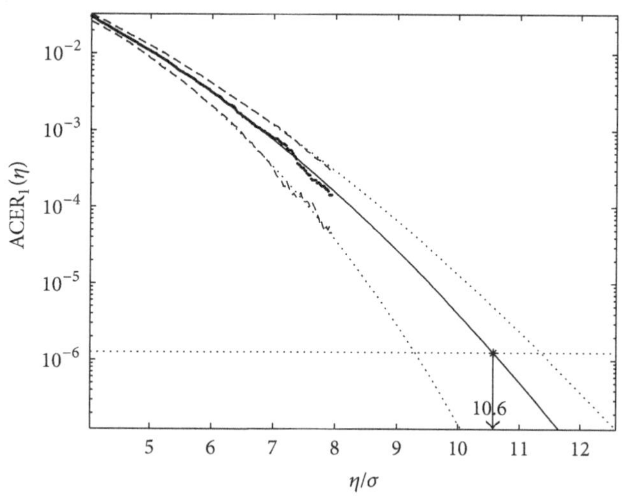

FIGURE 10: Hekkingen wind speed statistics, 14 years hourly data. $\hat{\varepsilon}_1(\eta)$ (∗); optimized curve fit (—); empirical 95% confidence band (- -); optimized confidence band (···). Tail marker $\eta_1 = 23$ m/s $= 4.02\sigma$ ($\sigma = 5.72$ m/s).

Also, for the Hekkingen data, the ACER plot of Figure 9 indicates that the ACER functions coalesce in the far tail. Hence, for the practical prediction of a 100-year value, one may use the first ACER function.

In Figure 10 is shown the results of parametric estimation of the return value and its 95% CI for 14 years of hourly maxima. The predicted 100-year return speed is $\eta^{100\,\mathrm{yr}} = 60.47$ m/s with 95% confidence interval (53.1, 64.9). Equation (22) has been used also for this example with $R = 14$.

Figure 11 presents POT predictions for different threshold numbers based on MLE. The POT prediction is $\eta^{100\,\mathrm{yr}} = 53.48$ m/s at threshold $n = 183$, while the bootstrapped

TABLE 3: Predicted 100-year return period levels for Nordøyan Fyr weather station by the ACER method for different degrees of conditioning, annual maxima, and POT methods, respectively.

Method	Spec	$\hat{\eta}^{100\,\mathrm{yr}}$, m/s	95% CI ($\hat{\eta}^{100\,\mathrm{yr}}$), m/s
ACER, various k	1	60.47	(53.1, 64.9)
	2	62.23	(53.3, 70.0)
	4	63.03	(53.0, 74.5)
	24	60.63	(51.3, 70.7)
	48	60.44	(51.3, 77.0)
	72	58.06	(51.2, 66.4)
	96	59.19	(52.0, 68.3)
Annual maxima	MM	58.10	(50.8, 67.3)
	GL	60.63	(53.0, 70.1)
POT	—	53.48	(48.9, 57.0)

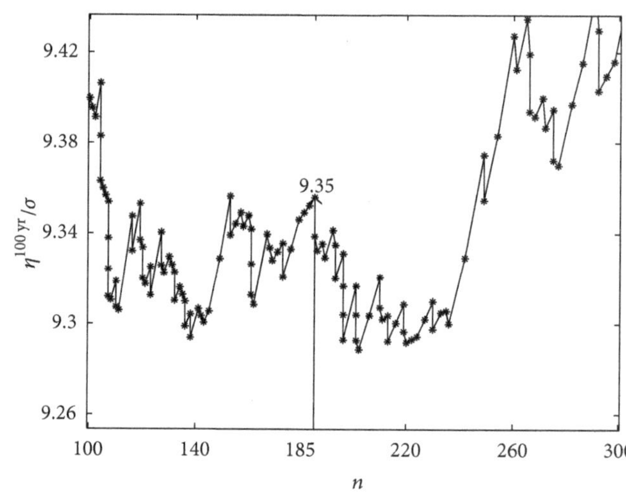

FIGURE 11: The point estimate $\eta^{100\,\mathrm{yr}}$ of the 100-year return level based on 14 years hourly data as a function of the number n of data points above threshold. $\sigma = 5.72$ m/s.

95% confidence interval is found to be (48.9, 57.0) m/s based on 10,000 generated samples. It is interesting to observe the unstable characteristics of the predictions over a range of threshold values, while they are quite stable on either side of this range giving predictions that are more in line with the results from the other two methods.

Figure 12 presents a Gumbel plot based on the 14 yearly extremes extracted from the 14 years of hourly data. The Gumbel prediction based on the method of moments (MM) is $\eta_{\mathrm{MM}}^{100\,\mathrm{yr}} = 58.10$ m/s, with a parametric bootstrapped 95% confidence interval equal to (50.8, 67.3) m/s. Prediction based on the Gumbel-Lieblein BLUE method (GL) is $\eta_{\mathrm{GL}}^{100\,\mathrm{yr}} = 60.63$ m/s, with a parametric bootstrapped 95% confidence interval equal to (53.0, 70.1) m/s.

In Table 3, the 100-year return period values for the Hekkingen station are listed together with the predicted 95% confidence intervals for all methods.

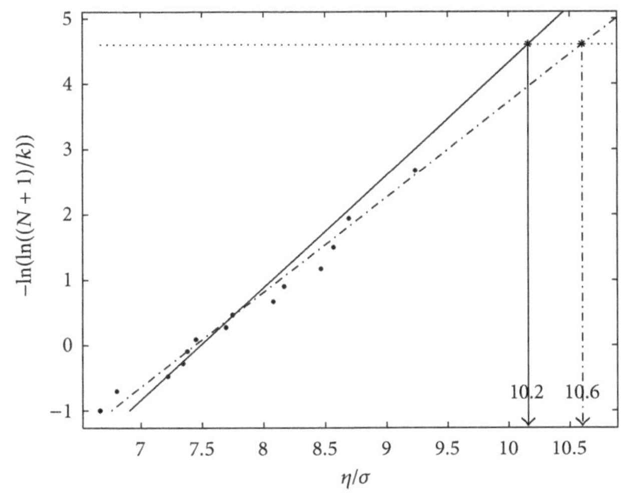

FIGURE 12: Hekkingen wind speed statistics, 14 years of hourly data. Gumbel plot of yearly extremes. Lines are fitted by the method of moments—solid line (—) and the Gumbel-Lieblein BLUE method—dash-dotted lite (- · · -). $\sigma = 5.72$ m/s.

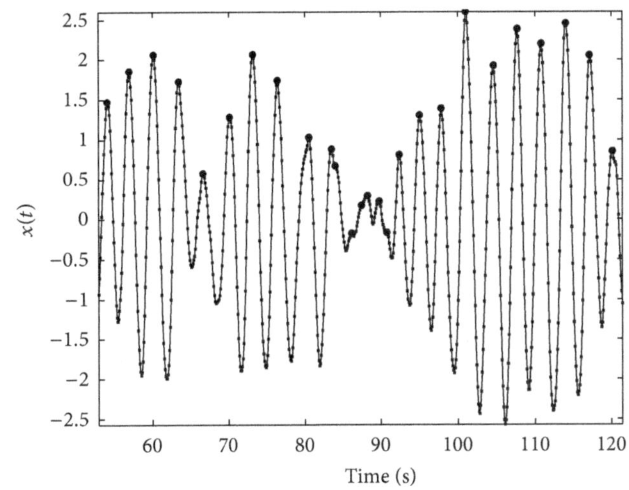

FIGURE 13: Part of the narrow-band response time series of the linear oscillator with fully sampled and peak values indicated.

10. Extreme Value Prediction for a Narrow Band Process

In engineering mechanics, a classical extreme response prediction problem is the case of a lightly damped mechanical oscillator subjected to random forces. To illustrate this prediction problem, we will investigate the response process of a linear mechanical oscillator driven by a Gaussian white noise. Let $X(t)$ denote the displacement response; the dynamic model can then be expressed as, $\ddot{X}(t) + 2\zeta\omega_e\dot{X}(t) + \omega_e^2 X(t) = W(t)$, where ζ = relative damping, ω_e = undamped eigenfrequency, and $W(t)$ = a stationary Gaussian white noise (of suitable intensity). By choosing a small value for ζ, the response time series will exhibit narrow band characteristics, that is, the spectral density of the response process $X(t)$ will assume significant values only over a narrow range of frequencies. This manifests itself by producing a strong beating of the response time series, which means that the size of the response peaks will change slowly in time, see Figure 13. A consequence of this is that neighbouring peaks are strongly correlated, and there is a conspicuous clumping of the peak values. Hence the problem with accurate prediction, since the usual assumption of independent peak values is then violated.

Many approximations have been proposed to deal with this correlation problem, but no completely satisfactory solution has been presented. In this section, we will show that the ACER method solves this problem efficiently and elegantly in a statistical sense. In Figure 14 are shown some of the ACER functions for the example time series. It may be verified from Figure 13 that there are approximately 32 sample points between two neighbouring peaks in the time series. To illustrate a point, we have chosen to analyze the time series consisting of all sample points. Usually, in practice, only the time series obtained by extracting the peak values would be used for the ACER analysis. In the present case, the first ACER function is then based on assuming that all the sampled data points are independent, which is obviously completely wrong. The second ACER function, which is based on counting each exceedance with an immediately preceding nonexceedance, is nothing but an upcrossing rate. Using this ACER function is largely equivalent to assuming independent peak values. It is now interesting to observe that the 25th ACER function can hardly be distinguished from the second ACER function. In fact, the ACER functions after the second do not change appreciably until one starts to approach the 32nd, which corresponds to hitting the previous peak value in the conditioning process. So, the important information concerning the dependence structure in the present time series seems to reside in the peak values, which may not be very surprising. It is seen that the ACER functions show a significant change in value as a result of accounting for the correlation effects in the time series. To verify the full dependence structure in the time series, it is necessary to continue the conditioning process down to at least the 64th ACER function. In the present case, there is virtually no difference between the 32nd and the 64th, which shows that the dependence structure in this particular time series is captured almost completely by conditioning on the previous peak value. It is interesting to contrast the method of dealing with the effect of sampling frequency discussed here with that of [49].

To illustrate the results obtained by extracting only the peak values from the time series, which would be the approach typically chosen in an engineering analysis, the ACER plots for this case is shown in Figure 15. By comparing results from Figures 14 and 15, it can be verified that they are in very close agreement by recognizing that the second ACER function in Figure 14 corresponds to the first ACER function in Figure 15, and by noting that there is a factor of approximately 32 between corresponding ACER functions in the two figures. This is due to the fact that the time series of peak values contains about 32 times less data than the original time series.

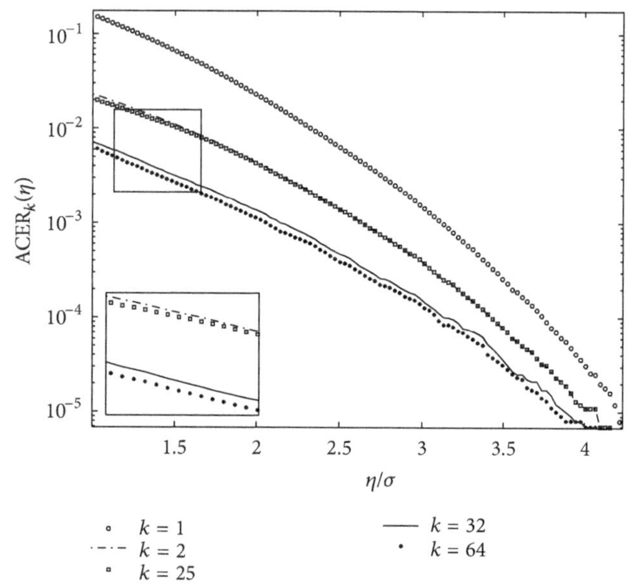

FIGURE 14: Comparison between ACER estimates for different degrees of conditioning for the narrow-band time series.

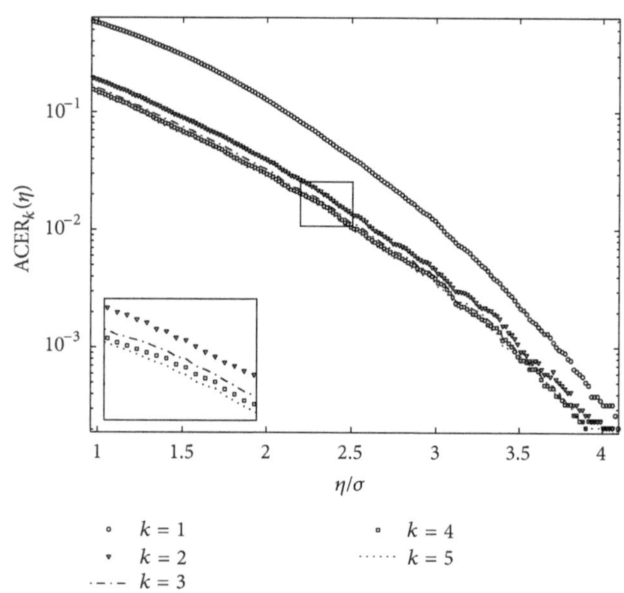

FIGURE 15: Comparison between ACER estimates for different degrees of conditioning based on the time series of the peak values, compared with Figure 13.

11. Concluding Remarks

This paper studies a new method for extreme value prediction for sampled time series. The method is based on the introduction of a conditional average exceedance rate (ACER), which allows dependence in the time series to be properly and easily accounted for. Declustering of the data is therefore avoided, and all the data are used in the analysis. Significantly, the proposed method also aims at capturing to some extent the subasymptotic form of the extreme value distribution.

Results for wind speeds, both synthetic and measured, are used to illustrate the method. An estimation problem related to applications in mechanics is also presented. The validation of the method is done by comparison with exact results (when available), or other widely used methods for extreme value statistics, such as the Gumbel and the peaks-over-threshold (POT) methods. Comparison of the various estimates indicate that the proposed method provides more accurate results than the Gumbel and POT methods.

Subject to certain restrictions, the proposed method also applies to nonstationary time series, but it cannot directly predict for example, the effect of climate change in the form of long-term trends in the average exceedance rates extending beyond the data. This must be incorporated into the analysis by explicit modelling techniques.

As a final remark, it may be noted that the ACER method as described in this paper has a natural extension to higher dimensional distributions. The implication is that, it is then possible to provide estimates of for example, the exact bivariate extreme value distribution for a suitable set of data [50]. However, as is easily recognized, the extrapolation problem is not as simply dealt with as for the univariate case studied in this paper.

Acknowledgment

This work was supported by the Research Council of Norway (NFR) through the Centre for Ships and Ocean Structures (CeSOS) at the Norwegian University of Science and Technology.

References

[1] S. Coles, *An Introduction to Statistical Modeling of Extreme Values*, Springer Series in Statistics, Springer, London, UK, 2001.

[2] J. Beirlant, Y. Goegebeur, J. Teugels, and J. Segers, *Statistics of Extremes*, Wiley Series in Probability and Statistics, John Wiley & Sons, Chichester, UK, 2004.

[3] A. C. Davison and R. L. Smith, "Models for exceedances over high thresholds," *Journal of the Royal Statistical Society. Series B. Methodological*, vol. 52, no. 3, pp. 393–442, 1990.

[4] R.-D. Reiss and M. Thomas, *Statistical Analysis of Extreme Values*, Birkhäuser, Basel, Switzerland, 3rd edition, 1997.

[5] P. Embrechts, C. Klüppelberg, and T. Mikosch, *Modelling Extremal Events*, vol. 33 of *Applications of Mathematics (New York)*, Springer, Berlin, Germany, 1997.

[6] M. Falk, J. Hüsler, and R.-D. Reiss, *Laws of Small Numbers: Extremes and Rare Events*, Birkhäuser, Basel, Switzerland, 2 nd edition, 2004.

[7] A. Naess and O. Gaidai, "Monte Carlo methods for estimating the extreme response of dynamical systems," *Journal of Engineering Mechanics*, vol. 134, no. 8, pp. 628–636, 2008.

[8] A. Naess, O. Gaidai, and S. Haver, "Efficient estimation of extreme response of drag-dominated offshore structures by Monte Carlo simulation," *Ocean Engineering*, vol. 34, no. 16, pp. 2188–2197, 2007.

[9] R. L. Smith, "The extremal index for a Markov chain," *Journal of Applied Probability*, vol. 29, no. 1, pp. 37–45, 1992.

[10] S. G. Coles, "A temporal study of extreme rainfall," in *Statistics for the Environment 2-Water Related Issues*, V. Barnett and K. F. Turkman, Eds., chapter 4, pp. 61–78, John Wiley & Sons, Chichester, UK, 1994.

[11] R. L. Smith, J. A. Tawn, and S. G. Coles, "Markov chain models for threshold exceedances," *Biometrika*, vol. 84, no. 2, pp. 249–268, 1997.

[12] S. Yun, "The extremal index of a higher-order stationary Markov chain," *The Annals of Applied Probability*, vol. 8, no. 2, pp. 408–437, 1998.

[13] S. Yun, "The distributions of cluster functionals of extreme events in a *d*th-order Markov chain," *Journal of Applied Probability*, vol. 37, no. 1, pp. 29–44, 2000.

[14] J. Segers, "Approximate distributions of clusters of extremes," *Statistics & Probability Letters*, vol. 74, no. 4, pp. 330–336, 2005.

[15] G. S. Watson, "Extreme values in samples from m-dependent stationary stochastic processes," *Annals of Mathematical Statistics*, vol. 25, pp. 798–800, 1954.

[16] M. R. Leadbetter, G. Lindgren, and H. Rootzén, *Extremes and Related Properties of Random Sequences and Processes*, Springer Series in Statistics, Springer, New York, NY, USA, 1983.

[17] D. E. Cartwright, "On estimating the mean energy of sea waves from the highest waves in a record," *Proceedings of the Royal Society of London. Series A*, vol. 247, pp. 22–28, 1958.

[18] A. Naess, "On the long-term statistics of extremes," *Applied Ocean Research*, vol. 6, no. 4, pp. 227–228, 1984.

[19] G. Schall, M. H. Faber, and R. Rackwitz, "Ergodicity assumption for sea states in the reliability estimation of offshore structures," *Journal of Offshore Mechanics and Arctic Engineering*, vol. 113, no. 3, pp. 241–246, 1991.

[20] E. H. Vanmarcke, "On the distribution of the first-passage time for normal stationary random processes," *Journal of Applied Mechanics*, vol. 42, no. 1, pp. 215–220, 1975.

[21] M. R. Leadbetter, "Extremes and local dependence in stationary sequences," *Zeitschrift für Wahrscheinlichkeitstheorie und Verwandte Gebiete*, vol. 65, no. 2, pp. 291–306, 1983.

[22] T. Hsing, "On the characterization of certain point processes," *Stochastic Processes and Their Applications*, vol. 26, no. 2, pp. 297–316, 1987.

[23] T. Hsing, "Estimating the parameters of rare events," *Stochastic Processes and Their Applications*, vol. 37, no. 1, pp. 117–139, 1991.

[24] M. R. Leadbetter, "On high level exceedance modeling and tail inference," *Journal of Statistical Planning and Inference*, vol. 45, no. 1-2, pp. 247–260, 1995.

[25] C. A. T. Ferro and J. Segers, "Inference for clusters of extreme values," *Journal of the Royal Statistical Society. Series B*, vol. 65, no. 2, pp. 545–556, 2003.

[26] C. Y. Robert, "Inference for the limiting cluster size distribution of extreme values," *The Annals of Statistics*, vol. 37, no. 1, pp. 271–310, 2009.

[27] A. Naess and O. Gaidai, "Estimation of extreme values from sampled time series," *Structural Safety*, vol. 31, no. 4, pp. 325–334, 2009.

[28] P. E. Gill, W. Murray, and M. H. Wright, *Practical Optimization*, Academic Press Inc, London, UK, 1981.

[29] W. Forst and D. Hoffmann, *Optimization—Theory and Practice*, Springer Undergraduate Texts in Mathematics and Technology, Springer, New York, NY, USA, 2010.

[30] N. R. Draper and H. Smith, *Applied Regression Analysis*, Wiley Series in Probability and Statistics: Texts and References Section, John Wiley & Sons, New York, NY, USA, 3rd edition, 1998.

[31] D. C. Montgomery, E. A. Peck, and G. G. Vining, *Introduction to Linear Regression Analysis*, Wiley Series in Probability and Statistics: Texts, References, and Pocketbooks Section, Wiley-Interscience, New York, NY, USA, Third edition, 2001.

[32] A. Naess, *Estimation of Extreme Values of Time Series with Heavy Tails*, Preprint Statistics No. 14/2010, Department of Mathematical Sciences, Norwegian University of Science and Technology, Trondheim, Norway, 2010.

[33] E. F. Eastoe and J. A. Tawn, "Modelling the distribution of the cluster maxima of exceedances of subasymptotic thresholds," *Biometrika*, vol. 99, no. 1, pp. 43–55, 2012.

[34] J. A. Tawn, "Discussion of paper by A. C. Davison and R. L. Smith," *Journal of the Royal Statistical Society. Series B*, vol. 52, no. 3, pp. 393–442, 1990.

[35] A. W. Ledford and J. A. Tawn, "Statistics for near independence in multivariate extreme values," *Biometrika*, vol. 83, no. 1, pp. 169–187, 1996.

[36] J. E. Heffernan and J. A. Tawn, "A conditional approach for multivariate extreme values," *Journal of the Royal Statistical Society. Series B*, vol. 66, no. 3, pp. 497–546, 2004.

[37] Numerical Algorithms Group, *NAG Toolbox for Matlab*, NAG, Oxford, UK, 2010.

[38] E. J. Gumbel, *Statistics of Extremes*, Columbia University Press, New York, NY, USA, 1958.

[39] K. V. Bury, *Statistical Models in Applied Science*, Wiley Series in Probability and Mathematical Statistics, John Wiley & Sons, New York, NY, USA, 1975.

[40] B. Efron and R. J. Tibshirani, *An Introduction to the Bootstrap*, vol. 57 of *Monographs on Statistics and Applied Probability*, Chapman and Hall, New York, NY, USA, 1993.

[41] A. C. Davison and D. V. Hinkley, *Bootstrap Methods and Their Application*, vol. 1 of *Cambridge Series in Statistical and Probabilistic Mathematics*, Cambridge University Press, Cambridge, UK, 1997.

[42] J. Pickands, III, "Statistical inference using extreme order statistics," *The Annals of Statistics*, vol. 3, pp. 119–131, 1975.

[43] A. Naess and P. H. Clausen, "Combination of the peaks-over-threshold and bootstrapping methods for extreme value prediction," *Structural Safety*, vol. 23, no. 4, pp. 315–330, 2001.

[44] N. J. Cook, *The Designer's Guide to Wind Loading of Building Structures*, Butterworths, London, UK, 1985.

[45] N. J. Cook, "Towards better estimation of extreme winds," *Journal of Wind Engineering and Industrial Aerodynamics*, vol. 9, no. 3, pp. 295–323, 1982.

[46] A. Naess, "Estimation of long return period design values for wind speeds," *Journal of Engineering Mechanics*, vol. 124, no. 3, pp. 252–259, 1998.

[47] J. P. Palutikof, B. B. Brabson, D. H. Lister, and S. T. Adcock, "A review of methods to calculate extreme wind speeds," *Meteorological Applications*, vol. 6, no. 2, pp. 119–132, 1999.

[48] O. Perrin, H. Rootzén, and R. Taesler, "A discussion of statistical methods used to estimate extreme wind speeds," *Theoretical and Applied Climatology*, vol. 85, no. 3-4, pp. 203–215, 2006.

[49] M. E. Robinson and J. A. Tawn, "Extremal analysis of processes sampled at different frequencies," *Journal of the Royal Statistical Society. Series B*, vol. 62, no. 1, pp. 117–135, 2000.

[50] A. Naess, *A Note on the Bivariate ACER Method*, Preprint Statistics No. 01/2011, Department of Mathematical Sciences, Norwegian University of Science and Technology, Trondheim, Norway, 2011.

16

A Survey Design for a Sensitive Binary Variable Correlated with Another Nonsensitive Binary Variable

Jun-Wu Yu,[1] Yang Lu,[2] and Guo-Liang Tian[3]

[1] *School of Mathematics and Computational Science, Hunan University of Science and Technology, Xiangtan, Hunan 411201, China*
[2] *School of Mathematics and Statistics, Central China Normal University, Wuhan, Hubei 430079, China*
[3] *Department of Statistics and Actuarial Science, The University of Hong Kong, Pokfulam Road, Hong Kong*

Correspondence should be addressed to Guo-Liang Tian; gltian@hku.hk

Academic Editor: Zhidong Bai

Tian et al. (2007) introduced a so-called hidden sensitivity model for evaluating the association of two sensitive questions with binary outcomes. However, in practice, we sometimes need to assess the association between one sensitive binary variable (e.g., whether or not a drug user, the number of sex partner being $\leqslant 1$ or >1, and so on) and one nonsensitive binary variable (e.g., good or poor health status, with or without cervical cancer, and so on). To address this issue, by sufficiently utilizing the information contained in the non-sensitive binary variable, in this paper, we propose a new survey scheme, called combination questionnaire design/model, which consists of a main questionnaire and a supplemental questionnaire. The introduction of the supplemental questionnaire which is indeed a design of direct questioning can effectively reduce the noncompliance behavior since more respondents will not be faced with the sensitive question. Likelihood-based inferences including maximum likelihood estimates via the expectation-maximization algorithm, asymptotic confidence intervals, and bootstrap confidence intervals of parameters of interest are derived. A likelihood ratio test is provided to test the association between the two binary random variables. Bayesian inferences are also discussed. Simulation studies are performed, and a cervical cancer data set in Atlanta is used to illustrate the proposed methods.

1. Introduction

Warner [1] introduced a randomized response technique to obtain truthful answers to questions with sensitive attributes. Using the Warner design, Kraemer [2] derived a bivariate correlation between an attribute with polytomous responses and an attribute with normally distributed responses. Fox and Tracy [3] derived estimation of the Pearson product moment correlation coefficient between two sensitive questions by assuming that randomized response observations can be treated as individual-level scores that are contaminated by random measurement error. Edgell et al. [4] considered the correlation between two sensitive questions using the unrelated question design or the additive constants design. Christofides [5] presented a randomized response technique with two randomization devices to estimate the proportion of individuals having two sensitive characteristics at the same time. Kim and Warde [6] considered a multinomial randomized response model which can handle untruthful responses. They also derived the Pearson product moment correlation estimator which may be used to quantify the linear relationship between two variables when multinomial response data are observed according to a randomized response procedure. However, all these randomized response procedures make use of one or two randomizing devices which (i) entail extra costs in both efficiency and complexity, (ii) increase the cognitive load of randomized response techniques, and (iii) allow for new sources of error, such as misunderstanding the randomized response procedures or cheating on the procedures [7].

From the perspective of incomplete categorical data design, Tian et al. [8] proposed a nonrandomized response model (called the hidden sensitivity model) for assessing the association of two sensitive questions with binary outcomes. To protect respondents' privacy and to avoid the use of any randomization device, they utilized a non-sensitive question in the questionnaire to indirectly obtain respondents' answers to the two sensitive questions. In the hidden sensitivity model, they implicitly assumed that all respondents are willing to follow the design instructions. In other words, the noncompliance behavior will not occur. However, in practice, we sometimes need to assess the association between one sensitive binary variable (e.g., whether or not a drug user, the number of sex partner being $\leqslant 1$ or >1, and so on) and one nonsensitive binary variable (e.g., good or poor health status, with or without cervical cancer, and so on). To our knowledge, the survey design for addressing this issue and corresponding statistical analysis methods are not available. Although we could directly adopt the hidden sensitivity model, the information contained in the nonsensitive binary variable cannot be utilized in the design. Intuitively, such information can be used to enhance the degree of privacy protection, so that more respondents will not be faced with the sensitive question. The major objective of this paper is to propose a new survey design and to develop corresponding statistical methods for analyzing sensitive data collected by this technique.

The rest of the paper is organized as follows. In Section 2, without using any randomizing device, we propose a survey scheme, called combination questionnaire design/model, which consists of a main questionnaire and a supplemental questionnaire. Likelihood-based inferences including maximum likelihood estimates via an *expectation–maximization* (EM) algorithm, asymptotic confidence intervals, and bootstrap confidence intervals of parameters of interest are derived in Section 3. A likelihood ratio test is also provided to test the association between the two binary random variables. In Section 4, we discuss Bayesian inferences when prior information on parameters is available. In Section 5, two simulation studies are performed to compare the efficiency of the proposed combination questionnaire model with that of the existing hidden sensitivity model of Tianet al. [8] (i.e., the main questionnaire only). A cervical cancer data set in Atlanta is used in Section 6 to illustrate the proposed methods. A discussion and an appendix on the mode of a group Dirichlet density and a sampling method from it are also presented.

2. The Survey Design

Assume that X is a sensitive binary random variable, Y is a non-sensitive binary random variable, and they are correlated. Let $\{X = 1\}$ denote the sensitive class (e.g., $X = 1$ if a respondent is a drug user) and $\{X = 0\}$ denote the non-sensitive class (e.g., $X = 0$ if a respondent is not a drug user). Furthermore, let both $\{Y = 1\}$ (e.g., $Y = 1$ if a respondent receives at least some college training) and $\{Y = 0\}$ (e.g., $Y = 0$ if a respondent graduates at most from some high school) be non-sensitive classes. Define $\boldsymbol{\theta} = (\theta_1, \ldots, \theta_4)^\top$, where $\theta_1 = \Pr(X = 0, Y = 0)$, $\theta_2 = \Pr(X = 1, Y = 0)$, $\theta_3 = \Pr(X = 1, Y = 1)$, and $\theta_4 = \Pr(X = 0, Y = 1)$, then $\boldsymbol{\theta} \in \mathbb{T}_4$, where

$$\mathbb{T}_n \,\hat{=} \left\{ (x_1, \ldots, x_n)^\top : x_i \geqslant 0,\ i = 1, \ldots, n,\ \sum_{i=1}^{n} x_i = 1 \right\} \quad (1)$$

Table 1: The main questionnaire for the combination questionnaire model.

Category	$W = 1$	$W = 2$	$W = 3$
I: $\{X = 0,\ Y = 0\}$	Block 1: —	Block 2: —	Block 3: —
II: $\{X = 1,\ Y = 0\}$	Category II: Please put a tick in Block 2		
III: $\{X = 1,\ Y = 1\}$	Category III: Please put a tick in Block 3		
IV: $\{X = 0,\ Y = 1\}$	Block 4: —		

Note: Only $\{X = 1\}$ is a sensitive class, while $\{X = 0\}$, $\{Y = 0\}$, and $\{Y = 1\}$ are nonsensitive classes.

Table 2: The supplemental questionnaire for the combination questionnaire model.

Category	
$\{Y = 0\}$	Please put a tick in Block 5: — if you belong to $\{Y = 0\}$
$\{Y = 1\}$	Please put a tick in Block 6: — if you belong to $\{Y = 1\}$

Note: Both the $\{Y = 0\}$ and $\{Y = 1\}$ are nonsensitive classes.

denotes the n-dimensional closed simplex in $\mathbb{R}^n$. The objective is to make inferences on $\boldsymbol{\theta}$, $\theta_x \,\hat{=}\, \Pr(X = 1) = \theta_2 + \theta_3$, $\theta_y \,\hat{=}\, \Pr(Y = 1) = \theta_4 + \theta_3$ and the odds ratio $\psi = \theta_1\theta_3/(\theta_2\theta_4)$.

The survey scheme consists of a main questionnaire and a supplemental questionnaire. To design the main questionnaire which is to be assigned to group 1 with n respondents (n is specified by the investigators), we first introduce a non-sensitive question (say, Q_W) with three possible answers. Assume that W is a non-sensitive variate with trichotomous outcomes associated with the Q_W and W is independent of both X and Y. Define $p_i = \Pr(W = i)$ for $i = 1, 2, 3$. For example, let $W = 1\ (2, 3)$ denote that a respondent was born in January–April (May–August, September–December), and thus we could assume that $p_i \approx 1/3$. The main questionnaire is shown in Table 1, under which each respondent is asked to answer the non-sensitive question Q_W.

On the one hand, since Category I (i.e., $\{X = 0, Y = 0\}$) and Category IV (i.e., $\{X = 0, Y = 1\}$) are non-sensitive to each respondent, it is reasonable to assume that a respondent is willing to provide his/her truthful answer by putting a tick in Block i ($i = 1, \ldots, 4$) according to his/her true status. On the other hand, Category II (i.e., $\{X = 1, Y = 0\}$) and Category III (i.e., $\{X = 1, Y = 1\}$) are usually sensitive to respondents. In this case, if a respondent belongs to Category II (III), he/she is designed to put a tick in Block 2 (3).

The supplemental questionnaire is designed as shown in Table 2, under which m respondents (m is also specified by the investigators) in group 2 are asked to put a tick in Block 5 or Block 6 depending on their true status; that is, $\{Y = 0\}$ or $\{Y = 1\}$. Since both the $\{Y = 0\}$ and $\{Y = 1\}$ are non-sensitive classes, the supplemental questionnaire is in fact a design of direct questioning. Therefore, we call this design the combination questionnaire model.

Table 3 shows the cell probabilities $\{\theta_i\}_{i=1}^4$, the observed frequencies $\{n_i\}_{i=1}^4$ and the unobservable frequencies $\{Z_i\}_{i=1}^3$, for the main questionnaire. The observed frequency n_2 is the sum of the frequency of respondents belonging to Block 2 and

TABLE 3: Cell probabilities, observed and unobservable counts for the main questionnaire.

Category	$W = 1$	$W = 2$	$W = 3$	Total
I: $\{X = 0, Y = 0\}$	$p_1\theta_1$	$p_2\theta_1$	$p_3\theta_1$	$\theta_1(Z_1)$
II: $\{X = 1, Y = 0\}$				$\theta_2(Z_2)$
III: $\{X = 1, Y = 1\}$				$\theta_3(Z_3)$
IV: $\{X = 0, Y = 1\}$		$\theta_4(n_4)$		$\theta_4(n_4)$
Total	$p_1\theta_1(n_1)$	$p_2\theta_1 + \theta_2(n_2)$	$p_3\theta_1 + \theta_3(n_3)$	$1(n)$

Note: $n = \sum_{i=1}^{4} n_i$, $Z_1 = n-(Z_2+Z_3)-n_4$, where (Z_2, Z_3) are unobservable.

TABLE 4: Cell probabilities and observed counts for the supplemental questionnaire.

Category		Total
$\{Y = 0\}$	Block 5: —	$\theta_1 + \theta_2(m_0)$
$\{Y = 1\}$	Block 6: —	$\theta_3 + \theta_4(m_1)$
Total		$1\ (m)$

Note: $m = m_0 + m_1$.

the frequency of those belonging to Category II. The observed frequency n_3 is the sum of the frequency of respondents belonging to Block 3 and the frequency of those belonging to Category III. Note that $n = \sum_{i=1}^{4} n_i = \sum_{i=1}^{3} Z_i + n_4$, we have $Z_1 = n-(Z_2+Z_3)-n_4$. Thus, only Z_2 and Z_3 are unobservable. Table 4 shows the cell probabilities and observed counts for the supplemental questionnaire.

Remark 1. The design of the main questionnaire is similar to that of the hidden sensitivity model of Tian et al. [8], while the design of the supplemental questionnaire is indeed a design of direct questioning since both the $\{Y = 0\}$ and $\{Y = 1\}$ are non-sensitive classes. Table 1 shows that Categories II and III are two sensitive subclasses. Therefore, putting a tick in Block 2 or Block 3 implies that the respondent could be suspected with the sensitive attribute. Let $\theta_1 = \cdots = \theta_4 \approx 0.25$ (see Table 3); then around half of the, say, $2n$ respondents will be suspected with the sensitive attribute if only the main questionnaire is employed. However, besides the main questionnaire (with n respondents), if the supplemental questionnaire (see Table 2) with $m = n$ respondents is also used, then only half of the n respondents will be suspected with the sensitive attribute. In other words, in the proposed combination questionnaire model, the information of the non-sensitive binary variable Y can be used to enhance the degree of privacy protection, so that more respondents will not be faced with the sensitive question. This is why we introduce the supplemental questionnaire besides the main questionnaire.

Remark 2. In practice, to simplify the design itself, we suggest that both the sample size n in the main questionnaire and the sample size m in the supplemental questionnaire should be fixed in advance rather than the total sample size $N = n + m$ being fixed. In this way, survey data can be collected in two independent groups, resulting in a relatively simpler statistical analysis. In addition, the interviewees are randomly assigned to either the group 1 or group 2.

3. Likelihood-Based Inferences

In this section, *maximum likelihood estimates* (MLEs) of the $\boldsymbol{\theta}$ and the odds ratio ψ are derived by using the *EM* algorithm. In addition, asymptotic confidence intervals and the bootstrap confidence intervals of an arbitrary function of $\boldsymbol{\theta}$ are also provided. Finally, a likelihood ratio test is presented for testing the association between the two binary random variables.

3.1. MLEs via the EM Algorithm. A total of $N = n + m$ respondents are classified into two groups by a randomization approach such that n respondents answer the questions in the main questionnaire and m respondents answer the questions in the supplemental questionnaire. Let $Y_{\text{obs},M} = \{n;\ n_1, \ldots, n_4\}$ denote the observed counts collected in the main questionnaire (see Table 3), where $\sum_{i=1}^{4} n_i = n$. The likelihood function of $\boldsymbol{\theta}$ based on $Y_{\text{obs},M}$ is

$$L(\boldsymbol{\theta} \mid Y_{\text{obs},M}) = \binom{n}{n_1, n_2, n_3, n_4} \times (p_1\theta_1)^{n_1} \left\{\prod_{i=2}^{3} (p_i\theta_1 + \theta_i)^{n_i}\right\} \theta_4^{n_4}. \tag{2}$$

Let $Y_{\text{obs},S} = \{m; m_0, m_1\}$ denote the observed counts gathered in the supplemental questionnaire (see Table 4), where $m_0 + m_1 = m$. The likelihood function of $\boldsymbol{\theta}$ based on $Y_{\text{obs},S}$ is

$$L(\boldsymbol{\theta} \mid Y_{\text{obs},S}) = \binom{m}{m_0} (\theta_1 + \theta_2)^{m_0} (\theta_3 + \theta_4)^{m_1}. \tag{3}$$

Let $Y_{\text{obs}} = \{Y_{\text{obs},M}, Y_{\text{obs},S}\}$. Since $Y_{\text{obs},M}$ and $Y_{\text{obs},S}$ are independent, the observed-data likelihood function of $\boldsymbol{\theta} \in \mathbb{T}_4$ is

$$L_{\text{CQ}}(\boldsymbol{\theta} \mid Y_{\text{obs},S}) \propto \theta_1^{n_1} \left\{\prod_{i=2}^{3} (p_i\theta_1 + \theta_i)^{n_i}\right\} \theta_4^{n_4} \times (\theta_1 + \theta_2)^{m_0} (\theta_3 + \theta_4)^{m_1}, \tag{4}$$

where the subscript "CQ" denotes the "combination questionnaire" model.

Since the corresponding cell probabilities to the observed counts n_2 and n_3 in the group 1 are in the form of summation (i.e., $p_i\theta_1 + \theta_i$, $i = 2, 3$), we cannot obtain the explicit expressions of the MLEs of $\boldsymbol{\theta}$ from the score equations of (4). By treating the observed counts n_2 and n_3 as incomplete data, we use the EM algorithm [9] to find the MLE $\widehat{\boldsymbol{\theta}}$ of $\boldsymbol{\theta}$. The counts $\{Z_i\}_{i=1}^{3}$ in Table 3 can be viewed as missing data. Briefly, Z_1, Z_2, Z_3, and n_4 represent the counts of the respondents belonging to Categories I, II, III, and IV, respectively. Thus, we denote the latent data by $Y_{\text{mis}} = \{z_2, z_3\}$ and the complete data by $Y_{\text{com}} = \{Y_{\text{obs}}, Y_{\text{mis}}\}$. Note that all $\{p_i\}$ are known. Consequently, the complete-data likelihood function for $\boldsymbol{\theta}$ is

$$L_{\text{CQ}}(\boldsymbol{\theta} \mid Y_{\text{COM}}) \propto \left(\prod_{i=1}^{3} \theta_i^{z_i}\right) \theta_4^{n_4} \times (\theta_1 + \theta_2)^{m_0} (\theta_3 + \theta_4)^{m_1}, \tag{5}$$

where $z_1 = n - (z_2 + z_3) - n_4$.

By treating $\{\theta_i\}_{i=1}^4$ as random variables, we note that the complete-data likelihood function (5) has the density form of a grouped Dirichlet distribution [10]. Ng et al. [11] derived the mode of a grouped Dirichlet density with explicit expressions (see the appendix). Hence, from (A.4) and (A.5), the complete-data MLEs for $\boldsymbol{\theta}$ are given by

$$\begin{aligned} \theta_1 &= 1 - \theta_2 - \theta_3 - \theta_4, \\ \theta_2 &= \frac{z_2}{N}\left(1 + \frac{m_0}{n - z_3 - n_4}\right), \\ \theta_3 &= \frac{z_3}{N}\left(1 + \frac{m_1}{z_3 + n_4}\right), \\ \theta_4 &= \frac{n_4}{N}\left(1 + \frac{m_1}{z_3 + n_4}\right). \end{aligned} \tag{6}$$

Given Y_{obs} and $\boldsymbol{\theta}$, Z_i follows the binomial distribution with parameters n_i and $\theta_i/(p_i\theta_1 + \theta_i)$; that is,

$$Z_i \mid (Y_{\text{obs}}, \boldsymbol{\theta}) \sim \text{Binomial}\left(n_i, \frac{\theta_i}{p_i\theta_1 + \theta_i}\right), \quad i = 2, 3. \tag{7}$$

Therefore, the E-step of the EM algorithm computes the following conditional expectations:

$$E(Z_i \mid Y_{\text{obs}}, \boldsymbol{\theta}) = \frac{n_i\theta_i}{p_i\theta_1 + \theta_i}, \quad i = 2, 3, \tag{8}$$

and the M-step updates (6) by replacing z_2 and z_3 with previous conditional expectations.

Remark 3. Based on the observed-data likelihood function (4), we could use the Newton-Raphson algorithm to find the MLEs of $\boldsymbol{\theta}$. However, it is well known that the Newton-Raphson algorithm does not necessarily increase the log likelihood, leading even to divergence sometimes [12, page 172]. In addition, the Newton–Raphson algorithm is sensitive to the initial values. One advantage of using the EM algorithm in the current situation is that both the E- and M-step have closed-form expressions. More importantly, the EM algorithm and the data augmentation algorithm of Tanner and Wong [13] share the same data augmentation structure in the Bayesian settings (see Section 4 for more details).

3.2. Asymptotic Confidence Intervals. Let $\boldsymbol{\theta}_{-4} = (\theta_1, \theta_2, \theta_3)^{\top}$. The asymptotic variance-covariance matrix of the MLE $\widehat{\boldsymbol{\theta}}_{-4}$ is then given by $\mathbf{I}_{\text{obs}}^{-1}(\widehat{\boldsymbol{\theta}}_{-4})$, where

$$\mathbf{I}_{\text{obs}}(\boldsymbol{\theta}_{-4}) = -\frac{\partial^2 \ell_{\text{CQ}}(\boldsymbol{\theta} \mid Y_{\text{obs}})}{\partial\boldsymbol{\theta}_{-4}\partial\boldsymbol{\theta}_{-4}^{\top}} \tag{9}$$

denotes the observed information matrix and $\ell_{\text{CQ}}(\boldsymbol{\theta} \mid Y_{\text{obs}}) = \log L_{\text{CQ}}(\boldsymbol{\theta} \mid Y_{\text{obs}})$ is the observed-data log-likelihood function. From (4), we have

$$\begin{aligned} \ell_{\text{CQ}}(\boldsymbol{\theta} \mid Y_{\text{obs}}) = {} & n_1 \log\theta_1 + n_2 \log(p_2\theta_1 + \theta_2) \\ & + n_3 \log(p_3\theta_1 + \theta_3) \\ & + n_4 \log(1 - \theta_1 - \theta_2 - \theta_3) \\ & + m_0 \log(\theta_1 + \theta_2) + m_1 \log(1 - \theta_1 - \theta_2). \end{aligned} \tag{10}$$

It is easy to show that

$$\begin{aligned} &\frac{\partial \ell_{\text{CQ}}(\boldsymbol{\theta} \mid Y_{\text{obs}})}{\partial\theta_1} \\ &\quad = \frac{n_1}{\theta_1} - \frac{n_4}{\theta_4} + \sum_{i=2}^{3} \frac{n_i p_i}{p_i\theta_1 + \theta_i} + \frac{m_0}{\theta_1 + \theta_2} - \frac{m_1}{1 - \theta_1 - \theta_2}, \\ &\frac{\partial \ell_{\text{CQ}}(\boldsymbol{\theta} \mid Y_{\text{obs}})}{\partial\theta_2} \\ &\quad = -\frac{n_4}{\theta_4} + \frac{n_2}{p_2\theta_1 + \theta_2} + \frac{m_0}{\theta_1 + \theta_2} - \frac{m_1}{1 - \theta_1 - \theta_2}, \\ &\frac{\partial \ell_{\text{CQ}}(\boldsymbol{\theta} \mid Y_{\text{obs}})}{\partial\theta_3} = -\frac{n_4}{\theta_4} + \frac{n_3}{p_3\theta_1 + \theta_3}, \\ &-\frac{\partial^2 \ell_{\text{CQ}}(\boldsymbol{\theta} \mid Y_{\text{obs}})}{\partial\theta_1^2} = \frac{n_1}{\theta_1^2} + \frac{n_4}{\theta_4^2} + \sum_{i=2}^{3} \frac{n_i p_i^2}{(p_i\theta_1 + \theta_i)^2} + \phi \\ &-\frac{\partial^2 \ell_{\text{CQ}}(\boldsymbol{\theta} \mid Y_{\text{obs}})}{\partial\theta_2^2} = \frac{n_4}{\theta_4^2} + \frac{n_2}{(p_2\theta_1 + \theta_2)^2} + \phi \\ &-\frac{\partial^2 \ell_{\text{CQ}}(\boldsymbol{\theta} \mid Y_{\text{obs}})}{\partial\theta_3^2} = \frac{n_4}{\theta_4^2} + \frac{n_3}{(p_3\theta_1 + \theta_3)^2} \\ &-\frac{\partial^2 \ell_{\text{CQ}}(\boldsymbol{\theta} \mid Y_{\text{obs}})}{\partial\theta_1\partial\theta_2} = \frac{n_4}{\theta_4^2} + \frac{n_2 p_2}{(p_2\theta_1 + \theta_2)^2} + \phi \\ &-\frac{\partial^2 \ell_{\text{CQ}}(\boldsymbol{\theta} \mid Y_{\text{obs}})}{\partial\theta_1\partial\theta_3} = \frac{n_4}{\theta_4^2} + \frac{n_3 p_3}{(p_3\theta_1 + \theta_3)^2} \\ &-\frac{\partial^2 \ell_{\text{CQ}}(\boldsymbol{\theta} \mid Y_{\text{obs}})}{\partial\theta_2\partial\theta_3} = \frac{n_4}{\theta_4^2}, \end{aligned} \tag{11}$$

where

$$\phi = \frac{m_0}{(\theta_1 + \theta_2)^2} + \frac{m_1}{(1 - \theta_1 - \theta_2)^2}. \tag{12}$$

Hence, the observed information matrix can be expressed as

$$\mathbf{I}_{\text{obs}}(\boldsymbol{\theta}_{-4}) = \text{diag}\left(\frac{n_1}{\theta_1^2}, 0, 0\right) + \frac{n_4}{\theta_4^2} \times \mathbf{1}_3\mathbf{1}_3^{\top} + \mathbf{A}, \tag{13}$$

where

$$\mathbf{A} = \begin{pmatrix} \sum_{i=2}^{3} \frac{n_i p_i^2}{(p_i\theta_1+\theta_i)^2} + \phi, & \frac{n_2 p_2}{(p_2\theta_1+\theta_2)^2} + \phi, & \frac{n_3 p_3}{(p_3\theta_1+\theta_3)^2} \\ \frac{n_2 p_2}{(p_2\theta_1+\theta_2)^2} + \phi, & \frac{n_2}{(p_2\theta_1+\theta_2)^2} + \phi, & 0 \\ \frac{n_3 p_3}{(p_3\theta_1+\theta_3)^2}, & 0, & \frac{n_3}{(p_3\theta_1+\theta_3)^2} \end{pmatrix}. \tag{14}$$

Let $\text{se}(\widehat{\theta}_i)$ denote the standard error of $\widehat{\theta}_i$ for $i = 1, 2, 3$. Note that $\text{se}(\widehat{\theta}_i)$ can be estimated by the square root of the ith diagonal element of $\mathbf{I}_{\text{obs}}^{-1}(\widehat{\boldsymbol{\theta}}_{-4})$. We denote the estimated value of $\text{se}(\widehat{\theta}_i)$ by $\widehat{\text{se}}(\widehat{\theta}_i)$. Thus, a 95% normal-based asymptotic confidence interval for θ_i can be constructed as

$$\left[\widehat{\theta}_i - 1.96 \times \widehat{\text{se}}\left(\widehat{\theta}_i\right), \widehat{\theta}_i + 1.96 \times \widehat{\text{se}}\left(\widehat{\theta}_i\right)\right], \quad i = 1, 2, 3. \tag{15}$$

Let $\vartheta = h(\boldsymbol{\theta}_{-4})$ be an arbitrary differentiable function of $\boldsymbol{\theta}_{-4}$. For example, $\theta_4 = 1 - \sum_{i=1}^{3} \theta_i$ and the odds ratio $\psi = \theta_1\theta_3/(\theta_2\theta_4)$. The delta method (e.g., [14, page 34]) can be used to approximate the standard error of $\widehat{\vartheta} = h(\widehat{\boldsymbol{\theta}}_{-4})$, and a 95% normal-based asymptotic confidence interval for ϑ is given by

$$\left[\widehat{\vartheta} - 1.96 \times \widehat{\text{se}}\left(\widehat{\vartheta}\right), \widehat{\vartheta} + 1.96 \times \widehat{\text{se}}\left(\widehat{\vartheta}\right)\right], \tag{16}$$

where

$$\widehat{\text{se}}\left(\widehat{\vartheta}\right) = \left\{ \left(\frac{\partial \vartheta}{\partial \boldsymbol{\theta}_{-4}}\right)^{\top} \mathbf{I}_{\text{obs}}^{-1}(\boldsymbol{\theta}_{-4}) \left(\frac{\partial \vartheta}{\partial \boldsymbol{\theta}_{-4}}\right)\Bigg|_{\boldsymbol{\theta}_{-4} = \widehat{\boldsymbol{\theta}}_{-4}} \right\}^{1/2}. \tag{17}$$

3.3. Bootstrap Confidence Intervals. When the normal-based asymptotic confidence interval like (15) is beyond the low bound zero or the upper bound one, the bootstrap approach [15] can be used to construct the bootstrap confidence interval of $\vartheta = h(\boldsymbol{\theta}_{-4})$. Based on the obtained MLE $\widehat{\boldsymbol{\theta}}$, we independently generate

$$(n_1^*, \dots, n_4^*)^{\top} \sim \text{Multinomial}\left(n; p_1\widehat{\theta}_1, p_2\widehat{\theta}_1 + \widehat{\theta}_2, p_3\widehat{\theta}_1 + \widehat{\theta}_3, \widehat{\theta}_4\right), \tag{18}$$

$$m_0^* \sim \text{Binomial}\left(m, \widehat{\theta}_1 + \widehat{\theta}_2\right). \tag{19}$$

Having obtained $Y_{\text{obs},M}^* = \{n; n_1^*, \dots, n_4^*\}$ and $Y_{\text{obs},S}^* = \{m; m_0^*, m_1^*\}$, where $m_1^* = m - m_0^*$, we can calculate the bootstrap replication $\widehat{\vartheta}^* = h(\boldsymbol{\theta}_{-4}^*)$ based on $Y_{\text{obs}}^* = \{Y_{\text{obs},M}^*, Y_{\text{obs},S}^*\}$ via the *EM* algorithm specified by (6) and (8). Independently repeating this process G times, we obtain G bootstrap replications $\{\widehat{\vartheta}_g^*\}_{g=1}^G$. Consequently, a $(1-\alpha)100\%$ bootstrap confidence interval for ϑ is given by

$$\left[\widehat{\vartheta}_L, \widehat{\vartheta}_U\right], \tag{20}$$

where $\widehat{\vartheta}_L$ and $\widehat{\vartheta}_U$ are the $100(\alpha/2)$ and $100(1-\alpha/2)$ percentiles of $\{\widehat{\vartheta}_g^*\}_{g=1}^G$, respectively.

3.4. The Likelihood Ratio Test for Testing Association. The likelihood ratio statistic can be used to test whether the two binary random variables X and Y are independent/correlated. The corresponding null and alternative hypotheses are [16, page 45]

$$H_0\colon \psi = 1 \text{ against } H_1\colon \psi \neq 1. \tag{21}$$

The likelihood ratio statistic is defined by

$$\Lambda = -2\left\{\ell_{\text{CQ}}\left(\widehat{\boldsymbol{\theta}}_R \mid Y_{\text{obs}}\right) - \ell_{\text{CQ}}\left(\widehat{\boldsymbol{\theta}} \mid Y_{\text{obs}}\right)\right\}, \tag{22}$$

where $\widehat{\boldsymbol{\theta}}_R$ denotes the restricted MLE of $\boldsymbol{\theta}$ under H_0, $\widehat{\boldsymbol{\theta}}$ denotes the MLE of $\boldsymbol{\theta}$, which can be obtained by the *EM* algorithm specified by (6) and (8), and $\ell_{\text{CQ}}(\boldsymbol{\theta} \mid Y_{\text{obs}}) = \log L_{\text{CQ}}(\boldsymbol{\theta} \mid Y_{\text{obs}})$.

To find the restricted MLE $\widehat{\boldsymbol{\theta}}_R$, we also employ the *EM* algorithm. Under $H_0\colon \theta_1\theta_3 = \theta_2\theta_4$, we have

$$\begin{aligned} \theta_1 &= (1-\theta_x)(1-\theta_y), \\ \theta_2 &= \theta_x(1-\theta_y), \\ \theta_3 &= \theta_x\theta_y, \\ \theta_4 &= (1-\theta_x)\theta_y. \end{aligned} \tag{23}$$

In other words, under H_0 we only have two free parameters θ_x and θ_y. Having obtained the restricted MLEs $\widehat{\theta}_{x,R}$ and $\widehat{\theta}_{y,R}$, we can compute the restricted MLE $\widehat{\boldsymbol{\theta}}_R = (\widehat{\theta}_{1,R}, \dots, \widehat{\theta}_{4,R})^{\top}$ from (23) by

$$\begin{aligned} \widehat{\theta}_{1,R} &= \left(1-\widehat{\theta}_{x,R}\right)\left(1-\widehat{\theta}_{y,R}\right), \\ \widehat{\theta}_{2,R} &= \widehat{\theta}_{x,R}\left(1-\widehat{\theta}_{y,R}\right), \\ \widehat{\theta}_{3,R} &= \widehat{\theta}_{x,R}\widehat{\theta}_{y,R}, \\ \widehat{\theta}_{4,R} &= \left(1-\widehat{\theta}_{x,R}\right)\widehat{\theta}_{y,R}. \end{aligned} \tag{24}$$

In what follows, we consider the computation of the restricted MLEs $\widehat{\theta}_{x,R}$ and $\widehat{\theta}_{y,R}$. Now, the complete-data likelihood function (5) becomes

$$\begin{aligned} L_{\text{CQ}}&\left(\theta_x, \theta_y \mid Y_{\text{com}}, H_0\right) \\ &\propto \left\{(1-\theta_x)(1-\theta_y)\right\}^{z_1}\left\{\theta_x(1-\theta_y)\right\}^{z_2} \\ &\quad \times \left(\theta_x\theta_y\right)^{z_3}\left\{(1-\theta_x)\theta_y\right\}^{n_4}\left(1-\theta_y\right)^{m_0}\theta_y^{m_1} \\ &= \theta_x^{z_2+z_3}(1-\theta_x)^{z_1+n_4}\theta_y^{z_3+n_4+m_1}\left(1-\theta_y\right)^{z_1+z_2+m_0} \end{aligned} \tag{25}$$

so that the restricted MLEs of θ_x and θ_y based on the complete-data are given by

$$\theta_x = \frac{z_2+z_3}{n}, \qquad \theta_y = \frac{z_3+n_4+m_1}{N}, \tag{26}$$

respectively. Thus, the M-step of the *EM* algorithm calculates (26), and the E-step computes the conditional expectations given in (8), where $\{\theta_i\}$ are defined in (23). Finally, under H_0, Λ asymptotically follows chi-squared distribution with one degree of freedom.

4. Bayesian Inferences

To derive the posterior mode of $\boldsymbol{\theta}$, we employ the *EM* algorithm again. The latent data $Y_{\text{mis}} = \{z_2, z_3\}$ are the same as those in Section 3.1. Based on the complete-data likelihood function (5), if the Dirichlet distribution Dirichlet $(a_1, \ldots, a_4)$ is adopted as the prior distribution of $\boldsymbol{\theta}$, then the complete-data posterior distribution is a *grouped Dirichlet* (GD) distribution with the formal definition given by (A.2); that is,

$$f(\boldsymbol{\theta} \mid Y_{\text{obs}}, Y_{\text{mis}}) = \text{GD}_{4,2,2}(\boldsymbol{\theta}_{-4} \mid \mathbf{a}, \mathbf{b}), \tag{27}$$

where $\mathbf{a} = (a_1 + z_1, a_2 + z_2, a_3 + z_3, a_4 + n_4)^{\top}$, $\mathbf{b} = (m_0, m_1)^{\top}$, and $z_1 = n - (z_2 + z_3) - n_4$. The conditional predictive distribution is

$$f(Y_{\text{mis}} \mid Y_{\text{obs}}, \boldsymbol{\theta}) = \prod_{i=2}^{3} \text{Binomial}\left(z_i \,\middle|\, n_i, \frac{\theta_i}{p_i\theta_1 + \theta_i}\right). \tag{28}$$

Therefore, the *M*-step of the *EM* algorithm is to calculate the complete-data posterior mode:

$$\begin{aligned}
\theta_1 &= 1 - \theta_2 - \theta_3 - \theta_4, \\
\theta_2 &= \frac{a_2 - 1 + z_2}{a_+ - 4 + N}\left(1 + \frac{m_0}{a_1 + a_2 - 2 + n - z_3 - n_4}\right), \\
\theta_3 &= \frac{a_3 - 1 + z_3}{a_+ - 4 + N}\left(1 + \frac{m_1}{a_3 + a_4 - 2 + z_3 + n_4}\right), \\
\theta_4 &= \frac{a_4 - 1 + n_4}{a_+ - 4 + N}\left(1 + \frac{m_1}{a_3 + a_4 - 2 + z_3 + n_4}\right),
\end{aligned} \tag{29}$$

where $a_+ = \sum_{i=1}^{4} a_i$, and the *E*-step is to replace $\{z_i\}_{i=2}^{3}$ by the conditional expectations given by (8).

In addition, based on (27) and (28), the data augmentation algorithm of Tanner and Wong [13] can be used to generate posterior samples of $\boldsymbol{\theta}$. A sampling method from (27) is given in the appendix.

5. Simulation Studies

In this section, two simulation studies are conducted to compare the efficiency of the proposed combination questionnaire model with that of the hidden sensitivity model of Tian et al. [8] (i.e., the main questionnaire model only), where (p_1, p_2, p_3) are assumed to be $(1/3, 1/3, 1/3)$ and $p_i \hat{=} \Pr(W = i)$ for $i = 1, 2, 3$. In the first simulated example, let the total sample size in the combination questionnaire model be the same as the sample size in the hidden sensitivity model. In the second example, the sample size for the main questionnaire in the combination questionnaire model is assumed to be equal to the sample size in the hidden sensitivity model.

In the first simulated example, let a total of $N = n + m = 50 + 50 = 100$ participants be interviewed by using the combination questionnaire model. The true values of $\{\theta_i\}_{i=1}^{4}$ are listed in the second column of Table 5. We first generate

$$(n_1, \ldots, n_4)^{\top} \sim \text{Multinomial}(n; p_1\theta_1, p_2\theta_1 + \theta_2, p_3\theta_1 + \theta_3, \theta_4), \tag{30}$$

and $m_0 \sim$ Binomial $(m, \theta_1 + \theta_2)$ so that $m_1 = m - m_0$. The *EM* algorithm (6) and (8) is used to calculate the MLEs of $\{\theta_i\}_{i=1}^{4}$. We repeated this experiment 1000 times. The average MLEs of $\{\theta_i\}_{i=1}^{4}$ are reported in the third column of Table 5. The corresponding bias, variance, and *mean square error* (MSE) are displayed in the fourth, fifth, and sixth columns of Table 5. Next, let $N = n + m = 100 + 0 = 100$ participants be interviewed by using the hidden sensitivity model (i.e., the main questionnaire only). The corresponding results are reported in the last four columns of Table 5.

From Table 5, we can see that both the MLEs of $\{\theta_i\}_{i=1}^{4}$ in the combination questionnaire model and the hidden sensitivity model are very close to their true values, while the MSEs of $\{\hat{\theta}_i\}_{i=1}^{4}$ in the combination questionnaire model are slightly larger than those in the hidden sensitivity model. These numerical results are not surprising since in the hidden sensitivity model, Tian et al. [8] implicitly assumed that all respondents must strictly follow the design instructions. In other words, the noncompliance behavior will not occur. The introduction of the supplemental questionnaire in the combination questionnaire model can effectively reduce the non-compliance behavior since more respondents will not be faced with the sensitive question, while the cost for introducing such a supplemental questionnaire is that we definitely lose a little of efficiency.

In the second simulated example, we assume that a total of $N = n + m = 100 + 100 = 200$ participants are interviewed by using the combination questionnaire model, while only $N = n + m = 100 + 0 = 100$ are interviewed by using the hidden sensitivity model. We repeat this experiment 1000 times. The corresponding results are reported in Table 6.

From Table 6, we can see that the MSEs of $\{\hat{\theta}_i\}_{i=1}^{4}$ in the combination questionnaire model are smaller than those in the hidden sensitivity model. In addition, by comparing the fifth columns in Tables 5 and 6, we can see that the precisions of $\{\hat{\theta}_i\}_{i=1}^{4}$ for the proposed combination questionnaire model in the second simulated example are significantly improved when compared with those in the first simulated example.

6. Analyzing Cervical Cancer Data in Atlanta

Williamson and Haber [17] reported a study which examined the relationship between disease status of cervical cancer and the number of sex partners and other risk factors. Cases were 20–79-year-old women of Fulton or DeKalb county in Atlanta, Georgia. They were diagnosed and were ascertained to have invasive cervical cancer. Controls were randomly chosen from the same counties and the same age ranges. Table 7 gives the cross-classification of number of sex partners ("few, 0–3" or "many, ⩾4", denoted by $X = 0$ or $X = 1$) and disease status (control or case, denoted by $Y = 0$ or $Y = 1$). Generally, a sizable proportion (13.5% in this example) of the responses would be missing because of the sensitive question about the number of sex partners in a telephone interview. The objective is to examine if association exists between the number of sex partners and disease status of cervical cancer.

For the purpose of illustration, we presume that $\{X = 0\}$ is non-sensitive although the number 0–3 of sex partners

TABLE 5: Comparison of the efficiency of the proposed combination questionnaire model with that of the hidden sensitivity model of Tian et al. [8] (i.e., the main questionnaire model only) with $(p_1, p_2, p_3) = (1/3, 1/3, 1/3)$.

Parameter	True value	The combination questionnaire model ($N = n + m = 50 + 50$)				The hidden sensitivity model ($N = n + m = 100 + 0$)			
		MLE	Bias	Variance	MSE	MLE	Bias	Variance	MSE
θ_1	0.10	0.1013	0.0013	0.0113	0.0113	0.1010	0.0010	0.0064	0.0064
θ_2	0.50	0.4985	−0.0015	0.0089	0.0089	0.4975	−0.0025	0.0041	0.0041
θ_3	0.20	0.2008	0.0008	0.0037	0.0037	0.2001	0.0001	0.0029	0.0029
θ_4	0.20	0.1994	−0.0006	0.0026	0.0026	0.2014	0.0014	0.0016	0.0016
θ_1	0.20	0.1957	−0.0043	0.0090	0.0090	0.1985	−0.0015	0.0055	0.0055
θ_2	0.30	0.2999	−0.0001	0.0070	0.0070	0.2999	−0.0001	0.0034	0.0034
θ_3	0.40	0.4038	0.0038	0.0040	0.0040	0.4025	0.0025	0.0037	0.0037
θ_4	0.10	0.1006	0.0006	0.0017	0.0017	0.0991	−0.0009	0.0009	0.0009
θ_1	0.30	0.3021	0.0021	0.0107	0.0107	0.3026	0.0026	0.0081	0.0081
θ_2	0.25	0.2483	−0.0017	0.0077	0.0077	0.2499	−0.0001	0.0039	0.0039
θ_3	0.15	0.1474	−0.0026	0.0042	0.0042	0.1476	−0.0024	0.0032	0.0032
θ_4	0.30	0.3021	0.0021	0.0033	0.0033	0.2999	−0.0001	0.0021	0.0021

Note: $\text{Bias}(\hat{\theta}_i) = E(\hat{\theta}_i) - \theta_i$ and $\text{MSE}(\hat{\theta}_i) = [\text{Bias}(\hat{\theta}_i)]^2 + \text{Var}(\hat{\theta}_i)$.

TABLE 6: Comparison of the efficiency of the proposed combination questionnaire model with that of the hidden sensitivity model of Tian et al. [8] (i.e., the main questionnaire model only) with $(p_1, p_2, p_3) = (1/3, 1/3, 1/3)$.

Parameter	True value	The combination questionnaire model ($N = n + m = 100 + 100$)				The hidden sensitivity model ($N = n + m = 100 + 0$)			
		MLE	Bias	Variance	MSE	MLE	Bias	Variance	MSE
θ_1	0.10	0.1003	0.0003	0.0030	0.0030	0.1013	0.0013	0.0064	0.0064
θ_2	0.50	0.5015	0.0015	0.0029	0.0029	0.5019	0.0019	0.0041	0.0041
θ_3	0.20	0.1986	−0.0014	0.0017	0.0017	0.1969	−0.0031	0.0028	0.0028
θ_4	0.20	0.1995	−0.0005	0.0013	0.0013	0.1999	−0.0001	0.0016	0.0016
θ_1	0.20	0.2007	0.0007	0.0041	0.0041	0.2049	0.0049	0.0057	0.0057
θ_2	0.30	0.2989	−0.0011	0.0033	0.0033	0.2965	−0.0035	0.0034	0.0034
θ_3	0.40	0.4007	0.0007	0.0020	0.0020	0.3992	−0.0008	0.0037	0.0037
θ_4	0.10	0.0997	−0.0003	0.0009	0.0009	0.0994	−0.0006	0.0009	0.0009
θ_1	0.30	0.3032	0.0032	0.0054	0.0054	0.2965	−0.0036	0.0079	0.0079
θ_2	0.25	0.2488	−0.0012	0.0039	0.0039	0.2501	0.0001	0.0038	0.0038
θ_3	0.15	0.1481	−0.0019	0.0021	0.0021	0.1531	0.0031	0.0032	0.0032
θ_4	0.30	0.3000	0.0000	0.0017	0.0017	0.3003	0.0003	0.0021	0.0021

Note: $\text{Bias}(\hat{\theta}_i) = E(\hat{\theta}_i) - \theta_i$ and $\text{MSE}(\hat{\theta}_i) = [\text{Bias}(\hat{\theta}_i)]^2 + \text{Var}(\hat{\theta}_i)$.

TABLE 7: Cervical cancer data from Williamson and Haber [17].

Number of sex partners	Disease status of cervical cancer	
	$Y = 0$ (control)	$Y = 1$ (case)
$X = 0$ (few, 0–3)	165 (r_1, θ_1)	103 (r_4, θ_4)
$X = 1$ (many, ⩾4)	164 (r_2, θ_2)	221 (r_3, θ_3)
Missing	43 $(r_{12}, \theta_1 + \theta_2)$	59 $(r_{34}, \theta_3 + \theta_4)$

Note: The observed counts and the corresponding cell probabilities are in parentheses. X is a sensitive binary variate and Y is a non-sensitive binary variate.

is somewhat sensitive for some respondents. To illustrate the proposed design and approaches, we let $W = 1\ (2,3)$ if a respondent was born in January–April (May–August, September–December). It is then reasonable to assume that $p_k = \Pr(W = k) \approx 1/3$ for $k = 1, 2, 3$, and W is independent of the sensitive binary variate X and the the non-sensitive binary variate Y. For the ideal situation (i.e., no sampling errors), the observed counts from the main questionnaire as shown in Tables 1 and 3 would be $n_1 = r_1/3 = 55$, $n_2 = 55 + r_2 = 219$, $n_3 = 55 + r_3 = 276$, $n_4 = r_4 = 103$; that is,

$$Y_{\text{obs},M} = \{n; n_1, \ldots, n_4\} = \{653; 55, 219, 276, 103\}. \quad (31)$$

On the other hand, we can view the missing data in Table 7 as the observed counts from the supplemental questionnaire as shown in Table 2. From Table 4, we have $m_0 = r_{12} = 43$ and $m_1 = r_{34} = 59$; that is, $Y_{\text{obs},S} = \{m; m_0, m_1\} = \{102; 43, 59\}$. Therefore, we obtain the observed data $Y_{\text{obs}} = \{Y_{\text{obs},M}, Y_{\text{obs},S}\}$.

TABLE 8: MLEs and 95% confidence intervals of $\boldsymbol{\theta}$ and ψ.

Parameter	MLE	std	95% asymptotic CI	95% bootstrap CI
θ_1	0.23793	0.02937	[0.18036, 0.29551]	[0.18006, 0.29879]
θ_2	0.24916	0.02261	[0.20484, 0.29348]	[0.20375, 0.29359]
θ_3	0.35196	0.02247	[0.30790, 0.39601]	[0.30731, 0.39666]
θ_4	0.16095	0.01434	[0.13284, 0.18905]	[0.13400, 0.18908]
ψ	2.08830	0.43971	[1.22646, 2.95011]	[1.35361, 3.18593]

6.1. Likelihood-Based Inferences. Using $\boldsymbol{\theta}^{(0)} = \mathbf{1}_4/4$ as the initial values, the *EM* algorithm in (6) and (8) converged in 29 iterations. The resultant MLEs of $\boldsymbol{\theta}$ and ψ are given in the second column of Table 8. From (13) and (14), the asymptotic variance-covariance matrix of the MLEs $\widehat{\boldsymbol{\theta}}_{-4}$ is

$$\mathbf{I}_{\text{obs}}^{-1}\left(\widehat{\boldsymbol{\theta}}_{-4}\right) = \begin{pmatrix} 0.00086302 & -0.000442447 & -0.000384452 \\ -0.00044245 & 0.000511253 & -0.000010026 \\ -0.00038445 & -0.000010026 & 0.000505241 \end{pmatrix}. \quad (32)$$

The estimated standard errors of $\widehat{\theta}_i$ $(i = 1, 2, 3)$ are square roots of the main diagonal elements of the above matrix. From (17), the estimated standard errors of $\widehat{\theta}_4 = 1-\widehat{\theta}_1-\widehat{\theta}_2-\widehat{\theta}_3$ and $\widehat{\psi} = \widehat{\theta}_1\widehat{\theta}_3/(\widehat{\theta}_2\widehat{\theta}_4)$ are given by

$$\widehat{\text{se}}\left(\widehat{\theta}_4\right) = \left\{\mathbf{1}_3^{\top}\, \mathbf{I}_{\text{obs}}^{-1}\left(\widehat{\boldsymbol{\theta}}_{-4}\right) \mathbf{1}_3\right\}^{1/2}, \qquad \widehat{\text{se}}\left(\widehat{\psi}\right) = \left\{\boldsymbol{\alpha}^{\top}\mathbf{I}_{\text{obs}}^{-1}\left(\widehat{\boldsymbol{\theta}}_{-4}\right)\boldsymbol{\alpha}\right\}^{1/2}, \quad (33)$$

respectively, where

$$\boldsymbol{\alpha} = \left(\frac{\widehat{\theta}_3\left(1-\widehat{\theta}_2-\widehat{\theta}_3\right)}{\widehat{\theta}_2\widehat{\theta}_4^2}, \frac{\widehat{\theta}_1\widehat{\theta}_3\left(\widehat{\theta}_2-\widehat{\theta}_4\right)}{\left(\widehat{\theta}_2\widehat{\theta}_4\right)^2}, \frac{\widehat{\theta}_1\left(1-\widehat{\theta}_1-\widehat{\theta}_2\right)}{\widehat{\theta}_2\widehat{\theta}_4^2}\right)^{\top}. \quad (34)$$

These estimated standard errors are listed in the third column of Table 8. From (15) and (16), we can obtain the 95% asymptotic confidence intervals of $\boldsymbol{\theta}$ and ψ, which are showed in the fourth column of Table 8.

Based on (18) and (19), we generate $G = 10{,}000$ bootstrap samples. The corresponding 95% bootstrap confidence intervals of $\boldsymbol{\theta}$ and ψ are displayed in the last column of Table 8.

To test the null hypothesis H_0: $\psi = 1$ against H_1: $\psi \neq 1$, we need to obtain the restricted MLE $\widehat{\boldsymbol{\theta}}_R$. Using

$$\left(\theta_x^{(0)}, \theta_y^{(0)}\right)^{\top} = (0.5, 0.5)^{\top}, \quad (35)$$

as the initial values, the *EM* algorithm in (26) converged to

$$\widehat{\theta}_{x,R} = 0.64807, \qquad \widehat{\theta}_{y,R} = 0.52948, \quad (36)$$

in 19 iterations. From (24), the restricted MLEs of $\boldsymbol{\theta}$ are obtained as

$$\widehat{\boldsymbol{\theta}}_R = (0.16559, 0.30493, 0.34314, 0.18634)^{\top}. \quad (37)$$

The log-likelihood ratio statistic Λ is equal to 12.469 and the P-value is 0.0004137. Since this P-value is far less than 0.05, the H_0 is rejected at the 0.05 level of significance. Thus, we can conclude that there is an association between sex partners and cervical cancer status based on the current data. This conclusion is identical to that from the two 95% confidence intervals of the odds ratio as shown in Table 8, where both confidence intervals exclude the value 1.

TABLE 9: Posterior modes and estimates of parameters for the cervical cancer data.

Parameter	Posterior mode	Bayesian mean	Bayesian std	95% Bayesian credible interval
θ_1	0.23793	0.24025	0.02934	[0.18574, 0.30050]
θ_2	0.24916	0.24789	0.02251	[0.20365, 0.29192]
θ_3	0.35196	0.35029	0.02246	[0.30621, 0.39432]
θ_4	0.16095	0.16155	0.01431	[0.13448, 0.19042]
ψ	2.08830	2.14538	0.45838	[1.39399, 3.18678]

6.2. Bayesian Inferences. When Dirichlet (1, 1, 1, 1) (i.e., the uniform distribution on $\mathbb{T}_4$) is adopted as the prior distribution of $\boldsymbol{\theta}$, the posterior modes of $\boldsymbol{\theta}$are equal to the corresponding MLEs. Using $\boldsymbol{\theta}^{(0)} = \mathbf{1}_4/4$ as the initial values, we employ the data augmentation algorithm to generate 1,000,000 posterior samples of $\boldsymbol{\theta}$ and discard the first half of the samples. The Bayesian estimates of $\boldsymbol{\theta}$ and ψ are given in Table 9. Since the lower bound of the Bayesian credible interval of the ψ is larger than 1, we believe that there is an association between the number of sexual partners and cervical cancer status.

The posterior densities of the $\{\theta_i\}_{i=1}^4$ and ψ estimated by a kernel density smoother are plotted in Figures 1 and 2.

7. Discussion

In this paper, we develop a general framework of design and analysis for the combination questionnaire model, which consists of a main questionnaire and a supplemental questionnaire. In fact, the main questionnaire (see Table 1) is a generalization of the nonrandomized triangular model [18] and is then a special case of the multicategory triangular model [19]. The supplemental questionnaire (see Table 2) is a design of direct questioning. The introduction of the supplemental questionnaire can effectively reduce the noncompliance behavior since more respondents will not be faced with the sensitive question, while the cost for introducing such a supplemental questionnaire is that we definitely lose a little of efficiency. The combination questionnaire model can be used to gather information to evaluate the association between one sensitive binary variable and one non-sensitive binary variable.

We note that the proposed combination questionnaire model has one limitation in applications; that is, it cannot be

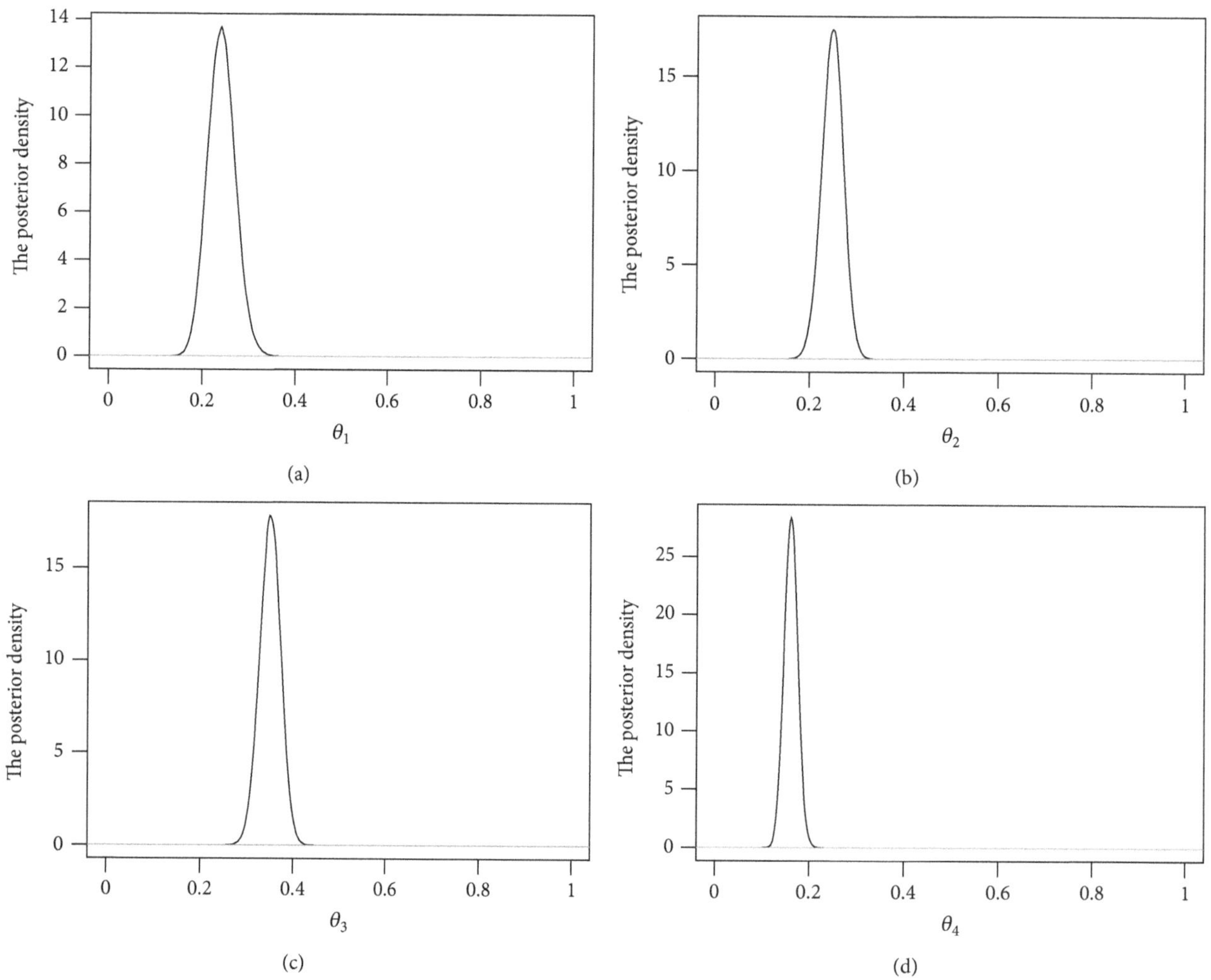

FIGURE 1: The posterior densities of the $\{\theta_i\}_{i=1}^4$ estimated by a kernel density smoother based on the second 500,000 posterior samples of $\boldsymbol{\theta}$generated by the data augmentation algorithm when the prior distribution is Dirichlet $(1, 1, 1, 1)$. (a) The posterior density of θ_1; (b) The posterior density of θ_2; (c) The posterior density of θ_3; (d) The posterior density of θ_4.

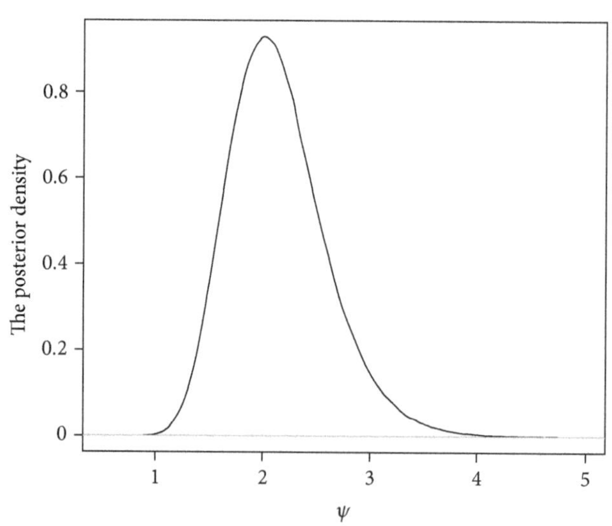

FIGURE 2: The posterior density of the odds ratio ψ estimated by a kernel density smoother based on the second 500,000 posterior samples of $\boldsymbol{\theta}$ generated by the data augmentation algorithm when the prior distribution is Dirichlet $(1, 1, 1, 1)$.

applied to situation where two categories $\{X = 0\}$ and $\{X = 1\}$ are sensitive like income. For example, let $X = 0$ if his/her annual income is \$25,000 or less and $X = 1$ if his/her annual income is more than \$25,000. For such cases, it is worthwhile to develop new designs to address this issue. One way is to replace the main questionnaire in Table 1 by a four-category parallel model (see [20, Section 4.1]). The other way is to employ the parallel model [21] to collect the information on X and to employ the design of direct questioning to collect information on Y then we could use the logistic regression to estimate the odds ratio, which is one of our further researches.

Appendix

The Mode of a Group Dirichlet Density and a Sampling Method from a GDD

Let

$$\mathbb{V}_{n-1} = \left\{ (x_1, \ldots, x_{n-1})^\top : x_i \geqslant 0, \; i = 1, \ldots, n-1, \sum_{i=1}^{n-1} x_i \leqslant 1 \right\} \quad \text{(A.1)}$$

denote the $(n-1)$-dimensional open simplex in $\mathbb{R}^{n-1}$. A random vector $\mathbf{x} = (X_1, \ldots, X_n)^\top \in \mathbb{T}_n$ is said to follow a *group Dirichlet distribution* (GDD) with two partitions, if the density of $\mathbf{x}_{-n} = (X_1, \ldots, X_{n-1})^\top \in \mathbb{V}_{n-1}$ is

$$\text{GD}_{n,2,s}(x_{-n} \mid \mathbf{a}, \mathbf{b}) = c_{\text{GD}}^{-1} \left(\prod_{i=1}^{n} x_i^{a_i - 1} \right) \left(\sum_{i=1}^{s} x_i \right)^{b_1} \left(\sum_{i=s+1}^{n} x_i \right)^{b_2}, \quad \text{(A.2)}$$

where $x_{-n} = (x_1, \ldots, x_{n-1})^\top$, $\mathbf{a} = (a_1, \ldots, a_n)^\top$ is a positive parameter vector, $\mathbf{b} = (b_1, b_2)^\top$ is a nonnegative parameter vector, s is a known positive integer less than n, and the normalizing constant is given by

$$c_{\text{GD}} = B(a_1, \ldots, a_s) B(a_{s+1}, \ldots, a_n) \times B\left(\sum_{i=1}^{s} a_i + b_1, \sum_{i=s+1}^{n} a_i + b_2 \right). \quad \text{(A.3)}$$

We write $\mathbf{x} \sim \text{GD}_{n,2,s}(\mathbf{a}, \mathbf{b})$ on $\mathbb{T}_n$ or $\mathbf{x}_{-n} \sim \text{GD}_{n,2,s}(\mathbf{a}, \mathbf{b})$ on $\mathbb{V}_{n-1}$ to distinguish the two equivalent representations.

If $a_i \geqslant 1$, then the mode of the grouped Dirichlet density (A.2) is given by [11, 22]

$$\hat{x}_i = \frac{a_i - 1}{\sum_{j=1}^{n} (a_j - 1) + b_1 + b_2} \left\{ 1 + \frac{b_1}{\sum_{j=1}^{s} (a_j - 1)} \right\}, \quad i = 1, \ldots, s, \quad \text{(A.4)}$$

$$\hat{x}_i = \frac{a_i - 1}{\sum_{j=1}^{n} (a_j - 1) + b_1 + b_2} \left\{ 1 + \frac{b_2}{\sum_{j=s+1}^{n} (a_j - 1)} \right\}, \quad i = s+1, \ldots, n. \quad \text{(A.5)}$$

The following procedure can be used to generate i.i.d. samples from a GDD [11]. Let

(1) $\mathbf{y}^{(1)} \sim$ Dirichlet $(a_1, \ldots, a_s)$ on $\mathbb{T}_s$;

(2) $\mathbf{y}^{(2)} \sim$ Dirichlet $(a_{s+1}, \ldots, a_n)$ on $\mathbb{T}_{n-s}$;

(3) $R \sim \text{Beta}(\sum_{i=1}^{s} a_i + b_1, \sum_{i=s+1}^{n} a_i + b_2)$; and

(4) $\mathbf{y}^{(1)}, \mathbf{y}^{(2)}$ and R are mutually independent.

Define

$$\mathbf{x}^{(1)} = R \times \mathbf{y}^{(1)}, \qquad \mathbf{x}^{(2)} = (1 - R) \times \mathbf{y}^{(2)}. \quad \text{(A.6)}$$

Then,

$$\mathbf{x} = \begin{pmatrix} \mathbf{x}^{(1)} \\ \mathbf{x}^{(2)} \end{pmatrix} \sim \text{GD}_{n,2,s}(\mathbf{a}, \mathbf{b}) \quad \text{(A.7)}$$

on $\mathbb{T}_n$, where $\mathbf{a} = (a_1, \ldots, a_n)^\top$ and $\mathbf{b} = (b_1, b_2)^\top$.

Acknowledgments

The authors are grateful to the Editor, an Associate Editor, and two anonymous referees for their helpful comments and suggestions which led to the improvement of the paper. They also would like to thank Miss Yin LIU of The University of Hong Kong for her help in the simulation studies. Jun-Wu Yu's research was partially supported by a Grant (no. 09BTJ012) from the National Social Science Foundation of China and a Grant from Hunan Provincial Science and Technology Department (no. 11JB1176). Partial work was done when Jun-Wu Yu visited The University of Hong Kong. Guo-Liang Tian's research was partially supported by a Grant (HKU 779210M) from the Research Grant Council of the Hong Kong Special Administrative Region.

References

[1] S. L. Warner, "Randomized response: a survey technique for eliminating evasive answer bias," *Journal of the American Statistical Association*, vol. 60, no. 309, pp. 63–69, 1965.

[2] H. C. Kraemer, "Estimation and testing of bivariate association using data generated by the randomized response technique," *Psychological Bulletin*, vol. 87, no. 2, pp. 304–308, 1980.

[3] J. A. Fox and P. E. Tracy, "Measuring associations with randomized response," *Social Science Research*, vol. 13, no. 2, pp. 188–197, 1984.

[4] S. E. Edgell, S. Himmelfarb, and D. J. Cira, "Statistical efficiency of using two quantitative randomized response techniques to estimate correlation," *Psychological Bulletin*, vol. 100, no. 2, pp. 251–256, 1986.

[5] T. C. Christofides, "Randomized response technique for two sensitive characteristics at the same time," *Metrika*, vol. 62, no. 1, pp. 53–63, 2005.

[6] J. M. Kim and W. D. Warde, "Some new results on the multinomial randomized response model," *Communications in Statistics: Theory and Methods*, vol. 34, no. 4, pp. 847–856, 2005.

[7] G. J. L. M. Lensvelt-Mulders, J. J. Hox, P. G. M. van der Heijden, and C. J. M. Maas, "Meta-analysis of randomized response research: thirty-five years of validation," *Sociological Methods & Research*, vol. 33, no. 3, pp. 319–348, 2005.

[8] G. L. Tian, J. W. Yu, M. L. Tang, and Z. Geng, "A new non-randomized model for analysing sensitive questions with binary outcomes," *Statistics in Medicine*, vol. 26, no. 23, pp. 4238–4252, 2007.

[9] A. P. Dempster, N. M. Laird, and D. B. Rubin, "Maximum likelihood from incomplete data via the EM algorithm," *Journal of the Royal Statistical Society B*, vol. 39, no. 1, pp. 1–38, 1977.

[10] G.-L. Tian, K. W. Ng, and Z. Geng, "Bayesian computation for contingency tables with incomplete cell-counts," *Statistica Sinica*, vol. 13, no. 1, pp. 189–206, 2003.

[11] K. W. Ng, M. L. Tang, M. Tan, and G. L. Tian, "Grouped Dirichlet distribution: a new tool for incomplete categorical data analysis," *Journal of Multivariate Analysis*, vol. 99, no. 3, pp. 490–509, 2008.

[12] D. R. Cox and D. Oakes, *Analysis of Survival Data*, Monographs on Statistics and Applied Probability, Chapman & Hall, London, UK, 1984.

[13] M. A. Tanner and W. H. Wong, "The calculation of posterior distributions by data augmentation," *Journal of the American Statistical Association*, vol. 82, no. 398, pp. 528–550, 1987.

[14] M. A. Tanner, *Tools for Statistical Inference: Methods for the Exploration of Posterior Distributions and Likelihood Functions*, Springer Series in Statistics, Springer, New York, NY, USA, 3rd edition, 1996.

[15] B. Efron and R. J. Tibshirani, *An Introduction to the Bootstrap*, vol. 57 of *Monographs on Statistics and Applied Probability*, Chapman & Hall, New York, NY, USA, 1993.

[16] A. Agresti, *Categorical Data Analysis*, Wiley Series in Probability and Statistics, Wiley-Interscience, New York, NY, USA, 2nd edition, 2002.

[17] G. D. Williamson and M. Haber, "Models for three-dimensional contingency tables with completely and partially cross-classified data," *Biometrics*, vol. 50, no. 1, pp. 194–203, 1994.

[18] J. W. Yu, G. L. Tian, and M. L. Tang, "Two new models for survey sampling with sensitive characteristic: design and analysis," *Metrika*, vol. 67, no. 3, pp. 251–263, 2008.

[19] M. L. Tang, G. L. Tian, N. S. Tang, and Z. Q. Liu, "A new non-randomized multi-category response model for surveys with a single sensitive question: design and analysis," *Journal of the Korean Statistical Society*, vol. 38, no. 4, pp. 339–349, 2009.

[20] Y. Liu and G. L. Tian, "Multi-category parallel models in the design of surveys with sensitive questions," *Statistics and Its Interface*, vol. 6, no. 1, pp. 137–149, 2013.

[21] G. L. Tian, "A new non-randomized response model: the parallel model," *Statistica Neerlandica*. In press.

[22] K. W. Ng, G. L. Tian, and M. L. Tang, *Dirichlet and Related Distributions: Theory, Methods and Applications*, Wiley Series in Probability and Statistics, John Wiley & Sons, Chichester, UK, 2011.

Bayesian Estimation of the Scale Parameter of Inverse Weibull Distribution under the Asymmetric Loss Functions

Farhad Yahgmaei, Manoochehr Babanezhad, and Omid S. Moghadam

Department of Statistics, Faculty of Sciences, Golestan University, Gorgan 49138-15739, Golestan, Iran

Correspondence should be addressed to Manoochehr Babanezhad; mbaba22@yahoo.com

Academic Editor: Aera Thavaneswaran

This paper proposes different methods of estimating the scale parameter in the inverse Weibull distribution (IWD). Specifically, the maximum likelihood estimator of the scale parameter in IWD is introduced. We then derived the Bayes estimators for the scale parameter in IWD by considering quasi, gamma, and uniform priors distributions under the square error, entropy, and precautionary loss functions. Finally, the different proposed estimators have been compared by the extensive simulation studies in corresponding the mean square errors and the evolution of risk functions.

1. Introduction

It is well known that the Weibull distribution is one of the most popular distributions in the lifetime data analyzing. The main reason is that one can create a wide variety of shapes with varying levels of its parameters. Therefore, during the past decades, extensive work has been done on this distribution in both the frequentist and Bayesian points of view; see, for example, the excellent reviews by Johnson et al. [1] and Kundu [2]. However, the Weibull distribution has two parameters, and in many practical applications, one or both of them might be unknown. To estimate them, we may use common approaches (see, e.g., Nordman and Meeker [3]). Moreover, it is clear through the distribution of Weibull that the Weibull probability density function (PDF) can be decreasing (or increasing) or unimodal, depending on the shape of distribution parameters. Due to the flexibility of the Weibull PDF, the inverse Weibull distribution (IWD) has been extensively employed in situation where a monotone data set is available (REF). Furthermore, if the empirical studies indicate that the Weibull PDF might be unimodal, then the inverse Weibull distribution (IWD) may be an appropriate model (Kundu [2]).

As a definition, if a positive random variable $Y > 0$ has the Weibull distribution with the following PDF:

$$f_Y(y;\alpha,\beta) = \alpha\beta y^{\beta-1} e^{-\alpha y^{\beta}}, \quad (1)$$

then the random variable $X = 1/Y$ has the IWD with the PDF of the following form:

$$f_X(x;\alpha,\beta) = \frac{\alpha\beta}{x^{\beta+1}} e^{-\alpha x^{-\beta}}, \quad (2)$$

where $\alpha > 0$ is called scale parameter and $\beta > 0$ is called shape parameter of this family. It also follows from (2) that the cumulative distribution function of X can be obtained:

$$F_X(x;\alpha,\beta) = e^{-\alpha x^{-\beta}}. \quad (3)$$

IWD plays an important role in many applications, including the dynamic components of diesel engines and several data sets such as the times to breakdown of an insulating fluid subject to the action of a constant tension (see Drapella [4], Jiang et al. [5], and Nelson [6] for more practical applications). For instance, Calabria and Pulcini [7] provide an interpretation of the IWD in the context of the load-strength relationship for a component. Maswadah [8] has fitted IWD to the flood data reported in Dumonceaux and Antle [9] (for more details see, e.g., Murthy et al. [10]).

The aim of this paper is to propose the different methods of estimation of the scale parameter for the inverse Weibull distribution (IWD). In the next section, we obtain the maximum likelihood estimator of the α scale parameter in IWD, when the shape parameter $\beta > 0$ is known. We also discuss the procedures to obtain the Bayes estimators for the quasi prior, gamma prior, and for uniform prior under square error, entropy, and precautionary loss functions for the scale parameter in IWD. In Section 3, we compare the maximum likelihood estimator and the Bayes estimators which are obtained in Section 2 based on their considered risk functions. The last section of the paper includes a discussion.

2. Estimation of the Scale Parameter α

In a situation where the shape parameter $\beta > 0$ is known in IWD, we can obtain the maximum likelihood estimator of the scale parameter $\alpha > 0$. Suppose that $X_1, \ldots, X_r$ is a random sample of size r, extracted from the density function defined in (2); then the likelihood function of α for fixed value of β is given by

$$L(\alpha) = \frac{(\alpha\beta)^r}{\prod_{i=1}^{r} x_i^{\beta+1}} e^{-\alpha \sum_{i=1}^{r} x_i^{-\beta}}. \quad (4)$$

By taking the natural logarithm on (4), we will obtain

$$l(\alpha) = r \ln(\alpha\beta) - (\beta+1) \sum_{i=1}^{r} \ln x_i - \alpha \sum_{i=1}^{r} x_i^{-\beta}, \quad (5)$$

and by taking derivative on (5) and setting with zero, the maximum likelihood estimator can be obtained as the following form:

$$\widehat{\alpha}_{\text{mle}} = \frac{r}{\sum_{i=1}^{r} x_i^{-\beta}}. \quad (6)$$

2.1. The Bayes Estimator. We now derive the Bayes estimator of the scale parameter α in IWD when the shape parameter β is known. We consider three different prior distributions and three different loss functions.

(a) The Quasi Prior. When there is no more information about the distribution parameter, one may use the quasi density as given by

$$\pi_1(\alpha) = \frac{1}{\alpha^d}; \quad \alpha > 0, \ d > 0. \quad (7)$$

The quasi-prior leads to a diffuse prior for a case where $d = 0$ and to a noninformative prior for a case where $d = 1$.

(b) The Gamma Prior. It is assumed that the scale parameter has a gamma prior distribution with the shape and scale parameters as c and d, respectively, when it has the following PDF:

$$\pi_2(\alpha) = \frac{d^c}{\Gamma(c)} \alpha^{c-1} e^{-d\alpha}, \quad \alpha > 0, \ c, d > 0. \quad (8)$$

Note that the gamma prior is one of the most considerable priors, which researchers often use. Note also that the gamma prior is a conjugate prior family.

(c) The Uniform Prior. It is assumed that the scale parameter has a uniform distribution over a finite range $[0, k]$, when it has the following form

$$\pi_3(\alpha) = \begin{cases} \dfrac{1}{k} & 0 < \alpha < k \\ 0 & \text{otherwise} \end{cases} \quad (9)$$

for all $k > 0$. Bayesian estimators are optimal decisions and are often obtained under a specific prior distribution and loss function. Suppose that $\widehat{\alpha}$ is an estimate of $\widehat{\alpha}$.

(i) The Square Error Loss Function. A commonly used loss function is the square error loss function (SLF)

$$L(\widehat{\alpha}, \alpha) = (\widehat{\alpha} - \alpha)^2, \quad (10)$$

which is a symmetric loss function that assigns equal losses to overestimation and underestimation. The SLF is often used because it does not need extensive numerical computation. However, several authors have recognized the inappropriateness of using an SLF in several applications (Calabria and Pulcini [11], Basu and Ebrahimi [12], Berger [13], and Norström [14]). For instance, Basu and Ebrahimi [12] derive Bayes estimators of the mean lifetime and the reliability function in the exponential life testing model. Instead, the loss functions that they used are asymmetric to reflect that, in most situations of interest, overestimating is more harmful than underestimating. Due to this, we use various asymmetric loss functions as follows.

(ii) The Entropy Loss Function. In many practical situations, it appears to be more realistic to express the loss in terms of the ratio $\widehat{\alpha}/\alpha$. In this case, Calabria and Pulcini [7] point out that a useful asymmetric loss function is the entropy loss function (ELS):

$$L(\delta) \propto \left[\delta^p - p \ln(\delta) - 1\right], \quad (11)$$

where $\delta = \widehat{\alpha}/\alpha$ and $p > 0$, whose minimum occurs at $\widehat{\alpha} = \alpha$. Also, the loss function $L(\delta)$ has been used in Dey et al. [15] and Dey and Liu [16], in the original form having $p = 1$. Thus, $L(\delta)$ can be written as

$$L(\delta) = b\left[\delta - \ln(\delta) - 1\right], \quad b > 0. \quad (12)$$

(iii) The Precautionary Loss Function. Norström [14] introduced an alternative asymmetric loss function and also presented a general class of precautionary loss functions as a special case. These loss functions approach infinitely near the origin to prevent underestimation, thus giving conservative estimators, especially when low failure rates are being estimated. These estimators are very useful when underestimation may lead to serious consequences. A very

useful and simple asymmetric precautionary loss function (PLF) is

$$L(\hat{\alpha}, \alpha) = \frac{(\hat{\alpha} - \alpha)^2}{\hat{\alpha}}. \tag{13}$$

2.2. The Bayes Estimator under $\pi_1(\alpha)$. Now, we obtain the Bayes estimators for parameter α for the quasi-prior density under square error, entropy, and precautionary loss functions. The posterior PDF of α is obtained as

$$\pi_1(\alpha \mid x) = \frac{\left(\sum_{i=1}^{r} x_i^{-\beta}\right)^{r-d+1}}{\Gamma(r-d+1)} \alpha^{r-d} e^{-\alpha \sum_{i=1}^{r} x_i^{-\beta}}, \tag{14}$$
$$\alpha > 0, \ r > d-1,$$

which is a gamma family with parameters $(r-d+1, \sum_{r}^{i=1} x_i^{-\beta})$.

The Bayes estimator under the square error loss function can clearly be obtained as

$$\hat{\alpha}_s = \frac{r-d+1}{\sum_{i=1}^{r} x_i^{-\beta}}, \tag{15}$$

the Bayes estimator under the entropy loss function by

$$\hat{\alpha}_e = \frac{r-d}{\sum_{i=1}^{r} x_i^{-\beta}}, \tag{16}$$

and the Bayes estimator under the precautionary loss function by

$$\hat{\alpha}_p = \frac{[(r-d+2)(r-d+1)]^{1/2}}{\sum_{i=1}^{r} x_i^{-\beta}}. \tag{17}$$

It is clear that the maximum likelihood estimator $\hat{\alpha}_{\text{mle}}$ is a special case of the Bayes estimator under square error loss function by $d = 1$. Therefore, the risk functions of $\hat{\alpha}_{\text{mle}}$ and $\hat{\alpha}_s$ are the same when $d = 1$.

2.2.1. The Risk Functions. The risk functions of the estimators $\hat{\alpha}_s$, $\hat{\alpha}_e$, and $\hat{\alpha}_p$, relative to SLF, are denoted by $R_s(\hat{\alpha}_s)$, $R_s(\hat{\alpha}_e)$, and $R_s(\hat{\alpha}_p)$, respectively, and are given by

$$\begin{aligned} R_s(\hat{\alpha}_s) &= \alpha^2 \left[\frac{(r-d+1)^2}{(r-1)(r-2)} - \frac{2(r-d+1)}{r-1} + 1\right], \\ R_s(\hat{\alpha}_e) &= \alpha^2 \left[\frac{(r-d)^2}{(r-1)(r-2)} - \frac{2(r-d)}{r-1} + 1\right], \\ R_s(\hat{\alpha}_p) &= \alpha^2 \left[\frac{(r-d+2)(r-d+1)}{(r-1)(r-2)} \right. \\ &\quad \left. - \frac{2[(r-d+2)(r-d+1)]^{1/2}}{r-1} + 1\right]. \end{aligned} \tag{18}$$

The risk functions of the estimators $\hat{\alpha}_s$, $\hat{\alpha}_e$, and $\hat{\alpha}_p$, relative to the entropy loss function, are denoted by $R_e(\hat{\alpha}_s)$, $R_e(\hat{\alpha}_e)$, and $R_e(\hat{\alpha}_p)$, respectively, and are given by

$$\begin{aligned} R_e(\hat{\alpha}_s) &= b\left[\frac{r-d+1}{r-1} + \ln\alpha - 1 - E_e(\ln\hat{\alpha}_s)\right], \\ R_e(\hat{\alpha}_e) &= b\left[\frac{r-d}{r-1} + \ln\alpha - 1 - E_e(\ln\hat{\alpha}_e)\right], \\ R_e(\hat{\alpha}_p) &= b\left[\frac{(r-d+2)(r-d+1)]^{1/2}}{r-1} \right. \\ &\quad \left. + \ln\alpha - 1 - E_e(\ln\hat{\alpha}_p)\right]. \end{aligned} \tag{19}$$

The risk functions of the estimators $\hat{\alpha}_s$, $\hat{\alpha}_e$, and $\hat{\alpha}_p$, relative to the precautionary loss function, are denoted by $R_p(\hat{\alpha}_s)$, $R_p(\hat{\alpha}_e)$, and $R_p(\hat{\alpha}_p)$, respectively, and are given by

$$\begin{aligned} R_p(\hat{\alpha}_s) &= \alpha\left[\frac{r-d+1}{r-1} + \frac{r}{r-d+1} - 2\right], \\ R_p(\hat{\alpha}_e) &= \alpha\left[\frac{r-d}{r-1} + \frac{r}{r-d} - 2\right], \\ R_p(\hat{\alpha}_p) &= \alpha\left[\frac{[(r-d+2)(r-d+1)]^{1/2}}{r-1} \right. \\ &\quad \left. + \frac{r}{[(r-d+2)(r-d+1)]^{1/2}} - 2\right]. \end{aligned} \tag{20}$$

2.3. The Bayes Estimator under $\pi_2(\alpha)$. The gamma density is the natural conjugate prior for the parameter α with respect to IWD. Using (4), the posterior distribution is obtained by

$$\pi_2(\alpha \mid x) = \frac{\left(d + \sum_{i=1}^{r} x_i^{-\beta}\right)^{r+c}}{\Gamma(r+c)} \alpha^{r+c-1} e^{-\alpha(d+\sum_{i=1}^{r} x_i^{-\beta})}, \tag{21}$$
$$\alpha > 0, \ r+c > 0,$$

which is again a gamma family of parameters $(r+c, d+\sum_{r}^{i=1} x_i^{-\beta})$. Thus, the Bayes estimators of α under the square loss function are given by

$$\hat{\alpha}_s = \frac{r+c}{d+\sum_{i=1}^{r} x_i^{-\beta}}. \tag{22}$$

The Bayes estimator of α under entropy loss function is given by

$$\hat{\alpha}_e = \frac{r+c-1}{d+\sum_{i=1}^{r} x_i^{-\beta}} \tag{23}$$

and the Bayes estimator under the precautionary loss function by

$$\hat{\alpha}_p = \frac{[(r+c+1)(r+c)]^{1/2}}{d+\sum_{i=1}^{r} x_i^{-\beta}}. \tag{24}$$

2.3.1. The Risk Functions. The risk functions of the estimators $\hat{\alpha}_s$, $\hat{\alpha}_e$, and $\hat{\alpha}_p$, relative to SLF, are denoted by $R_s(\hat{\alpha}_s)$, $R_s(\hat{\alpha}_e)$, and $R_s(\hat{\alpha}_p)$, respectively, and are given by

$$\begin{aligned}R_s(\hat{\alpha}_s) &= E(\hat{\alpha}_s - \alpha)^2 = \alpha^2 - 2\alpha(r+c)G(1) + (r+c)^2 G(2),\\ R_s(\hat{\alpha}_e) &= E(\hat{\alpha}_e - \alpha)^2 = \alpha^2 - 2\alpha(r+c-1)G(1) + (r+c-1)^2 G(2),\\ R_s(\hat{\alpha}_p) &= E(\hat{\alpha}_p - \alpha)^2 = \alpha^2 - 2\alpha[(r+c+1)(r+c)]^{1/2}G(1) + (r+c+1)(r+c)G(2),\end{aligned} \tag{25}$$

where

$$G(s) = \int_o^\infty \frac{1}{(d+z)^s}\frac{\alpha^r}{\Gamma(r)} z^{r-1} e^{-\alpha z} dz, \quad z = \sum_{i=1}^r x_i^{-\beta}. \tag{26}$$

Similarly, the risk functions of the estimators $\hat{\alpha}_s$, $\hat{\alpha}_e$, and $\hat{\alpha}_p$, relative to the entropy loss function, are denoted by $R_e(\hat{\alpha}_s)$, $R_e(\hat{\alpha}_e)$, and $R_e(\hat{\alpha}_p)$, respectively, and are given by

$$\begin{aligned}R_e(\hat{\alpha}_s) &= b\left[\frac{r+c}{\alpha}G(1) + \ln\alpha - 1 - E_e(\ln\hat{\alpha}_s)\right],\\ R_e(\hat{\alpha}_e) &= b\left[\frac{r+c-1}{\alpha}G(1) + \ln\alpha - 1 - E_e(\ln\hat{\alpha}_e)\right],\\ R_e(\hat{\alpha}_p) &= b\left[\frac{[(r+c+1)(r+c)]^{1/2}}{\alpha}G(1) + \ln\alpha - 1 - E_e(\ln\hat{\alpha}_p)\right].\end{aligned} \tag{27}$$

The risk functions of the estimators $\hat{\alpha}_s$, $\hat{\alpha}_e$, and $\hat{\alpha}_p$, relative to the precautionary loss function, are denoted by $R_p(\hat{\alpha}_s)$, $R_p(\hat{\alpha}_e)$, and $R_p(\hat{\alpha}_p)$, respectively, and are given by

$$\begin{aligned}R_p(\hat{\alpha}_s) &= (r+c)G(1) + \frac{\alpha^2}{r+c}\left(d+\frac{r}{\alpha}\right) - 2\alpha,\\ R_p(\hat{\alpha}_e) &= (r+c-1)G(1) + \frac{\alpha^2}{r+c-1}\left(d+\frac{r}{\alpha}\right) - 2\alpha,\\ R_p(\hat{\alpha}_p) &= [(r+c+1)(r+c)]^{1/2}G(1) + \frac{\alpha^2}{[(r+c+1)(r+c)]^{1/2}}\left(d+\frac{r}{\alpha}\right) - 2\alpha.\end{aligned} \tag{28}$$

2.4. The Bayes Estimator under $\pi_3(\alpha)$. Under $\pi_3(\alpha)$, using (4), the posterior distribution is obtained by

$$\pi_3(\alpha \mid x) = \frac{1}{kB}\alpha^r e^{-\alpha\sum_{i=1}^r x_i^{-\beta}}, \quad 0 < \alpha < k, \tag{29}$$

where

$$B(k) = \frac{1}{k}\int_0^k \alpha^r e^{-\alpha\sum_{i=1}^r x_i^{-\beta}} d\alpha. \tag{30}$$

The Bayes estimators of α under the square loss function are given by

$$\hat{\alpha}_s = \frac{1}{\sum_{i=1}^r x_i^{-\beta}}\left[(r+1) - \frac{k^r}{B(k)}e^{-k\sum_{i=1}^r x_i^{-\beta}}\right], \tag{31}$$

the Bayes estimator under the entropy loss function by

$$\hat{\alpha}_e = \frac{rB(k)}{k^{r-1}e^{-k\sum_{i=1}^r x_i^{-\beta}} + B(k)\sum_{i=1}^r x_i^{-\beta}}, \tag{32}$$

and the Bayes estimator under the precautionary loss function by

$$\hat{\alpha}_p = \frac{1}{\sum_{i=1}^r x_i^{-\beta}}\left[(r+2)(r+1) - \frac{k^r e^{-k\sum_{i=1}^r x_i^{-\beta}}}{B(k)} \times \left(k\sum_{i=1}^r x_i^{-\beta} + (r+2)\right)\right]^{1/2}. \tag{33}$$

In this case, there is no closed-form solution to obtain the risk functions of the latter estimators. Therefore, we employ the importance sampling technique for constructing the Bayes estimators and obtaining risk functions which is presented in next section.

3. Comparisons

This section presents the comparison of the various estimators obtained by the use of different methods in Sections 2 and 3 on the basis of their risks. In the previous section, the risk function of the estimators is computed under SLF, ELF, and PLF.

3.1. The Case of Quasi Prior. The Bayes estimators are seen to depend upon the parameters of prior distributions. In Figure 1, we have plotted the ratio of the risk functions to α^2, that is,

$$A_1 = \frac{R_s(\hat{\alpha}_s)}{\alpha^2}, \quad A_2 = \frac{R_s(\hat{\alpha}_e)}{\alpha^2}, \quad A_3 = \frac{R_s(\hat{\alpha}_p)}{\alpha^2}, \tag{34}$$

for the Bayes estimators $\hat{\alpha}_s$, $\hat{\alpha}_e$, and $\hat{\alpha}_p$, respectively, under the square error loss function, as given in (18), for $r = 5(5)20$ and $d = 0.5(0.5)4.5$.

In Figure 2, we have plotted the ratio of the risk functions to α, that is,

$$B_1 = \frac{R_p(\hat{\alpha}_s)}{\alpha}, \quad B_2 = \frac{R_p(\hat{\alpha}_e)}{\alpha}, \quad B_3 = \frac{R_p(\hat{\alpha}_p)}{\alpha}, \tag{35}$$

for the Bayes estimators $\hat{\alpha}_s$, $\hat{\alpha}_e$, and $\hat{\alpha}_p$, respectively, under the precautionary loss function, as given in (20), for $r = 5(5)20$ and $d = 0.5(0.5)4.5$.

It is important to mention here that the scales on y-axis of the graphs are not the same and they vary from figure to figure. From Figures 1 and 2, we see that none of the estimators uniformly dominates any other. We therefore recommend that the estimators to be chosen according to the values of d when quasi density is used as the prior distribution, and this choice in turn depends on the situations at hand.

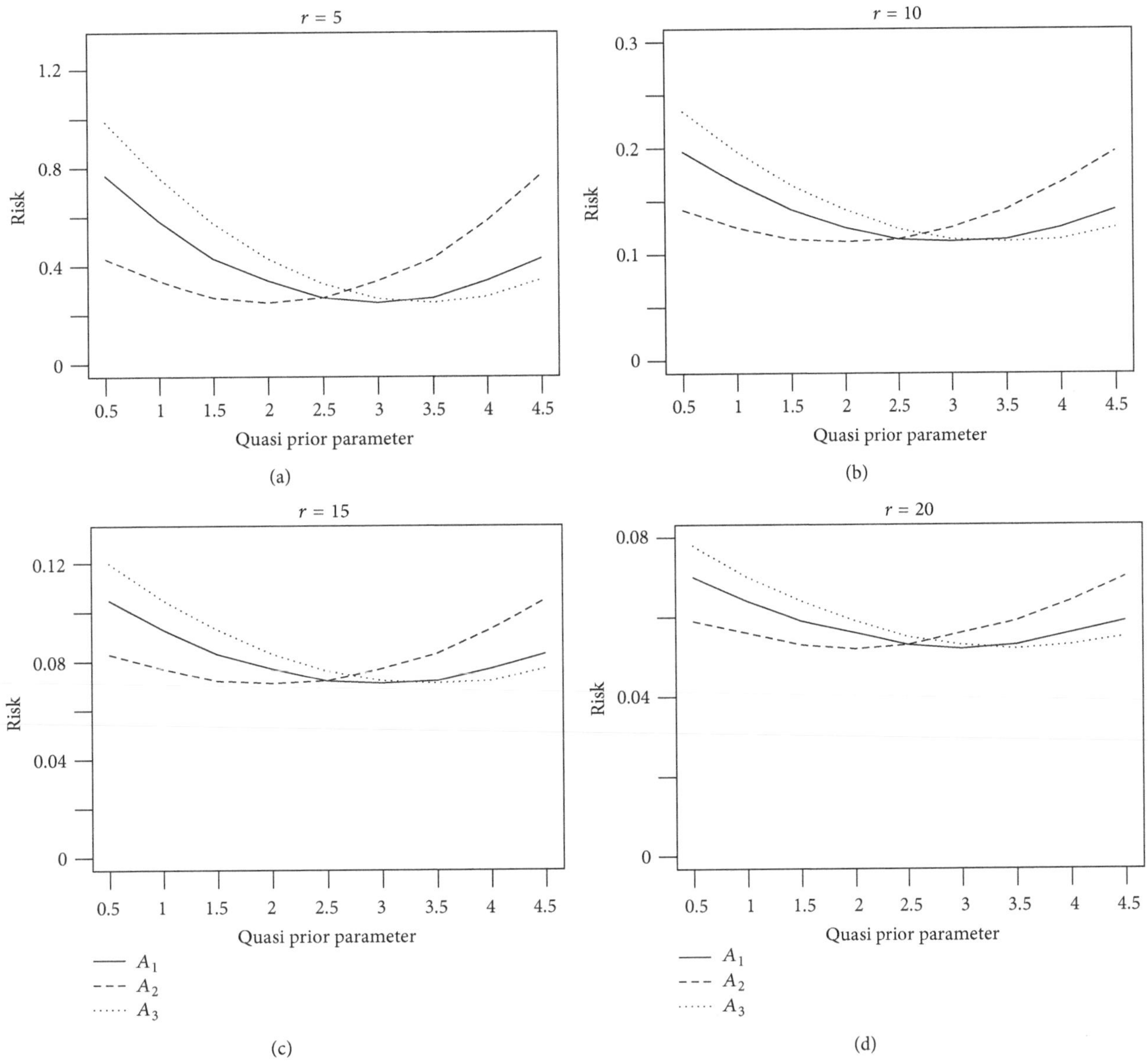

FIGURE 1: The evolution of risk ratio to α^2 when $r = 5, 10, 15$, and 20.

3.2. The Case of Gamma Prior. The risk functions under the gamma prior are dependent on the population parameter α, which is not separable. Therefore, a comparison could only be made by using numerical techniques. Random samples of different size are generated, and the estimators obtained in Sections 2 and 3 are compared in the following steps.

Algorithm 1. Consider the following.

Step 1. For given values ($c = 2, d = 3, \alpha = 4.055$), we generate prior (8).

Step 2. By using the value $\alpha = 4.055$ from Step 1 and true value $\beta = 2$, we select the sample size $r = 10, 20, 30$, and 40. We then generate the likelihood function (4).

Step 3. The MLE and different Bayes estimators of α are computed through Step 3.

Step 4. Steps 1 to 3 are repeated 1000 times, and the mean square error (MSE) for each estimator is computed.

Table 1 given herein shows the mean square error (MSE) of the different estimators based on 1000 runs of Monte Carlo simulation.

From Table 1, we see that the estimators are consistent in MSE of the all considered cases. As expected, the Bayes estimators are doing better than the maximum likelihood estimators. Also, the Bayes estimators under precautionary loss function are doing better than the all other estimators.

3.3. The Case of Uniform Prior. Since the risk functions of estimators cannot be obtained in a closed form, we propose to use the Gibbs sampling technique to generate MCMC

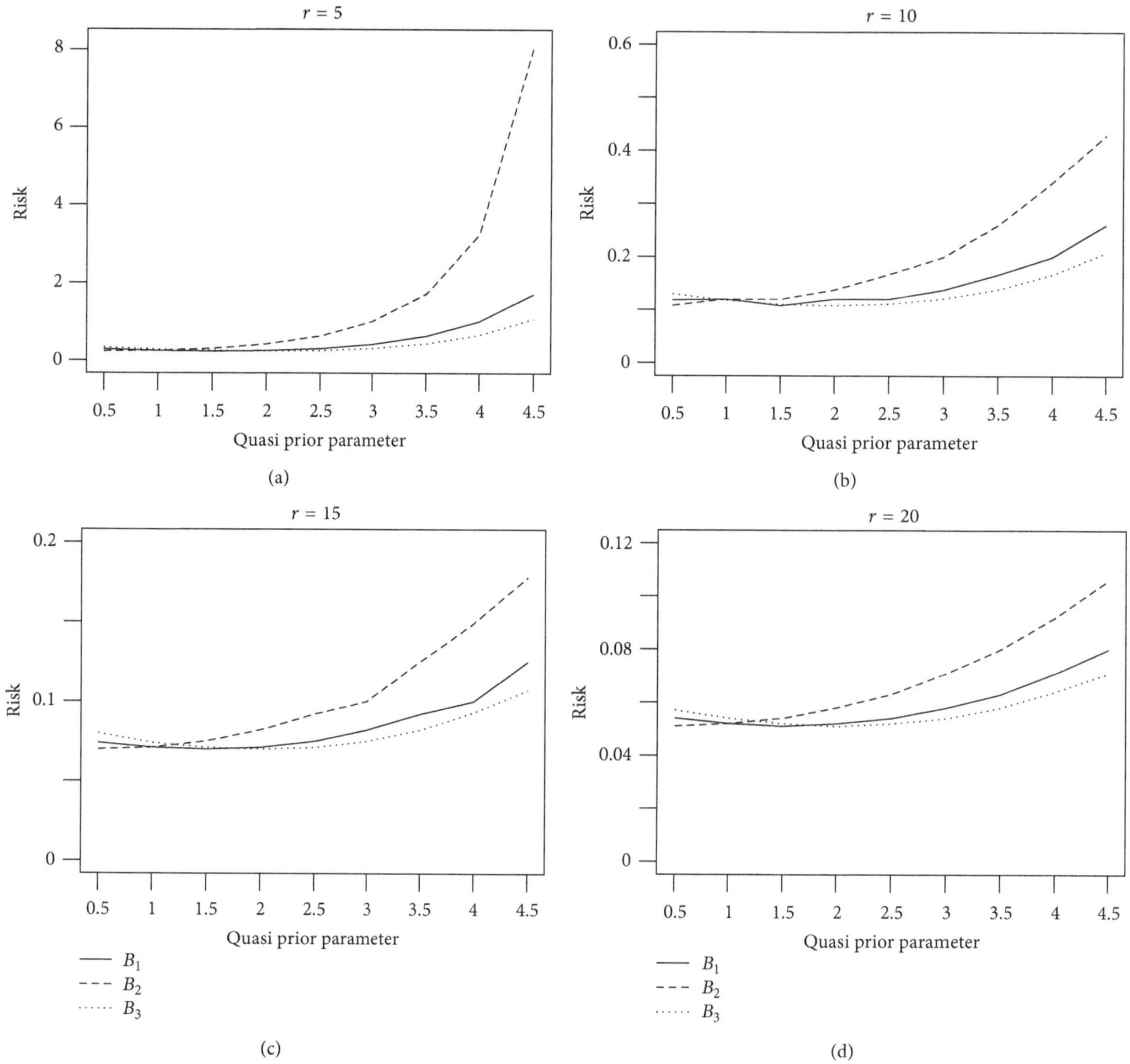

FIGURE 2: The evolution of risk ratio to α when $r = 5, 10, 15$, and 20.

TABLE 1: Mean square error (MSE) of the different estimators for α.

r	$\widehat{\alpha}_s$	$\widehat{\alpha}_e$	$\widehat{\alpha}_p$	$\widehat{\alpha}_{\text{mle}}$
10	0.03605	0.03609	0.03604	0.03612
20	0.02684	0.02686	0.02683	0.02687
30	0.00772	0.00774	0.00771	0.00776
40	0.00049	0.00049	0.00048	0.00052

samples and then use importance sampling technique for constructing the Bayes estimators.

Now, we provide an algorithm to draw MCMC samples from the posterior distribution (29). Since

$$\frac{\text{Gamma}\left(r, \sum_{i=1}^{r} x_i^{-\beta}\right)}{B} \geq \pi_3\left(\alpha \mid x\right), \tag{36}$$

where B have been defined in (30), it is possible to use the acceptance rejection method to generate samples from $\pi_3(\alpha \mid x)$, by using gamma generation, and we use Algorithm 2 in what follows to generate Gibbs sample from the posterior density function of α.

Algorithm 2. Consider the following.

Step 1. Generate α from the Gamma $(r, \sum_{i=1}^{r} x_i^{-\beta})$ and U from the Uniform (0,1).

Step 2. If $U \leq (\Gamma(r)/(\sum_{i=1}^{r} x_i^{-\beta})^r)(\alpha/k)$, then accept α; otherwise, go back to Step 1.

Step 3. Generate $\alpha_1, \ldots, \alpha_N$.

TABLE 2: Mean (MSEs) of the different estimators for α.

r	$\beta = 2$		$\beta = 2.5$		$\beta = 3$		$\beta = 4$	
	$\widehat{\alpha}_s$	$\widehat{\alpha}_p$	$\widehat{\alpha}_s$	$\widehat{\alpha}_p$	$\widehat{\alpha}_s$	$\widehat{\alpha}_p$	$\widehat{\alpha}_s$	$\widehat{\alpha}_p$
10	0.4987 (0.0910)	0.5723 (0.1002)	0.5036 (0.0878)	0.5841 (0.0944)	0.4991 (0.0845)	0.5769 (0.0889)	0.4981 (0.0847)	0.5738 (0.0904)
20	0.4961 (0.0863)	0.5746 (0.0932)	0.4946 (0.8610)	0.5751 (0.0925)	0.5047 (0.0843)	0.5745 (0.0891)	0.4995 (0.0838)	0.5781 (0.0897)
30	0.4966 (0.0853)	0.5768 (0.0908)	0.4983 (0.8481)	0.5772 (0.0910)	0.5006 (0.0839)	0.5819 (0.0879)	0.4989 (0.0833)	0.5766 (0.0890)
50	0.5012 (0.0828)	0.5787 (0.0900)	0.4885 (0.0835)	0.5675 (0.0897)	0.4966 (0.0835)	0.5801 (0.0874)	0.5008 (0.0831)	0.5792 (0.0884)
80	0.5046 (0.0826)	0.5741 (0.0880)	0.5048 (0.0821)	0.5805 (0.0878)	0.4982 (0.0827)	0.5738 (0.0869)	0.5006 (0.0827)	0.5786 (0.0883)

Step 4. Obtain the Bayes estimate of α under the square error loss function as the posterior mean, that is,

$$\widehat{\alpha}_s = \widehat{E}(\alpha \mid x) = \frac{1}{N}\sum_{i=1}^{N}\alpha_i. \quad (37)$$

Step 5. Obtain the Bayes estimator α under the precautionary loss function as follows:

$$\widehat{\alpha}_p = \left[\frac{1}{N}\sum_{i=1}^{N}\alpha_i^2\right]^{1/2}. \quad (38)$$

Step 6. Obtain the mean square error α.

In order to compare the proposed Bayes estimators with the corresponding Bayes estimators, we perform a Monte Carlo simulation study of 1000 using different sample sizes $r = 10, 20, 30, 50$, and 80. The IWD samples were generated from (2) for all combinations of $\alpha = 0.5$ and $\beta = 2, 2.5, 3$, and 4. For the uniform prior, we have considered $k = 1$. In this case, we have chosen the hyperparameters in such a way that the prior mean becomes the expected value of the corresponding parameter. The averages and mean square errors (MSE) in parentheses of estimators of $\widehat{\alpha}_s$ and $\widehat{\alpha}_p$ are presented in Table 2.

It is clear from Table 2 that the proposed Bayes estimators perform very well for r and the estimators is consistent in MSE of the all considered case. Also, the Bayes estimators $\widehat{\alpha}_s$ under square loss function are doing better than the Bayes estimators under precautionary loss function, that is, $\widehat{\alpha}_p$.

4. Conclusion

In this paper, we have proposed the classical and the Bayesian approaches to estimate the scale parameter for inverse Weibull distribution, when the shape parameter was known [12]. Bayes estimators are often obtained using both symmetric and asymmetric loss functions ([11, 12]). In view of this, we have obtained and then compared the different Bayes estimators corresponding to the different loss functions. To compare the considered estimators, extensive simulation studies have been performed. The results show that, in the case of quasi-prior, none of the estimators uniformly dominates any other. Therefore, it might recommend that the estimators be chosen according to the value of d, when quasi-density is used as the prior distribution. This choice in turn depends on the situations at hand. It appears to be clear from this study that the Bayes method of estimation for gamma prior is superior to the MLE method. Also, in the case of gamma prior, the Bayes estimators related to precautionary loss function have the smallest MSE as compared with the Bayes estimators related to square error loss function or the Bayes estimators under entropy loss function or the MLEs. Furthermore, in the case of uniform prior, the Bayes estimators under square error loss function are doing better than the Bayes estimators under precautionary loss function.

References

[1] N. L. Johnson, S. Kotz, and N. Balakrishnan, *Continuous Univariate Distribution*, John Wiley & Sons, New York, NY, USA, 2nd edition, 1995.

[2] D. Kundu, "Bayesian inference and life testing plan for the Weibull distribution in presence of progressive censoring," *Technometrics*, vol. 50, no. 2, pp. 144–154, 2008.

[3] D. J. Nordman and W. Q. Meeker, "Weibull prediction intervals for a future number of failures," *Technometrics*, vol. 44, no. 1, pp. 15–23, 2002.

[4] A. Drapella, "The complementary Weibull distribution: unknown or just forgotten?" *Quality and Reliability Engineering International*, vol. 9, pp. 383–385, 1993.

[5] R. Jiang, D. N. P. Murthy, and P. Ji, "Models involving two inverse Weibull distributions," *Reliability Engineering and System Safety*, vol. 73, no. 1, pp. 73–81, 2001.

[6] W. B. Nelson, *Applied Life Data Analysis*, John Wiley & Sons, New York, NY, USA, 1982.

[7] R. Calabria and G. Pulcini, "On the maximum likelihood and least-squares estimation in the Inverse Weibull distributions," *Statistical Application*, vol. 2, no. 1, pp. 53–66, 1990.

[8] M. Maswadah, "Conditional confidence interval estimation for the inverse Weibull distribution based on censored generalized order statistics," *Journal of Statistical Computation and Simulation*, vol. 73, no. 12, pp. 887–898, 2003.

[9] R. Dumonceaux and C. E. Antle, "Discrimination between the lognormal and Weibull distribution," *Technometrics*, vol. 15, pp. 923–926, 1973.

[10] D. N. P. Murthy, M. Xie, and R. Jiang, *Weibull Model*, John Wiley & Sons, New York, NY, USA, 2004.

[11] R. Calabria and G. Pulcini, "An engineering approach to Bayes estimation for the Weibull distribution," *Microelectronics Reliability*, vol. 34, no. 5, pp. 789–802, 1994.

[12] A. P. Basu and N. Ebrahimi, "Bayesian approach to life testing and reliability estimation using asymmetric loss function," *Journal of Statistical Planning and Inference*, vol. 29, no. 1-2, pp. 21–31, 1992.

[13] J. O. Berger, *Statistical Decision Theory, Foundation, Concepts and Method*, Springer, New York, NY, USA, 1985.

[14] J. G. Norström, "The use of precautionary loss functions in risk analysis," *IEEE Transactions on Reliability*, vol. 45, no. 3, pp. 400–403, 1996.

[15] D. K. Dey, M. Ghosh, and C. Srinivasan, "Simultaneous estimation of parameters under entropy loss," *Journal of Statistical Planning and Inference*, vol. 15, pp. 347–363, 1987.

[16] D. K. Dey and P.-S. L. Liu, "On comparison of estimators in a generalized life model," *Microelectronics Reliability*, vol. 32, no. 1-2, pp. 207–221, 1992.

18

A Note on Functional Averages over Gaussian Ensembles

Gabriel H. Tucci[1] and Maria V. Vega[2]

[1] *Bell Laboratories, Alcatel-Lucent, 600 Mountain Avenue, Murray Hill, NJ 07974, USA*
[2] *William Paterson University, Mathematics Department, 300 Pompton Road, Wayne, NJ 07470, USA*

Correspondence should be addressed to Gabriel H. Tucci; gabrieltucci@gmail.com

Academic Editor: Aera Thavaneswaran

We find a new formula for matrix averages over the Gaussian ensemble. Let $\mathbf{H}$ be an $n \times n$ Gaussian random matrix with complex, independent, and identically distributed entries of zero mean and unit variance. Given an $n \times n$ positive definite matrix $\mathbf{A}$ and a continuous function $f : \mathbb{R}^+ \to \mathbb{R}$ such that $\int_0^\infty e^{-\alpha t}|f(t)|^2 dt < \infty$ for every $\alpha > 0$, we find a new formula for the expectation $\mathbb{E}[\mathrm{Tr}(f(\mathbf{HAH}^*))]$. Taking $f(x) = \log(1 + x)$ gives another formula for the capacity of the MIMO communication channel, and taking $f(x) = (1 + x)^{-1}$ gives the MMSE achieved by a linear receiver.

1. Introduction

Random matrix theory was introduced to the theoretical physics community by Wigner in his work on nuclear physics in the 1950s [1, 2]. Since that time, the subject is an important and active research area in mathematics, and it finds applications in fields as diverse as the Riemann conjecture, physics, chaotic systems, multivariate statistics, wireless communications, signal processing, compressed sensing, and information theory. In the last decades, a considerable amount of work has emerged in the communications and information theory on the fundamental limits of communication channels that make use of results in random matrix theory [3–5]. For this reason, computing averages over certain matrix ensembles becomes extremely important in many situations. To be more specific, consider the well-known case of the single user MIMO channel with multiple transmitting and receiving antennas. Denoting the number of transmitting antennas by t and the number of receiving antennas by r, the channel model is

$$\mathbf{y} = \mathbf{H}\mathbf{u} + \mathbf{n}, \tag{1}$$

where $\mathbf{u} \in \mathbb{C}^t$ is the transmitted vector, $\mathbf{y} \in \mathbb{C}^r$ is the received vector, $\mathbf{H}$ is a $r \times t$ complex matrix, and $\mathbf{n}$ is the zero mean complex Gaussian vector with independent, equal variance entries. We assume that $\mathbb{E}(\mathbf{nn}^*) = \mathbf{I}_r$, where $(\cdot)^*$ denotes the complex conjugate transpose. It is reasonable to put a power constraint

$$\mathbb{E}(\mathbf{u}^*\mathbf{u}) = \mathbb{E}(\mathrm{Tr}(\mathbf{uu}^*)) \le P, \tag{2}$$

where P is the total transmitted power. The signal to noise ratio, denoted by snr, is defined as the quotient of the signal power and the noise power and in this case is equal to P/r.

Recall that if $\mathbf{A}$ is an $n \times n$ Hermitian matrix, then there exists $\mathbf{U}$ unitary and $\mathbf{D} = \mathrm{diag}(d_1, \ldots, d_n)$ such that $\mathbf{A} = \mathbf{UDU}^*$. Given a continuous function f, we define $f(\mathbf{A})$ as

$$f(\mathbf{A}) = \mathbf{U}\,\mathrm{diag}(f(d_1), \ldots, f(d_n))\,\mathbf{U}^*. \tag{3}$$

Naturally, the simplest example is the one where $\mathbf{H}$ has independent and identically distributed (i.i.d.) Gaussian entries, which constitutes the canonical model for the single user narrow band MIMO channel. It is known that the capacity of this channel is achieved when $\mathbf{u}$ is a complex Gaussian zero mean and covariance snr $\mathbf{I}_t$ vector (see e.g., [4, 6]). For the fast fading channel, assuming the statistical channel state information at the transmitter, the ergodic capacity is given by

$$\mathbb{E}[\log\det(\mathbf{I}_r + \mathrm{snr}\,\mathbf{HH}^*)] = \mathbb{E}[\mathrm{Tr}\log(\mathbf{I}_r + \mathrm{snr}\,\mathbf{HH}^*)], \tag{4}$$

where in the last equality we use the fact that $\operatorname{Tr}\log(\cdot) = \log\det(\cdot)$. We refer the reader to [6] or [4] for more details on this.

Another important performance measure is the minimum mean square error (MMSE) achieved by a linear receiver, which determines the maximum achievable output signal to interference and noise ratio (SINR). For an input vector $\mathbf{x}$ with i.i.d. entries of zero mean and unit variance, the MSE at the output of the MMSE receiver is given by

$$\min_{\mathbf{M}\in\mathbb{C}^{t\times r}} \mathbb{E}\left[\|\mathbf{x}-\mathbf{M}\mathbf{y}\|^2\right] = \mathbb{E}\left[\operatorname{Tr}\left(\mathbf{I}_t + \operatorname{snr}\mathbf{H}^*\mathbf{H}\right)^{-1}\right], \quad (5)$$

where the expectation on the left-hand side is over both the vectors $\mathbf{x}$ and the random matrices $\mathbf{H}$, while the right-hand side is over $\mathbf{H}$ only. We refer to [4] for more details on this.

There is a big literature and history of work on averages over Gaussian ensembles; see, for instance, [3–14] and references therein. In [6], the capacity of the Gaussian channel was computed as an improper integral. This integral is difficult to compute, and asymptotic and simulation results are provided. In [8, 11, 12, 14, 15], several asymptotic results for large complex Gaussian random matrices are studied in connection with wireless communication and information theory. In [8], many aspects of correlated Gaussian matrices are addressed, and in particular, the capacity of Rayleigh channel was computed as the number of antennas increases to infinity. The books [3–5] are excellent introductions to random matrix theory and their applications to physics and information theory. In [10], the spectral eigenvalue distribution for a random infinite d-regular graph was computed.

The typical approach in computing averages over random matrices is to consider the asymptotic behavior as the size of the matrix increases to infinity. In this work, we contribute to this area by providing a unified framework to express the ergodic mutual information, the MSE at the output of the MMSE decoder, and other types of functionals of a single user MIMO channel, when the number of transmitting and receiving antennas is equal and finite. We do not rely on asymptotic results as the number of antennas increases. The results shown in this work are new and novel to the best knowledge of the authors and they were not discovered before.

In Section 2, we present some preliminaries in Schur polynomials that are later used in this work. In Section 3, we prove the main result of the paper, Theorem 2. This theorem provides a new formula for the expectation

$$\mathbb{E}\left[\operatorname{Tr}\left(f\left(\mathbf{H}\mathbf{A}\mathbf{H}^*\right)\right)\right], \quad (6)$$

where $\mathbf{A}$ is a positive definite matrix and f a continuous function such that

$$\int_0^\infty e^{-\alpha t}\left|f(t)\right|^2 dt < \infty, \quad (7)$$

for every $\alpha > 0$. Notice that, as previously stated, taking $f(x) = \log(1+x)$ gives another formula for the capacity of the MIMO communication channel, and taking $f(x) = (1+x)^{-1}$ gives the MMSE achieved by a linear receiver. We also discuss some applications and present some examples.

2. Schur Polynomials Preliminaries

A symmetric polynomial is a polynomial $P(x_1, x_2, \ldots, x_n)$ in n variables such that if any of the variables are interchanged one obtains the same polynomial. Formally, P is a symmetric polynomial if for any permutation σ of the set $\{1, 2, \ldots, n\}$ one has

$$P\left(x_{\sigma(1)}, x_{\sigma(2)}, \ldots, x_{\sigma(n)}\right) = P\left(x_1, x_2, \ldots, x_n\right). \quad (8)$$

Symmetric polynomials arise naturally in the study of the relation between the roots of a polynomial in one variable and its coefficients, since the coefficients can be given by a symmetric polynomial expressions in the roots. Symmetric polynomials also form an interesting object by themselves. The resulting structures, and in particular the ring of symmetric functions, are of great importance in combinatorics and in representation theory (see e.g., [16–19] for more on details on this topic).

The Schur polynomials are certain symmetric polynomials in n variables. This class of polynomials is very important in representation theory since they are the characters of irreducible representations of the general linear groups. The Schur polynomials are indexed by partitions. A partition of a positive integer n, also called an integer partition, is a way of writing n as a sum of positive integers. Two partitions that differ only by the order of their summands are considered to be equal. Therefore, we can always represent a partition λ of a positive integer n as a nonincreasing sequence of n nonnegative integers d_i such that

$$\sum_{i=1}^{n} d_i = n \quad \text{with } d_1 \geq d_2 \geq d_3 \geq \cdots \geq d_n \geq 0. \quad (9)$$

Notice that some of the d_i could be zero. Integer partitions are usually represented by the so-called Young diagrams (also known as Ferrers' diagrams). A Young diagram is a finite collection of boxes, or cells, arranged in left-justified rows, with the row lengths weakly decreasing (each row has the same or shorter length than its predecessor). Listing the number of boxes on each row gives a partition λ of a nonnegative integer n, the total number of boxes of the diagram. The Young diagram is said to be of shape λ, and it carries the same information as that partition. For instance, later we can see the Young diagram corresponding to the partition $(5, 4, 1)$ of the number 10:

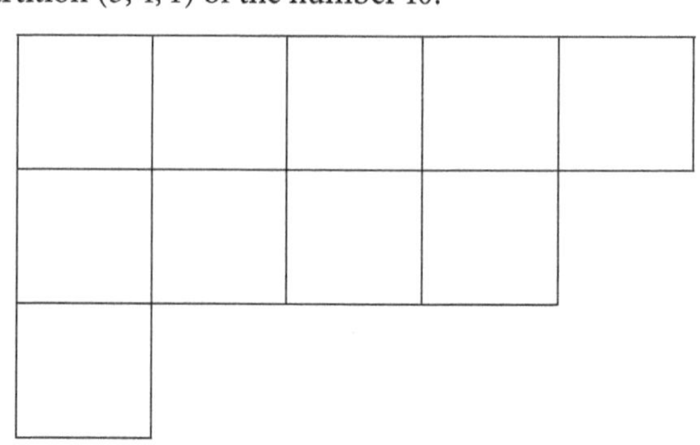

(10)

Given a partition λ of n

$$n = d_1 + d_2 + \cdots + d_n : d_1 \geq d_2 \geq \cdots \geq d_n \geq 0, \quad (11)$$

the following functions are alternating polynomials (in other words they change sign under any transposition of the variables):

$$a_{(d_1,\ldots,d_n)}(x_1,\ldots,x_n) = \det \begin{bmatrix} x_1^{d_1} & x_2^{d_1} & \cdots & x_n^{d_1} \\ x_1^{d_2} & x_2^{d_2} & \cdots & x_n^{d_2} \\ \vdots & \vdots & \ddots & \vdots \\ x_1^{d_n} & x_2^{d_n} & \cdots & x_n^{d_n} \end{bmatrix} = \sum_{\sigma\in S_n} \epsilon(\sigma)\, x_{\sigma(1)}^{d_1}\cdots x_{\sigma(n)}^{d_n}, \tag{12}$$

where S_n is the permutation group of the set $\{1,2,\ldots,n\}$. Since they are alternating, they are all divisible by the Vandermonde determinant

$$\Delta(x_1,\ldots,x_n) = \prod_{1\le j<k\le n} (x_j - x_k). \tag{13}$$

The Schur polynomial associated with λ is defined as the ratio

$$s_\lambda(x_1,x_2,\ldots,x_n) = \frac{a_{(d_1+n-1,d_2+n-2,\ldots,d_n+0)}(x_1,\ldots,x_n)}{\Delta(x_1,\ldots,x_n)}. \tag{14}$$

This is a symmetric function because the numerator and denominator are both alternating and a polynomial since all alternating polynomials are divisible by the Vandermonde determinant (see [16, 18, 19] for more details here). For instance,

$$\begin{aligned} s_{(2,1,1)}(x_1,x_2,x_3) &= x_1x_2x_3(x_1+x_2+x_3), \\ s_{(2,2,0)}(x_1,x_2,x_3) &= x_1^2x_2^2 + x_1^2x_3^2 + x_2^2x_3^2 + x_1^2x_2x_3 \\ &\quad + x_1x_2^2x_3 + x_1x_2x_3^2. \end{aligned} \tag{15}$$

Another definition we need for the next section is the so-called hook length, hook (x), of a box x in a Young diagram of shape λ. This is defined as the number of boxes that is in the same row to the right of it plus those boxes in the same column below it, plus one (for the box itself). As an example, later we show the hook lengths of the partition (5, 4, 1). The product of the hook's length of a partition is the product of the hook lengths of all the boxes in the partition:

7	5	4	3	1
5	3	2	1	
1				

(16)

We recommend the interested reader to consult [16, 18, 19] for more details and examples on this topic.

3. Averages over Gaussian Ensembles

Let M_n be the set of all $n\times n$ complex matrices and $\mathbf{U}_n$ the set of $n\times n$ unitary complex matrices. Let $d\mathbf{H}$ be the Lebesgue measure on M_n, and let

$$d\nu(\mathbf{H}) = \pi^{-n^2}\exp(-\operatorname{trace}(\mathbf{H}^*\mathbf{H}))\, d\mathbf{H} \tag{17}$$

be the Gaussian measure on M_n. This is the induced measure by the Gaussian random matrix with complex independent and identically distributed entries with zero mean and unit variance in the set of matrices, when this is represented as a Euclidean space of dimension $2n^2$. Note that this probability measure is left and right invariant under unitary multiplication (i.e., $d\nu(\mathbf{HU}) = d\nu(\mathbf{UH}) = d\nu(\mathbf{H})$ for every unitary $\mathbf{U}$). The following theorem can be found on page 447 of [18].

Theorem 1 (see [18]). *For all Hermitian $n\times n$ matrices $\mathbf{A},\mathbf{B}$ and every partition λ*

$$\int_{M_n} s_\lambda(\mathbf{AH}^*\mathbf{BH})\, d\nu(\mathbf{H}) = h(\lambda)\, s_\lambda(\mathbf{A})\, s_\lambda(\mathbf{B}), \tag{18}$$

where $h(\lambda)$ is the product of the hook lengths of λ.

Denote by $(m-k,1^k)$ the partition $(m-k,1,1,\ldots,1)$ with k ones. It is a well-known fact in matrix theory (see [16] or [18]) that for every Hermitian $n\times n$ matrix $\mathbf{A}$ and for every integer m

$$\operatorname{Tr}(\mathbf{A}^m) = \sum_{k=0}^{n-1} (-1)^k s_{(m-k,1^k)}(\mathbf{A}). \tag{19}$$

Note that for the case $1\le m<n$, even though the sum is up to the $n-1$ term, all the terms between $\min\{n,m\}$ and $n-1$ are zero. In particular,

(i) $\operatorname{Tr}(\mathbf{A}) = s_{(1)}(\mathbf{A})$,

(ii) $\operatorname{Tr}(\mathbf{A}^2) = s_{(2,0)}(\mathbf{A}) - s_{(1,1)}(\mathbf{A})$,

(iii) $\operatorname{Tr}(\mathbf{A}^3) = s_{(3,0)}(\mathbf{A}) - s_{(2,1)}(\mathbf{A}) + s_{(1,1,1)}(\mathbf{A})$,

(iv) $\operatorname{Tr}(\mathbf{A}^4) = s_{(4,0)}(\mathbf{A}) - s_{(3,1)}(\mathbf{A}) + s_{(2,1,1)}(\mathbf{A}) - s_{(1,1,1,1)}(\mathbf{A})$.

The constant $s_{(m-k,1^k)}(\mathbf{I}_p)$ is equal to

$$s_{(m-k,1^k)}(\mathbf{I}_p) = \frac{(m+p-(k+1))!}{k!\,(p-(k+1))!\,(m-(k+1))!m} \tag{20}$$

(see [18] for a proof of this formula). Therefore,

$$\frac{s_{(m-k,1^k)}(\mathbf{I}_p)}{s_{(m-k,1^k)}(\mathbf{I}_n)} = \frac{(m+p-(k+1))!}{(m+n-(k+1))!}\cdot\frac{(n-(k+1))!}{(p-(k+1))!}. \tag{21}$$

For every $\alpha>0$, let us define the following class of functions:

$$L_\alpha^2 := \left\{ f:\mathbb{R}^+ \longrightarrow \mathbb{R} : \text{measurable such that} \int_0^\infty e^{-\alpha t}\,|f(t)|^2 dt < \infty \right\}. \tag{22}$$

This is a Hilbert space with respect to the inner product $\langle f, g\rangle_\alpha = \int_0^\infty e^{-\alpha t} f(t)g(t)dt$. Moreover, polynomials are dense with respect to this norm (see Chapter 10 in [20]). Let $\mathscr{A}_\alpha$ be the set of continuous functions in L^2_α, and let $\mathscr{A}$ be the intersection of all the $\mathscr{A}_\alpha$,

$$\mathscr{A} = \bigcap_{\alpha>0} \mathscr{A}_\alpha. \tag{23}$$

Note that the family $\mathscr{A}$ is a very rich family of functions. For instance, all functions that do not grow faster than polynomials belong to this family. In particular, $f(t) = \log(1+t) \in \mathscr{A}$.

Theorem 2. *Let* $\mathbf{A}$ *be an* $n \times n$ *positive definite matrix, and let* $\{d_1, \ldots, d_n\}$ *be the set of eigenvalues of* $\mathbf{A}$. *Assume that all the eigenvalues are different. Then, for every* $f \in \mathscr{A}$ *one has that*

$$\int_{M_n} \operatorname{Tr}\left(f\left(\mathbf{H}^*\mathbf{A}\mathbf{H}\right)\right) d\nu(\mathbf{H}) = \frac{1}{\det(\Delta(\mathbf{D}))} \sum_{k=0}^{n-1} \det(\mathbf{T}_k), \tag{24}$$

where $\Delta(\mathbf{D})$ *is the Vandermonde matrix associated with the matrix* $\mathbf{D} = \operatorname{diag}(d_1, \ldots, d_n)$ *and* $\mathbf{T}_k$ *is the matrix constructed by replacing the* $(k+1)$ *row of* $\Delta(\mathbf{D})$ ($\{d_i^{n-(k+1)}\}_{i=1}^n$) *by*

$$\frac{1}{(n-(k+1))!} \{f_k(d_i)\}_{i=1}^n, \tag{25}$$

where

$$f_k(x) := \int_0^\infty e^{-t} (tx)^{n-(k+1)} f(tx)\, dt. \tag{26}$$

Proof. First, we will prove the theorem for polynomials. Let p and q be two polynomials. It is clear that

$$\operatorname{Tr}\left((p+q)\left(\mathbf{H}^*\mathbf{A}\mathbf{H}\right)\right) = \operatorname{Tr}\left(p\left(\mathbf{H}^*\mathbf{A}\mathbf{H}\right)\right) + \operatorname{Tr}\left(q\left(\mathbf{H}^*\mathbf{A}\mathbf{H}\right)\right) \tag{27}$$

and $(p+q)_k = p_k + q_k$ for every $k = 0, \ldots, n-1$. Therefore, both sides of (24) are linear, and it is enough to prove the Theorem for the case $p(x) = x^m$ with $m \geq 0$. Using Theorem 1 and (19), we see that for every positive definite $n \times n$ matrix $\mathbf{A}$, the average

$$\begin{aligned} &\int_{M_n} \operatorname{Tr}\left(\left(\mathbf{H}^*\mathbf{A}\mathbf{H}\right)^m\right) d\nu(\mathbf{H}) \\ &\quad = \sum_{k=0}^{n-1} (-1)^k h(\lambda_k)\, s_{\lambda_k}(\mathbf{A})\, s_{\lambda_k}(\mathbf{I}_n), \end{aligned} \tag{28}$$

where λ_k is the partition $(m-k, 1^k)$. It is well known (see [16]) that for every partition $\lambda = (\lambda_1, \ldots, \lambda_n)$

$$s_\lambda(\mathbf{I}_n) = \prod_{1\leq i\leq j\leq n} \frac{\lambda_i - \lambda_j + j - i}{j-i}. \tag{29}$$

Therefore, we can deduce that

$$s_{\lambda_k}(\mathbf{I}_n) = \frac{1}{m} \cdot \frac{(m+n-(k+1))!}{k!\,(n-(k+1))!\,(m-(k+1))!}. \tag{30}$$

We can see by direct examination that the hook length of the partition λ_k is equal to

$$h(\lambda_k) = k!\,(m-(k+1))!m. \tag{31}$$

Hence,

$$s_{\lambda_k}(\mathbf{I}_n)\, h(\lambda_k) = \frac{(m+n-(k+1))!}{(n-(k+1))!}. \tag{32}$$

Since $\mathbf{A}$ is a positive definite matrix, by the spectral theorem there exists $\mathbf{U}$ unitary and $\mathbf{D} = \operatorname{diag}(d_1, \ldots, d_n)$ diagonal such that $\mathbf{A} = \mathbf{U}\mathbf{D}\mathbf{U}^*$. Note that the d_i are the eigenvalues of $\mathbf{A}$. By definition of the Schur polynomials,

$$s_{\lambda_k}(\mathbf{A}) = s_{\lambda_k}(\mathbf{D}) = \frac{\det(\mathbf{S}_k)}{\det(\Delta(\mathbf{D}))}, \tag{33}$$

where $\Delta(\mathbf{D})$ is the Vandermonde matrix associated with the sequence $\{d_i\}_{i=1}^n$ and $\mathbf{S}_k$ is a matrix whose ith column is equal to

$$\begin{pmatrix} d_i^{n-1+m-k} \\ d_i^{n-2+1} \\ d_i^{n-3+1} \\ \vdots \\ d_i^{n-(k+1)+1} \\ d_i^{n-(k+2)} \\ \vdots \\ d_i^{n-(n-1)} \\ 1 \end{pmatrix}. \tag{34}$$

It is easy to see that after k transpositions of the rows of the matrix $\mathbf{S}_k$, we obtain a new matrix $\mathbf{H}_k$ whose ith column is equal to

$$\begin{pmatrix} d_i^{n-1} \\ d_i^{n-2} \\ d_i^{n-3} \\ \vdots \\ d_i^{n-k} \\ d_i^{n+m-(k+1)} \\ d_i^{n-(k+2)} \\ \vdots \\ d_i^{n-(n-1)} \\ 1 \end{pmatrix}. \tag{35}$$

This matrix is equal to the matrix $\Delta(\mathbf{D})$ except for the $(k+1)$ row, $\{d_i^{n-(k+1)}\}_{i=1}^n$, which is substituted by the row $\{d_i^{n+m-(k+1)}\}_{i=1}^n$. Note also that

$$\det(\mathbf{S}_k) = (-1)^k \det(\mathbf{H}_k). \tag{36}$$

Therefore,

$$\int_{M_n} \operatorname{Tr}\left(\left(\mathbf{H}^*\mathbf{AH}\right)^m\right) d\nu(\mathbf{H}) = \frac{1}{\det(\Delta(\mathbf{D}))}\sum_{k=0}^{n-1}\frac{(m+n-(k+1))!}{(n-(k+1))!}\cdot \det(\mathbf{H}_k). \quad (37)$$

Using the fact that $\int_0^\infty e^{-t}t^p\,dt = p!$ and the definition of $p_k(x)$ for the case $p(x) = x^m$, we see that

$$p_k(x) := \int_0^\infty e^{-t}(tx)^{n+m-(k+1)}dt = (m+n-(k+1))!x^{m+n-(k+1)}. \quad (38)$$

Therefore, our claim holds and we have proven the result for all polynomials. Now consider $f \in \mathscr{A}$, and let β be the maximum eigenvalue; that is, $\beta = \max\{d_1, \ldots, d_n\}$. Define $\alpha = 1/\beta$. Since $f \in \mathscr{A}$, then $f \in \mathscr{A}_\alpha$, and let $\{p^{(r)}\}_{r\geq 1}$ be a sequence of polynomials such that $\|f - p^{(r)}\|_\alpha \to 0$. Let $\mathbf{T}_k^{(n)}$ be the matrix constructed by replacing the $(k+1)$ row of $\Delta(\mathbf{D})$ $(\{d_i^{n-(k+1)}\}_{i=1}^n)$ by

$$\frac{1}{(n-(k+1))!}\left\{p_k^{(r)}(d_i)\right\}_{i=1}^n, \quad (39)$$

where

$$p_k^{(r)}(x) := \int_0^\infty e^{-t}(tx)^{n-(k+1)}p^{(r)}(tx)\,dt. \quad (40)$$

Let $\mathbf{T}_k$ be the matrix constructed by replacing the $(k+1)$ row of $\Delta(\mathbf{D})$ by

$$\frac{1}{(n-(k+1))!}\{f_k(d_i)\}_{i=1}^n, \quad (41)$$

where

$$f_k(x) := \int_0^\infty e^{-t}(tx)^{n-(k+1)}f(tx)\,dt. \quad (42)$$

To prove that (24) holds, it is enough to prove that

$$\det\left(\mathbf{T}_k^{(n)}\right) \longrightarrow \det(\mathbf{T}_k) \quad (43)$$

as $n \to \infty$ for every $k = 0, 1, \ldots, n-1$. For this, it is enough to prove that $p_k^{(r)}(d_i) \to f_k(d_i)$ for every k and every $i = 1, 2, \ldots, n$. Note that

$$\begin{aligned}
&\left|f_k(d_i) - p_k^{(r)}(d_i)\right| \\
&= \int_0^\infty e^{-t}(td_i)^{n-(k+1)}\left|f(td_i) - p^{(r)}(td_i)\right| dt \\
&\leq d_i^{n-(k+1)}\sqrt{(2(n-(k+1)))!} \\
&\quad\cdot\left(\int_0^\infty e^{-t}\left|f(td_i) - p^{(r)}(tdi)\right|^2 dt\right)^{1/2} \\
&= d_i^{n-(k+3/2)}\sqrt{(2(n-(k+1)))!} \\
&\quad\cdot\left(\int_0^\infty e^{-t/d_i}\left|f(t) - p^{(r)}(t)\right|^2 dt\right)^{1/2},
\end{aligned} \quad (44)$$

where we use Cauchy-Schwartz for the second inequality and change of variable for the last one. Now, by construction, the sequence $\{p^{(r)}\}$ satisfies

$$\lim_{n\to\infty}\left\|f - p^{(r)}\right\|_\alpha^2 = \lim_{n\to\infty}\int_0^\infty e^{-\alpha t}\left|f(t) - p^{(r)}(t)\right|^2 dt = 0 \quad (45)$$

and $\alpha \leq d_i^{-1}$. Hence, we see that

$$\lim_{n\to\infty}\left|f_k(d_i) - p_k^{(r)}(d_i)\right| = 0 \quad (46)$$

finishes the proof. □

Remark 3. We would like to observe that the case when not all the eigenvalues are different can be treated as previously by perturbing the original eigenvalues and applying a subsequent limit. We present an instance of this situation in Corollary 6.

As a consequence, we have a new formula for the capacity of the MIMO communication channel and for the MMSE described in the introduction.

Corollary 4. *Let* $\mathbf{A}$ *be as in Theorem 2. Then,*

$$\int_{M_n}\operatorname{Tr}\left(\log\left(\mathbf{I}_n + \mathbf{H}^*\mathbf{AH}\right)\right) d\nu(\mathbf{H}) = \frac{1}{\det(\Delta(\mathbf{D}))}\sum_{k=0}^{n-1}\det(\mathbf{T}_k), \quad (47)$$

where $\mathbf{T}_k$ *is the matrix constructed by replacing the* $(k+1)$ *row of* $\Delta(\mathbf{D})(\{d_i^{n-(k+1)}\}_{i=1}^n)$ *by*

$$\left\{\frac{1}{(n-(k+1))!}\int_0^\infty e^{-t}(td_i)^{n-(k+1)}\log(1+td_i)\,dt\right\}_{i=1}^n. \quad (48)$$

Corollary 5. *Let* $\mathbf{A}$ *be as in Theorem 2. Then,*

$$\int_{M_n}\operatorname{Tr}\left(\left(\mathbf{I}_n + \mathbf{H}^*\mathbf{AH}\right)^{-1}\right) d\nu(\mathbf{H}) = \frac{1}{\det(\Delta(\mathbf{D}))}\sum_{k=0}^{n-1}\det(\mathbf{T}_k), \quad (49)$$

where $\mathbf{T}_k$ *is the matrix constructed by replacing the* $(k+1)$ *row of* $\Delta(\mathbf{D})(\{d_i^{n-(k+1)}\}_{i=1}^n)$ *by*

$$\left\{\frac{1}{(n-(k+1))!}\int_0^\infty e^{-t}(td_i)^{n-(k+1)}(1+td_i)^{-1}dt\right\}_{i=1}^n. \quad (50)$$

As an application, let us compute explicitly the two-dimensional case for the capacity.

Corollary 6. *Let* $\mathbf{A}$ *be a Hermitian* 2×2 *matrix with eigenvalues* d_1 *and* d_2*. If* $d_1 \neq d_2$*, then*

$$\int_{M_2}\operatorname{Tr}\left(\log\left(\mathbf{I}_2 + \mathbf{H}^*\mathbf{AH}\right)\right) d\nu(\mathbf{H}) = \frac{f_0(d_1) - f_0(d_2) + d_1 f_1(d_2) - d_2 f_1(d_1)}{d_1 - d_2}, \quad (51)$$

where $f_0(d_i) = \int_0^\infty e^{-t} t d_i \log(1+td_i)\,dt$ *and* $f_1(d_i) = \int_0^\infty e^{-t}\log(1+td_i)\,dt$. *If* $d_1 = d_2 = d$, *then*

$$\int_{M_2} \operatorname{Tr}\left(\log\left(\mathbf{I}_2 + d\cdot \mathbf{H}^*\mathbf{H}\right)\right) d\nu(\mathbf{H}) = \int_0^\infty e^{-t}\left[(1+t)\log(1+td) + \frac{td(t-1)}{1+td}\right] dt. \quad (52)$$

Proof. The case $d_1 \neq d_2$ is a direct application of Theorem 2 for $n = 2$ and $f(x) = \log(1+x)$. For the case $d_1 = d_2 = d$, then both the top and the bottom vanish, and we have to take the limit of $d_1 = d + \epsilon$ and $d_2 = d$ as $\epsilon \to 0$. More precisely,

$$\begin{aligned}\lim_{\epsilon\to 0}\frac{f_0(d+\epsilon) - f_0(d)}{\epsilon} &= \int_0^\infty e^{-t}\left[t\log(1+td) + \frac{t^2 d}{1+td}\right] dt,\\ \lim_{\epsilon\to 0}\frac{(d+\epsilon) f_1(d) - d f_1(d+\epsilon)}{\epsilon} &= \int_0^\infty e^{-t}\left[(1+d)\log(1+td) - \frac{td}{1+td}\right] dt.\end{aligned} \quad (53)$$

Putting all the pieces together, we finish the proof. □

Analogously, we can compute explicitly the moments for the two-dimensional case.

Theorem 7. *Let* $\mathbf{A}$ *be a Hermitian* 2×2 *matrix with eigenvalues* d_1 *and* d_2 *and* $m \geq 1$. *If* $d_1 \neq d_2$, *then*

$$\int_{M_2} \operatorname{Tr}\left(\left(\mathbf{H}^*\mathbf{A}\mathbf{H}\right)^m\right) d\nu(\mathbf{H}) = m!\left((m+1)\frac{d_1^{m+1} - d_2^{m+1}}{d_1 - d_2} + \frac{d_1 d_2^m - d_2 d_1^m}{d_1 - d_2}\right). \quad (54)$$

If $d_1 = d_2 = d$, *then*

$$\int_{M_2} \operatorname{Tr}\left(\left(\mathbf{H}^*\mathbf{A}\mathbf{H}\right)^m\right) d\nu(\mathbf{H}) = m!\left(m^2 + m + 2\right) d^m. \quad (55)$$

4. Conclusion

Using results on random matrix theory and representation theory, in particular Schur polynomials, we prove a new formula for the average of functionals over the Gaussian ensemble. In particular, this gives another formula for the capacity of the MIMO Gaussian channel and the MMSE achieved by a linear receiver.

References

[1] E. P. Wigner, "Characteristic vectors of bordered matrices with infinite dimensions," *Annals of Mathematics*, vol. 62, pp. 548–564, 1955.

[2] E. P. Wigner, "On the distribution of the roots of certain symmetric matrices," *Annals of Mathematics*, vol. 67, pp. 325–327, 1958.

[3] M. L. Mehta, *Random Matrices*, vol. 142, Academic Press, 3rd edition, 2004.

[4] S. Verdu and A. M. Tulino, *Random Matrix Theory and Wireless Communications*, Now Publishers, 2004.

[5] G. W. Anderson, A. Guionnet, and O. Zeitouni, *An Introduction to Random Matrices*, vol. 118, Cambridge University Press, Cambridge, UK, 2010.

[6] E. Telatar, "Capacity of multi-antenna Gaussian channels," *European Transactions on Telecommunications*, vol. 10, no. 6, pp. 585–595, 1999.

[7] S. H. Simon, A. L. Moustakas, and L. Marinelli, "Capacity and character expansions: moment-generating function and other exact results for MIMO correlated channels," *IEEE Transactions on Information Theory*, vol. 52, no. 12, pp. 5336–5351, 2006.

[8] T. Ratnarajah, R. Vaillancourt, and M. Alvo, "Complex random matrices and Rayleigh channel capacity," *Communications in Information and Systems*, vol. 3, no. 2, pp. 119–138, 2003.

[9] T. L. Marzetta and B. M. Hochwald, "Capacity of a mobile multiple-antenna communication link in Rayleigh flat fading," *IEEE Transactions on Information Theory*, vol. 45, no. 1, pp. 139–157, 1999.

[10] B. D. McKay, "The expected eigenvalue distribution of a large regular graph," *Linear Algebra and its Applications*, vol. 40, pp. 203–216, 1981.

[11] J. W. Silverstein and Z. D. Bai, "On the empirical distribution of eigenvalues of a class of large-dimensional random matrices," *Journal of Multivariate Analysis*, vol. 54, no. 2, pp. 175–192, 1995.

[12] M. Debbah, W. Hachem, P. Loubaton, and M. de Courville, "MMSE analysis of certain large isometric random precoded systems," *IEEE Transactions on Information Theory*, vol. 49, no. 5, pp. 1293–1311, 2003.

[13] V. Marchenko and L. Pastur, "Distribution of eigenvalues for some sets of random matrices," *Mathematics of the USSR-Sbornik*, vol. 1, pp. 457–483, 1967.

[14] A. Edelman, *Eigenvalues and condition numbers of random matrices [Ph.D. thesis]*, Department Mathematics, MIT, 1989.

[15] J. W. Silverstein and S.-I. Choi, "Analysis of the limiting spectral distribution of large-dimensional random matrices," *Journal of Multivariate Analysis*, vol. 54, no. 2, pp. 295–309, 1995.

[16] W. Fulton and J. Harris, *Representation Theory*, vol. 129, Springer, New York, NY, USA, 1991.

[17] R. J. Muirhead, *Aspects of Multivariate Statistical Theory*, John Wiley & Sons Inc., New York, NY, USA, 1982.

[18] I. G. Macdonald, *Symmetric Functions and Hall Polynomials*, The Clarendon Press Oxford University Press, New York, NY, USA, 2nd edition, 1995.

[19] B. E. Sagan, *The Symmetric Group: Representations, Combinatorial Algorithms, and Symmetric Functions*, vol. 203, Springer, New York, NY, USA, 2nd edition, 2001.

[20] B. Spain and M. Smith, *Functions of Mathematical Physics*, Von Nostrand Reinhold Company, 1970.

Permissions

The contributors of this book come from diverse backgrounds, making this book a truly international effort. This book will bring forth new frontiers with its revolutionizing research information and detailed analysis of the nascent developments around the world.

We would like to thank all the contributing authors for lending their expertise to make the book truly unique. They have played a crucial role in the development of this book. Without their invaluable contributions this book wouldn't have been possible. They have made vital efforts to compile up to date information on the varied aspects of this subject to make this book a valuable addition to the collection of many professionals and students.

This book was conceptualized with the vision of imparting up-to-date information and advanced data in this field. To ensure the same, a matchless editorial board was set up. Every individual on the board went through rigorous rounds of assessment to prove their worth. After which they invested a large part of their time researching and compiling the most relevant data for our readers. Conferences and sessions were held from time to time between the editorial board and the contributing authors to present the data in the most comprehensible form. The editorial team has worked tirelessly to provide valuable and valid information to help people across the globe.

Every chapter published in this book has been scrutinized by our experts. Their significance has been extensively debated. The topics covered herein carry significant findings which will fuel the growth of the discipline. They may even be implemented as practical applications or may be referred to as a beginning point for another development.

The editorial board has been involved in producing this book since its inception. They have spent rigorous hours researching and exploring the diverse topics which have resulted in the successful publishing of this book. They have passed on their knowledge of decades through this book. To expedite this challenging task, the publisher supported the team at every step. A small team of assistant editors was also appointed to further simplify the editing procedure and attain best results for the readers.

Our editorial team has been hand-picked from every corner of the world. Their multi-ethnicity adds dynamic inputs to the discussions which result in innovative outcomes. These outcomes are then further discussed with the researchers and contributors who give their valuable feedback and opinion regarding the same. The feedback is then collaborated with the researches and they are edited in a comprehensive manner to aid the understanding of the subject.

Apart from the editorial board, the designing team has also invested a significant amount of their time in understanding the subject and creating the most relevant covers. They scrutinized every image to scout for the most suitable representation of the subject and create an appropriate cover for the book.

The publishing team has been involved in this book since its early stages. They were actively engaged in every process, be it collecting the data, connecting with the contributors or procuring relevant information. The team has been an ardent support to the editorial, designing and production team. Their endless efforts to recruit the best for this project, has resulted in the accomplishment of this book. They are a veteran in the field of academics and their pool of knowledge is as vast as their experience in printing. Their expertise and guidance has proved useful at every step. Their uncompromising quality standards have made this book an exceptional effort. Their encouragement from time to time has been an inspiration for everyone.

The publisher and the editorial board hope that this book will prove to be a valuable piece of knowledge for researchers, students, practitioners and scholars across the globe.

List of Contributors

Sanjay Kumar Singh, Umesh Singh and Vikas Kumar Sharma
Department of Statistics and DST-CIMS, Banaras Hindu University, Varanasi 221005, India

Philippe Bernardoff
Universit´e de Pau et des Pays de l'Adour, Avenue de l'Universit´e, 64012 Pau, France

Florent Chatelain
Gipsa-Lab, D´epartement Images et Signal, 961 rue de la Houille Blanche, BP 46, 38402 Saint Martin d'H´eres, France

Jean-Yves Tourneret
Universite de Toulouse, IRIT/ENSEEIHT/TESA, 2 rue Charles Camichel, BP 7122, 31071 Toulouse Cedex 7, France

Evgueni Gordienko, Andrey Novikov and J. Ruiz de Chávez
Department of Mathematics, Autonomous Metropolitan University, Iztapalapa, San Rafael Atlixco 186, Col. Vicentina, C.P. 09340, Mexico City, DF, Mexico

Matthijs J. Warrens
Unit of Methodology and Statistics, Institute of Psychology, Leiden University, P.O. Box 9555, 2300 RB Leiden, The Netherlands

Chris Bambey Guure and Samuel Bosomprah
Department of Biostatistics, School of Public Health, University of Ghana, Legon Accra, Ghana

Chrysoula Ganatsiou
Department of Civil Engineering, Section of Mathematics and Statistics, School of Technological Sciences, University of Thessaly, Pedion Areos, 38334 Volos, Greece

Thomas L. Toulias and Christos P. Kitsos
Technological Educational Institute of Athens, Department of Mathematics, Ag. Spyridonos & Palikaridi Street, 12210 Egaleo, Athens, Greece

Francisco Louzada
Department of Applied Mathematics and Statistics, ICMC, University of S˜ao Paulo, 13560-970 Sao Carlos, SP, Brazil

Vitor Marchi
Department of Statistics, Federal University of S˜ao Paulo, 13565-905 S˜ao Carlos, SP, Brazil

James Carpenter
London School of Hygiene and Tropical Medicine, University of London, Keppel Street, LondonWC1E 7HT, UK

Mengyuan Xu
Statistics Genomics Unit, NIH/NIMH (National Institutes of Health/National Institute of Mental Health), Bethesda, MD 20892, USA

Wayne A. Woodward and Henry L. Gray
Department of Statistical Science, Southern Methodist University, Dallas, TX 75205, USA

David Shilane
Department of Health Research and Policy, Stanford University, Stanford, CA 94305, USA

Derek Bean
Department of Statistics, University of California, Berkeley, CA 94720, USA

Zhengjia Chen
Department of Biostatistics and Bioinformatics, Emory University, Atlanta, GA 30322, USA

Xinjia Chen
Department of Electrical Engineering, Southern University and A&M College, Baton Rouge, LA 70813, USA

Robert W. Chen
Department of Mathematics, University of Miami, Coral Gables, FL 33124, USA

Larry A. Shepp
Department of Statistics, Wharton School, University of Pennsylvania, Philadelphia, PA 19104, USA

Justín Ju-Chen Yang
Department of Statistics, Harvard University, 1 Oxford Street, Cambridge, MA 02138-2901, USA

Trent Gaugler
Department of Statistics, Carnegie Mellon University, 5000 Forbes Avenue, 132B Baker Hall, Pittsburgh, PA 15213, USA

Michael G. Akritas
Department of Statistics, Penn State University, 325Thomas Building, University Park, PA 16802, USA

Akbar Biglarian, Enayatollah Bakhshi, Mehdi Rahgozar and Masoud Karimloo
Department of Biostatistics, University of Social Welfare and Rehabilitation Sciences (USWRS), Tehran 1985713834, Iran

Ahmad Reza Baghestani
Department of Biostatistics, Faculty of Paramedical Sciences, Shahid Beheshti University ofMedical Sciences, Tehran 1971653313, Iran

Mahmood Reza Gohari
Hospital Management Research Center, Tehran University of Medical Sciences (TUMS), Tehrann 1996713883, Iran

A. Naess
Department of Mathematical Sciences and CeSOS, Norwegian University of Science and Technology, 7491 Trondheim, Norway

O. Gaidai
Norwegian Marine Technology Research Institute, 7491 Trondheim, Norway

O. Karpa
Centre for Ships and Ocean Structures (CeSOS), Norwegian University of Science and Technology, 7491 Trondheim, Norway

Jun-Wu Yu
School of Mathematics and Computational Science, Hunan University of Science and Technology, Xiangtan, Hunan 411201, China

Yang Lu
School of Mathematics and Statistics, Central China Normal University, Wuhan, Hubei 430079, China

Guo-Liang Tian
Department of Statistics and Actuarial Science, The University of Hong Kong, Pokfulam Road, Hong Kong

Farhad Yahgmaei, Manoochehr Babanezhad and Omid S. Moghadam
Department of Statistics, Faculty of Sciences, Golestan University, Gorgan 49138-15739, Golestan, Iran

Gabriel H. Tucci
Bell Laboratories, Alcatel-Lucent, 600 Mountain Avenue, Murray Hill, NJ 07974, USA

Maria V. Vega
William Paterson University, Mathematics Department, 300 Pompton Road, Wayne, NJ 07470, USA